THE MITES OF STORED FOOD AND HOUSES

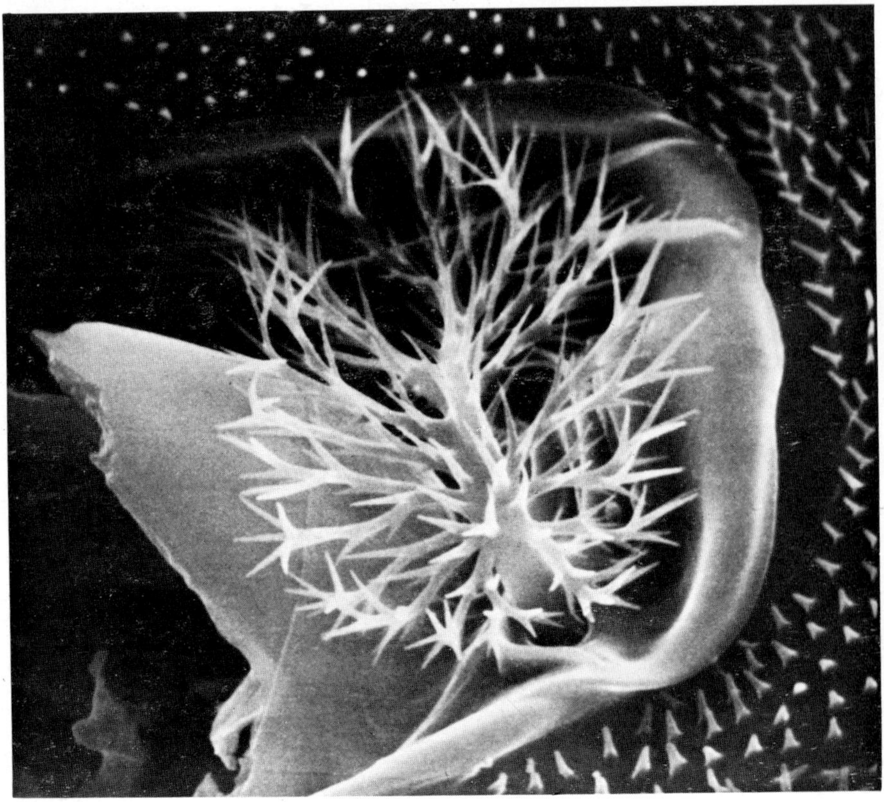

Glycyphagus sp. Supracoxal seta (X4,800) (by courtesy of D. A. Griffiths).

'Only those who are not acquainted with the Acarina can suppose that none is beautiful'
A. D. Michael, 1883

MINISTRY OF
AGRICULTURE, FISHERIES AND FOOD

Technical Bulletin 9

THE MITES OF STORED FOOD AND HOUSES

A M HUGHES* PHD

Biology Department, Royal Free Hospital
School of Medicine (University of London) and
Pest Infestation Control Laboratory
Ministry of Agriculture, Fisheries and Food

**Present address*
Cox Green Cottage
Rudgwick, Horsham
Sussex RH12 3AA

LONDON
HER MAJESTY'S STATIONERY OFFICE

First published 1961
Second edition 1976

Any reference to proprietary products in this bulletin should not be construed as an official endorsement of those products, nor is any criticism implied of products which are not mentioned.

ISBN 0 11 240909 1*

Contents

ACKNOWLEDGEMENTS

This bulletin is based on work carried out for the Pest Infestation Laboratory of the Ministry of Agriculture, Fisheries and Food, and is intended for those whose work concerns mites in food. Many of these mites are also found occasionally in houses and it is hoped that it may be of use to Public Health Officials.

I should like to take this opportunity of expressing my gratitude for their co-operation to the Inspectors and Scientific Staff of the Pest Infestation Control Laboratory, in particular to Dr. D. A. Griffiths whose help is always so readily given. I should also like to thank Professor G. Owen Evans for reading and commenting on the section concerned with the Mesostigmata and Mr. D. Macfarlane for helping with the Cryptostigmata. My thanks are also due to Dr. D. A. Griffiths and Dr. E. Lindquist for allowing me to reproduce their photographs and drawings, and to Mr. A. M. Cunnington for Plate IV, figure 1.

A.M.H.

February 1973

The subclass Acari

The mites comprise a large group of Arthropoda, belonging to the subclass Acari of the class Arachnida. They are about as closely related to the insects as birds are to mammals, and they are easily distinguished from them by the possession of eight, instead of six legs in the adult stage. As a group, they show considerable variation in external and internal structure, in habitat and mode of living, so that a satisfying definition is hard to frame. Apart from smallness of size, their most conspicuous character is a reduction in the segmentation which is so characteristic of the typical arthropod.

In an insect, the body is divided into three main regions: a head composed of six fused segments and bearing the complex of mouth parts, the three-segmented thorax whose appendages are the six walking legs and also the wings, and an abdomen where the segmentation is clearly visible. In a mite, however, the limits of head, thorax and abdomen cannot be determined, and, in the vast majority of cases, little trace of segmentation is visible, either during the course of development or in the adult stage. This loss of segmentation leads one to regard the mites as highly specialized forms of Arthropoda, and members of a group which diverged from the main line of evolution at an early stage in the history of the phylum.

Owing to the difficulty of identifying a head, thorax and abdomen on a mite, the following nomenclature is now used, instead of the older terms of cephalothorax and abdomen (Fig. 1).

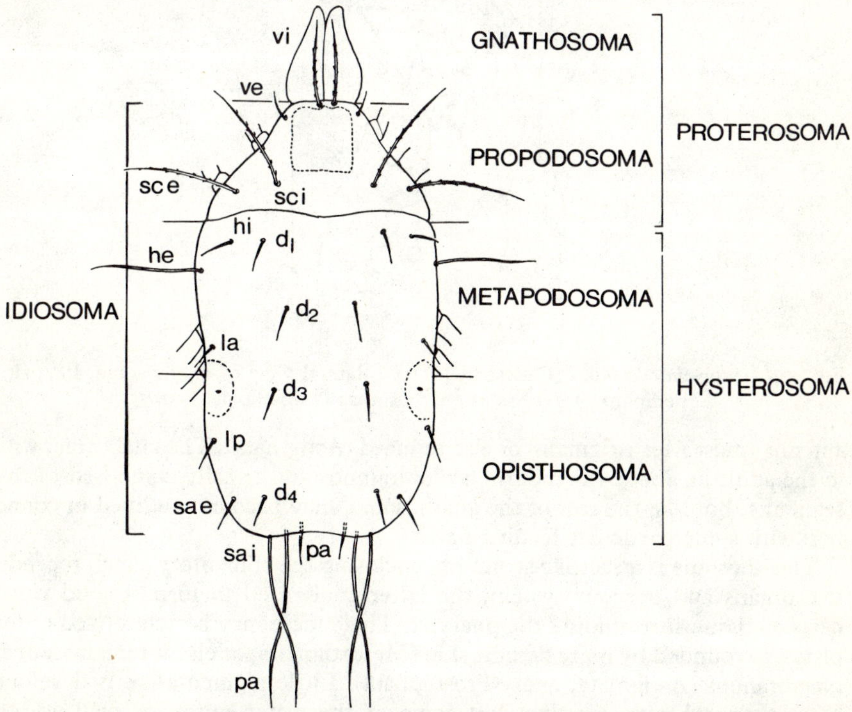

FIG. 1. *Acarus siro* L. ♀. Dorsal view. Setae of the idiosoma: *v e* and *v i* external and internal vertical; *sc e* and *sc i* external and internal scapular; *h e* and *h i* external and internal humeral; *l a* and *l p* anterior and posterior lateral; d_1 to d_4 dorsals; *sa e* and *sa i* external and internal sacral; *p a* post anal.

Segments	Region		
Segments bearing mouth and its appendages	Gnathosoma	Proterosoma	Idiosoma
Segments bearing first and second pairs of legs	Propodosoma	Proterosoma	Idiosoma
Segments bearing third and fourth pairs of legs	Metapodosoma	Hysterosoma	Idiosoma
Segments posterior to fourth pair of legs	Opisthosoma	Hysterosoma	Idiosoma

The gnathosoma (see T. Hughes, 1959) is attached to the idiosoma by an arthrodial membrane, allowing movement to occur. It can be withdrawn within the idiosoma, the opening into which it is retracted being known as the camerostome. The gnathosoma is formed as a result of the fusion of at least three segments, and consists of little more than the remains of the tergites and sternites of these segments, and of the two pairs of appendages: chelicerae and pedipalps, with their associated musculature. In the more primitive species, the tergites are well-developed and project forward as a tectum or 'epistome' lying above the chelicerae and forming the dorsal part of the gnathosoma. The ventral and lateral regions, however, are largely composed of the pedipalpal coxae which become fused with both sternites and tergites (Fig. 2). The external ramus of the pedipalp is produced to form a palp, which may be long and then function as an

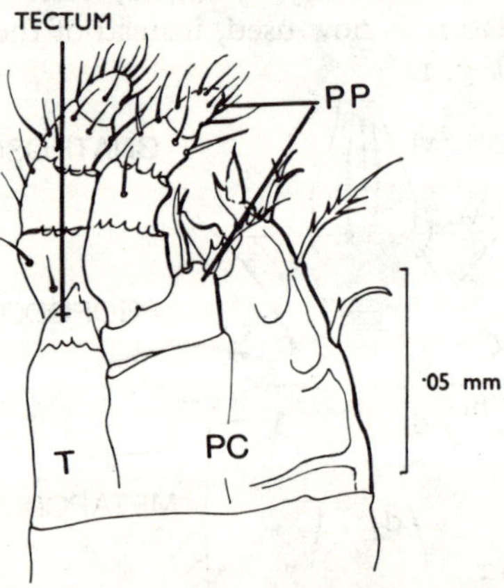

FIG. 2. *Leiodinychus krameri* (Canestrini). Dorso-lateral view of gnathosoma. PP palp of pedipalp; T tergite of gnathosoma; PC pedipalpal coxa.

antenna (most Mesostigmata) or else reduced (Astigmata). The chelicerae, which in the adult lie above the mouth, are uniramous and usually composed of three segments, but like the rest of the gnathosoma, may become modified in connection with some particular feeding habit.

The idiosoma is a sac-like structure, enclosing the alimentary canal, reproductive organs and nervous system, the latter condensed to form a solid ring of nervous tissue surrounding the pharynx. The cuticle may be sclerotized to form plates surrounded by more flexible skin (Mesostigmata) or else it remains soft and membranous (Astigmata, many Prostigmata). Little pigmentation is developed, other than a brown tanning, but some of the water mites, as well as other terrestrial Prostigmata, are brightly coloured. Sensory setae are often found projecting from the surface of the body, and simple eyes may be present in the anterior region.

With the exception of a degenerate group of plant parasites—the Tetrapodili (Prostigmata)—eight legs are present in the adult, each consisting of six segments (Fig. 3): coxa, trochanter, femur, genu, tibia and tarsus joined together by

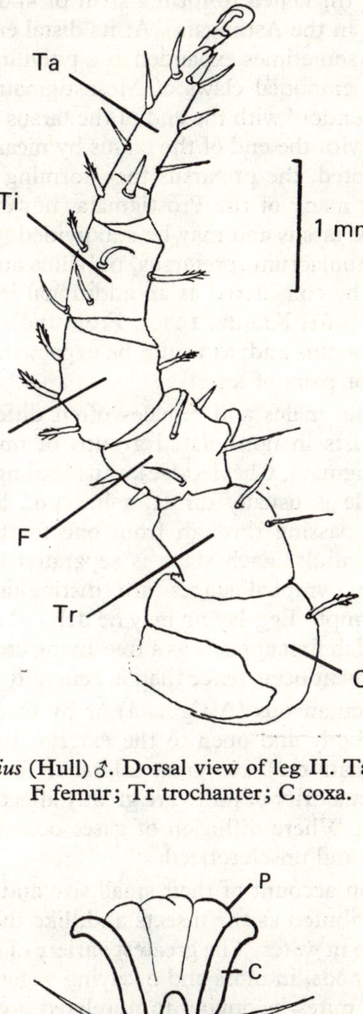

FIG. 3. *Macrocheles matrius* (Hull) ♂. Dorsal view of leg II. Ta tarsus; Ti tibia; G genu; F femur; Tr trochanter; C coxa.

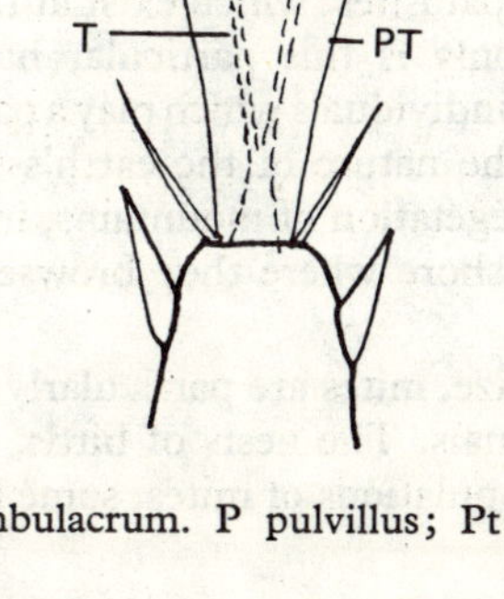

FIG. 4. *Macrocheles* sp. Ambulacrum. P pulvillus; Pt pretarsus; C claw; T tendon.

arthrodial membranes. Of these, the femur may be divided to form a basi- and telofemur (some Tarsonemini) or the number of segments may be reduced, the coxa becoming incorporated into the ventral surface of the body to form a plate. This plate is sometimes partially or completely sclerotized to form an epimeron, or its anterior margin is thickened to form a strut or apodeme to which the leg muscles are attached (as in the Astimgata). At its distal end, the leg may remain soft to form a pretarsus, sometimes expanded as a pulvillus, on which is inserted one or more claws (the empodial claws of Mesostigmata and Astigmata); this claw is connected by a 'tendon' with the end of the tarsus (Fig. 4). Alternatively, the claw may articulate with the end of the tarsus by means of a pair of sclerites, between which it is pivoted, the pretarsus then forming a membranous sheath enveloping the claw. In many of the Prostigmata, however, the claw is joined directly to the end of the tarsus and may be subdivided (the ambulacral claw of Tetranychidae). The ambulacrum (pretarsus, pulvillus and claws) is also known as the apotele and may be considered as an additional leg segment (see Evans, Sheals and Macfarlane, 1961, Krantz, 1970). From the surface of the legs arise various setae and sense organs and, as would be expected, these tend to be more numerous on the anterior pairs of legs.

The sexes are separate, males and females often differing widely from one another. A tendency exists in non-related groups of mites for heteromorphic males to be formed (Astigmata, Cheyletidae), thus leading to much confusion in taxonomy. The life cycle is usually direct, a hexapod larva emerging from a heavily-yolked egg and passing through from one to three octopod nymphal stages before becoming adult. Each stage is separated by a resting period or diapause and the three nymphal stages are distinguished as protonymph, deutonymph and tritonymph. Egg-laying may be delayed and it is then the larva, nymph or even adult which first appears as a free-living creature. Parthenogenesis is probably of more frequent occurrence than is generally supposed.

Respiration is either cutaneous (Astigmata) or by means of tracheae, which branch throughout the body and open to the exterior by stigmata or spiracles (Mesostigmata, Prostigmata etc). No great tolerance for dry surroundings has been developed and the majority of mites are greatly affected by a lowering of the water content of the air. Where diffusion of gases occurs through the cuticle, this usually remains soft and unsclerotized.

Although unnoticed on account of their small size and dislike of strong light, mites are as widely distributed as the insects and, like them, are found in large numbers on land and also in water. The greatest variety of terrestrial forms occurs in the surface litter of woods, in moss and decaying vegetation and in the upper layers of the soil. Here, mites belonging to unrelated groups, feed on decaying leaves and wood, on mosses, lichens and fungal mycelium. Accompanying the scavengers are predators and parasites, which exist at the expense of other mites and small arthropods. Not only is this particular habitat rich in number of species, but also in number of individuals which play a part comparable with that of the earthworm in changing the nature of the earth's surface. Many species of mites also live in the sparse vegetation of mountains, in fresh or salt water, or in the intertidal zone of the seashore where they browse on decaying seaweed or prey on other small animals.

On account of their small size, mites are particularly well adapted to link their lives with those of other animals. The nests of birds, rodents, insectivores and social insects all have large populations of mites, some living on minute particles of food or on the general organic detritus accumulated by any animal household. From sharing an animal's house to feeding on the animal itself is only a short step, and many mites are adapted to scraping the surface of mammalian hair, burrowing into the skin, sucking blood and tissue fluid or even living inside the body. Much evidence has been accumulated recently of the part that species of Acari play as vectors in transmitting pathogenic organisms.

Classification

Many attempts have been made to draw up a satisfactory scheme of classification, but as Krantz (1970) points out: acarology is, in fact, in a state of systematic turmoil similar to that experienced by entomologists a century ago. The foundations of a classification were laid down by Kramer (1877) and modified by subsequent authors (Michael, 1883, Canestrini, 1891, Berlese, 1899, Oudemans, 1906a, Vitzthum, 1929, Trägårdh, 1932, Baker and Wharton, 1952).

Evans *et al.*, 1961 divide the subclass Acari into seven orders: Notostigmata, Tetrastigmata, Mesostigmata, Metastigmata, Cryptostigmata, Astigmata and Prostigmata. Grandjean (1935a) examined specimens from these orders with polarized light, and noticed that the axes of most of the setae of mites belonging to the Prostigmata, Astigmata and Cryptostigmata were birefringent and have a core of actinochitin, whereas in the remaining groups they are all optically inactive. Acari with setae containing actinochitin (Prostigmata, Astigmata and Cryptostigmata), he united in the Actinochitinosi (=Actinochaeta Evans *et al.*, 1961); acari with setae lacking actinochitin, he placed in the Anactinochitinosi (=Anactinochaeta Evans *et al*, 1961).

The order Notostimgata With, 1903 contains a single, widely distributed genus *Opilioacarus* of doubtful relationship. This is placed in a group by itself on account of its segmented hysterosoma and eight dorsal stigmata. It is free-living and found under stones in the Mediterranean region, Central Asia, North and South America.

The order Tetrastigmata Evans *et al.*, 1961 also contains a single genus of wide distribution, found in the Seychelles, Mauritius, Australia, New Zealand, New Guinea and Sri Lanka. Most of the species of *Holothyrus* are large, up to 7 mm in length, well-sclerotized and with a pair of stigmata opening lateral to coxae III, which are connected to elongate peritremes (see p. 281). In addition a pair of pores, opening posterior to the stigmata, is connected to a system of air sacs. The mites are free-living, found under stones and are probably predaceous on other small animals.

Species belonging to the Metastigmata Canestrini, 1891 are more familiarly known as ticks. They are all parasitic on Vertebrata and many are of medical and veterinary importance. The ticks are the largest of the mites, a fully-gorged female sometimes attaining a length of almost an inch. The surface of the body is covered with a leathery integument, thickened in some to form a dorsal shield, which may cover part, or the whole of the idiosoma. The stigmata are surrounded by a perforated plate and are not connected to tubular peritremes and, in connexion with the mode of life, the mouth parts are modified for piercing the host tissue and sucking blood.

Representatives of the Astigmata (p. 7), Mesostigmata (p. 277), Prostigmata (p. 229) and occasionally, the Cryptostigmata (p. 223) are found in stored food and also in houses. Species belonging to the Astigmata are the most important economically, since they consume the food and constitute the dominant group.

Stored product mites (adults)

Key to the orders

1. With a long, whip-like solenidion (Fig. 13) arising from the dorsal extremity of tibiae I and II and usually extending beyond the end of the tarsus (exception: Histiostomidae p. 214 2

— Without such a whip-like solenidion 3

2. Well-sclerotized mites with a pair of prominent sensillae or pseudostigmatic organs (Fig. 263) arising from the dorsal surface of the propodosoma near its posterior marign .... Cryptostigmata, p. 223

— Feebly-sclerotized mites, without prominent pseudostigmatic organs .... Astigmata, p. 7

3. Stigmata easily seen, usually situated on the sides of the idiosoma and associated with a tubular peritreme (Fig. 313) .... Mesostigmata, p. 277

— Stigmata difficult to see, usually situated on or at the base of the gnathosoma and sometimes associated with a peritreme (Fig. 266) .. Prostigmata p. 229

Astigmata Canestrini, 1891

The Astigmata possess a pallid, soft cuticle varying in colour from pearly white to yellow brown. This cuticle may become pigmented in certain species (*Gohieria*, many hypopal forms), but it never becomes hard or densely sclerotized. Its surface may be smooth, rough or finely grooved; such grooving is characteristic of parasites, but is not unknown among free-living species. It may be that the possession of such a cuticle is a prerequisite for the development of a parasitic mode of life.

According to T. Hughes (1959), the cuticle is composed of a thin exocuticle lying directly above the epidermis which, except at ecdysis, is a thin meshwork of branched cells. At certain regions of the body, namely, the dorsal region of the propodosoma and also the coxal regions or epimera, the cuticle is thicker and appears punctate on account of the presence of pore canals. The whole surface is covered by an epicuticle. A distinct chitinous endocuticle is missing, except where the exocuticle is thickened to form the ring-shaped sockets of sensory setae.

The cuticle is the main respiratory surface and must also play a part in whatever control mechanism there is regulating the loss and uptake of water. Knülle and Wharton (1964) believe that, above a critical equilibrium point, uptake of water vapour can be compared with water uptake in non-living hygroscopic substances. In the case of *Acarus siro*, this point lies at a relative humidity of 71 per cent and above this point, the mite freely gains or loses moisture according as to whether it is being transferred to a higher or lower relative humidity. Below this point, a shortening of life coincides with increase in water loss. A relative humidity of 71 per cent is likely to be encountered by *A. siro* under natural conditions and would be the condition prevailing in stored cereals with a moisture content of 14 per cent. On the other hand, as Solomon (1966) points out, *A. siro* can live at humidities down to 62 per cent, so presumably below the critical equilibrium point, the cuticle ceases to behave like a non-living hygroscopic system. The fact that when a mite is killed, there is a great increase in rate of cuticular water loss, indicates the existence of some energy consuming mechanism which tends to reduce water loss at relatively low humidities.

A. siro has been shown to have a very high resistance to carbon dioxide accumulation and to lack of oxygen (T. Hughes, 1943). High moisture content home-grown barley is sometimes stored in silos and in one example, it was found that sufficient carbon dioxide had been evolved to prevent the development of insect pests, but not to inhibit the development of the flour mite. This was present in numbers sufficient to make the barley unpalatable to cattle.

The gnathosoma, although very mobile, is only partially retractile within the body, and in life is usually held at an angle to the idiosoma, so that the tips of the chelicerae come in contact with the food. It may become concealed by a rostral prolongation of the dorsal surface of the propodosoma (*Gohieria*, Fig. 206). Typically, the dorsal surface of the gnathosoma is reduced to form a small lobe between the bases of the chelicerae which are thus exposed dorsally. Each chelicera (Figs. 5 and 6), is bilaterally compressed and consists of a large basal region prolonged anteriorly as the fixed digit, and articulating with a movable one. The shears work in a vertical plane and their edges are often toothed or serrated. On the internal side of the fixed digit is a conical spur and beyond it, the mandibular spine. The muscles moving the chelicera originate from the

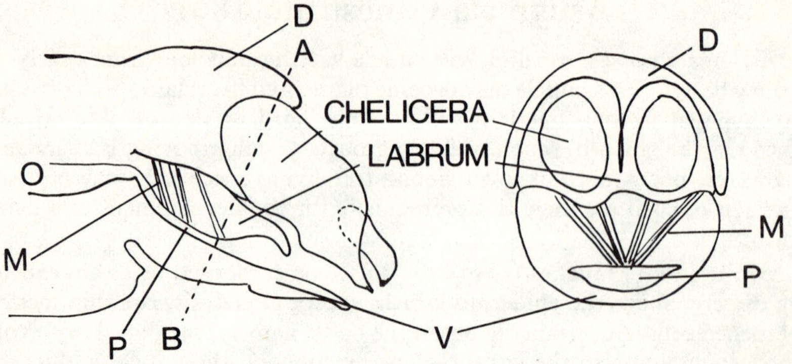

FIG. 5. Diagrams to show the arrangement of the mouth parts in an astigmatid mite in longitudinal section (*left*), and transverse section (*right*) at the level AB. D and V dorsal and ventral regions of gnathosoma; M dilator muscles of pharynx; P pharynx; O oesophagus.

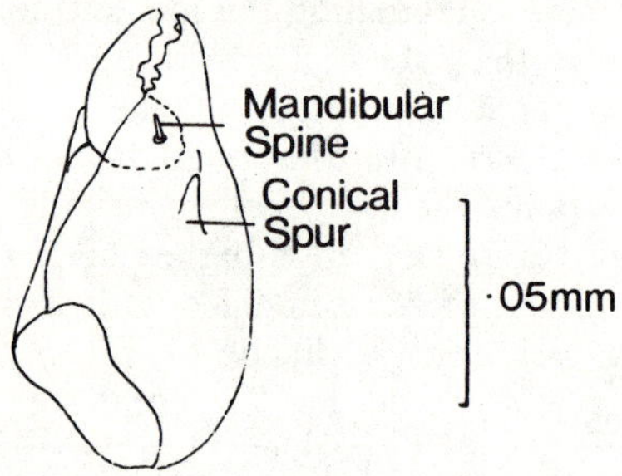

FIG. 6. *Acarus siro* L. Internal view of chelicera.

dorsal propodosomal plate and the two chelicerae can move independently of one another. Below the chelicerae lies the labrum (or true epistome): a hollow structure forming the roof of the mouth. The labrum continues backwards into the body as a plate, the lateral walls of which are continuous with the ventral part of the gnathosoma, and from it originate the dilator muscles of the pharynx (T. Hughes, 1953). The ventral part of the gnathosoma (the hypostome or infracapitulum) is formed mainly by the fused coxae of the pedipalps (Fig. 7) produced forward as a pair of internal lobes or malae, and a pair of external two-segmented palps. From the basal portion arise two pairs of setae. The palp is a flattened structure whose proximal segment bears a pair of setae and the distal, a single seta and an eccentric rod-shaped structure, possibly the vestige of a third segment or a sense organ. In some species, the mouth parts may become modified in connection with some particular way of life. For instance, in the Histiostomidae, p. 214, the mouth parts are adapted for the concentration of minute particles of food from a fluid medium (T. Hughes, 1953), and those of the Listrophoridae for clinging to hair and scraping its surface (T. Hughes, 1954). The Cytodytidae which inhabit the lungs of birds, have reduced mouth parts.

The idiosoma is usually an oval structure, showing little or no traces of segmentation, apart from a transverse constriction dividing it into propodosoma and hysterosoma. Sometimes the posterior edge of the body is lobed and this is found occasionally in the males of free-living species (*Thyreophagus*, *Histiogaster*),

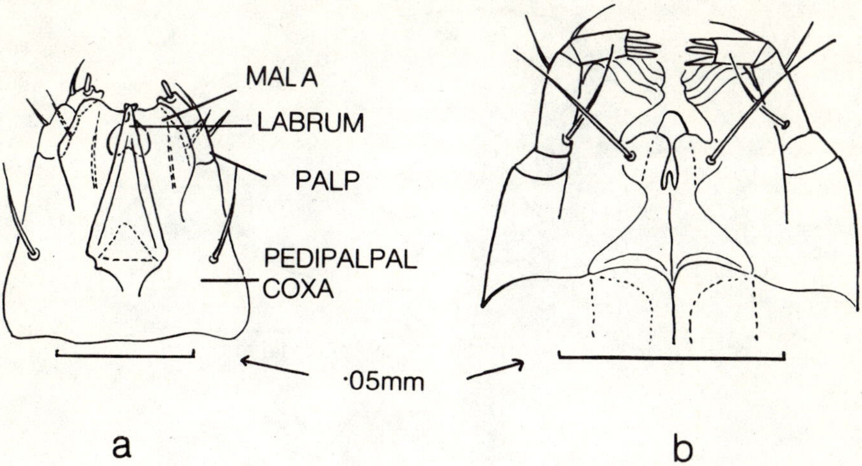

FIG. 7(a). *Acarus siro* L. Dorsal view of gnathosoma with the chelicerae removed; (b) *Lepidoglyphus destructor* (Schrank). Ventral view of gnathosoma.

and more often in ectoparasites (Proctophyllodidae, Psoroptidae). A pair of longitudinal grooves extending the length of the idiosoma, marks the position of the dorso-ventral muscles and, since the cuticle is inextensible, these grooves are particularly conspicuous in specimens which have been kept under dry conditions.

Arising from the idiosoma are a number of setae, which may be of various lengths and shapes (Fig. 8), but whose arrangement is constant throughout the group (Figs. 1 and 9). The vertical internal setae (*v i*) arise in the midline on the anterior dorsal edge of the propodosoma and project forward over the gnathosoma. On either side of the chelicerae, or more posterior in position, are the external verticals (*v e*). Across the posterior dorsal region of the propodosoma originate the internal (*sc i*) and external (*sc c*) scapulars arranged in a transverse line. On the anterior lateral edge of the hysterosoma between legs II and III, are

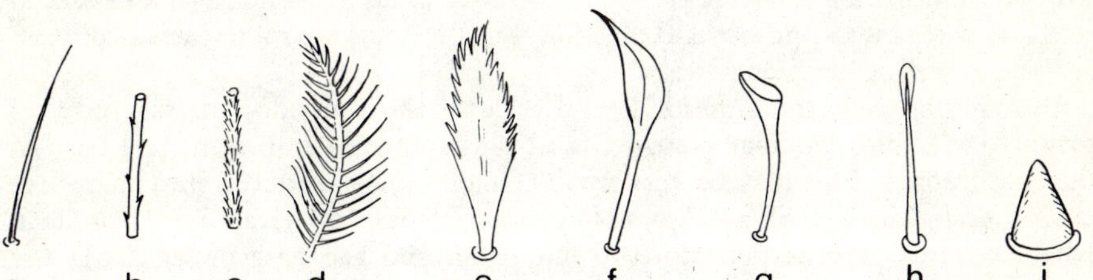

FIG. 8. Types of setae: (*a*) smooth or simple; (*b*) slightly pectinate; (*c*) pectinate; (*d*) bipectinate; (*e*) fringed; (*f*) leaf-shaped or falcate; (*g*) sucker-shaped; (*h*) spatulate; (*i*) spine.

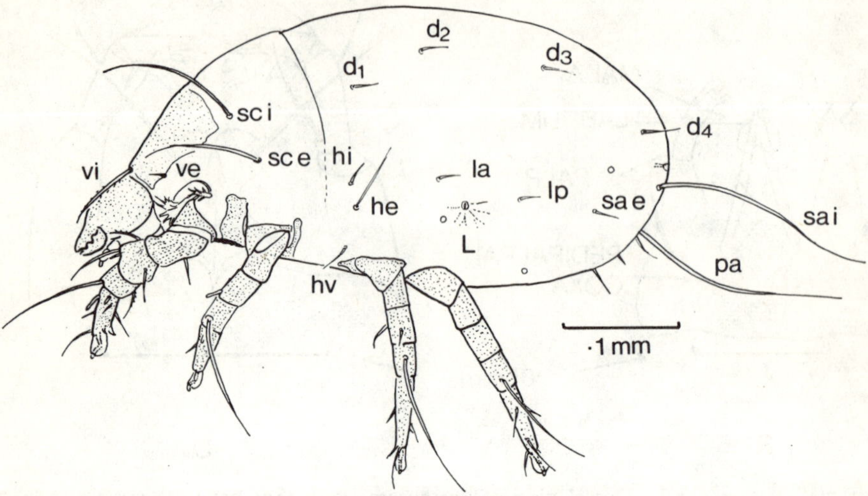

FIG. 9. *Acarus siro* L., ♀. Lateral view. Setae of the idiosoma: *v e* and *v i* external and internal vertical; *sc e* and *sc i* external and internal scapular; *h e* and *h i* external and internal humeral; d_1 to d_4 dorsals; *l a* and *l p* anterior and posterior lateral; *sa e* and *sa i* external and internal sacral; *p a* post anal. L latero-abdominal gland.

one to three pairs of humerals: the internal (*h i*), external (*h e*) and ventral (*h v*) according to their positions. The four pairs of dorsals (d_1 to d_4) are arranged in two longitudinal rows on the hysterosoma on either side of the midline. Two (*l a* and *l p*) or three (l_1 to l_3) pairs of laterals are inserted on the sides of the body, *l a* or l_1 being anterior to the opening of the latero-abdominal gland. From the dorsal posterior margin arise one or two pairs of sacrals: the internal (*sa i*) and external (*sa e*). These setae vary considerably in length and shape in different species; generally speaking, they tend to increase in length from anterior to posterior end of the body, and some may be reduced in length or absent altogether.

The ventral side of the body bears fewer setae and they are simpler in structure. Two pairs of coxal setae (*cx*) are inserted on the coxal regions of legs I and III, and three pairs of genitals (*g*) surround the genital opening (Fig. 10). The anal complex can be divided into two groups: one to two pairs of pre-anals (*pr a*) and one to five pairs of post-anals (*p a*). Sometimes the two groups are continuous with one another and they are then referred to as anals (*a*). The coxal and genital setae are constant in number and position, but the anals are more varied between species, as well as between sexes.

In many species, the anterior lateral edge of the propodosoma is produced forward as a thin cuticular plate—Grandjean's organ— embracing the base of the gnathosoma. This may be very small (*Otodectes*) or produced into flame-like extensions (*Acarus gracilis*, Fig. 11). From the base of Grandjean's organ a lateral sclerite extends posteriorly to curve around the base of leg I. At the posterior end of this sclerite is the supracoxal fossa, from which may arise the supracoxal seta; this may be a simple rod (*Caloglyphus berlesei*) or richly branched (*Glycyphagus* sp.). In many acarids, at the posterior end of the lateral sclerite and adjacent to the supracoxal seta, is a slit or pore, possibly the opening of

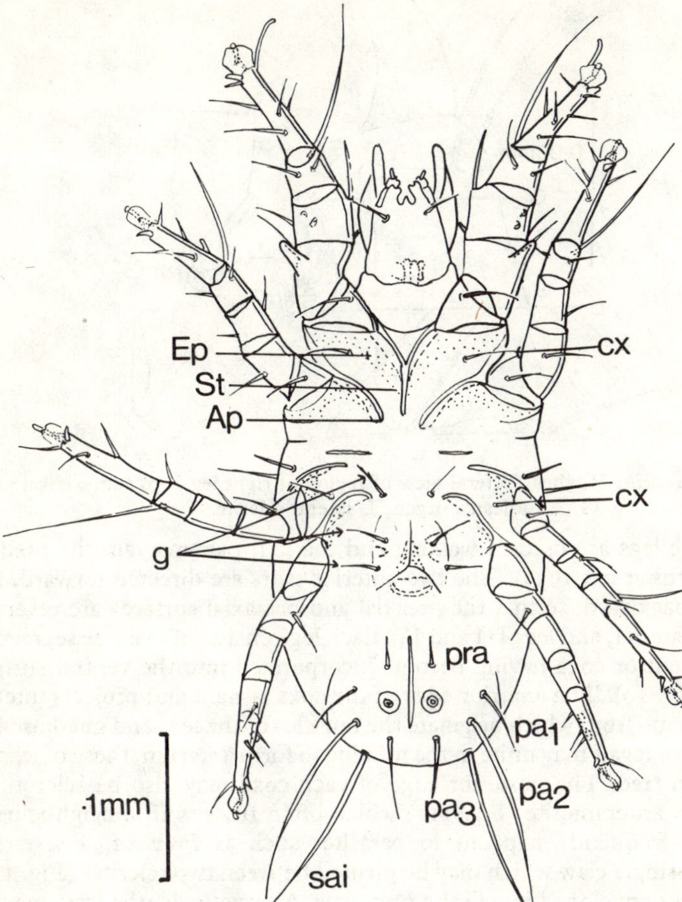

FIG. 10. *Acarus siro* L., ♂. Ventral view. Setae of the idiosoma: *p* a_1 to *p* a_3 post-anals; *pr a* pre-anal; *sa i* internal sacral; *cx* coxal; *g* genital. Ap apodeme; Ep epimeron; St sternum.

glands lying beneath the cuticle. Similar slits, the openings of the latero-abdominal or 'fat' glands (Fig. 9), leading into a sac lined with chitin, are also present on the sides of the body between the anterior and posterior lateral setae. The latero-abdominal glands, which are present in all stages of the life cycle, contain a highly refractile liquid which is colourless, sometimes yellow or brown, and red in the case of *Austroglycyphagus geniculatus* (Vitzthum). The lining of the sac is shed at ecdysis (Michael, 1901) and the opening in some mites can be partially closed by a cuticular flap (Brody and Wharton, 1970). Four pairs of circular rings are also found on the hysterosoma, one near the humeral setae, one near the anterior laterals, another near the posterior end of the body and the remaining pair on either side of the anus. These rings are well-developed in *Histiostoma laboratorium* R. Hughes which leads a semiaquatic existence, and it has been suggested that they function as osmoregulating organs (Perron, 1954).

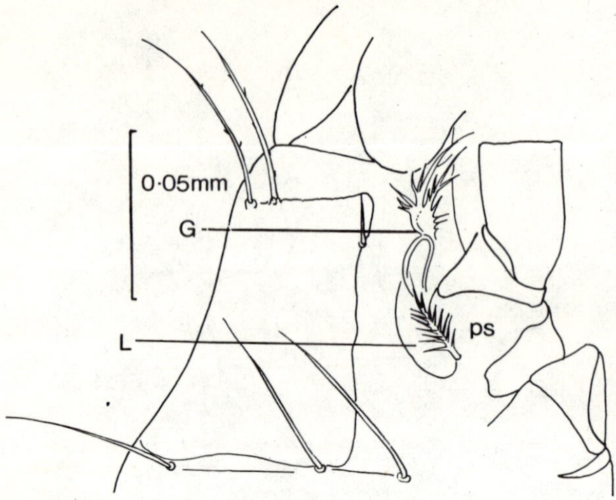

FIG. 11. *Acarus gracilis* Hughes. Lateral view of region of right leg I. *ps* supra-coxal seta; G Grandjean's organ; L lateral sclerite.

In life all the legs are used in walking and the first pair may also be used in feeding (Griffiths *et al*., 1959). The two anterior pairs are directed forward, the two posterior backward, so that the preaxial and postaxial surfaces are reversed between legs I and II, and legs III and IV. Each leg consists of five free segments, the basal segment or coxa having become incorporated into the ventral surface of the body (Fig. 10). The anterior edge of the coxa is rigid and projects internally as an apodeme from which originate the muscles of the legs and gnathosoma. The apodemes of legs I may unite in the midline to form a sterum, those of legs II to IV are often free. The posterior edge of each coxa may also be sclerotized and then forms an epimerite (Fig. 111) which often fuses with a neighbouring apodeme; this frequently happens in parasites such as *Freyana*. The tarsus terminates in a single claw which may be pivoted between two sclerites (Fig. 12a) and have its base enveloped in a fleshy pretarsus. Alternatively, the claw may be attached to the distal end of a short, or fleshy pretarsus and is then connected to the end of the tarsus by a pair of tendons (Fig. 12b). In *Lardoglyphus*, the

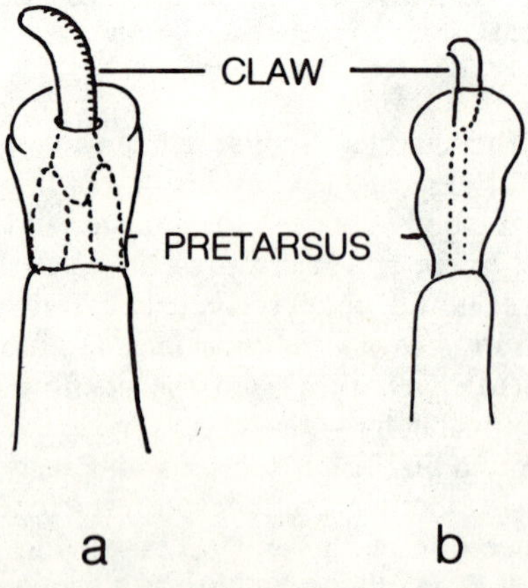

FIG. 12. Ambulacrum of (*a*) *Rhizoglyphus* sp., (*b*) *Glycyphagus* sp. The tarsal setae are not shown.

claw may have a biramous end. The claw is sometimes missing and the pretarsus may be elaborately folded to form an adhesive apparatus as in certain parasites.

Various hair-like outgrowths are inserted on the legs. They are most numerous on the tarsi and diminish in number from legs I to IV. Grandjean (1935 a, b) has classified these outgrowths into (a) true setae which are similar to those arising from the rest of the body, (b) solenidia and (c) the famulus. The true seta consists of a core of actinochitin surrounded by an additional layer which gives rise to any pectinations. The axis of the seta is swollen at its base and usually closed, the whole structure being inserted into a hole in the cuticle. Setae are to be found arising from any part of the body and, like the chelicerae and claws, are birefringent when viewed with polarized light.

The solenidion is a thin chitinous tube without a swollen base and open at its proximal end. It is never pectinate, but may appear to be striated owing to the presence of lines of slit-like pits; unlike the seta it is not birefringent. The famulus epsilon (ε) is found on tarsus I only; it is usually very small, often conical in shape and arises from a small pit which is homologous with the Haller's organ found in other Acari. It is always close to a solenidion, has a hollow core containing protoplasm and is birefringent.

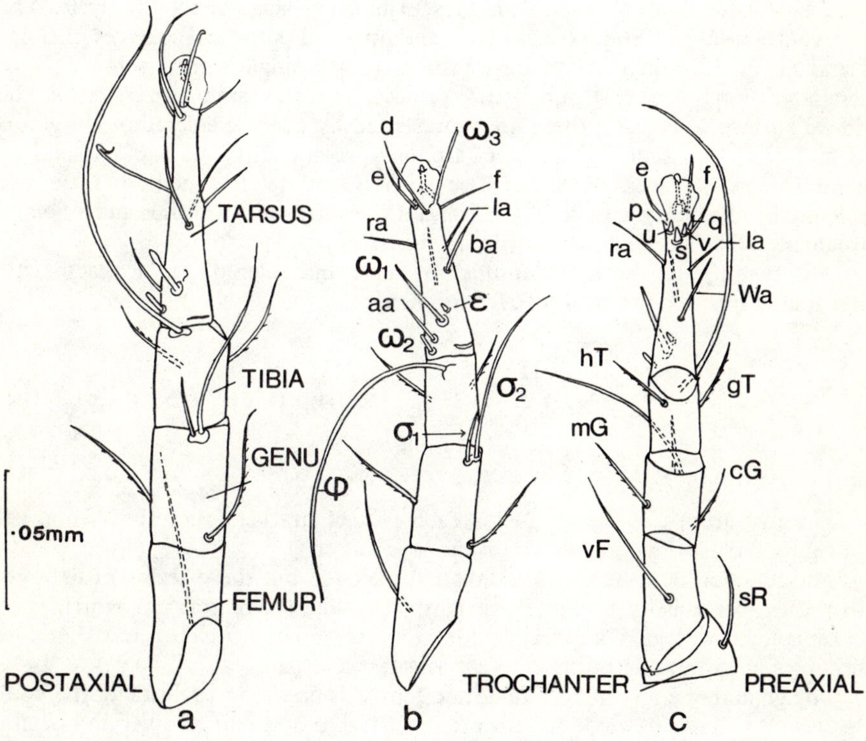

FIG. 13. Dorsal view of left leg I of (*a*) *Acarus siro* L., ♀. (*b*) *A. gracilis* Hughes ♀. (*c*) Ventral view of left leg I of *A. gracilis* Hughes ♀. ω_1 to ω_3 (omega$_1$ to omega$_3$) solenidia of tarsus; Φ (phi) solenidion of tibia; σ_1, σ_2 (sigma$_1$, sigma$_2$) solenidia of genu; ε (epsilon) famulus of tarsus; *aa, ba, d, e, f, la, ra, wa, p, q, s, u, v* setae and spines of tarsus; *gT, hT* setae of tibia; *cG, mG* setae of genu; *vF* seta of femur; *sR* seta of trochanter.

The arrangement and number of setae and solenidia have the same basic pattern in all the Astigmata, although certain setae and solenidia may be lost or occupy a different position; differences can also occur between the two sexes of the same species. Grandjean (1939a) has devised the following terminology which is here applied to *Acarus gracilis*. On leg I (Fig. 13) eight setae arise in a circle surrounding the base of the pretarsus: *d* is dorsal in position, *e* and *f* are post- and preaxial respectively, *p*, *q*, *u*, *v* and *s* are ventral and usually take the form of short, blunt spines, *q* and *v* being preaxial, *u* and *p* postaxial and *s* central. These setae and spines are present on all the tarsi. The solenidion omega$_3$ (ω_3) is a terminal cylindrical rod arising on tarsus I only; it is first developed in the last nymphal stage. The median region of tarsus I has a whorl of setae in which *ba* is dorsal, *wa* ventral and *la* and *ra* are pre- and postaxial respectively. These setae are also present on tarsus II, but III and IV bear two setae only: *r* and *w*. The basal group includes the dorsal blunt solenidion omega$_1$ (ω_1) which is present in all stages on tarsi I and II. It is often longer on tarsus II than tarsus I and is disproportionately long in the larva. The famulus is always closely associated with omega$_1$ on tarsus I. Omega$_2$ (ω_2) is a smaller solenidion, more proximal in position, which first appears in the protonymph and, like the seta *aa*, is restricted to tarsus I.

The whip-like solenidion phi (Φ) is inserted on the dorsal apex of each tibia and, except for that on tibia IV, is present in every stage of the life cycle. The two ventral setae *gT* and *hT* arise from the pre- and post-axial surfaces of tibiae I and II; on III and IV they are represented by a single ventral seta *kT*. Two solenidia sigma$_1$ (σ_1) and sigma$_2$ (σ_2) project from the same depression on the dorsal surface of genu I; these are represented by a single solenidion on genua II and III. Two setae *cG* and *mG* are also present on genua I and II and on genu III are replaced by a single seta *nG*. Genu IV has neither setae nor solenidia. The femora of legs I, II and IV each bear a ventral seta *vF*, and trochanters I to III each have a similar seta *sR*.

It is possible to show the number of setae and solenidia as a chaetotactic formula. Thus, for the adult of *Acarus gracilis*:

	Leg I	II	III	IV
Setae	(13.2.2.1.1)	(12.2.2.1.1)	(10.1.1.0.1)	(10.1.0.1.0)
Solenidia	(3.1.2.0.0)	(1.1.1.0.0)	(0.1.1.0.0)	(0.1.0.0.0.)

The five groups of figures inside each pair of brackets represent from left to right: tarsus, tibia, genu, femur and trochanter.

The function of setae and solenidia is unknown, but there seems little doubt that they are sensory organs. The long whip-like solenidion phi which is so conspicuous on legs I and II, in life projects straight forward and therefore occupies an admirable position to act as a tactile organ.

The genital opening in both male and female is on the ventral side of the body between the bases of the legs. It is often concealed by a pair of diverging genital folds on the internal side of which is a pair of blunt tubular structures—the so-called genital 'suckers' or sense organs. In some females (*Blomia*), an additional unpaired fold may also be present, closing the genital opening from behind. In the female the opening is large, to allow the passage of heavily-yolked eggs, and it may be longitudinal (most free-living forms Fig. 14) or transverse (many

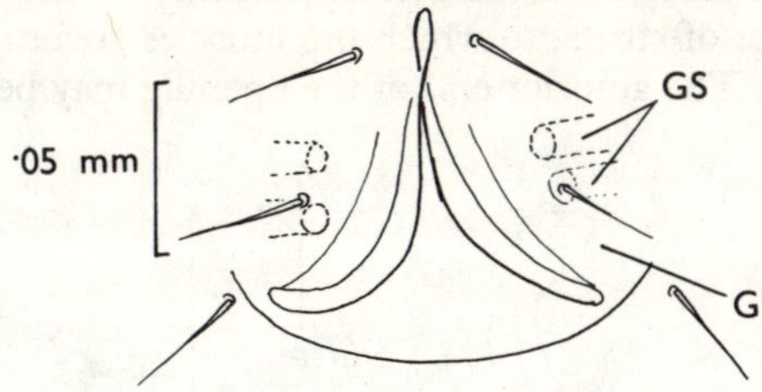

FIG. 14. *Acarus siro* L., ♀. Genital opening. GS genital sense organs; GF genital folds.

parasites). Sometimes the anterior end of the female opening is strengthened by a small crescentic epigynium (*Glycyphagus*), and this may become united with the sternum (*Carpoglyphus*) or with a circumgenital ring surrounding the opening of the oviduct (*Gohieria*). An additional opening, the bursa copulatrix, is also present at the posterior end of the body. It is usually a circular pore, more rarely at the end of a tubular structure (*Blomia*, *Glycyphagus*) and, internally, it leads eventually into a receptaculum seminis which communicates with the ovary (Fig. 15).

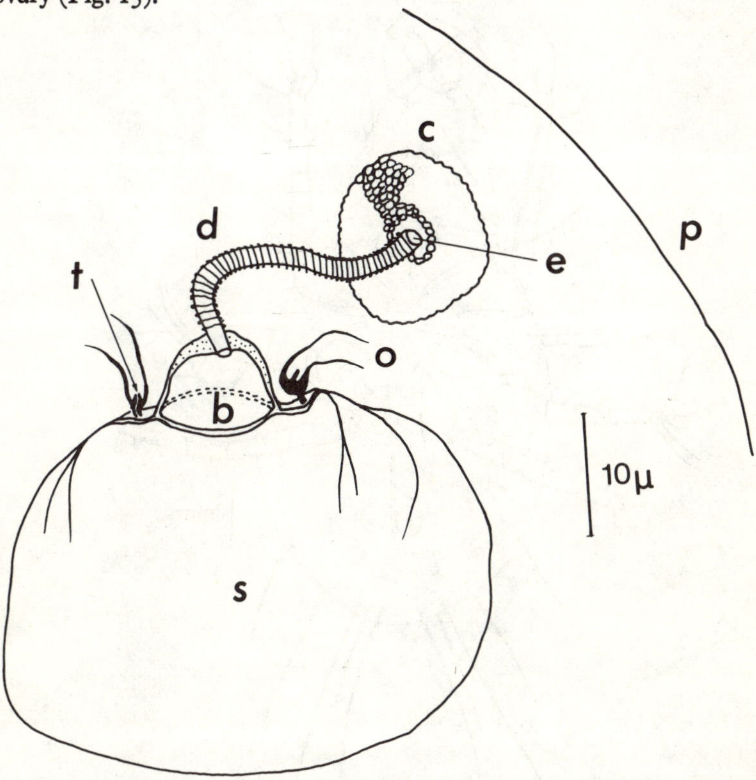

FIG. 15. *Acarus siro* L., ♀. Reproductive system. p posterior margin of body; c chitinised depression in body wall; e opening of expansible tube, or bursa copulatrix; d expansible tube; b chitinised bell-shaped structure leading into receptaculum seminis s; t chitinised opening to thin-walled oviduct o. (From Griffiths, 1969).

In the male, the vas deferens terminates in a chitinous tube or penis mounted on a complicated series of struts, to which the muscles governing its movement are attached (Fig. 16). The anterior end of the opening may be supported by an

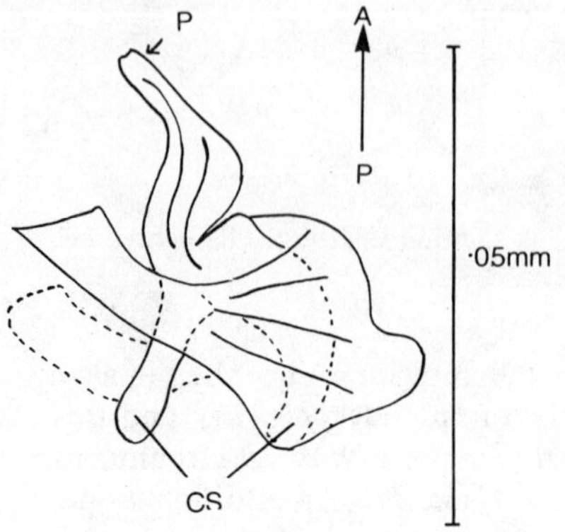

FIG. 16. *Suidasia medanensis* Oudemans. Lateral view of male genitalia. P penis; CS chitinous struts.

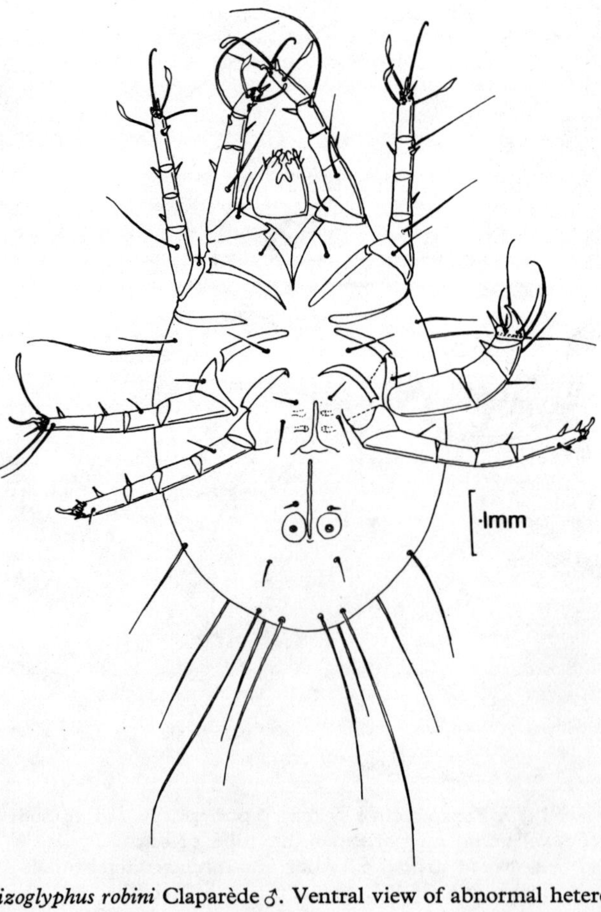

FIG. 17. *Rhizoglyphus robini* Claparède ♂. Ventral view of abnormal heteromorphic form.

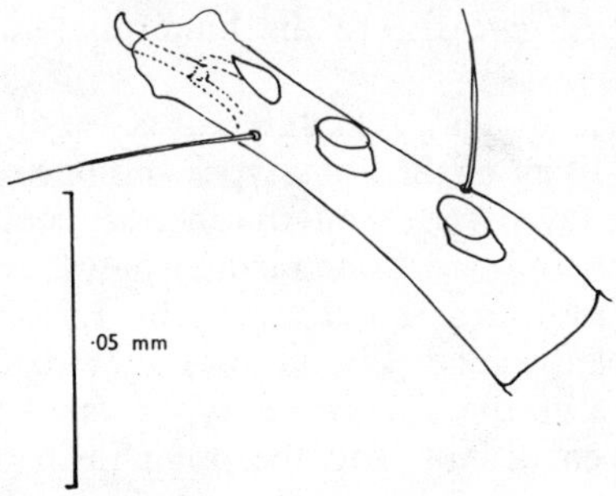

FIG. 18. *Tyrolichus casei* Oudemans, ♂. Dorsal view of tarsus IV.

epiandrium. Special copulatory organs are also developed in the male. These may take the form of large adhesive suckers on either side of the anal slit (Fig. 17) and a pair of smaller suckers on the fourth tarsus (Fig. 18) (*Acarus*, *Tyrophagus* etc.) or one only on tarsi I and II (*Calvolia*). A similar function is also performed by a modification in the shape of the legs, e.g., of legs I and II in *Glycyphagus ornatus* or legs III or IV in many parasites.

Sexual dimorphism is sometimes slight, or extreme cases may occur as in *Orycteroxenus dispar*, an inhabitant of moles' nests, where male and female bear no resemblance to each other either in size or structure.

Polymorphism amongst some of the males of the Astigmata was first recognized by Canestrini in 1888. He established the fact that *Rhizoglyphus* sp. can form two types of male: one resembling the female and one which has the third pair of legs enlarged and the tarsus of this leg replaced by a slightly curved claw. Michael (1901) noticed that this leg is not used in walking, but is held away from the ground and in *Caloglyphus anomalus* can be used as a weapon to kill young homomorphic males or even to fight amongst themselves (Woodring, 1969a). Abnormal specimens are sometimes seen where the hypertrophied leg and its associated heavily sclerotized apodemes are found on one side of the body only (Fig. 17). The most elaborate forms of these males are found among the bird parasites such as *Analges*, where the third pair of legs is almost as large as the rest of the body. In *Lardoglyphus* (p. 80) only the heteromorphic male is developed and the homomorphic is unknown.

Türk and Türk (1957) recognized four types of male in some species belonging to the genera *Caloglyphus*, *Rhizoglyphus* and *Schwiebia*: the homomorphic male resembling the non-gravid female in body shape and length of dorsal setae, the bimorphic which has a larger body and longer dorsal setae, the heteromorphic, resembling the homomorphic but with a modified third pair of legs and the pleomorphic which has the modified third pair of legs and a bimorphic body. Woodring (1969a) arranges the four types of male into two groups according to body shape and length of dorsal setae. The homotype males have a body shape and setal length similar to that of the female and the bimotypes, longer bodies and longer dorsal setae. Thus:

homotype male (Female type body)	{ homomorphic male heteromorphic male with modified legs III
bimotype male (Longer body)	{ bimorphic male pleomorphic male with modified legs III.

Some species, such as *Caloglyphus anomalus* can form all four types of male; others, such as *C. berlesei*, give rise to the bimorphic and pleomorphic forms only.

According to Woodring (1969a), two different types of polymorphism can be recognized: in the first, either of the homotypes and one of the bimotypes occur in a fixed ratio under a given set of environmental conditions. For instance, in *Caloglyphus berlesei*, there is a definite pleomorphic-homomorphic male ratio. When environmental conditions are changed, by raising the temperature or altering the character of the food, this ratio is altered. In the second type of polymorphism, any one of the four male types can be the dominant form irrespective of external conditions, and the other male type of that particular body shape (homotype or bimotype) rarely occurs. Thus in *Rhizoglyphus echinopus* (sensu Woodring), the homomorphic male is the dominant one and the heteromorphic is rarely found. Likewise in *Caloglyphus boharti*, the bimorphic male is predominant, and the pleomorphic male occurs only occasionally. Both types of polymorphism can occur in the same species; and in *C. anomalus* the pleomorphic and homomorphic males occur in a regular ratio, and heteromorphic and bimorphic males are found very rarely. Woodring also found that in *C. boharti*, pleomorphic male formation was inhibited by the presence of other mites which it was thought produced a volatile pheromone, the threshold pheromone level depending on the degree of separation of the developing immature stages.

The majority of the free-living Astigmata are oviparous, although egg-laying may be delayed until the larva, protonymph or even the hypopus is formed

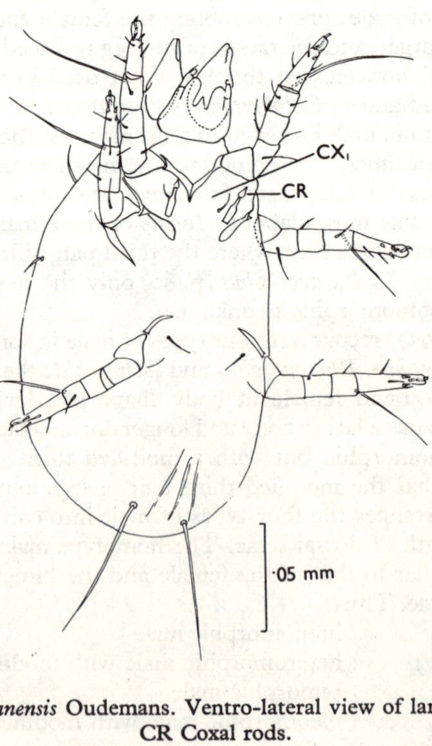

FIG. 19. *Suidasia medanensis* Oudemans. Ventro-lateral view of larva. CX_1 coxal region; CR Coxal rods.

within the parent body. The eggs are relatively large, owing to the presence of yolk, and covered with a smooth or sculptured shell. From the egg emerges a larva, differing from other stages by lacking the fourth pair of legs (Fig. 19) and, sometimes, by possessing coxal rods ('bruststiele', Claparèdè's organs) in the region of coxae I. These resemble solenidia, as they are distally-enlarged hollow rods. Their function is unknown. Similar coxal rods occur in the larvae of Bdellidae (Prostigmata) and it has been suggested by Grandjean (1938) that they are homologous with the genital sense organs ('suckers') found in the nymphs and adults. In the larval stage, the hysterosoma is not fully developed and certain setae or solenidia of the body (d_4, *lp*, genital and adanal) and legs (*sR* of trochanters I to III, omega$_2$ and omega$_3$ of tarsus I) are lacking. On the other hand, *sa* can be disproportionately long (*Glycyphagus* sp.) and may be used in dispersal by wind. The solenidion omega$_1$ is also large in comparison with the size of the larval tarsus and it may be that there is a certain minimum size below which its function becomes impaired. External genitalia are not visible in the larva.

After a period of activity, the larva becomes inert, and from this resting form emerges the protonymph (Fig. 20) which has acquired a fourth pair of legs and

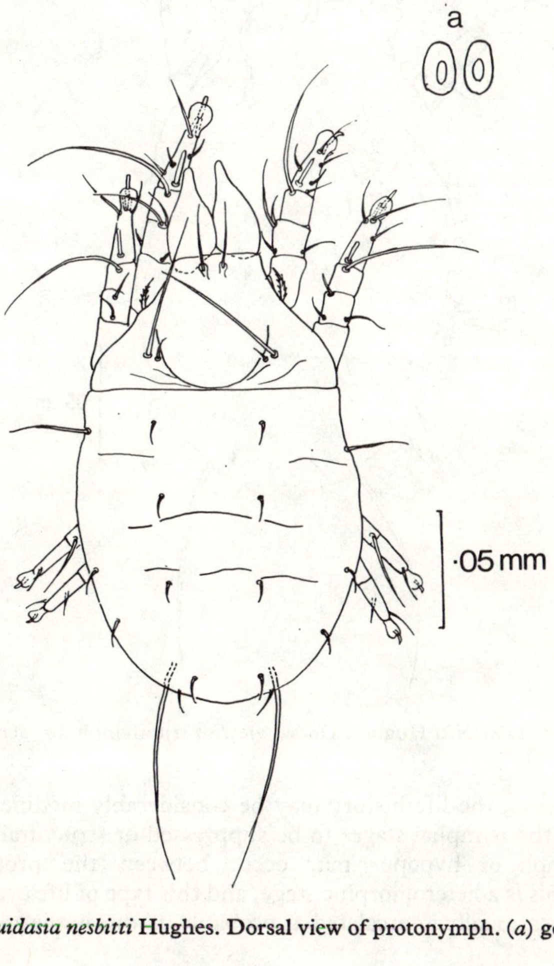

FIG. 20. *Suidasia nesbitti* Hughes. Dorsal view of protonymph. (*a*) genital region.

lost its coxal rods. This stage is characterized by possessing the rudiments of a genital opening, the rudiments of one pair of genital sense organs, one pair of genital setae and adanal setae when they exist in the adult. Setae d_4 and *lp* are also developed on the body and, except for the lack of a seta on trochanters I to III and a simplification of setae on legs IV, the chaetotaxy of legs I to III is the same as that of the adult. From the resting protonymph emerges the tritonymph (Fig. 21) closely resembling the adult. It differs in having a vestigial genital opening flanked by two pairs of sense organs, and in lacking some of the anal setae.

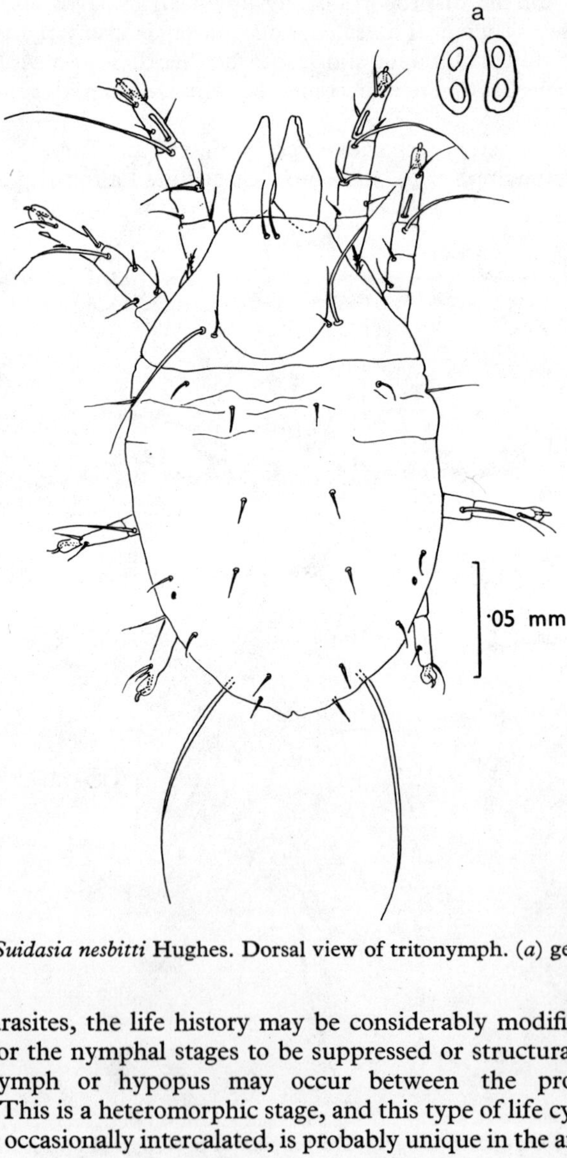

FIG. 21. *Suidasia nesbitti* Hughes. Dorsal view of tritonymph. (*a*) genital region.

Among parasites, the life history may be considerably modified and there is a tendency for the nymphal stages to be suppressed or structurally modified.

A deutonymph or hypopus may occur between the protonymph and tritonymph. This is a heteromorphic stage, and this type of life cycle in which an extra stage is occasionally intercalated, is probably unique in the animal kingdom.

The hypopus does not feed and serves either as a means of dispersal or of survival under adverse conditions. There are two kinds: the active ones move freely about and are adapted for clinging to other Arthropoda or to mammals; the inert are almost or completely devoid of movement and often remain within the protonymphal cuticle to await passively for more favourable living conditions. In some cases they are known to be resistant to desiccation (see p. 45). The distinction between the two types is not a rigid one and structurally they tend to grade into one another. Although hypopi only resemble the adults in certain details, those of related species are usually of the same type.

The active hypopus (Fig. 22) is yellow or brown in colour and has a hard,

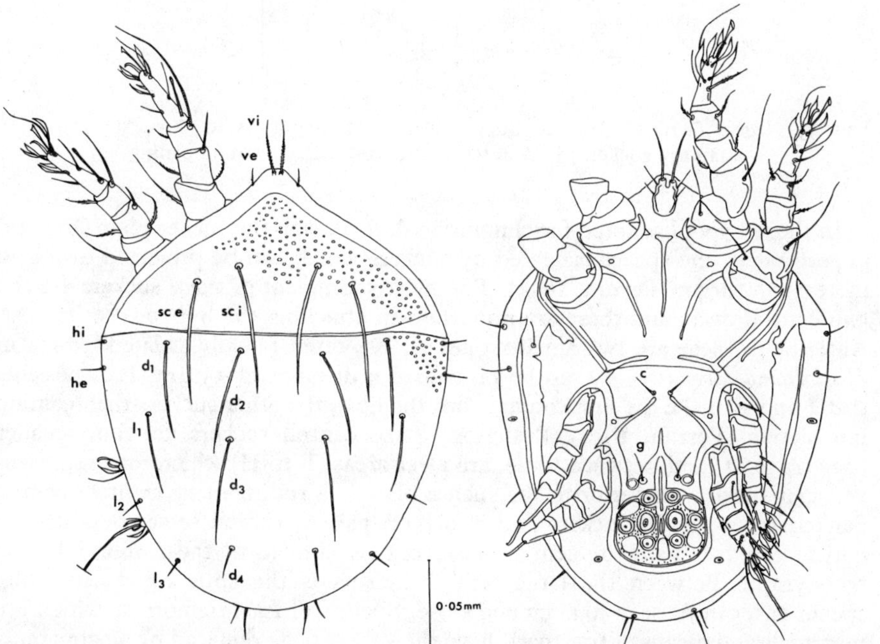

FIG. 22. *Acarus siro* L. Hypopus. (*a*) Dorsal view; (*b*) ventral view. Setae of the idiosoma: *v e*, *v i*, *sc e*, *sc i*, d_1 to d_4, *h e*, *h i*, l_1 to l_3. Genital seta: *g*. Coxal seta: *c* (From Griffiths, 1964).

rigid cuticle. The body is round or oval, and dorso-ventrally flattened, the ventral surface being concave and the dorsal convex, the curvature enabling the hypopus to fit more closely to the body of another arthropod. The dorsal surface is completely covered by propodosomal and hysterosomal shields. Two corneae may be present on the rostral region of the propodosoma (*Calvolia*). A mouth is lacking and the gnathosoma reduced to an unpaired plate whose anterior edge is bilobed. From each lobe arises a flagelliform seta. On the ventral surface of the body, the coxal plates are well-defined and bounded anteriorly and posteriorly by apodemes and epimerites respectively. In the opisthosomal region, there is a well-defined sucker plate forming part of the adhesive apparatus (Fig. 23).

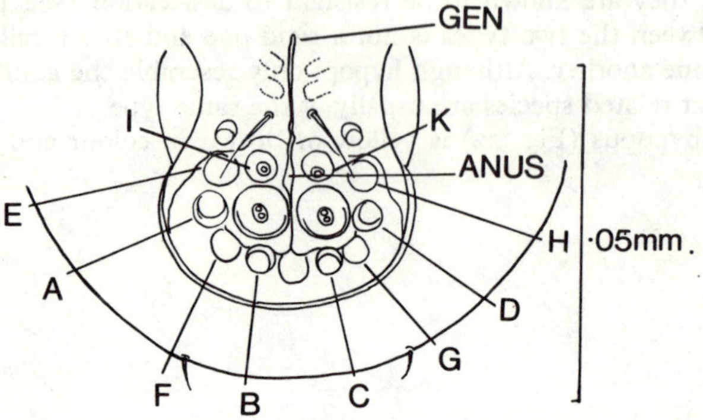

FIG. 23. *Acarus farris* Oudemans. Sucker plate of hypopus. A to D suckers; E to H auxiliary suckers; I, K anterior suckers; GEN genital opening.

In those hypopi adapted for clinging to Arthropoda, the sucker plate (Fig. 23) is perforated, the spaces occupied by suckers which can be projected owing to increase in turgor (Perron, 1954). The most prominent of these suckers are the two central ones, and these are important in attaching the hypopus to its host. Anterior to these are two smaller ones (I, K), often radially striated, which in *Histiostoma laboratorium*, can be protruded a distance of 15 μm. It is possible that I and K make the first contact with the host, the other suckers then coming into action (Perron, 1954). Posterior to the central suckers are four smaller ones (A to D) and between these, are clear areas (E to H) which may represent vestigial suckers. Anterior to the sucker plate is a rudimentary genital opening flanked by a pair of suckers and also by a pair of genital setae. Beneath the cuticle are two pairs of genital sense organs similar to those found in the tritonymph. Between the large central suckers is the anus which, in living specimens, can be seen to open and close. Species of Labidophorinae which use mammals as a means of transport, have the sucker plate replaced by an apparatus for clinging to mammalian hair. This consists of a pair of movable folds (Fig. 24) with a ridged internal surface, which covers two pairs of transversely grooved claspers. The whole organ functions as a vice which grips the hair securely.

The first two pairs of legs are better developed than the posterior which are almost completely tucked away beneath the body and, in the Histiostomidae are bent sharply towards the head. In some species, legs I and II in the live animal, make questing movements in the air, the body being supported by the remaining legs and the sucker plate. The chaetotaxy of the legs is different from that of the other stages, the differences not only involving a change in shape of setae, but also in the expression and suppression of some setae and solenidia. Tarsi I to III often bear setae widened to resemble leaves, or else expanded distally as small suckers (Fig. 123) and serving as part of the adhesive apparatus. Tarsus IV sometimes terminates in one or two long setae and is used for clinging to insects.

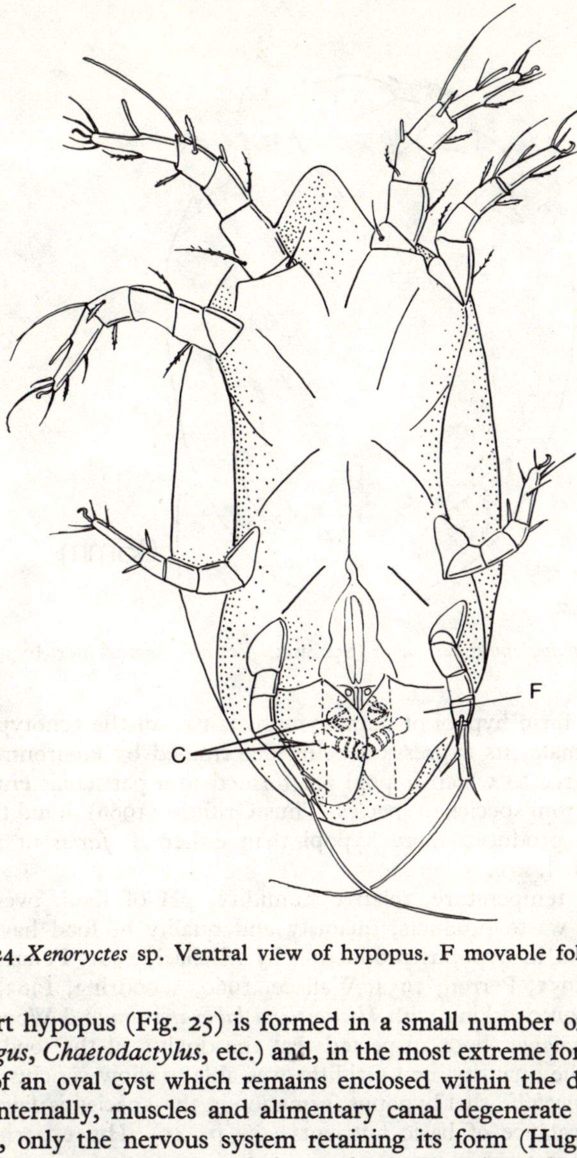

FIG. 24. *Xenoryctes* sp. Ventral view of hypopus. F movable fold; C claspers

An inert hypopus (Fig. 25) is formed in a small number of genera (*Acarus*, *Glycyphagus*, *Chaetodactylus*, etc.) and, in the most extreme form (*G. domesticus*) consists of an oval cyst which remains enclosed within the dry protonymphal cuticle. Internally, muscles and alimentary canal degenerate into a structureless mass, only the nervous system retaining its form (Hughes and Hughes, 1939). In the motile hypopus of *Acarus farris*, Boczek *et al.*, (1969) found that the internal structure was similar to that of *G. domesticus*, except that a muscular system is formed which operates the sucker plate and legs.

In the Hypodectidae, species of which are parasitic on birds, the hypopus has become the dominant stage of the life cycle. The hypopus of *Hypodectes propus* is found beneath the skin of pigeons and obtains its food from the tissue fluid of the host. Larva, protonymph and tritonymph are suppressed and the adult stage, which lives freely in the bird's nest, is only found during the nesting period. It is possible that the adults do not feed, since the chelicerae of the female are rudimentary (Fain, 1967a). It is not known how the hypopus manages to penetrate the host's skin.

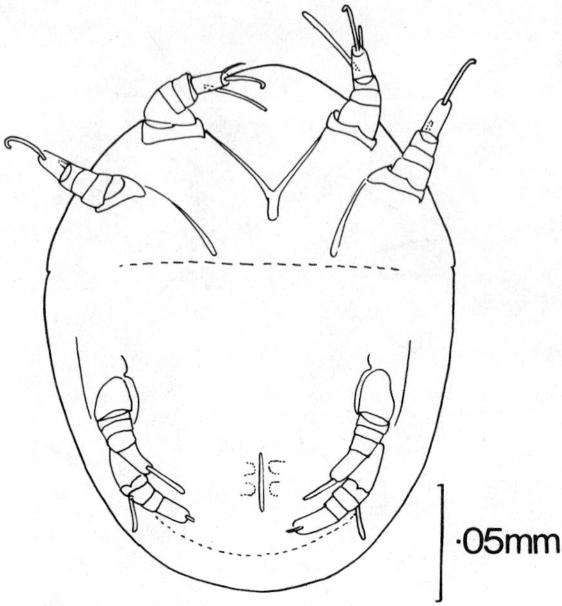

FIG. 25. *Lepidoglyphus destructor* (Schrank). Ventral view of inert hypopus.

The ability to form hypopi or deutonymphs is part of the genotype of many species of Astigmata, its expression being determined by environmental conditions. The degree to which hypopi are formed in a particular environment, however, varies from species to species. Thus Griffiths (1966) found that *Acarus immobilis* always produced more hypopi than either *A. farris* or a hypopus forming-strain of *A. siro*.

Conditions of temperature, relative humidity, pH of food, overcrowding, accumulation of waste products, quantity and quality of food have all been studied in relation to hypopus production by Michael, 1901, Polezhaev, 1938, 1940, Sorokin, 1951, Perron, 1954, Wallace, 1960, Woodring, 1963, Griffiths, 1966, 1969. Perron, working with *Histiostoma laboratorium* and Woodring with *Rhizoglyphus echinopus*, both suspected that the quality of the food was more important than the quantity, and Griffiths was able to show conclusively in the case of *Acarus immobilis*, that hypopus formation in this species is favoured by a scarcity and imbalance of basic nutrients (see p. 45). His experiments also indicated that ergosterol is necessary, possibly because it is a precursor of a hormone necessary for hypopus formation.

Hughes (1964) found that protonymphs from a starved culture of the hypopus-forming strain of *A. siro* lacked the neurosecretory cells present at ecdysis in the antero-dorsal region of the central nervous system, but similar cells were present in the so-called salivary gland. These cells appear at normal ecdysis during the first thirty hours of the protonymphal instar and Griffiths (1966) showed that it is during this time that the fate of the protonymph is fixed. Griffiths also showed that the developmental period of protonymphs which become deutonymphs is much extended, possibly to allow time for the accumulation of a food reserve.

Deutonymphs may appear only at a time when a particular vector is likely to be available. Robinson (1953) found that in *Hericia hericia* which lives in sap exuding from trees, the production of hypopi in May and early June coincides with the egg-laying activities of flies visiting the sap, and these flies are used as a means of dispersal.

Water is necessary for moulting to occur from the deutonymph to tritonymph and R. A. Baker (1964b) noticed that in *Histiostoma feroniarum* which lives in sewage bacteria beds, hypopi were present in large numbers where the active zoogloeal film had been allowed to dry out in the atmosphere. In *H. laboratorium* yeast must be present for moulting, although this need not be in contact with the hypopus. Thiamin and more particularly, riboflavin are also said to play an important part (Perron, 1954).

Working with *Caloglyphus boharti*, Cutcher and Woodring (1969) showed that hypopi will moult at temperatures of over 15 °C and a relative humidity above 97 per cent if they are isolated, but not if they are retained in old stock cultures. They also believe that moulting is affected by the salt concentration of the environment.

Chmielewski (1967a) compared the structure of the genital region of the different developmental stages of the Cryptostigmata and Astigmata. He assumes that the two groups are closely related and regards the hypopus as a heteromorphic deutonymph, in order to distinguish it from the succeeding stage, the homomorphic deutonymph. This nomenclature is not tenable, since the Astigmata and Cryptostigmata are no longer considered to be closely related to one another, and Chmielewski's homomorphic and heteromorphic deutonymphs are not alternative stages in the life cycle, but stages which always succeed one another.

The free-living Astigmata show the same wide range of habitat as is shown by other groups of mites. Some species (*Rhizoglyphus* sp., *Tyrophagus palmarum*, etc.) are found in soil and surface litter and feed on decaying plant remains. Some occur on the leaves of trees [some Saproglyphidae] and others such as *T. palmarum*, *Acarus farris*, *Tyrophagus longior*, *Mycetoglyphus fungivorus* are the inhabitants of grassland and pastures (Griffiths, 1960, Sheals, 1956) and may also be found in stacks of grain and hay sited in open fields. A few species are aquatic (Hyadesidae) and others live in the sap exudates from trees (*Hericia* sp.) or in the surface film of sewage beds.

Perhaps owing to the softness of the cuticle and their small size, many species are domestic in their habits and seek shelter in the nests of insects, birds and mammals where a choice of foods is available. The majority of these nest-dwelling species are scavengers and feed on fungi, remains of food, excrement and on desquamated skin scales or feathers. Many of these scavenging species can live on a wide variety of substances and can also live in human food stores. G. E. Woodroffe (1953) examined the nest contents of a number of species of birds and found six species of stored products mite together with a number of their predators. He is of the opinion that these mites act as scavengers in the nests. One of the mites (*Glycyphagus domesticus*) was not only frequently present in birds' nests, but was later found in a bat roost (Woodroffe, 1956) and is also commonly found in houses and in bee hives. Wasylik (1959) investigated the fauna of sparrows' (*Passer domesticus*) nests in Poland and found 11 species of Astigmata, of which 10 are common pests of stored food. He also examined the nests of *Passer montanus* and noticed that sexually mature mites predominate in the summer and nymphs in the autumn. The mites leave the nests in summer and autumn and it is suggested that at this time, infestation of food stores is likely to occur (Wasylik, 1964).

Grain stores which harbour stored product mites also provide a suitable habitat for many species of fungi and Machacek *et al.*, (from Sinha, 1964a) have isolated over 100 species of fungi from stored grain. Conflicting views have been expressed by early writers as to the relationship between the mites and fungi, but one relationship is well known to mycologists whose culture plates are accidentally infected by *Tyrophagus putrescentiae*. The mites walk across the plates, feeding on the fungi and leaving a trail of footprints marked by the germination of bacterial and fungal spores which they have carried on their bodies. In addition, Griffiths *et al.*, (1959) showed conclusively that *A siro* and *T. putrescentiae* can carry spores in the digestive tract as well as on the outside of the body; these spores are shed with the faeces and serve to distribute the fungi from one stack of grain to another. Griffiths also found that the mites were selective in their feeding (see p. 55) and when given a free choice, they sought out the preferred fungi and digested a greater percentage of spores of the preferred group than they did of the non-preferred fungi.

Sinha (1966a, 1968a), and Sinha and Mills (1968) cultured a number of species of stored product mites on potato-sucrose-agar plates of fungi and Actinomycetes originally obtained from grain. They found that species such as *T. putrescentiae*, *Tyrolichus casei*, *Acarus siro*, *A. gracilis* would feed and multiply on a wide variety of fungal cultures, whilst others (*Suidasia* sp., *Glycyphagus* sp.,) would only flourish on a few species. They also suggest that the presence of moulds on mite-infested food may accelerate the rate of reproduction of the mites (Sinha and Mills, 1968).

Competition can also occur between fungi and mites for the same substrate and most Acarologists are familiar with cultures which grow moulds rather than mites. Solomon *et al.*, (1964) was able to show that the presence of the xerophilic moulds *Sporendonema sebi* Fr. and *Eurotium restrictus* on wheat germ flakes caused populations of *A. siro* to decline owing to reduced reproduction, high mortality and retarded development. Griffiths *et al.*, (1959) found that *A. siro* is able to feed on and digest spores of *E. restrictus*, so that it is possible that the metabolism of the fungus leads to the accumulation of substances in the substrate having a deleterious effect on the mites. Indeed, Solomon found that resterilized food previously infected with *E. restrictus* was consumed by the mites with reluctance.

Heavy infestations of animal and human feeding material with *A. siro* are commonly encountered and conflicting opinions have been expressed as to their possible harmful effects. Szwabowicz *et al.*, (1957, 1958) fed chicks, ducks, pigs, sheep and horses with food contaminated with *A. siro*, but could find no difference in weight and condition between these animals and control ones. Post-mortem examination also failed to show any pathological changes. Parrish's work (personal communication from Brett, 1955), however, indicates that mite-infested food, if fed for long periods of time, is liable to produce degeneration of liver, kidney, adrenal gland and testicle in cockerels, rabbits and rats. He also found that one effect of the mites on the food is to reduce the available amounts of vitamins A and D. Laboratory rabbits, pigs and cockerels fed with clean cereals and dosed with many millions of mites were unaffected by them, whereas control animals fed on mite-contaminated food showed vitamin A and D deficiencies.

Chmielewski (1970) has shown that a small percentage of mites belonging to the species *Carpoglyphus lactis* and *Thyreophagus entomophagus* will survive passage through the alimentary canal of mice and the domestic sparrow, the eggs and immature stages being more resistant than the adults. These animals may therefore contribute to the dispersal of these mites.

Many attempts have been made to provide a system of classification for a group conspicuous for the uniformity of structure and close relationship of its members (Michael, 1901, Oudemans, 1924a and 1906a, Zachvatkin, 1941, Vitzthum, 1943, Nesbitt, 1945, Dubinin, 1953, Yunker, 1955, Hughes, 1961b). In this work a classification based on that devised by Zachvatkin is used and the order Astigmata is divided into a number of families. Of these, the following families include species which are associated with stored food or household premises: Acaridae, Glycyphagidae, Chortoglyphidae, Carpoglyphidae, Histiostomidae, Pyroglyphidae.

ASTIGMATA

Key to the families associated with stored food

Adults

1. Pedipalps with a flattened distal segment. One digit of the chelicera usually with a serrated edge. Ventral surface of idiosoma with four prominent chitinous rings ....Histiostomidae, p. 214

—. Pedipalps without a conspicuously flattened distal segment. Chelicerae chelate. Ventral surface of idiosoma without prominent chitinous rings...2

2. Without vertical setae in either sex. On tarsus I, omega$_1$ is displaced to the distal end of this segment (Fig. 226)...........Pyroglyphidae, p. 193

—. With at least one pair of vertical setae in both sexes (Fig. 9). On tarsus I, omega$_1$ arises from the base of the segment (Fig. 13)..................3

3. With a dorsal transverse groove dividing the propodosoma from the hysterosoma (Fig. 1). Claw attached to the end of the tarsus by paired sclerites surrounded by a short, cushion-like pretarsus (Fig. 12a). If the pretarsus is elongated, then the claw is bifid in the female...Acaridae, p. 28

—. Without a dorsal transverse groove dividing the propodosoma from the hysterosoma. Claw attached to the end of the pretarsus and sometimes joined to the end of the tarsus by two thin tendons (Fig. 12b).........4

4. Apodemes of legs I and II united with the sternum..Carpoglypidae, p. 186

—. Apodemes of legs II free...5

5. Genital opening of female large and covered by two sclerotized plates and situated between coxae III and IV (Fig. 213). Male with conspicuous anal suckers .... Chortoglyphidae, p. 182

—. Genital opening of female with genital plates not so strongly defined or, if so, then the opening is between coxae I and II. Male without anal suckers .... Glycyphagidae, p. 133

ASTIGMATA

Key to stages in life cycle

1. Without or with reduced appendages and usually enclosed within the protonymphal cuticle........Inert hypopus or deutonymph (Fig. 25)

—. With well-developed appendages.................................2

2. With six legs, sometimes with coxal rods (Fig. 19)...............Larva

—. With eight legs, without coxal rods...............................3

3. Chelicerae and pedipalps reduced to a bifid appendage. No mouth. With a collection of suckers on the ventral surface at the posterior end of the body ....Motile hypopus or deutonymph (Fig. 22)

— Chelicerae and pedipalps normally developed. Mouth present. No suckers on the ventral surface at the posterior end of the body.................4

4. With one pair of genital sense organs associated with a rudimentary genital opening.................................Protonymph (Fig. 20)

—. With two pairs of genital sense organs.............................5

5. Genital opening rudimentary. No genital folds....Tritonymph (Fig. 21)

—. Genital folds present.

6. Genital folds short. Penis supported by a series of chitinous struts ....Male (Fig. 10)

—. Genital folds usually long or genital opening covered by one or two plates. Opening to bursa copulatrix at posterior end of body....Female (Fig. 14)

ACARIDAE Ewing and Nesbitt, 1942

All the species are free-living, associated with insects or found in the nests of small mammals.

The dorsal side of the idiosoma is clearly divided by a transverse constriction into propodosoma and hysterosoma.

A dorsal propodosomal shield is usually present.

The cuticle is smooth, rough or thickened to form plates, not evenly wrinkled.

The body setae are usually smooth, sometimes sparsely pectinate, never strongly pectinate or leaf-like in appearance.

Claws are usually well developed and articulate with a pair of sclerites joined to the end of the tarsus; a fleshy pretarsus may be present surrounding the claw and sclerites; where the pretarsus is elongated, the claw is bifid in the female.

On tarsi I and II, the solenidion omega$_1$ arises from the base of the tarsus.

The female genital opening is a longitudinal slit covered by genital folds, each bearing a pair of sense organs on its internal surface.

Anal and two pairs of tarsal suckers are usually present in the male.

Acaridae

Key to the adults

1. External vertical setae *v e* arising near the anterior angles of the dorsal propodosomal shield at the same level as *v i*, or slightly posterior........ (Fig. 26a)...2

—. Seta *v e* rudimentary or absent or, when present, arising near the middle of the lateral edge of the propodosomal shield (Fig. 26b, c)................8a

2. On genu I, $sigma_1$ is more than three times longer than $sigma_2$; claws of female never bifid; femur of male enlarged and bearing ventrally a conical process (Fig. 32)....................................*Acarus*, p. 30

—. On genu I, $sigma_1$ is less than three times longer than $sigma_2$; femur of male not enlarged and without a ventral conical process.....................3

3. Female with bifid claws on all the legs (Fig. 89); male always heteromorphic with the third leg ending in two conspicuous processes......... ....*Lardoglyphus*, p. 80

—. Female without bifid claws, homomorphic males usually found, although heteromorphic ones also occur.....................................4

4. Internal scapular setae *sc i* longer than the external scapular setae *sc e*; chelicerae and legs poorly tanned...................................5

—. Setae *sc i* shorter than *sc e*; chelicerae and legs reddish-brown in colour ....*Aleuroglyphus ovatus* p. 77

5. Setae *v e* shorter than the genu, arising posterior to *v i* (Fig. 27) ....*Mycetoglyphus fungivorus*, p. 73

—. Setae *v e* as long as or longer than the genu, arising almost on a level with *v i* 6

6. Setae d_1 and *l a* about equal in length, shorter than d_3 and d_4............7

—. Seta *l a* four to six times longer than d_1..........*Tyrolichus casei*, p. 70

7. Seta *e* on the dorsal apex of tarsus I short and needle-shaped; five ventral terminal spines at the ends of the tarsi, of which the three central ones are thickened (Fig. 28a, b).............................*Tyrophagus*, p. 49

—. Seta *e* is usually spine-shaped; three ventral spines at the end of the tarsi (Fig. 28c, d)...................................*Tyroborus lini*, p. 68

8b. On tarsi I and II, *ba* is enlarged to form a stout conical spine and is situated close to $omega_1$ (Fig. 29)........................*Rhizoglyphus*, p. 115

—. On tarsi I and II, *ba* is a slender seta.................................9

9. Setae *sc i*, *h i*, d_1 to d_3 absent in the adult; male with a projecting shelf on the posterior dorsal margin of the hysterosoma ....*Thyreophagus entomophagus*, p. 123

—. A complete set of setae is present on the dorsal and lateral surface of the idiosoma; the posterior edge of the hysterosoma is not produced into a projecting shelf in the male...........................*Caloglyphus*, p. 88

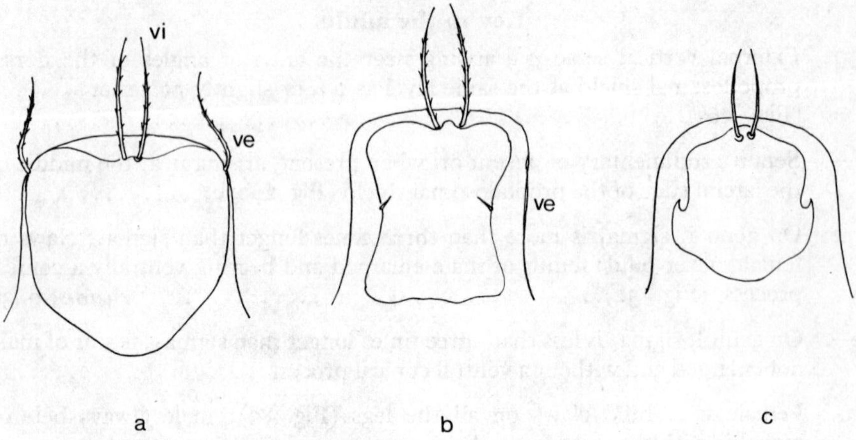

FIG. 26. Position of vertical setae *v e* in (*a*) *Tyroborus lini* Oudemans, (*b*) *Caloglyphus mycophagus* (Mégnin) and (*c*) *Thyreophagus entomophagus* (Laboulbène); *v e* and *v i* external and internal vertical setae.

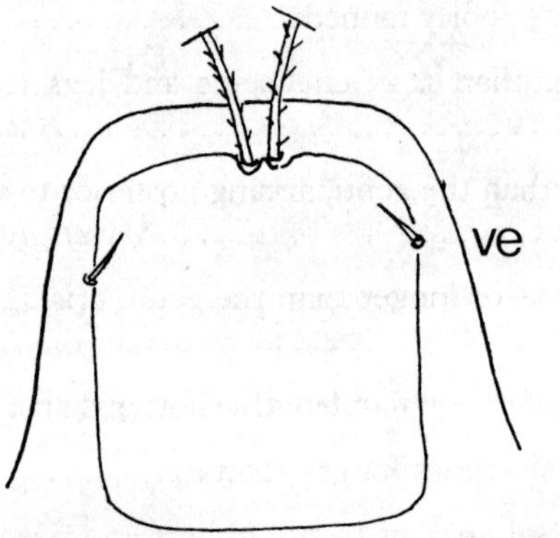

FIG. 27. Position of vertical setae in *Mycetoglyphus fungivorus* Oudemans.

ACARUS L., 1758

Tyroglyphus Latreille, 1796; *Aleurobius* Canestrini, 1888

Setae *v e* are less than half the length of *v i*. Setae d_1 and *l a* are always short. The solenidion sigma_1 on genu I is more than three times longer than sigma_2.

Sexual dimorphism is apparent, the male being distinguished by an enlarged first pair of legs, the femur of which bears a stout apophysis or process on its ventral surface; small thickenings of cuticle may also be present on the ventral side of genu I. The reproductive system of the female has a characteristic structure [see Fig. 15].

Type species: *Acarus siro* L, 1758.

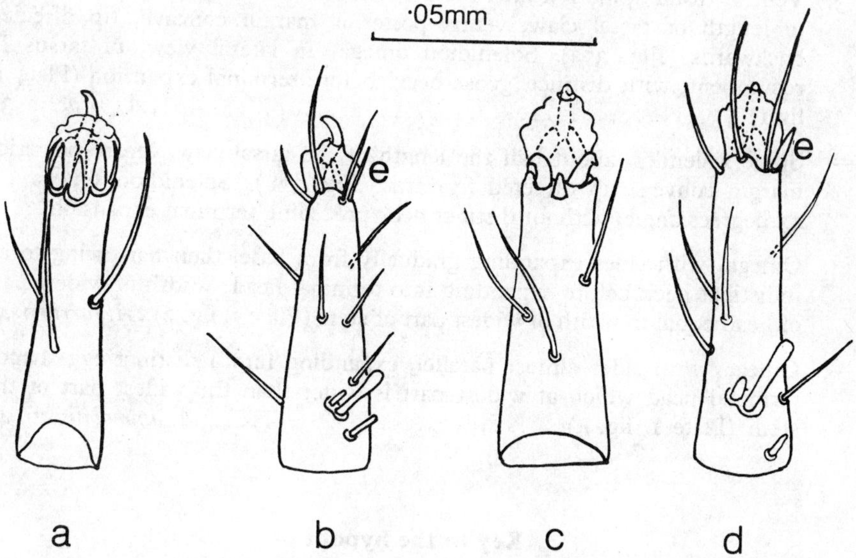

FIG. 28. *Tyrophagus perniciosus* Zachvatkin, ventral (*a*) and dorsal (*b*) views of tarsus I. *Tyroborus lini* Oudemans, ventral (*c*) and dorsal (*d*) views of tarsus I.

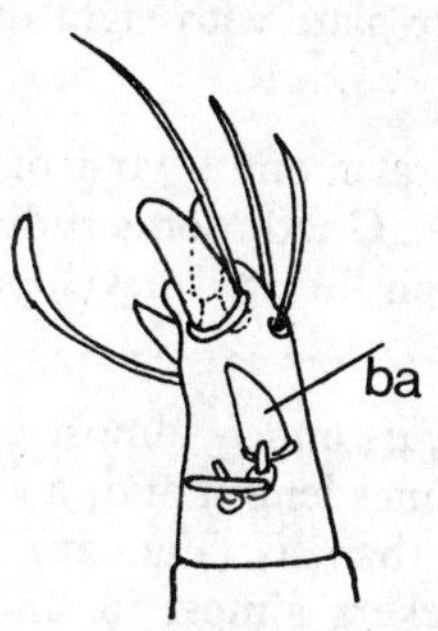

FIG. 29. *Rhizoglyphus robini* Claparède ♂. Dorsal view of right leg I to show spine *ba*.

ACARUS

Key to the adults [Griffiths, 1964a]

1. Dorsal setae d_2 not more than twice the length of d_1 2

—. Dorsal seta d_2 four to five times longer than d_1 ......*Acarus gracilis*, p. 46

2. Hysterosomal setae *h i*, *l a*, *l p* and d_1 to d_4 short. In particular, the length of dorsal setae d_2 or d_3 does not exceed the distance between its base and the base of the seta immediately posterior to it..........*A. siro* complex 3

—. Hysterosomal setae *h i*, *l a*, *l p* and dorsals longer; generally, in the majority of individuals of a given population, the length of d_2 and d_3 is greater than the distance between its base and the base of the seta immediately posterior to it.......................long seta species group (*see Griffiths, 1970)

*This includes *A. nidicolous*, *A. macrocoryne* and *A. chaetoxysilos:* Field species so far rarely encountered in stored product material.

3. Ventral distal spine *s* of tarsi I and II (not tarsus I of ♂) large, about equal to length of tarsal claw, ventro-posterior margin concave, tip directed backwards (Fig. 37a). Solenidion omega$_1$ in lateral view, of tarsus II recumbent, with distinct 'goose-neck' before terminal expansion (Plate 1, fig.1) . . . . *A. siro*, p. 33

—. Spine *s* slender, about half the length of the tarsal claw, ventro-posterior margin convex, tip directed forwards (Fig. 37b). Solenidion omega$_1$ at 45 degrees angle, without distinct neck preceding terminal expansion 4

4. Omega$_1$ with sides expanding gradually from base, then narrowing to an indistinct neck before expanding into terminal head. Width of widest part of head equal to width of widest part of stem (Plate 1, fig. 2) . *A. farris* p. 41

—. Omega$_1$ with sides almost parallel, expanding into a distinct egg-shaped terminal head which at widest part is wider than the widest part of the stem (Plate 1, fig. 4) . . . . *A. immobilis*, p. 43

Key to the hypopi

1. In dorso-ventral mount, terminal three or more segments of legs I and II extend beyond margin of body. Gnathosoma bears pair of long terminal flagelliform bristles. Sucker plate with eight distinct suckers . . . . (Motile hypopal form) 2

—. Legs short; in dorso-ventral mounts tarsi of legs I and II are the only segments completely visible. Gnathosoma rudimentary, flagelliform bristles not present. Only one pair of well-developed suckers on sucker plate (Inert hypopal form) 3

2. Hysterosomal setae d_1, d_2, d_3 and l_1 almost as long as scapulars (*sc i* and *sc e*). Setae *sc i* about 1.2 times length of d_1 and 1.5 times l_1. Setae d_1 and l_1 about three times longer than d_4 (Fig. 22a). Bases of genital setae and flanking pair of coxal suckers almost in line, distance between base of sucker and base of seta is less than width of setal base (Fig. 22b) . . . . *A. siro*, p. 35

—. Hysterosomal setae appreciably shorter than scapulars. Setae *sc i* about twice the length of d_1 and three times l_1. Setae d_1 and l_1 about equal in length to d_4 (Fig. 39a). Bases of genital setae just forward of coxal sucker bases. Distance between base of sucker and base of seta about equal to width of setal base (Fig. 39b) . . . . *A. farris*, p. 42

3. On sucker plate the central pair of suckers vestigial, anterior peripheral pair moderately well developed (Fig. 40b). All setae on tarsus III and IV shorter than length of tarsus, spine-like, never leaf-shaped (Fig. 41b). Omega$_1$ long, at least twice the length of tarsal claw (Fig. 41a) . . . . *A. immobilis*, p. 44

—. Central pair of suckers well developed, peripheral pairs vestigial (Fig. 45). All setae on tarsi III and IV longer than length of tarsus, leaf-like or pectinate (Fig. 46b). Omega$_1$ at least three times shorter than tarsal claw (Fig. 46a) . . . . *A. gracilis*, p. 48

ACARUS SIRO L., 1758

Acarus siro var. farinae L, 1758; *Aleurobius farinae var. africana* Oudemans, 1906.

MALE (Fig. 10). Length of idiosoma: 320 to 460 µm.

The body is colourless, the gnathosoma and legs varying in colour, according to diet and age, from pale yellow to reddish brown; the hind margin is smoothly rounded. The idiosomal setae are fine and some are sparsely pectinate, the pectinations being particularly visible on *v i* and the distal halves of *sc*.

The dorsal propodosomal shield is a broad plate extending back almost as far as the scapular setae. Setae *v i* extend almost to the tips of the chelicerae, *v e* are considerably shorter (less than one-quarter the length of *v i*) and arise close to the lateral corners of the propodosomal shield. Setae *sc* are about 25 per cent of the idiosoma in length and arranged in a transverse row; *sc e* are slightly shorter than *sc i*. The supracoxal seta (Fig. 30) is expanded basally and thickly pectinate. Grandjean's organ is a colourless fold of cuticle whose distal edge is drawn out into a number of filaments of unequal length.

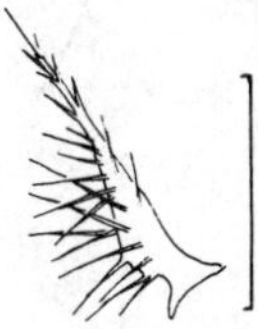

FIG. 30. *Acarus siro* L. Supracoxal seta.

The setae of the dorsal surface of the hysterosoma vary in length and thickness according to the nutritional quality of the food received in pre-adult stages. A poor diet produces adults with short, thin setae; a rich diet, longer thicker ones. However, in the vast majority of the populations examined by D. A. Griffiths (1970) (Fig. 31), the length of the dorsal setae d_1 and d_3 is less than the distance between the base of each seta and the base of the dorsal seta immediately posterior to it. The tips of the d_2 setae, however, when projected backwards in a straight line, do not often overlap the bases of d_3. Great variation in setal length can occur in the same population [see Griffiths, 1970]. The posterior margin of the body is fringed with two pairs of short setae—*sa e* and $p\ a_3$—and two pairs of long ones—*sa i* and $p\ a_2$, the long setae are curved and trail along the ground.

The apodemes of legs I are united in the mid-line to form a short sternum, those of legs II, III and IV are free.

The genital opening is between coxae IV; the lateral arms of the penis support diverge posteriorly and the penis is an arc-shaped tube with a blunt end. A pair of anal suckers lies on either side of the posterior end of the anus and anterior to these are the *pr a* setae.

The chelicerae (Fig. 6) are distinctly toothed; at the base of the fixed limb is a mandibular spine and, posterior to this, a conical spurlike outgrowth.

All the legs terminate in a well-developed pretarsus and stalked claw. Leg I (Fig. 32a) has the genu and femur enlarged and a spur projecting from the ventral side of the femur from which arises the seta *vF*. Two blunter thickenings of the cuticle are also present on the ventral side of the genu, but are not easily seen on lightly tanned specimens.

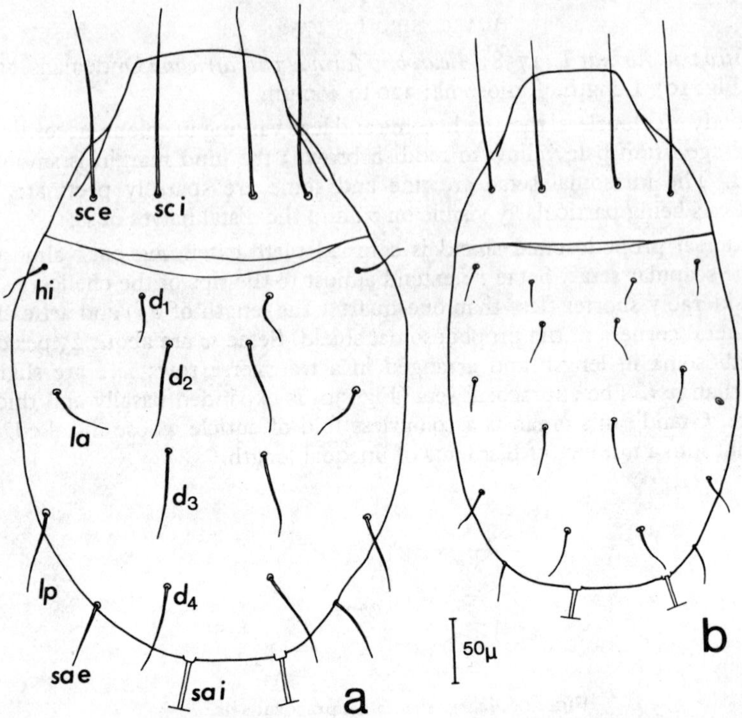

FIG. 31. *Acarus siro* L. Idiosoma of sibling ♀♀ illustrating variation in length of dorsal setae (d_1 to d_4), brought about by exposing pre-adult stages to (*a*) a rich diet and (*b*) a poor one. Setae of the idiosoma: *sc e, sc i, h i, l a, l p, sa e, sa i* (From Griffiths, 1970).

On tarsi I and II (Plate 1, fig. 1) omega$_1$ is recumbent and the angle between the dorsal surface of the tarsus and the anterior face of the solenidion rarely exceeds 45 °C. It is broadest at the base, then narrows before swelling out into a terminal expansion. The famulus (ε) is shaped like a minute hump, and arises from a small pit anterior to omega$_1$. At the distal end of tarsus I, *u* and *v* are united to form one large spine (Fig. 32a). The spine *s* on tarsi II to IV is large and, when the tarsus is viewed laterally, the length of the longest edge (the concave margin) is equal to that of the tarsal claw. On genu I, sigma$_1$ is more than three times as long as sigma$_2$. The suckers on tarsus IV of the male (Fig. 33a) are separate from one another by a distance equal to their diameter, and are nearer to the base than to the apex of the segment.

Griffiths (1966) has shown that the tarsal length in adults also varies according to the quality of the food received during the pre-adult period.

FEMALE (Fig. 1). Length of idiosoma: 350 to 650 µm.

The body is more oval than that of the male and the posterior margin is slightly indented in the region of the terminal bursa copulatrix.

The arrangement of the dorsal setae of the idiosoma is the same as that of the male, but the setae are usually more sparsely pectinate. Ventrally (Fig. 34) five pairs of setae surround the anal opening. Of these, a_2 are twice and a_3 almost four times the lengths of a_1, a_4 and a_5; *p* a_1 and *p* a_2 extend well beyond the posterior margin of the body.

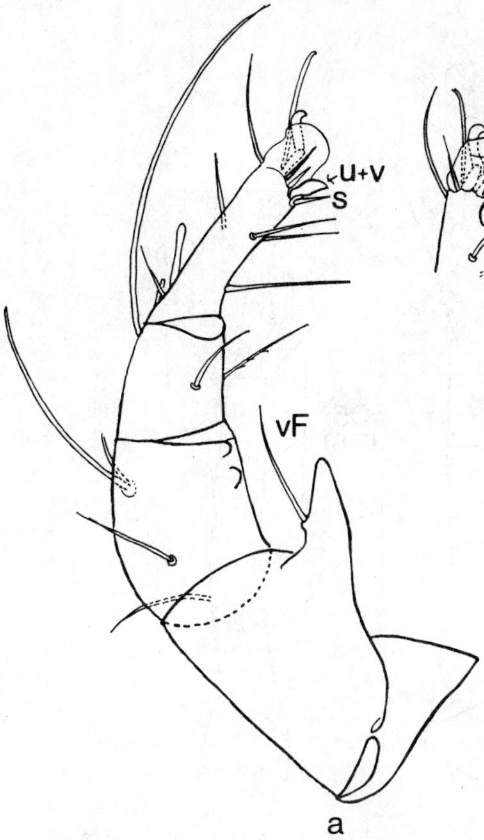

FIG. 32. *Acarus siro* L. ♂. (*a*) Internal view of right leg I of ♂. (*b*) Internal view of apex of tarsus I of ♀. Tarsal spines: *s*, *v*, *u* + *v*. Femoral seta: *vF*.

The genital opening lies between coxae III and IV. The bursa copulatrix opens into a narrow expansible tube which joins a sclerotized bell-shaped structure. This expands into a large thin-walled receptaculum which communicates with the oviducts by two openings (Fig. 15).

The first pair of legs is no wider than the remaining ones and no spur arises from the femur. The chaetotaxy is the same as that of the male, except that on tarsus I, the spines *u* and *v* are distinct and much narrower than *s* (Fig. 32b); *s* on all the legs, is large, with a backward curve (Fig. 37a).

HYPOPUS (Fig. 22a, b). Average length of 10 specimens: 230 µm (Griffiths, 1964a).

Only a motile form is developed and the strain which can give rise to it is rarely found.

The idiosoma is pinkish brown in colour, its dorsal surface arched and covered with scattered pits; the ventral side is concave. The propodosomal shield is distinctly separated from the hysterosomal and projects forward, almost completely concealing the gnathosoma. No eyes are present.

The *v i* setae are strongly pectinate, the *v e* much shorter. Supracoxal setae arise above coxae I and are almost the same length as *v i*. The bases of the four scapular setae traverse the propodosoma, *sc i* being slightly longer than *sc e*.

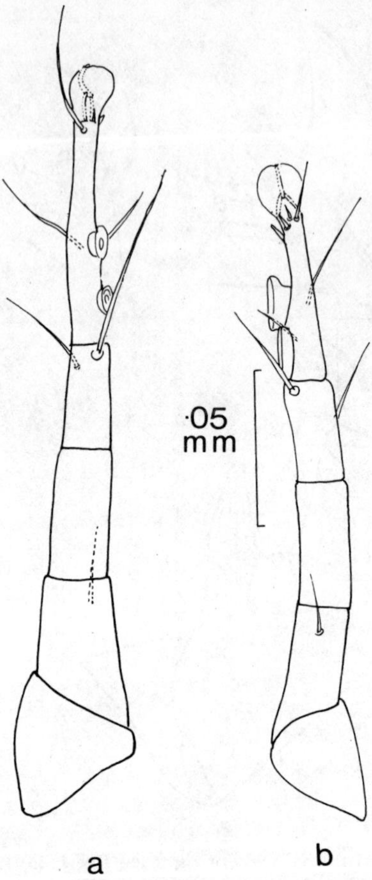

FIG. 33. Lateral view of leg IV of (*a*) *Acarus siro* L. ♂, (*b*) *A. gracilis* Hughes ♂.

These setae are flattened and slightly expanded in their middle regions and extend posteriorly beyond the junction of propodosoma and hysterosoma.

Of the hysterosomal setae, d_2 lie between the d_1 setae, and d_3 and d_4 are in linear series with d_2. Two pairs of humerals project from the sides of the body on the same transverse line as the bases of d_1 and d_2. Three pairs of laterals are present; l_1 is lateral to d_3, l_2 and l_3 arise from the body margin between l_1 and d_4. Setae d_2 and d_3 are almost as long as *sc i* and approximately equal to d_1 and l_1 in length. Setae d_1 and l_1 are three times longer than d_4. The latero-abdominal glands open on the sides of the body on a level with coxae IV. Ventrally, coxae II are well-defined laterally by epimerites II which are joined to apodemes III; the sternum and apodemes II are free and coxae III are only partially separated from one another in the midline. Apodemes IV are barely curved and do not meet; just in front of their termination arise the coxal setae. Coxae II, III and IV are distinctly thickened along their edges.

The genital setae arise almost in line with a pair of suckers which are anterior to the sucker plate; the setal base is separated from the edge of the sucker by a distance less than the width of the setal base. The sucker plate is small and well-removed from the posterior edge of the body, its large central suckers are surrounded by three pairs of peripheral suckers separated from one another by clear areas.

Legs I and II are almost completely visible when the animal is viewed from above, but only the terminal setae of III and IV project beyond the edge of the body. All four pairs of legs have well-developed claws and a reduced pretarsus. The chaetotaxy differs from that of other stages, the differences involving change in shape, as well as the expression or suppression of certain setae and solenidia.

Omega$_2$ and one sigma on leg I (Fig. 35a) and a sigma on leg III are not developed. The seta *l* on legs III and IV is found in the hypopus only. The ventral spine complex of the other stages is replaced by two expanded leaflike setae *vsc*. Omega$_1$ on tarsi I and II is longer and more slender than in other stages, it is slightly 'goose-necked' with a distinct terminal expansion. Omega$_3$ which is terminal in nymphal and adult stages, arises from the middle dorsal region of tarsus I. On tarsi I and II, seta *e* is expanded at its distal end into a sucker-like structure, on tarsus III it is leaflike and on tarsus IV it is about half the length of the idiosoma. Seta *f* is thin, transparent and leaflike on all the legs; *ra* or *r* is also leaflike on all the legs except on IV where it is simple; *la* or *l* is long and leafshaped on tarsi I to III; *wa* is broad, flattened and heavily pectinate; *w* is smooth on tarsus III, flattened and pectinate on tarsus IV. The solenidion phi exceeds the length of the tarsus on legs I, on tibia II it is about the same length as tarsus II.

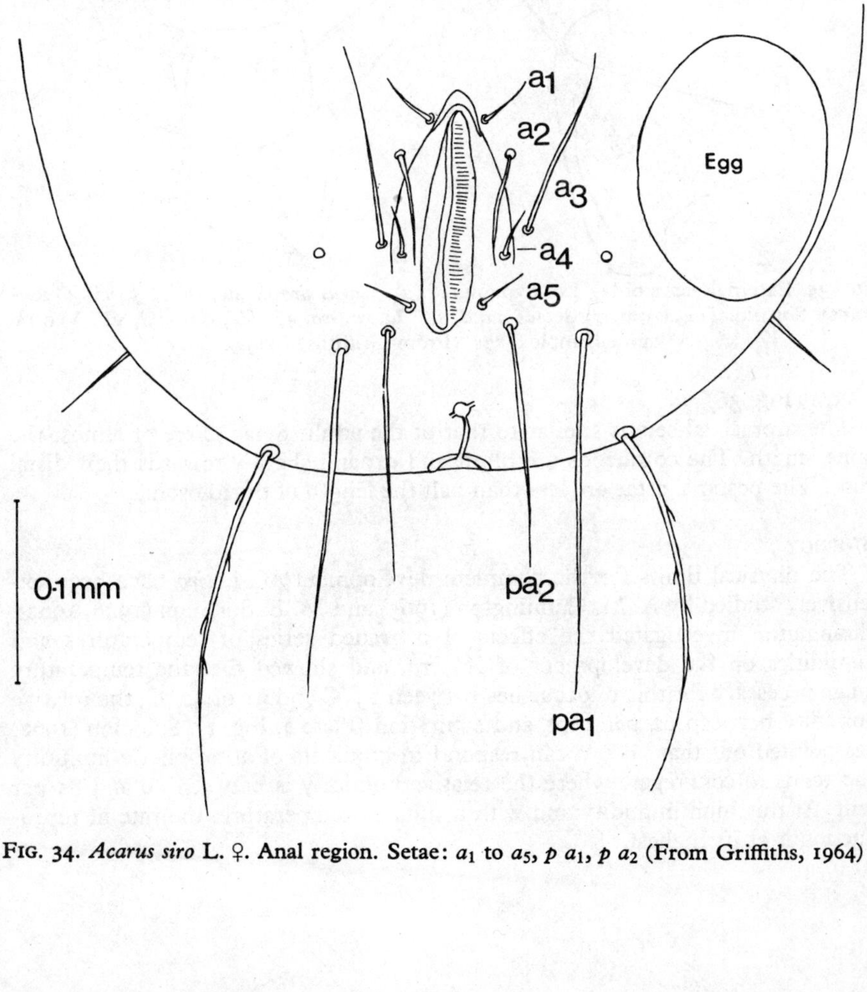

FIG. 34. *Acarus siro* L. ♀. Anal region. Setae: a_1 to a_5, $p\ a_1$, $p\ a_2$ (From Griffiths, 1964).

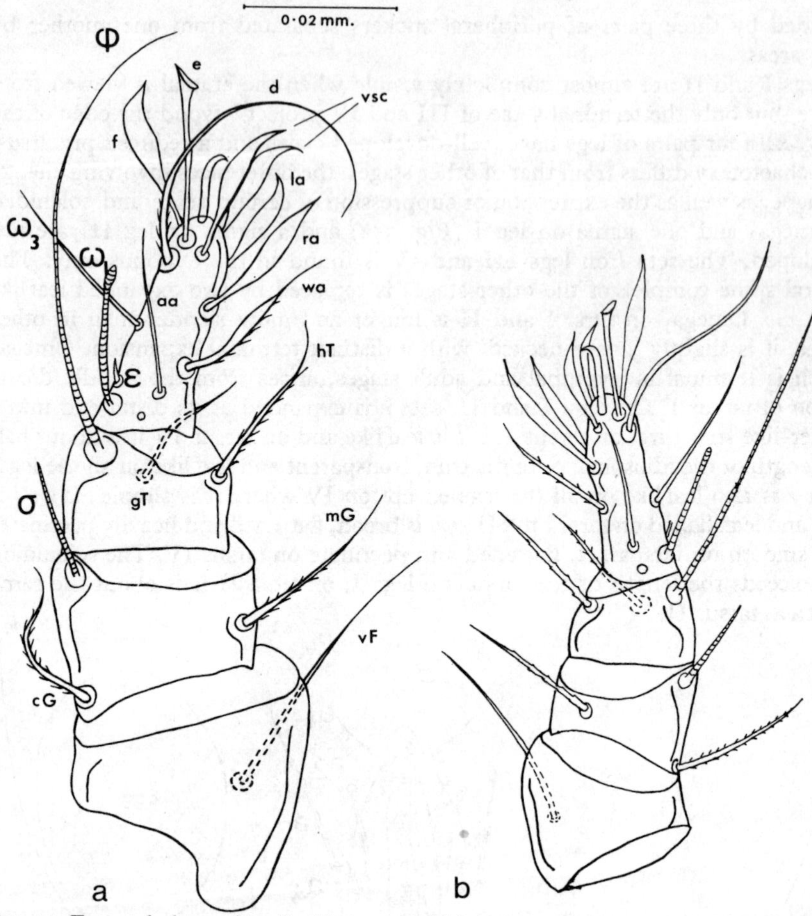

FIG. 35. External views of leg I of hypopus of (*a*) *Acarus siro* L. and (*b*) *A. farris* (Oudemans). Solenidia: ω_1, ω_3, Φ, σ; Setae: *aa, d, e, f, la, ra, wa, gT, hT, cG, mG, vF*. Ventral spine complex: *vsc*. (From Griffiths, 1964a).

LARVA (Fig. 36)

The supracoxal seta is similar to that of the adult. Setae *sc* are of almost the same length. The coxal rods are blunt and expand slightly towards their distal ends. The postanal setae are less than half the length of the idiosoma.

BIOLOGY

The physical limits for the complete development of *A. siro* have been extensively studied by A. M. Cunnington (1965) and M. E. Solomon (1946, 1962). Cunnington investigated the effects of a graded series of temperatures and humidities on the development of *A. siro*, and showed that the temperature range necessary for this to occur lies between 25 °C and 31 or 32 °C, the relative humidity between 62 per cent and saturation (Plate 4, Fig. 1). Solomon (1962) has pointed out that *A. siro* can respond to gradients of atmospheric humidity and tends to congregate where the relative humidity is between 80 and 85 per cent. At this high humidity and with a suitable temperature, the rate of reproduction is at its highest.

Solomon (1946) has also related mite numbers and grain damage with moisture content of grain. He found that damaged grain with a water content of about 14 per cent or over had the germ eaten away, whilst grain containing 13 per cent water or less was unaffected. Few mites were found in very damp grain as the germ had already been consumed and the mites had left the site because of overcrowding and food shortage.

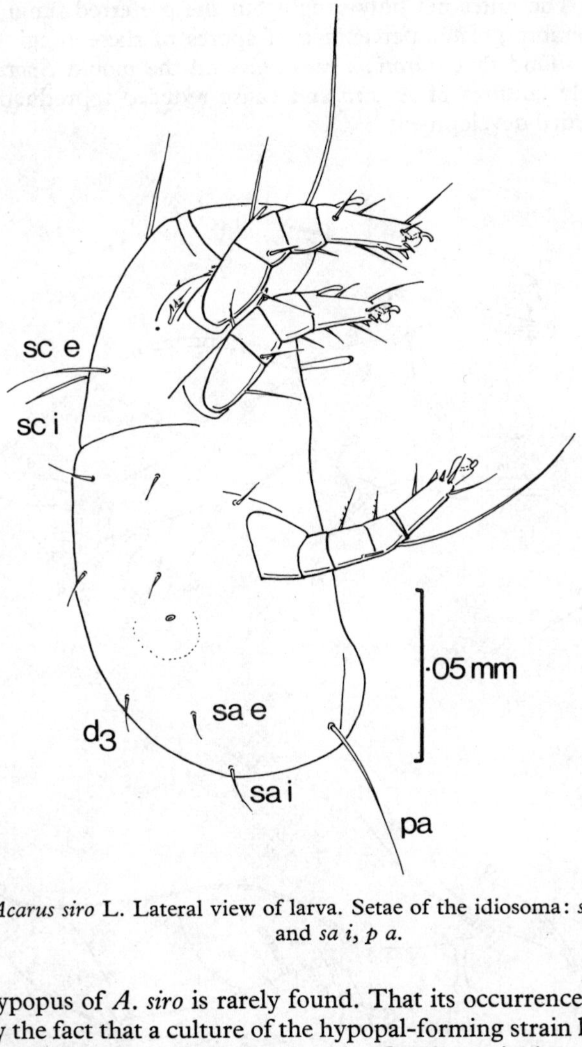

FIG. 36. *Acarus siro* L. Lateral view of larva. Setae of the idiosoma: *sc e* and *sc i*, d_3, *sa e* and *sa i*, *p a*.

The hypopus of *A. siro* is rarely found. That its occurrence is not sporadic is shown by the fact that a culture of the hypopal-forming strain kept by the author over a period of years always produced a few hypopi, the number increasing when the culture was starved. Non-hypopal forming strains fail to produce hypopi when the food becomes scarce.

Acarus siro is the most important pest of storage premises and is found more often on processed cereal products such as flour than on cereals such as whole grain, or on hay in farm stores (Griffiths, 1964a). It has also been recorded from cheese, the deep litter of broiler houses and in disused bee hives. On the surface of flour it occurs to the depth of a few inches. In 1965 in the Eastern region of

England, barley infested with *A. siro* was thought to be the cause of scouring in cattle. In 1968 Sinha and Mills recorded it as a pest of *Penicillium* sp. and in 1959 Griffiths *et al.* showed that when *A. siro* and *T. putrescentiae* were kept on grain contaminated with fungi belonging to the genera *Aspergillus* and *Eurotium*, they preferred to feed on species of *Eurotium*; thus *E. amstelodami*, *E. repens* and *E. ruber* which thrive in grain of moisture content of 13 to 15 per cent, were preferred to *A. candidus*, *A. ochraceus* and *A. flavus* which require a higher moisture content. The mites not only sought out the preferred fungi, but also digested a considerably greater percentage of spores of these fungi. Solomon *et al.* (1964) have found that *Eurotium restrictus* and the mould *Sporendonema sebi* affect adversely cultures of *A. siro* and cause reduced reproduction, high mortality and retarded development.

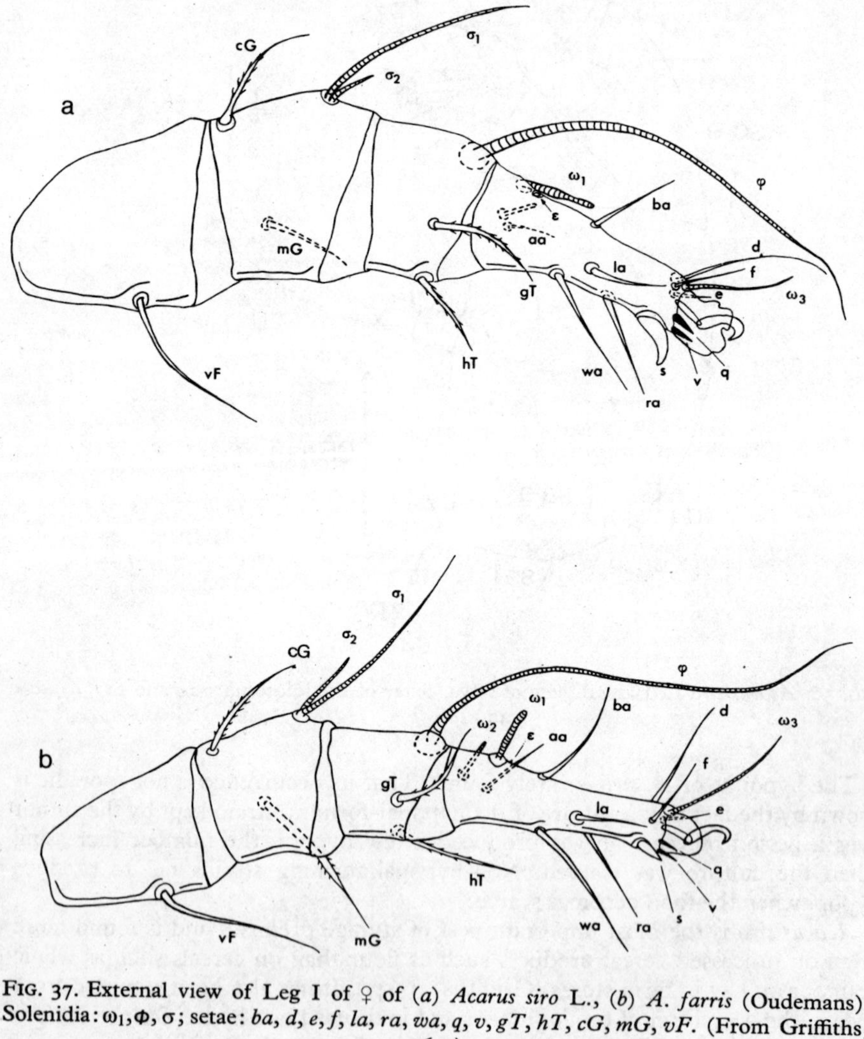

FIG. 37. External view of Leg I of ♀ of (*a*) *Acarus siro* L., (*b*) *A. farris* (Oudemans). Solenidia: ω_1, Φ, σ; setae: *ba, d, e, f, la, ra, wa, q, v, gT, hT, cG, mG, vF*. (From Griffiths, 1964a).

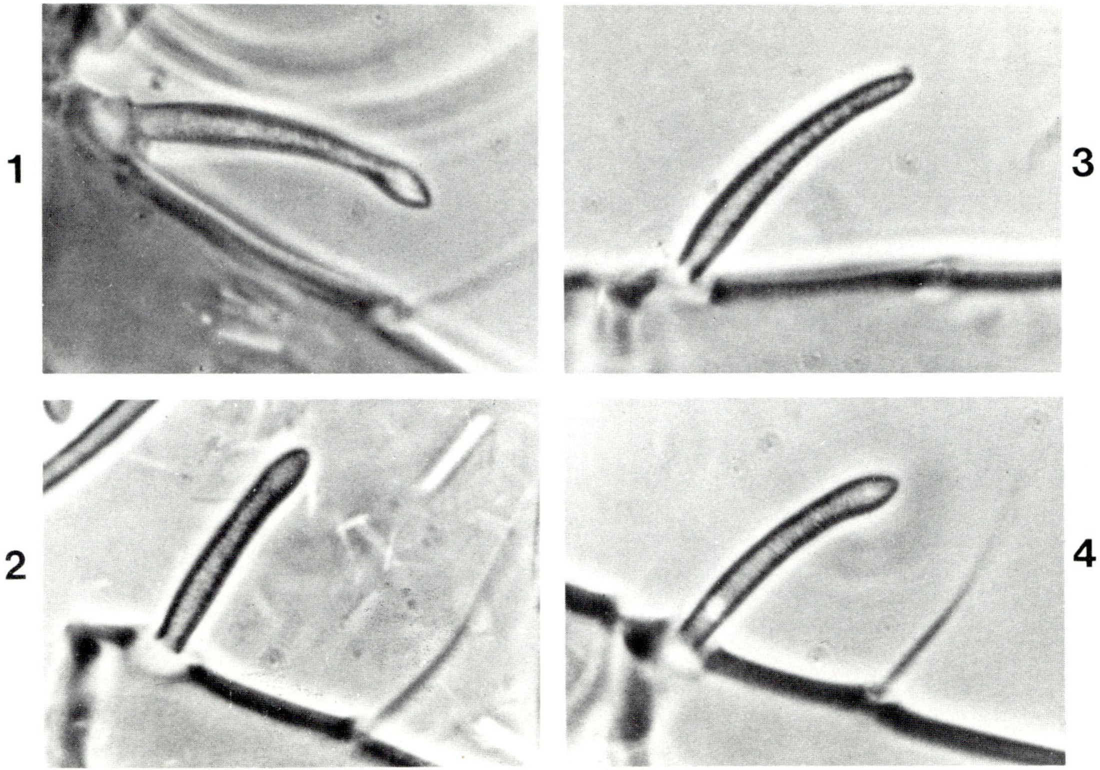

PLATE I

Lateral view of solenidion omega I situated on tarsus II of the females. Mag. x 2,250, Fig. 1 *Acurus siro* L. Fig. 2. *A. farris* (Oudemans), Fig. 3. *A. gracilis* Hughes. Fig. 4. *A. immobilis* Griffiths. (From Griffiths, 1964.)

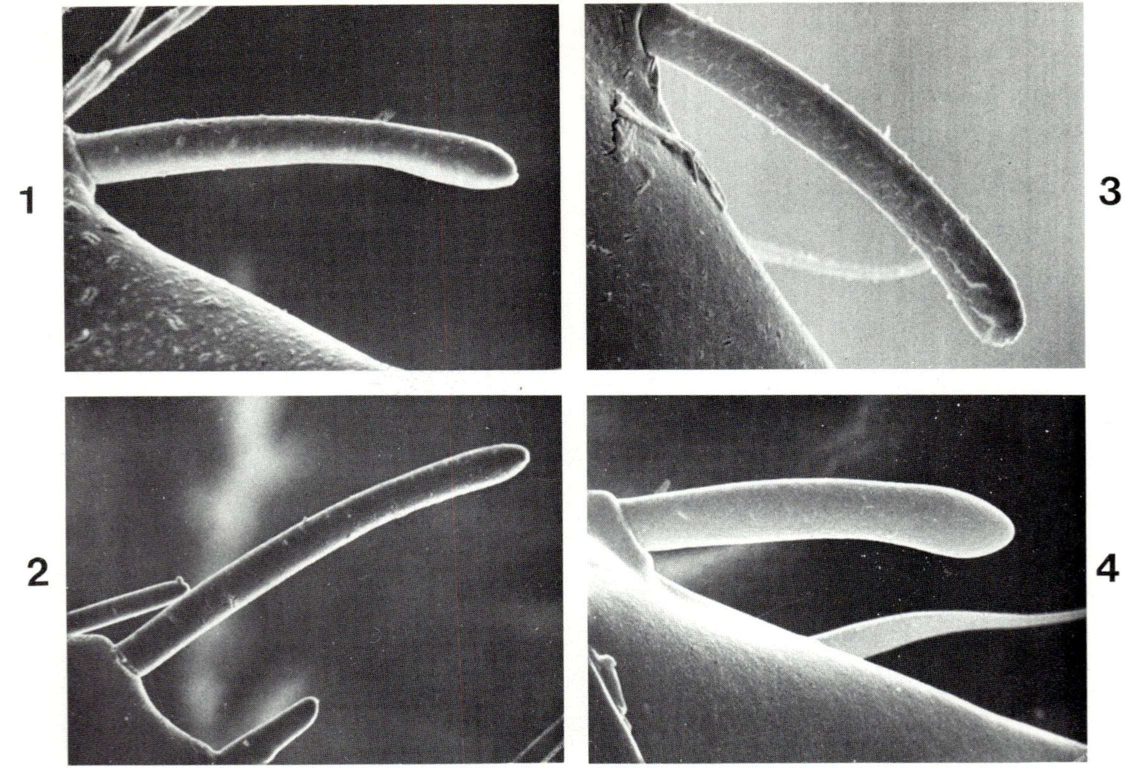

PLATE 2

Lateral view of solenidion omega I situated on tarsus II. Fig. 1. *Tyrophagus putrescentiae* (Schrank). Tarsus II (x 2,700). Fig. 2. *T. longior* (Gervais) Tarsus I (x 2,900). Fig. 3. *T. palmarum* Oudemans Tarsus II (x 3,000). Fig. 4. *T. perniciosus* Zachvatkin Tarsus II (x 3,100). (By courtesy of D. A. Griffiths.)

1

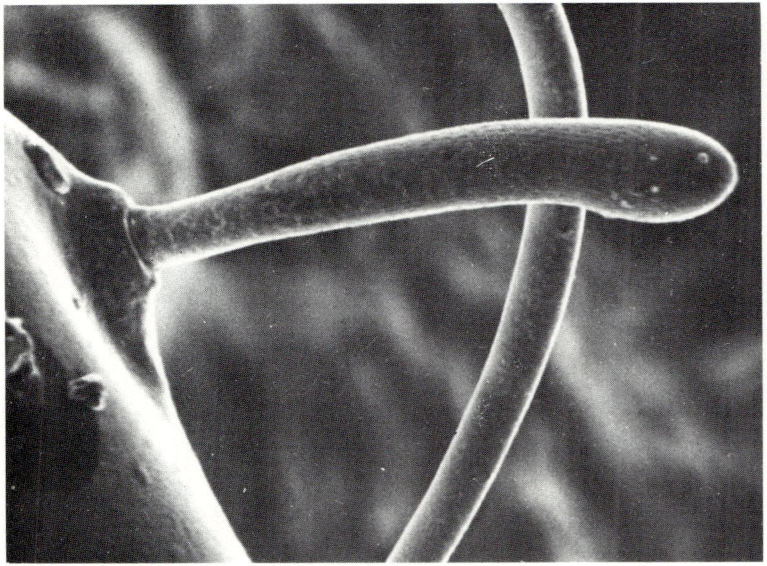

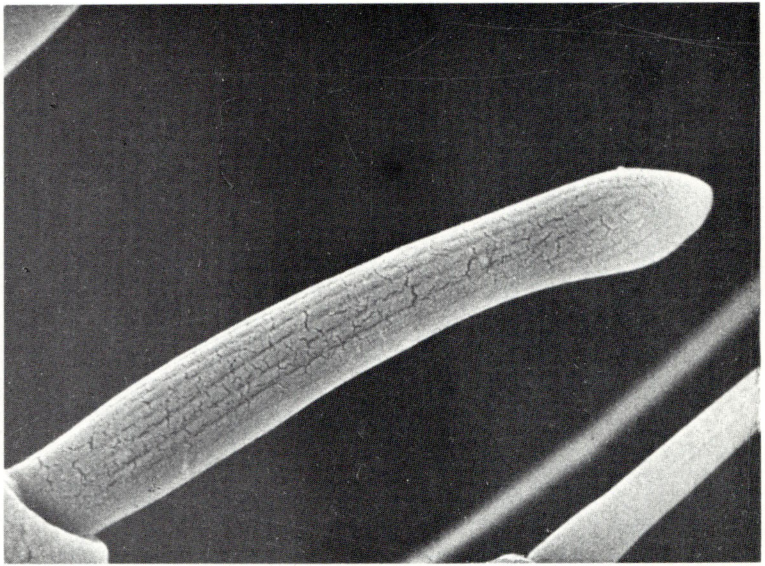

2

PLATE 3

Lateral view of solenidion omega I situated on tarsus II. Fig. 1. *Tyrophagus similis* Volgin Tarsus II (x 3,800). Fig. 2. *Tyrolichus casei* Oudemans Tarsus II (x 4,500). (By courtesy of D. A. Griffiths.)

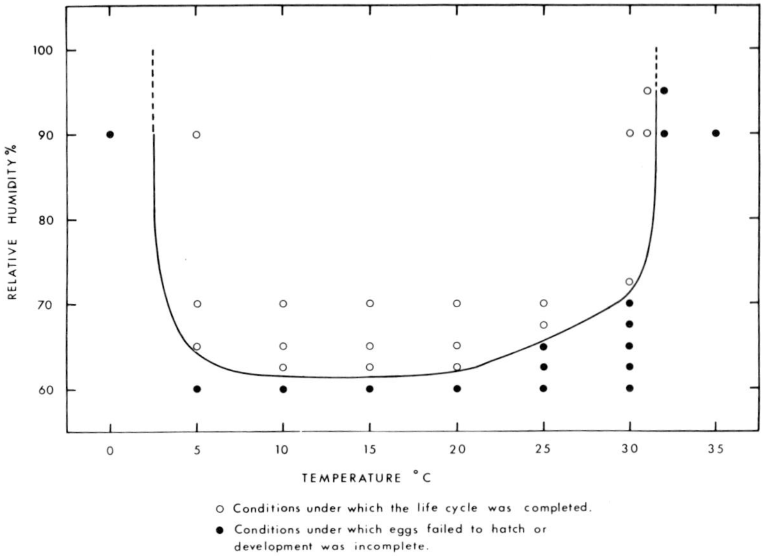

Fig. 1. Physical limits for complete development of the grain mite, *Acurus siro* L. (From Cunnington, 1965.)

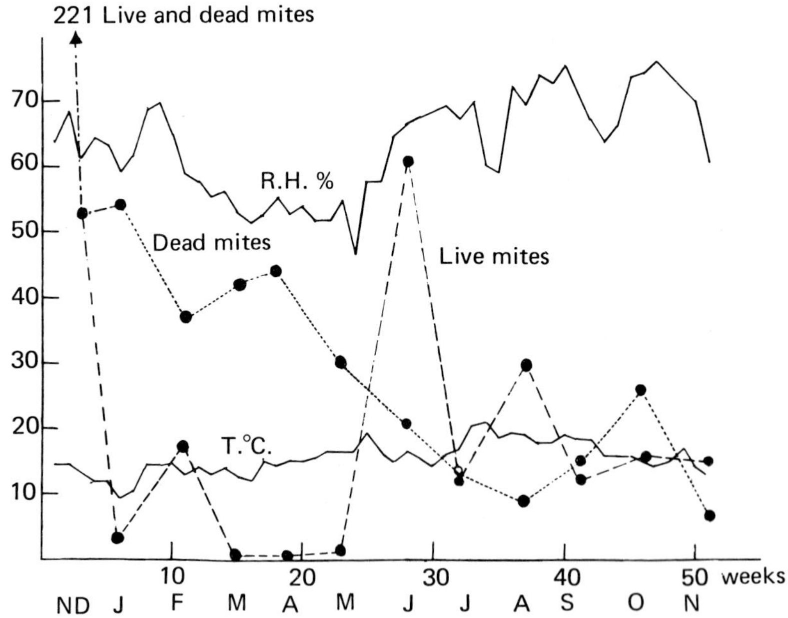

Fig. 2. Weekly averages of temperature (T) and relative humidity (RH) outside a bed between November, 1970 and November, 1971. The months of the year are indicated by their initials. (From Hughes and Maunsell, 1973.)

PLATE 4

ACARUS FARRIS (Oudemans, 1905)

Aleurobius farris Oudemans, 1905.

Generally speaking, *A. farris* is smaller, with more lightly tanned appendages than *A. siro*. Hybridization can occur to a limited extent between the two species (see D. A. Griffiths, 1964a).

MALE. Average length of the idiosoma of six specimens: 365 μm (D. A. Griffiths, 1964a).

This closely resembles the male of *A. siro* and differs in the following particulars:

On legs I and II, omega$_1$, when viewed laterally, increases slightly in diameter from the base upwards, then narrows almost imperceptibily before expanding a little into a terminal head (Plate 1, Fig. 2), the angle between its anterior margin and the dorsal surface of the tarsus is nearer to 90 degrees than 45 degrees; the spine *s* on tarsi II, III and IV of the male, and I to IV of the female (Fig. 37b) when viewed laterally, is about one-half to two-thirds the length of the tarsal claw and points anteriorly.

FEMALE (Fig. 38).

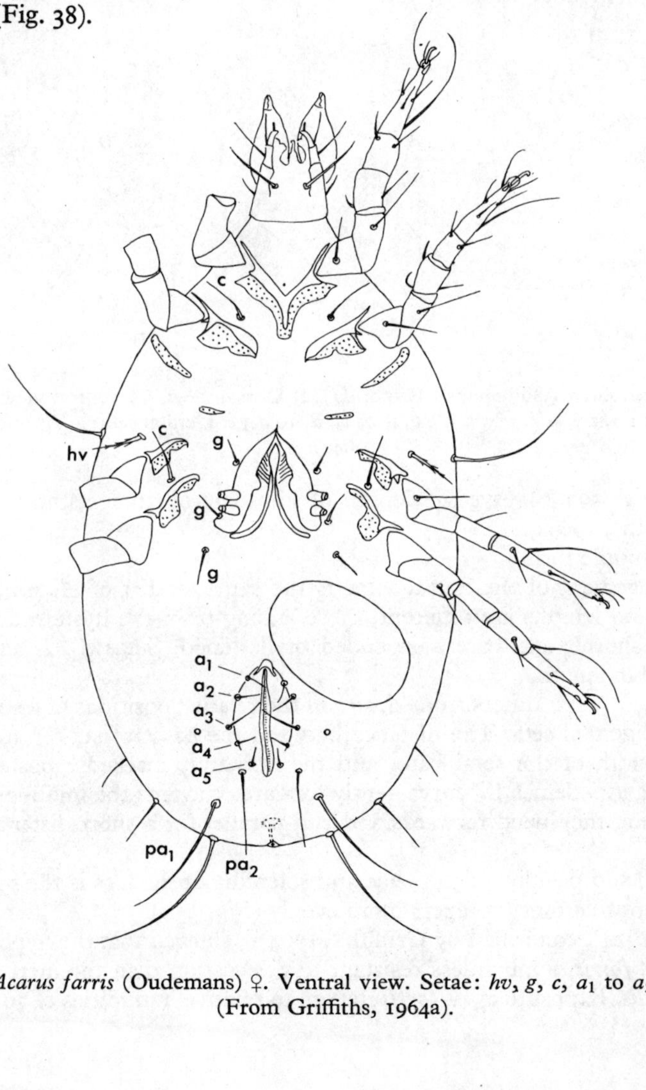

FIG. 38. *Acarus farris* (Oudemans) ♀. Ventral view. Setae: *hv*, *g*, *c*, a_1 to a_5, $p\ a_1$, $p\ a_2$. (From Griffiths, 1964a).

This is larger than the male. The anal setae differ from those of *A. siro* in that a_1, a_4 and a_5 are almost equal in length, a_2 are about one-third and a_3 are about twice as long (Fig. 38).

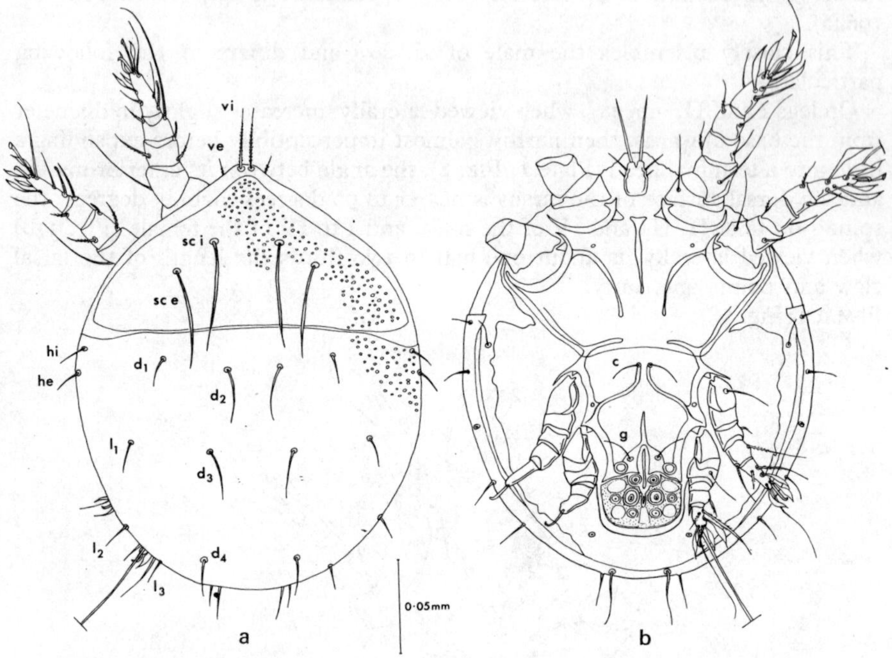

FIG. 39. *Acarus farris* (Oudemans). Hypopus. (*a*) Dorsal view, (*b*) ventral view. Setae of the idiosoma: *v e*, *v i*, *sc e*, *sc i*, *h i*, *h e*, l_1 to l_3, d_1 to d_4, *c*. Genital seta: *g*. (From Griffiths, 1964a).

HYPOPUS (Fig. 39a, b). Average length of idiosoma of 20 specimens: 240 μm (D. A. Griffiths, 1964a).

This is a motile form.

The arrangement of the dorsal setae is the same as that of *A. siro*, but the setal shape and lengths are different, those arising from the hysterosoma being appreciably shorter and barely expanded or flattened. Setae d_1, l_1 and d_4 are almost equal in length.

Ventrally, *A. farris* differs from *A. siro* in the relative positions of a sucker and the adjacent genital seta. The distance between the bases of sucker and seta is about the width of the setal base, and the sucker is markedly posterior and external to it. Apodemes IV curve gently forwards towards the midline, but just before meeting they bend forwards and run parallel for a short distance before ending.

The shape and position of the setae and solenidia of the legs is the same as in *A. siro*, except that omega$_1$ tapers more evenly (Fig. 35b).

Schulze [1924], confirmed by Griffiths (1964b), showed that the hypopus now known as *A. farris* is much less resistant to desiccation than the inert hypopus of *A. immobilis*. Exposure of *A. farris* hypopi to relative humidities of 70 per cent or below caused them to die quickly, whereas hypopi of *A. immobilis* exposed to this and lower humidities were unaffected. No doubt the power of rapid movement possessed by *A. farris* hypopi together with its habit of clinging to other mites and insects enables it to seek conditions of optimum humidity and temperature.

DISTRIBUTION

A. farris is a field species, but it has been recorded from barley, hay, cheese, poultry food, oats etc. It is found in birds' nests and the deep litter of poultry houses and is often particularly abundant on farms during August and September when it is associated with *Tyrophagus longior*. The hypopus may be found clinging to larger mites and to insects, such as the scatopsid flies which infest worm culture beds (M. Connell coll.). On one occasion it was present in bales of hay in such numbers that skin irritation was caused to men handling the bales (J. B. Pendlebury coll.).

DISTRIBUTION

England, Scotland, Wales, Netherlands, Germany, Kenya, U.S.A., Poland, Czechoslovakia (D. A. Griffiths, 1964a).

ACARUS IMMOBILIS Griffiths, 1964

The morphology of the adults, tritonymph, protonymph and larva are indistinguishable from the corresponding stages of *Acarus farris* except for a difference in shape of omega$_1$ on tarsi I and II of the adults. In a true lateral view, the sides of omega$_1$ are virtually parallel, expanding into a distinct egg-shaped terminal head (Plate 1, Fig. 4). Griffiths [1964b) has shown that the two species are genetically different.

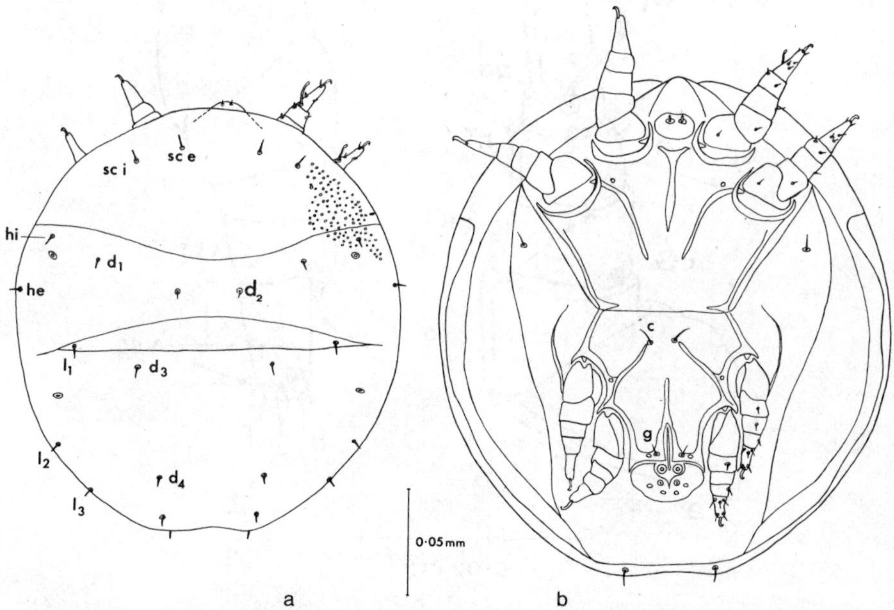

FIG. 40. *Acarus immobilis* Griffiths. Hypopus. (*a*) Dorsal view, (*b*) ventral view. Setae of the idiosoma: *sc e*, *sc i*, d_1, d_2, d_3, d_4, *h e*, *h i*, l_1 to l_3, *c*, *g*. (After Griffiths, 1964a).

HYPOPUS (Fig 40a, b). Average lengths of 10 specimens: 210 μm (D. A. Griffiths).

This is an inert form and was regarded as the hypopus of *Acarus siro* by Schulze (1923), Zachvatkin (1941) and Hughes (1955, 1961b). It is usually found free from the protonymphal skin, although the legs are capable of only feeble movement.

The idiosoma is broadly oval, white and translucent. The dorsal surface is arched and covered with shallow punctuations, whilst the ventral surface is concave. The division between propodosoma and hysterosoma is indicated by a transverse groove. The gnathosoma is considerably reduced, only represented by paired protuberances, each terminating in a blunt process.

The chaetotaxy of the dorsal surface resembles that of the motile forms of *A. siro* and *A. farris*, except that *v e*, d_2 and one pair of setae on the posterior margin of the hysterosoma are absent. All the setae are very short and difficult to see. The hysterosoma bears a pair of circular pores, one behind each of the *h i* setae, and a pair of glands on a level with coxae IV.

Ventrally, the coxal skeleton resembles more closely that of the motile hypopus of *A. siro*, since apodemes IV are straight rather than bow-shaped as in *A. farris*. The relationship of the genital setae to the adjacent sucker is also similar to that found in *A. siro*.

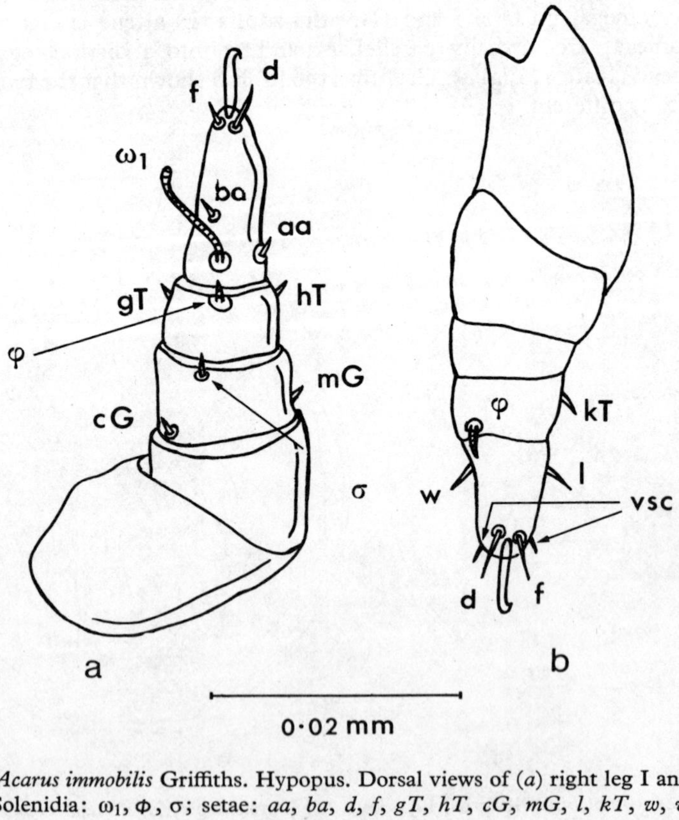

FIG. 41. *Acarus immobilis* Griffiths. Hypopus. Dorsal views of (*a*) right leg I and (*b*) right leg IV. Solenidia: ω_1, Φ, σ; setae: *aa, ba, d, f, gT, hT, cG, mG, l, kT, w, vsc*. (After Griffiths, 1964a).

On the legs the number and size of setae and solenidia are reduced compared with those of the motile hypopus of *A. siro* or *A. farris* (Fig. 41), although omega$_1$ is more than half the length of tarsi I or II, and has an egg-shaped terminal expansion. Solenidia sigma and phi are short and blunt on legs I. The ventral spine complex and *e* are absent from legs I and II, and seta *f* from legs II. The chaetotaxy of legs III and IV is the same as for the motile form, except that the long seta *e* is not present.

The hypopus is very resistant to desiccation and will withstand prolonged exposure to low relative humidities (Schulze, 1924, Griffiths 1964b).

By means of a series of exhaustive experiments, Griffiths [1966] has shown that hypopal formation in *A. immobilis* is favoured by malnutrition brought about by the exhaustion of an initially adequate food supply. His results also indicate that a minimum quantity of vitamin B and ergosterol is required which may be needed for the synthesis of a hormone controlling hypopus formation [T. E. Hughes, 1964]. Further experiments [1969] showed that when the basic food requirements were present in correct proportions and in an available form, hypopal production was suppressed. The most successful artifical diet for preventing the formation of hypopi was found to be one in which the ingredients resembled in nature and distribution those found in yeast. It is possible also that ribonucleic acid or its bases are also necessary for this 'direct' type of development.

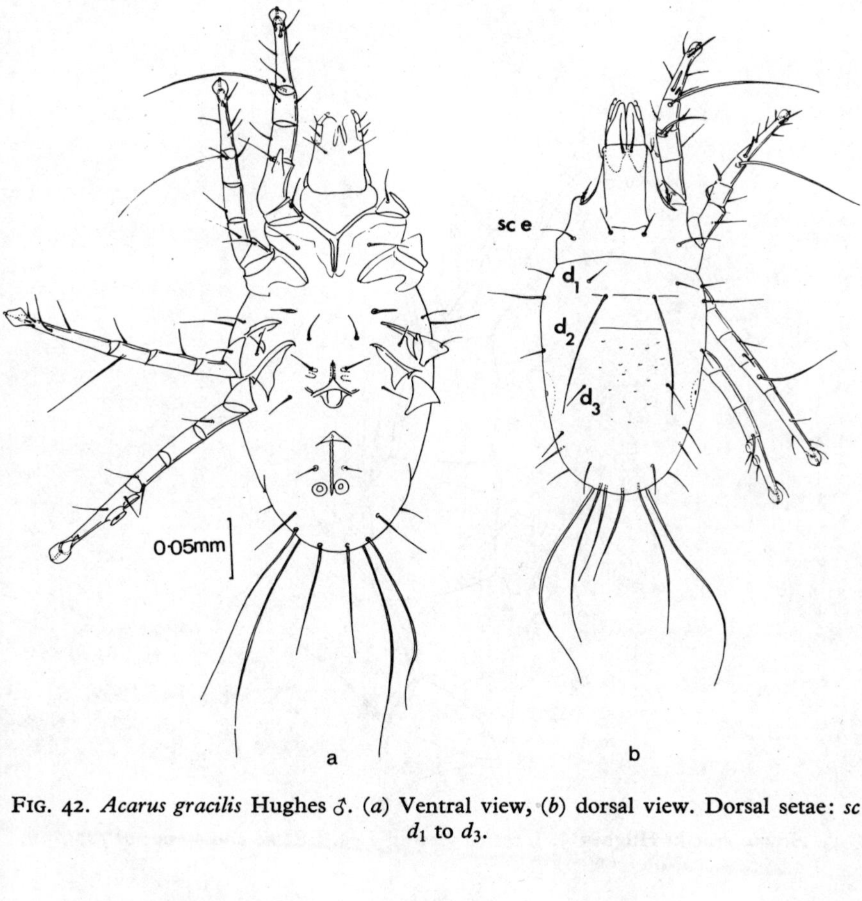

FIG. 42. *Acarus gracilis* Hughes ♂. (*a*) Ventral view, (*b*) dorsal view. Dorsal setae: *sc e*, d_1 to d_3.

DISTRIBUTION

A. immobilis is an out-of-doors species occurring mainly in birds' nests, humus and ground vegetation; it is found occasionally on unprocessed cereals in farms and storage premises, in grain residues and on cheese [Hughes, 1955]. So far it has mainly been recorded from temperate regions.

ACARUS GRACILIS Hughes, 1957

This species will neither mate nor breed with any of the three species of the *A. siro* complex (D. A. Griffiths, 1964b).

MALE (Fig. 42a, b). Length of idiosoma of six specimens: 280 to 360 μm.

This resembles the preceding species, but the cuticle is wrinkled and covered with minute papillae on the posterior region of the body.

The setae of the idiosoma are sparsely pectinate, their arrangement differs from that of the *A. siro* complex as follows: *sc* are short, about the same length as d_3; d_2 are more than four times as long as d_1, d_3, d_4, *h i*, *h v*, *l a*, *l p* and *sa e*; the distance between the insertions of d_2 and d_3 is more than four times that between d_1 and d_2. Setae *sa i* are long, about 70 per cent the length of the idiosoma.

Leg I (Fig. 43) has a ventral spur on the femur, but the femur and genu are less enlarged. On tarsi I and II omega$_1$ (Plate 1, Fig. 3) is long and tapering, so that the distance between its base and that of *ba* is less than the length of omega$_1$;

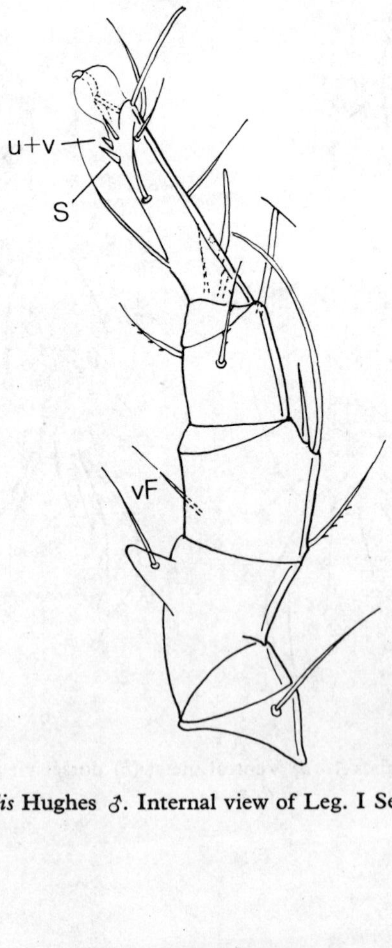

FIG. 43. *Acarus gracilis* Hughes ♂. Internal view of Leg. I Setae and spine: *vF*, *s*, *u+v*.

the famulus is more easily seen in *A. gracilis* than in *A. siro* as a small knob immediately distal to the base of omega$_1$. The ventral spine *s* is small as in *A. farris*. On tarsus IV (Fig. 33b) the tarsal copulatory suckers are large and fleshy and lie close together at the base of the segment, they are not separated by a space as in *A. siro*.

FEMALE (Fig. 44). Length of idiosoma: 200 to 250 µm.

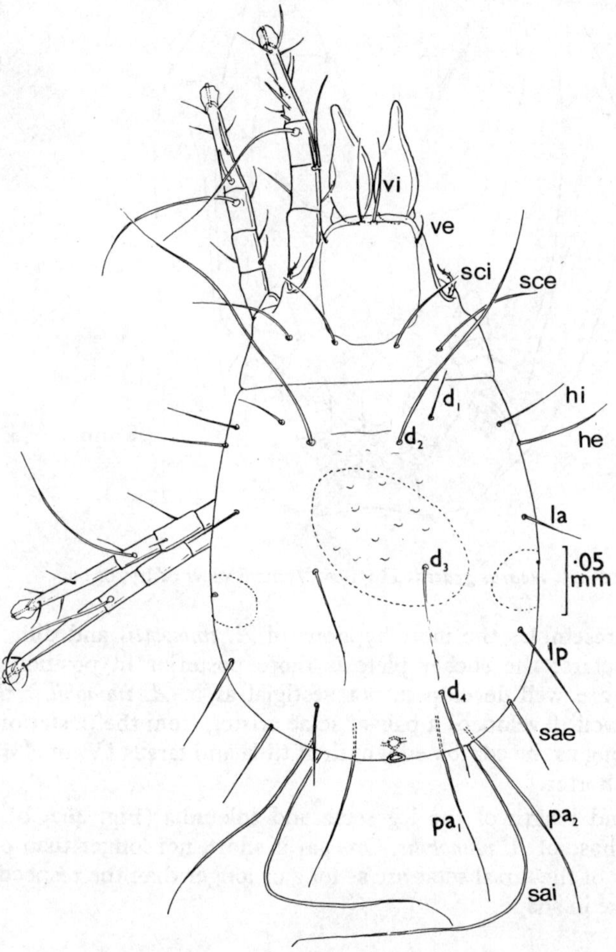

FIG. 44. *Acarus gracilis* Hughes ♀. Dorsal view. Setae of the idiosoma: *ve, v i, sc e, sc i, h e, h i,* d_1 to d_4, *l a, l p, sa e, sa i, p* a_1 *p* a_2.

The dorsal propodosomal shield is wider than that of the male, with a rounded posterior margin. The arrangement and lengths of the dorsal setae are also the same, except that d_3 are longer, more than twice as long as d_1. The setae of the anal region resemble more closely those of *A. farris* and *A. immobilis* although pa_1 are longer; they differ from *A. siro* in that a_3 are not more than twice as long as a_1 or a_2.

HYPOPUS (Fig. 45). Length of idiosoma: 200 to 250 µm.

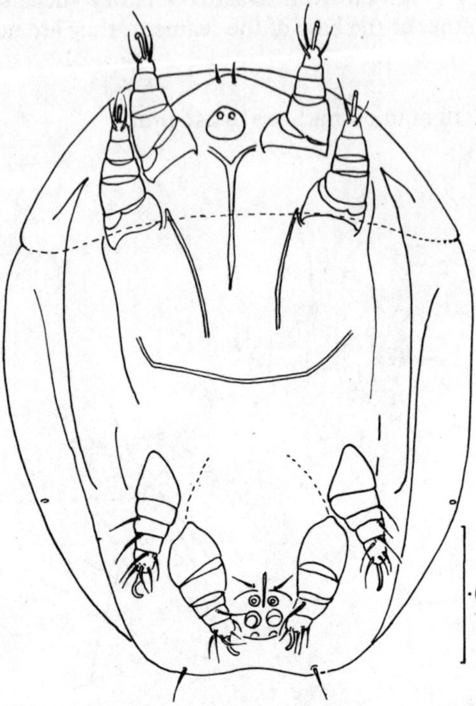

FIG. 45. *Acarus gracilis* Hughes. Ventral view of hypopus.

This closely resembles the inert hypopus of *A. immobilis* and differs in the following characters: the sucker plate is more posterior in position and the central suckers are well developed, not vestigial as in *A. immobilis*; the coxal skeleton is not well developed; a pair of setae arising from the posterior edge of the body is as long as the combined length of tibia and tarsus IV, in *A. immobilis* these setae are shorter.

The shape and length of the leg setae and solenidia (Fig. 46a, b) are also different from those of *A. immobilis*. Omega$_1$ is short, not longer than phi of the genu, and many of the tarsal setae are as long or longer than the respective tarsi, and often leaflike in shape.

BIOLOGY (A. M. Cunnington, pers. comm.).

The life cycle was completed in 20 to 21 days at 20°C and a relative humidity of 90 per cent when no hypopus was formed.

This species has mainly been found in bat roosts, inside houses and a stone tower (G. E. Woodroffe coll.), in old grain debris and a bird's nest, also in old nest residues beneath the tiled roof of a house also occupied by bats and rats (D. A. Griffiths coll.).

DISTRIBUTION

South-east England: Berkshire, Buckinghamshire and Hampshire; Argentina.

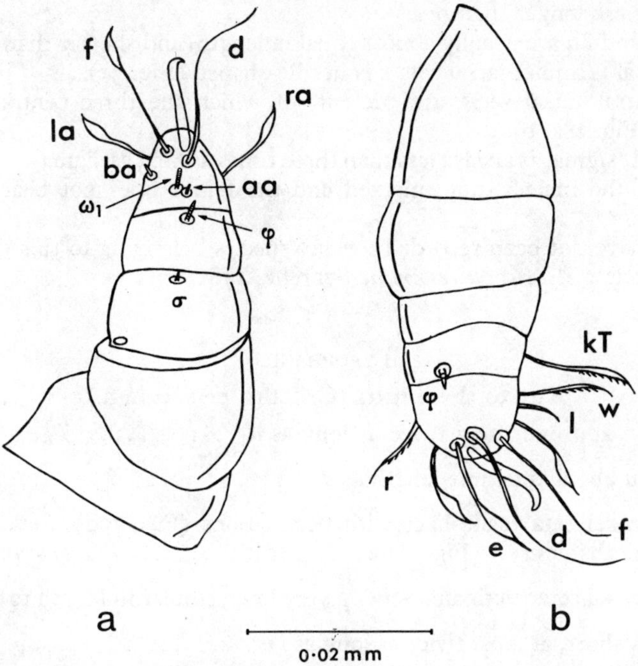

FIG. 46. *Acarus gracilis* Hughes. Hypopus. Dorsal views of (*a*) leg I and (*b*) leg IV., Solenidia: ω_1, Φ; setae: *aa, ba, d, e, f, la, l, ra, r, kT*. (After Griffiths, 1964a).

TYROPHAGUS OUDEMANS, 1924

Setae *v e* are longer than the length of the genu, and pectinate. They arise at almost the same level as *v i* (Fig. 47a) and curve downwards.

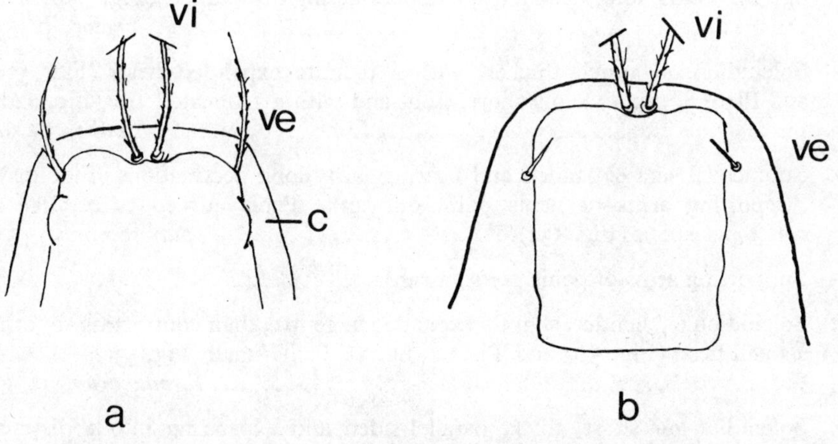

FIG. 47. Propodosomal region of (*a*) *Tyrophagus putrescentiae* (Schrank), (*b*) *Mycetoglyphus fungivorus* Oudemans. Vertical setae: *v e* and *v i*; c cornea.

Setae *sc i* are longer than *sc e*.

Setae d_1 and *l a* are usually almost equal in length, and shorter than d_3 and d_4.

The dorsal terminal tarsal seta *e* is needle-shaped (Fig. 51).

Five ventral tarsal setae are present, of which the three central ones are thickened (Fig. 28a, b).

On tibia I, $sigma_1$ is always less than three times as long as $sigma_2$.

Leg I of the male is not enlarged and the femur does not bear a ventral apophysis.

Hypopi have not been recorded for any species belonging to this genus.

Type species: *Acarus putrescentiae* Schrank, 1781.

Tyrophagus

Key to the adults (Griffiths, pers. comm.)

1. Seta *l a* approximately twice as long as d_1 . . . . *T. tropicus*, p. 66

—. Seta *l a* about the same length as d_1 2

2. Supracoxal seta blade-like, with pectinations (Fig. 59d). Seta *l p* much shorter than *sa i* (Fig. 66). *T. brevicrinatus*, p. 67

—. Supracoxal seta pectinate. Seta *l p* very long, similar in length to *sa i* 3

3. Seta d_2 short, at most twice as long as *l a* 4

—. Seta d_2 always more than twice as long as *l a* 6

4. With pigmented corneae present on the antero-lateral margins of the propodosomal shield (Fig. 58). Solenidion ω_1 like that of *T. putrescentiae*, possibly more slender (Fig. 60a) . . . . *T. neiswanderi*, p. 59

—. Without pigmented corneae. Supracoxal seta with short pectinations (fig. 49b) 5

5. Solenidion ω_1 long, slender, tapering towards the tip, terminating in a pointed end, or with a slightly expanded head (Fig. 51b and Plate 2, fig. 2). Penis long, slender and tapering, slightly curved (Fig. 53b) . . . . *T. longior*, p. 57

—. Solenidion ω_1 much thicker, with a definite expanded head (Fig. 51d and Plate 3, fig. 1) Penis short, stout and with a truncated tip (Fig. 62d) . . . . *T. similis*, p. 64

6. Supracoxal seta expanded and bearing fairly long pectinations (Fig. 49a). Supporting arms of penis point outwards. Penis curved twice, like a coffee pot spout (Fig. 53a) . . . . *T. putrescentiae*, p. 51

—. Supporting arms of penis point inwards 7

7. Solenidion ω_1 slender, slightly expanded at centre, then contracting to form a small head (Fig. 51c and Plate 2, fig. 3). Penis small (Fig. 53c) . . . . *T. palmarum*, p. 59

—. Solenidion ω_1 short, thick, parallel-sided and expanding into a distinct head (Fig. 60b and Plate 2, fig. 4). Penis large, truncate (Fig. 62c) . . . . *T. perniciosus*, p. 62

TYROPHAGUS PUTRESCENTIAE (Schrank, 1781)

Acarus putrescentiae Schrank, 1781; *Tyrophagus longior* var. *castellanii* Hirst, 1912; *T. noxius* Zachvatkin, 1941; *T. brauni* E. and F. Türk, 1957.

Also see Robertson, 1959, Samšiňák, 1962, Johnston and Bruce, 1965.

MALE (Fig. 48). Length of idiosoma: 280 to 350 μm.

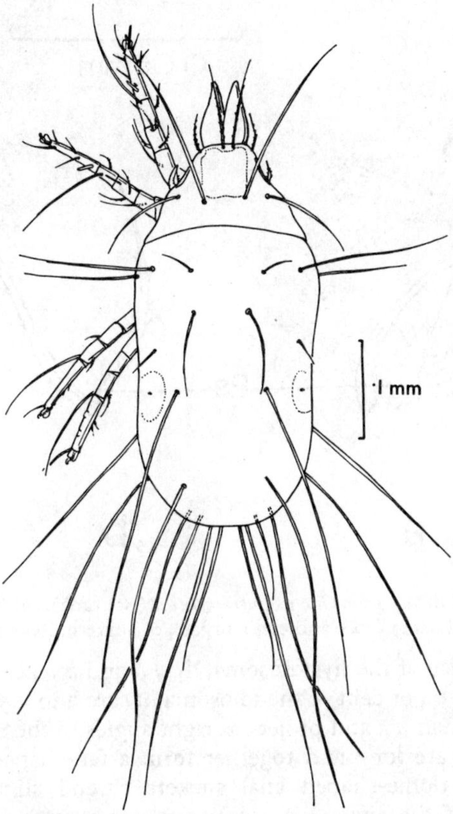

FIG. 48. *Tyrophagus putrescentiae* (Schrank) ♂. Dorsal view.

This is a small species with a smooth, shining cuticle, in which the degree of tanning of the appendages varies with the nature of the food. On flour, they are almost colourless, on cheese they are distinctly tanned. When alive, the body is more slender than in other species and the long setae do not project so stiffly from the surface of the body.

The propodosomal shield is often indistinct, depending on the tanning, and extends as far back as the scapular setae, with an almost straight posterior edge. A pair of colourless corneae is present on the side of the shield (Fig. 47a). Setae *v i* project beyond the tips of the chelicerae and like all the other long setae, are sparsely pectinate; *v e* arise slightly posterior to *v i* and are longer than the length of the genu. Setae *sc* are longer than the propodosoma, *sc i* being longer than *sc e*. The supracoxal seta (Fig. 49a) is flattened and has an expanded base from which arise stiff lateral projections, the expanded base is drawn out into a long thin point. The width of the base and length of the projections vary in different specimens and also with the angle at which the specimen is examined. As in other species of *Tyrophagus*, Grandjean's organ (Fig. 49a) has two main branches, one rod-shaped and one with an irregular outline. On the ventral surface, the apodemes are feebly tanned and the epimeral plates practically colourless; as in many other Acaridae, the anterior edge of apodemes I has an irregular outline.

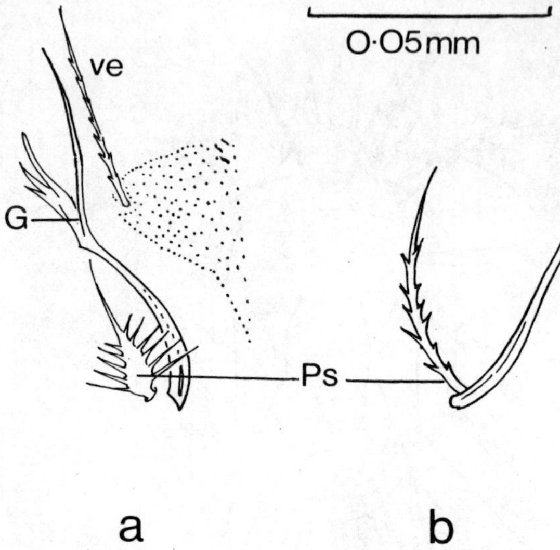

FIG. 49. Supracoxal seta of (*a*) *Tyrophagus putrescentiae* (Schrank), (*b*) *T. longior* (Gervais). Ps supracoxal seta; G Grandjean's organ; *v e* external vertical seta.

On the dorsal surface of the hysterosoma, d_1, *l a* and *h v* are short and almost equal in length (8 to 10 per cent of the idiosoma) d_2 are 2 to 3.5 times the length of d_1; *h i* are longer than *h e* and project at right angles to the sides of the body. The remaining setae are long and together form a fan-shaped 'train'. On the ventral surface, the dome-shaped anal suckers extend slightly beyond the posterior extremity of the anus (Fig. 50a); $p\ a_1$ are shorter and more slender

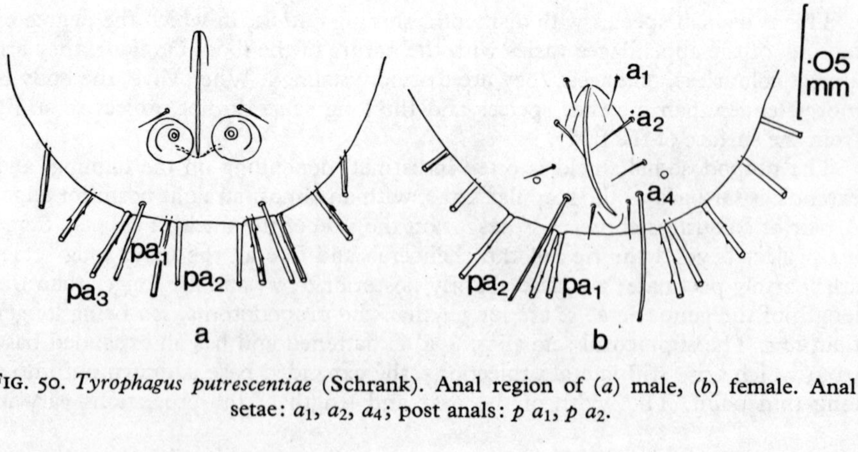

FIG. 50. *Tyrophagus putrescentiae* (Schrank). Anal region of (*a*) male, (*b*) female. Anal setae: a_1, a_2, a_4; post anals: $p\ a_1$, $p\ a_2$.

than $p\ a_2$ and $p\ a_3$, and arise from the end of the body. The body proportions and lengths of dorsal setae are variable within the species (Robertson, 1961).

Each chelicera is toothed and bears a spur-like outgrowth and mandibular spine. All the legs terminate in a stalked claw and well-developed pretarsus. The length of tarsus I does not exceed the combined length of the genu and tibia. On tarsus I omega$_1$ (Plate 2, fig. 1) is slightly expanded distally and arises close to the famulus; the origin of *aa* is distal to that of omega$_1$ (Fig. 51a). Omega$_3$

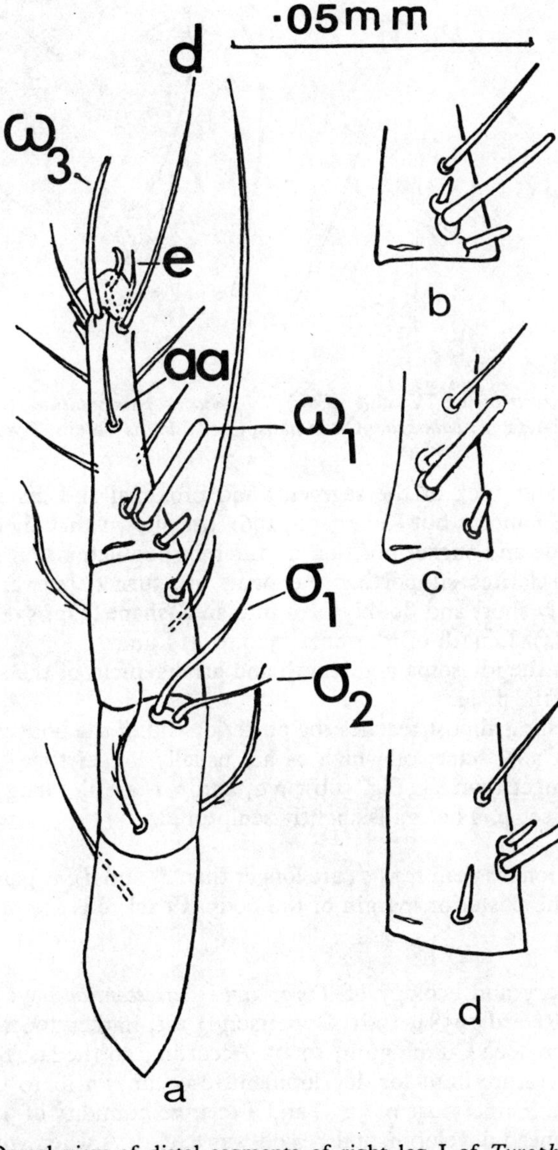

FIG. 51. (*a*) Dorsal view of distal segments of right leg I of *Tyrophagus putrescentiae* (Schrank). Base of dorsal surface of right leg I of (*b*) *T. longior*, (Gervais), (*c*) *T. palmarum* Oudemans, (*d*) *T. similis* Volgin. Solenidia: ω_1, ω_3, σ_1, σ_2; setae: *aa, d, e*.

and *d* extend well beyond the end of the claw and are longer than *e*; u, *v* and *s* are spines flanked by the slender setae *p* and *q*. On genu I $sigma_1$ is only slightly longer than $sigma_2$. The two suckers on tarsus IV (Fig. 52a) are equidistant

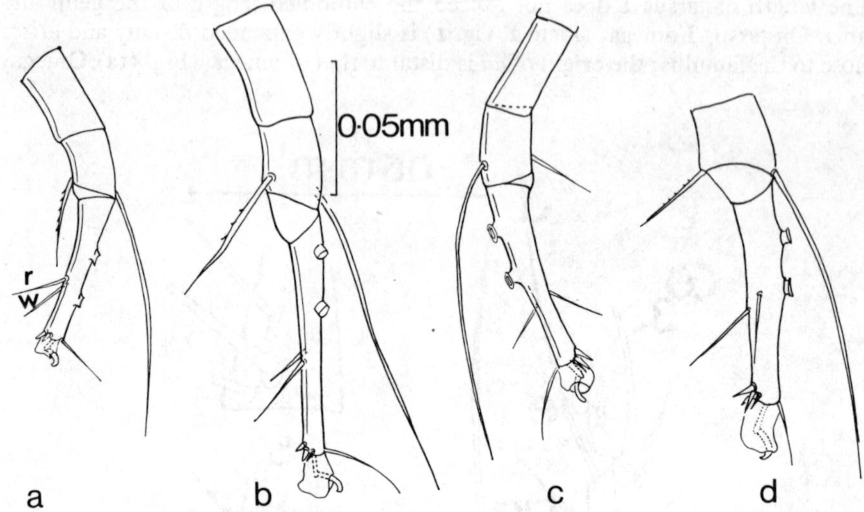

FIG. 52. Lateral view of leg IV of ♂ of (*a*) *Tyrophagus putrescentiae* (Schrank), (*b*) *T. longior* (Gervais), (*c*) *T. palmarum* Oudemans, (*d*) *T. similis* Volgin. Tarsal setae: *r*, *w*.

from the base and apex of the segment, and proximal and distal respectively to the two setae *r* and *w*, but Robertson, 1961 has shown that there is variation in the size, shape and sucker position in different populations.

The lateral sclerites supporting the penis are turned outwards, the penis itself is relatively short and doubly bent into an S-shape (Fig. 53a).

FEMALE (Fig. 54). Length of idiosoma: 320 to 415 µm.

The shape of the idiosoma and length and arrangement of the setae resemble closely those of the male.

The anal opening almost reaches the posterior end of the body; it is encircled by five pairs of anal setae, of which a_2 are usually longer than a_1, and a_4 are considerably longer than a_2 (Fig. 50b); $p\ a_1$ and $p\ a_2$ are also long and form part of the 'train' of setae. The egg is slightly sculptured.

LARVA

Setae *sc i* are longer than *sc e*; d_3 are longer than d_1 and d_2. A pair of long setae projects from the posterior margin of the body. Coxal rods are present, as well as coxal setae.

BIOLOGY

The life history and ecology of *Tyrophagus putrescentiae* have been studied extensively by Rivard 1958a, 1961, Robertson, 1961, Barker 1967a, Cunnington (1969) and others (see Cunnington, 1969). According to the last named author, the lower temperature limit for development to occur is 7 to 10°C, the highest 35 to 37°C. At a temperature of 32°C and a relative humidity of 90 per cent, he found that the mean developmental period was 8.46 days when wheat germ was used as food, whereas Barker (1967a) found that the most rapid development i.e., 21 days, occurred at 32.2°C and a relative humidity of 98 to 100 per cent, with

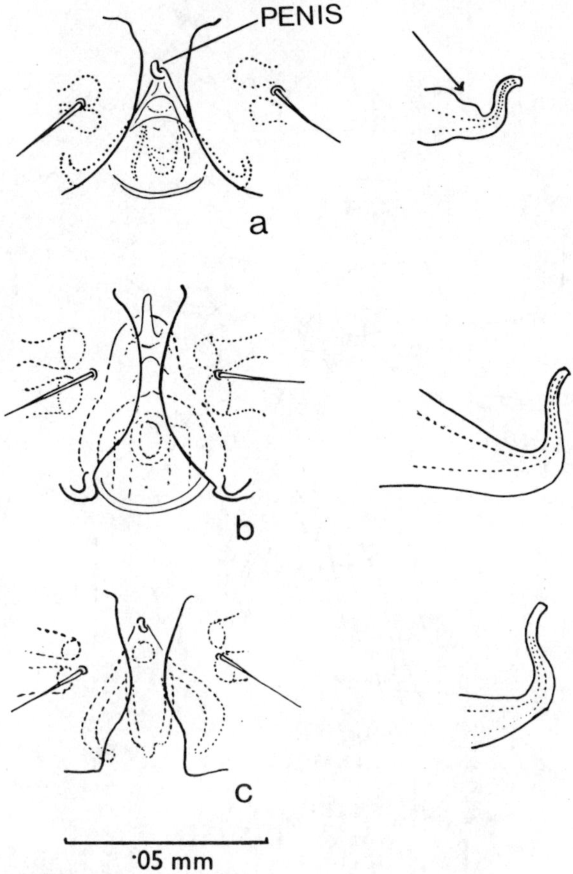

FIG. 53. Genital region and penis of (*a*) *Tyrophagus putrescentiae* (Schrank), (*b*) *T. longior* (Gervais), (*c*) *T. palmarum* Oudemans.

brewers yeast as substrate. He also found that 66 per cent of the adults were females. At 23°C and a relative humidity of 87 per cent and wheat germ as food, the life cycles was found to take from two to three weeks [Hughes, 1961]. As is the case with most of the acarid mites, development takes place most readily at high relative humidities, even as high as 100 per cent. The lowest limit that this species can tolerate is 60 per cent [Cunnington, 1969].

Ždárková (1971) tested the reactions of *T. putrescentiae* to the choice of a series of humidities ranging from 14.3 to 89 per cent. Below 22 per cent the mites appeared to be indifferent to changes in humidity, but between 22 per cent and 78 per cent, the higher humidity was always selected and, within this range, they were able to distinguish a 1 per cent difference in humidity.

T. putrescentiae has long been known as a pest of fungal cultures in Mycology Departments and indeed, Rivard [1958a and b] has reported rearing it on *Aspergillus* sp. Griffiths *et al.*, [1959] and Sinha and Mills [1968] have shown that it is attracted by and can complete its development on species of moulds belonging to the genera *Eurotium* and *Penicillium*, some species and strains being more attractive than others. Thus, according to Griffiths (1959), members

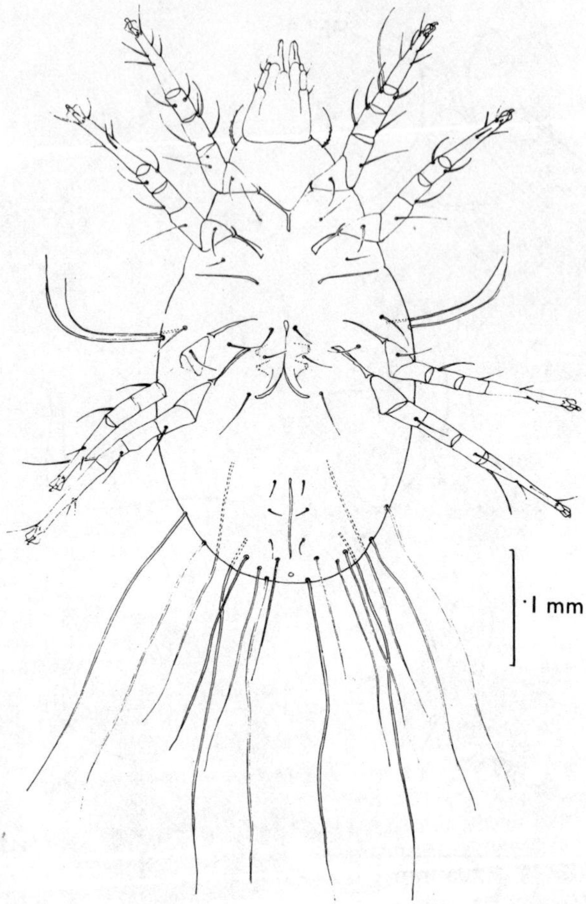

FIG. 54. *Tyrophagus putrescentiae* (Schrank) ♀. Ventral view.

of the greenish-yellow *Eurotium glaucus* group such as *E. amstelodami*, *E. repens* and *E. rubrum*, which thrive on grain of moisture content 13 to 15 per cent, are preferred to *Aspergillus candidus*, *A. ochraceus* and *A. flavus* which require a higher moisture content. Happily for the mites, the spores of the preferred fungi are more readily digested than those of the non-preferred group.

Sinha and Mills [1968] estimated the attractiveness of ten species of *Penicillium* to *T. putrescentiae* and *A. siro* by keeping these mites on potato-dextrose-agar cultures of the fungi and estimating their rate of increase. They found that *T. putrescentiae* was the more efficient fungivore and that its rate of increase was greatest on *Penicillium cyclopium*, least on *P. implicatum*. *P. cyclopium* occurs in soil, linseed cake, palm kernel cake, shelled corn in Canada, U.S.A., China and England. *C. implicatum* is cosmopolitan and has only been found in soil.

Like *A. siro*, *T. putrescentiae* can carry viable spores both outside and within its body and these are distributed rapidly throughout the substrate on which the mites are living [Griffiths, 1958]. That the mites are not completely dependent on fungi was shown by rearing them on sterile wheat germ. Under these conditions the F_2 generation took a little longer to reach maturity and the mites were slightly larger than their counterparts reared on mouldy germ.

In the temperate climate of this country, *T. putrescentiae* is most frequently found in large numbers in stored food with a high fat and protein content: dried eggs, ham, herring meal, copra, cheese and different kinds of nuts. Žďárková (1969) in a series of food preference tests, found that groundnuts, sunflower seed and rape seed were preferred to linseed, palm kernal, poppy seed and cotton seed, whereas soya had no attraction at all. The preferred nuts corresponded roughly to the degree of suitability of the material for their development. *T. putrescentiae* also occurs on dried bananas, wheat spillage, barley, tobacco and flour. Smith (1969) using the 'filth' test method of extraction (Kent Jones *et al.*, 1948) found that it was the species most frequently found in samples of flour collected at the bagging-off spouts of some 52 mills in England.

Robertson (1961) showed that, with *T. longior* and *T. palmarum*, *T. putrescentiae* is found on cheeses throughout the world, but the three species tend to occupy different, although overlapping geographical zones. *T. longior* is a cool to temperate form, *T. palmarum* a temperate species and *T. putrescentiae* tends to be found in subtropical and tropical regions. *T. putrescentiae* is attracted by the odour of brick or blue cheese and also by 1 to 5 per cent solutions of lactic acid. Other materials with a distinctive odour, such as cinnamaldehyde and anisaldehyde were also found to be attractive in low concentrations, but repellent in high (Žďárková, 1971).

DISTRIBUTION

Cosmopolitan

TYROPHAGUS LONGIOR (Gervais, 1844)

Tyroglyphus longior Gervais, 1844; *T. infestans* Berlese, 1884; *Tyrophagus tenuiclavus* Zachvatkin, 1941.

MALE (Fig. 55). Length of idiosoma: 330 to 535 µm.

A larger species, with more heavily tanned legs and chelicerae which sometimes have a faint reticulate pattern; the degree of tanning and size are, however, very variable.

The arrangement of the setae on the idiosoma and legs resembles that of the previous species. The principle differences are: the supracoxal seta (Fig. 49b) is curved, not thickened towards the base, with short lateral barbs of about equal length and d_2 are 1 to 1.3 to 2 times longer than d_1 and *l a*. On the legs, omega$_1$ of tarsi I and II (Fig. 51b and Plate 2, fig. 2) is long and gradually tapers towards its distal end. Tarsus IV is longer than the combined length of the genu and tibia and the two tarsal suckers are nearer to the base of the segment than the apex, *r* and *w* being distal to the suckers (Fig. 52b).

The lateral sclerites supporting the penis are turned inwards, the penis itself tapers towards its free end and is shaped like the spout of a teapot (Fig. 53b).

FEMALE. Length of idiosoma: 530 to 670 µm.

The general structure of the female is very similar to that of the male and only differs in secondary sexual characters.

OTHER STAGES

The larva closely resembles that of *T. putrescentiae.*

The egg (Fig. 56) differs from that of the previous species in being covered with a more distinct sculpturing.

BIOLOGY

The life cycle takes from two to three weeks at 23°C and a relative humidity of 87 per cent.

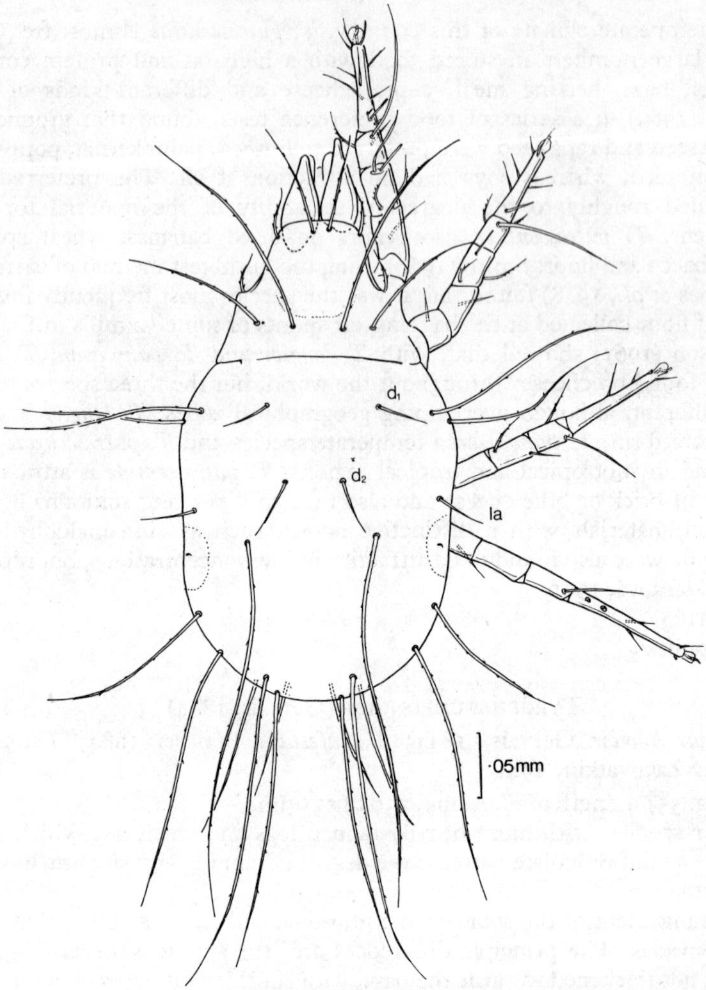

FIG. 55. *Tyrophagus longior* (Gervais) ♂. Dorsal view. Setae of the idiosoma: d_1, d_2, *l a*.

T. longior is found on cheese, particularly in the cool temperate regions of the world (Robertson, 1961), but it more commonly occurs in stored home-produced grain, hay and straw where it often increases to very large numbers by the end of the autumn. It is the dominant species in haystacks and grainstacks standing in the open field and is often associated with *Acarus farris* and the mesostigmatid mites *Alliphis halleri* and *Kleemannia plumosus* (Griffiths, 1960). It has also been found on barley in very large numbers in Hertfordshire (S. H. Beatson coll.), on a grass-cleaning plant in Hampshire (C. Wayman coll.), on sugar beet seed in Poland (W. Chmielewski, 1969) and in a sparrow's nest in Poland (Wasylik, 1959). It is the most common pest of stockfish in Iceland (G. Gigja, 1964) and has been recorded as attacking cucumber plants, beetroots, tomatoes and *Cyclamen* seed. Brady (1970) lists *T. longior* as one of the 10 species achieving the greatest density on poultry droppings in broiler houses.

DISTRIBUTION

Cosmopolitan.

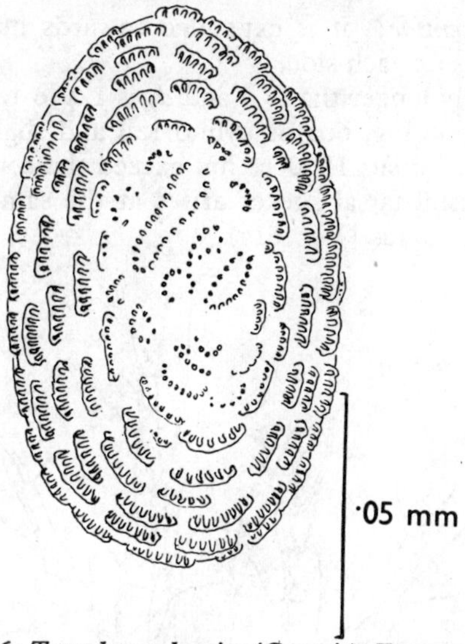

FIG. 56. *Tyrophagus longior* (Gervais). Egg.

TYROPHAGUS PALMARUM Oudemans, 1924

Tyrophagus perniciosus Zachvatkin, 1941 sensu Hughes, 1961.

MALE (Fig. 57). Length of idiosoma: 330 to 450 µm.

This species, in general appearance, closely resembles *T. longior*. The chief differences may be summarized as follows: d_2 are three to four times the length of *l a*; on tarsi I and II (Fig. 51c and Plate 2, fig. 3) omega$_1$ does not taper towards its distal end, but is cigar-shaped; tarsus IV is almost equal to the combined length of the genu and tibia, the distal sucker occupying a position in the middle of the segment (Fig. 52c). The genitalia and penis resemble those of *T. longior* (Fig. 53c), the penis is rather shorter in length.

FEMALE. Length of idiosoma: 350 to 550 µm.

The general structure of the female is very similar to that of the male.

BIOLOGY

The length of the life cycle is the same as that of the two preceding species. Attempts to cross this species with either *T. longior* or *T. putrescentiae* have always been unsuccessful.

T. palmarum is commonly found in the surface soil of grasslands and in stacks standing in open fields (Griffiths, 1960). It can occur in large numbers in hay, home-grown corn and barley, also in cheese, particularly in temperate zones (Robertson, 1961), old honey bee comb (D. C. Lee coll.) and in spiders' webs where it feeds on the bodies of dead insects. Wasylik (1963) records it from birds' nests.

DISTRIBUTION

Cosmopolitan.

TYROPHAGUS NEISWANDERI Johnston and Bruce, 1965

MALE (Fig. 58). Length of idiosoma of holotype: 413 µm (Johnston and Bruce, 1965).

The corneae, situated in the antero-lateral corners of the dorsal propodosomal shield, are distinctly pigmented (Fig. 58). The supracoxal seta (Fig. 59a) resembles that of *T. putrescentiae*; it is expanded towards its base and has about five lateral pectinations on each side.

Seta d_1 is slightly longer than *l a* and d_2 is 1.4 to 1.58 times the length of *l a*. On tarsus I omega$_1$ (Fig. 60a) is cylindrical and slightly curved. The famulus is short and stout. Tarsus IV does not exceed the combined length of the tibia and genu. The distal tarsal sucker arises at the same level as *r* and *w*: about midway along the tarsus (Fig. 61a).

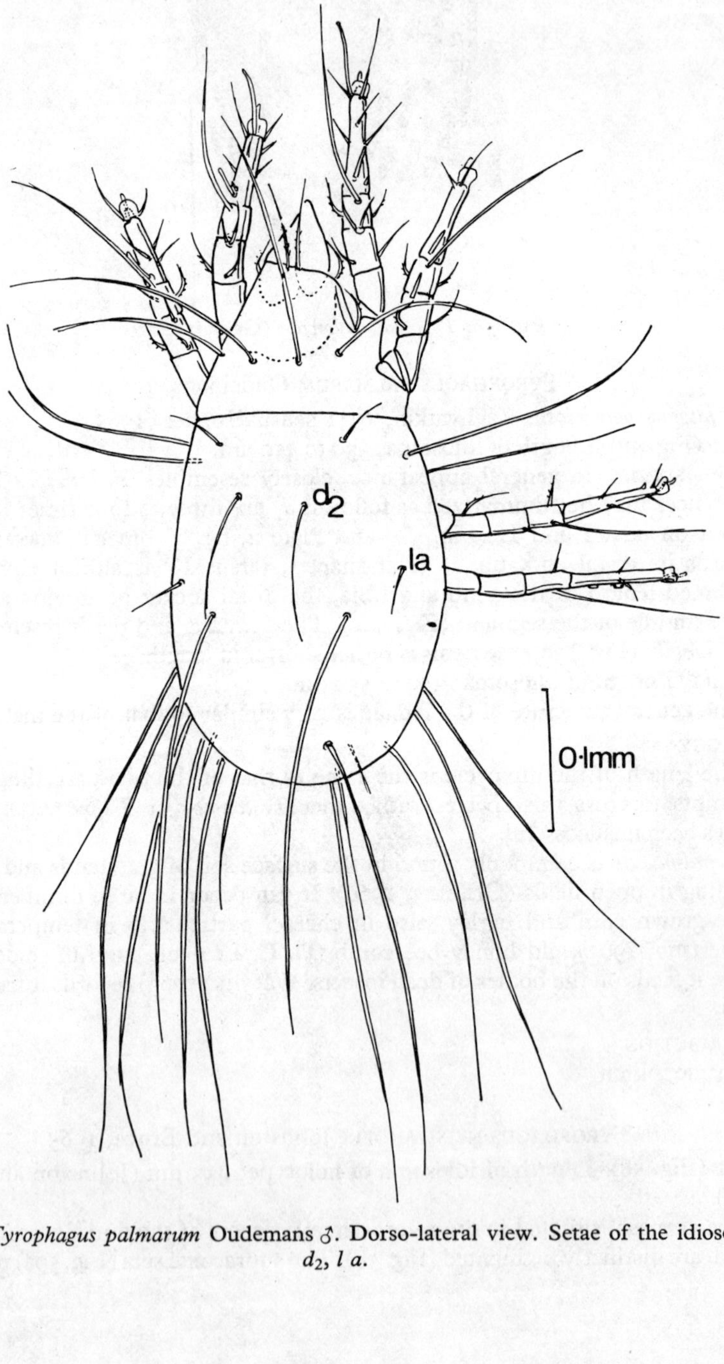

FIG. 57. *Tyrophagus palmarum* Oudemans ♂. Dorso-lateral view. Setae of the idiosoma: d_2, *l a*.

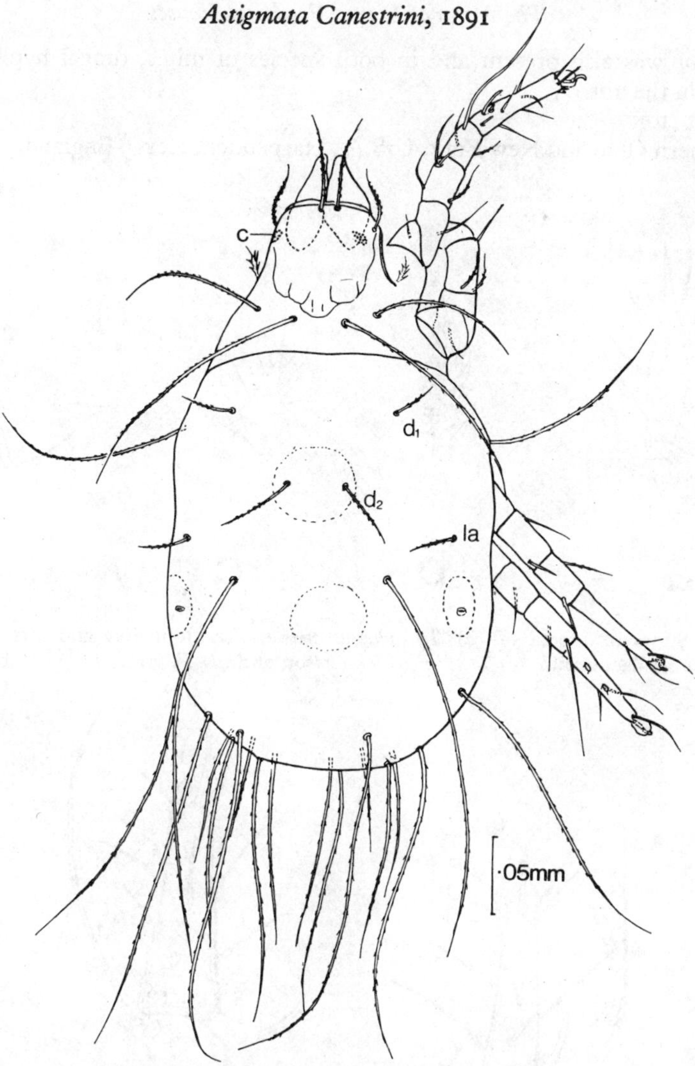

FIG. 58. *Tyrophagus neiswanderi* Johnston and Bruce ♂. Dorsal view. c cornea; setae of the idiosoma: d_1, d_2, *l a*.

The supporting sclerites of the penis are turned outwards. The penis (Fig. 62a) is curved twice and does not taper towards its distal end.

FEMALE

This resembles the male. Seta d_2 is longer than in the male, and is 1.7 times the length of *l a* (Johnston and Bruce, 1965).

The chorion of the egg is distinctly ornamented as in *T. longior*.

BIOLOGY

T. neiswanderi was originally found on greenhouse cucumber plants in Northern Ohio, U.S.A., where it was feeding on the foliage of the plants (Johnston and Bruce, 1965). It has also been found on the growing tips of chrysanthemum cuttings that had been damaged by noctuid moths. The chrysanthemums were imported from Kenya (J. C. Eames coll.). In the latter case,

T. longior was also present and in both species of mites, fungal hyphae were present in the gut.

DISTRIBUTION

Northern Ohio and New York, U.S.A., Harpenden, Herts, England.

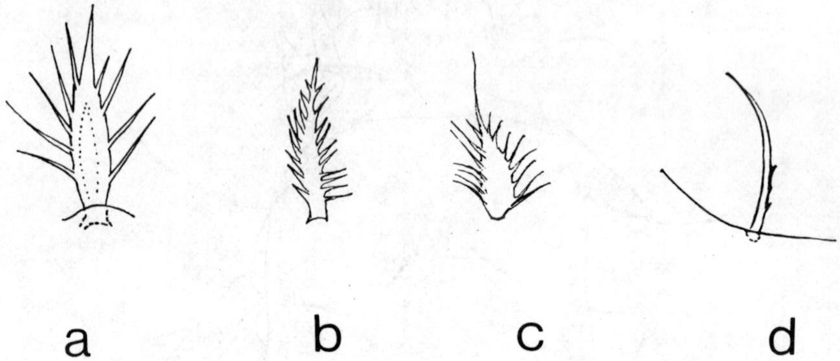

FIG. 59. Supracoxal seta of (*a*) *Tyrophagus neiswanderi* Johnston and Bruce, (*b*) *T. perniciosus* Zachvatkin, (*c*) *T. tropicus* Robertson and (*d*) *T. brevicrinatus* Robertson.

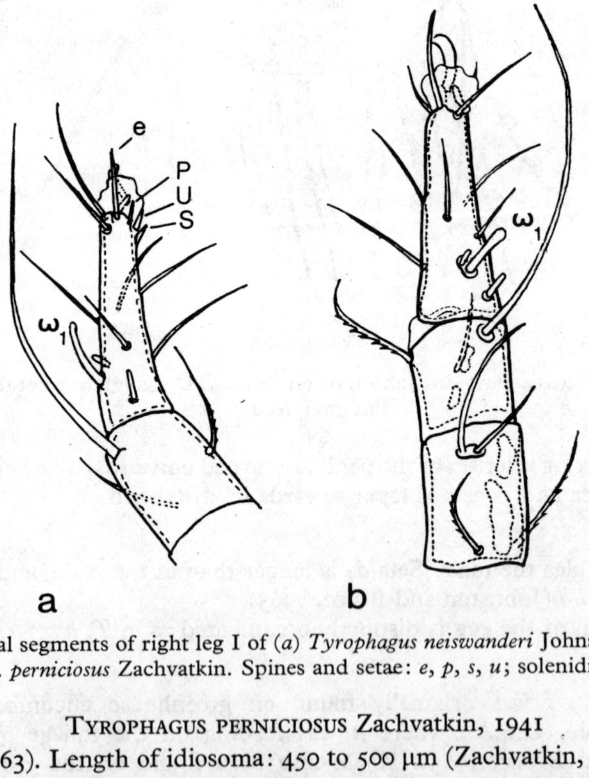

FIG. 60. Distal segments of right leg I of (*a*) *Tyrophagus neiswanderi* Johnston and Bruce, (*b*) *T. perniciosus* Zachvatkin. Spines and setae: *e*, *p*, *s*, *u*; solenidion: ω_1.

TYROPHAGUS PERNICIOSUS Zachvatkin, 1941

MALE (Fig. 63). Length of idiosoma: 450 to 500 μm (Zachvatkin, 1941).

The legs and gnathosoma are fairly well-sclerotized. The supracoxal seta (Fig. 59b) expands gradually towards the base, its lateral pectinations diminishing

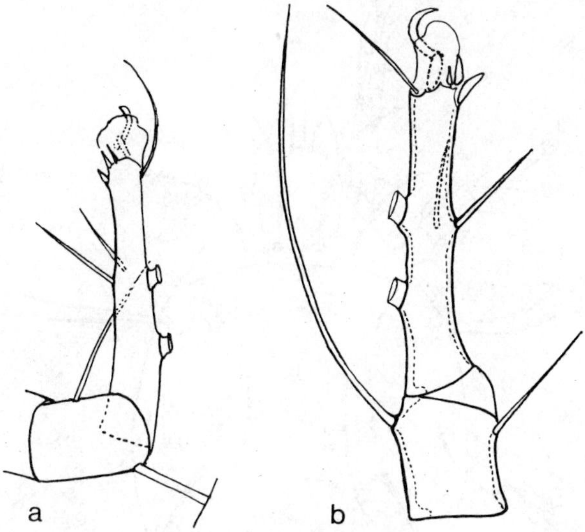

FIG. 61. Tibia and tarsus of (*a*) right leg IV of *Tyrophagus neiswanderi* Johnston and Bruce, (*b*) left leg IV of *T. perniciosus*.

in length from base to apex. Setae d_2 are considerably (2.5 to 4.5 times) longer than d_1. On tarsus I (Fig. 60b and Plate 2, fig. 4), omega$_1$ is stout and slightly expanded distally. On tarsus IV, the distal sucker arises at about the same level as the ventral setae (Fig. 61b).

The lateral sclerites supporting the penis are turned inwards (Fig. 62b); the penis (Fig. 62c) itself is large, curved into an arc and has a truncate end.

FEMALE. Length of idiosoma: 550 to 700 µm (Zachvatkin, 1941).

This resembles the male.

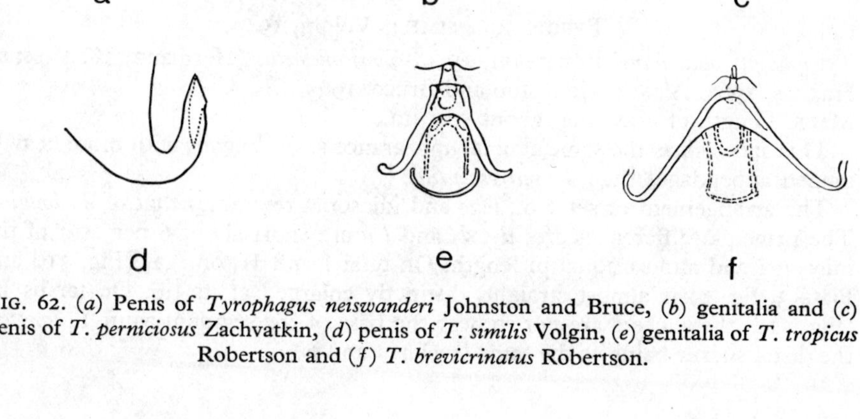

FIG. 62. (*a*) Penis of *Tyrophagus neiswanderi* Johnston and Bruce, (*b*) genitalia and (*c*) penis of *T. perniciosus* Zachvatkin, (*d*) penis of *T. similis* Volgin, (*e*) genitalia of *T. tropicus* Robertson and (*f*) *T. brevicrinatus* Robertson.

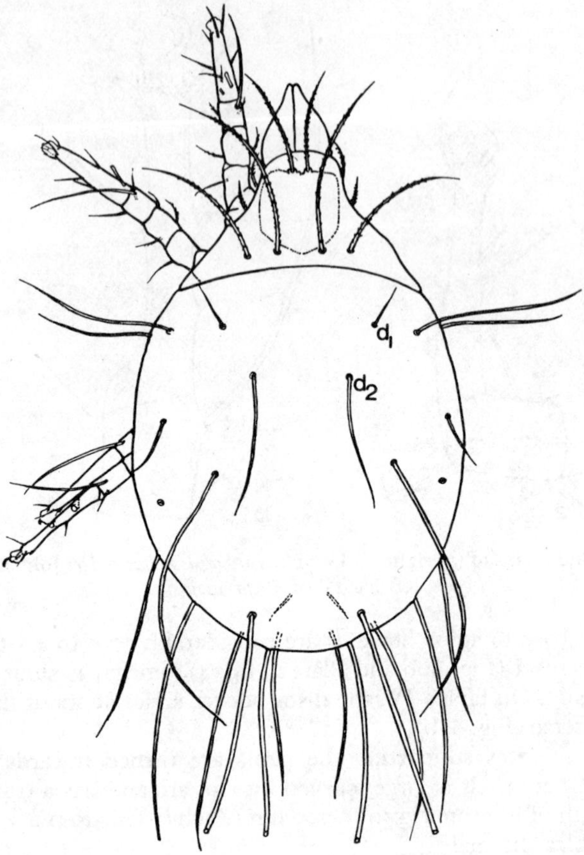

FIG. 63. *Tyrophagus perniciosus* Zachvatkin ♀. Dorsal view. Setae of the idiosoma: d_1, d_2.

BIOLOGY

T. perniciosus is found in the dust of granaries and warehouses, and is associated with oats and barley (Zachvatkin, 1941). It has also been recorded from cheese and the dirty seed found in a budgerigar cage.

DISTRIBUTION

England, Bulgaria, U.S.S.R., Australia.

TYROPHAGUS SIMILIS Volgin, 1949

Tyrophagus oudemansi Robertson, 1959; *T. dimidiatus* (Hermann, 1804) sensu Hughes, 1961. Also see Johnston and Bruce, 1965.

MALE. Length of idiosoma: about 500 µm.

This species has the same general appearance as *T. longior*, with more heavily tanned appendages than *T. putrescentiae*.

The arrangement of setae on legs and idiosoma resembles that of *T. longior*. The principle differences are: d_1, d_2 and *l a* are short about 6 per cent of the idiosoma and almost equal in length. On tarsi I and II, omega$_1$ (Fig. 51d and Plate 3, fig. 1) is almost straight, distinctly enlarged at its tip. On tarsus IV (Fig. 52d) the suckers are not so near the base of the segment as in *T. longior*, the distal sucker being at the same level as *r* and *w*.

The lateral sclerites supporting the penis are turned inwards as in *T. longior*, but the penis (Fig. 62d) does not taper distally and has a truncate end.

FEMALE (Fig. 64). Length of idiosoma: about 600 µm.

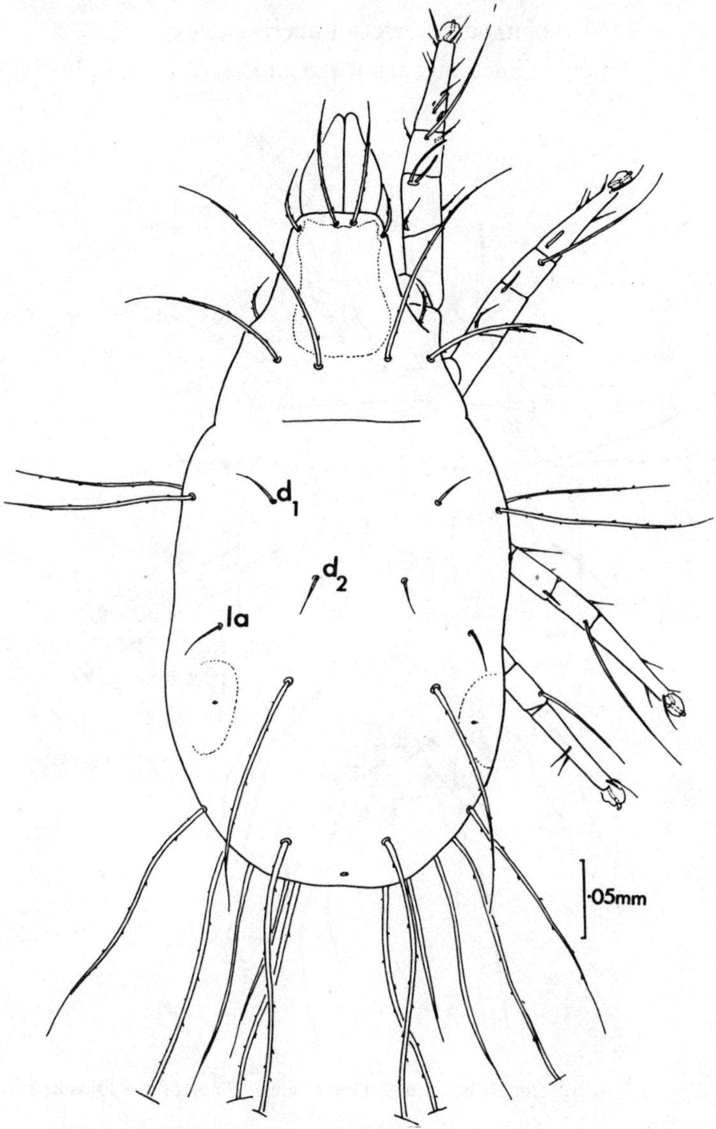

FIG. 64. *Tyrophagus similis* Volgin ♀. Dorsal view. Setae of the idiosoma: d_1, d_2, *l a*.

The general structure of the female is similar to that of the male.

BIOLOGY

T. similis is found on grassland, in soil and on plants such as *Phlox*, spinach, mushrooms etc. It has also been recorded from old hay [S. L. Tuxen coll. Hughes, 1961a], chaff (D. A. Griffiths coll.), in an eider duck nest and a nest of *Bombus ionellus* [S. L. Tuxen coll.].

DISTRIBUTION

British Isles, New Zealand, U.S.A., Belgium, Germany, Iceland, Australia. Holland.

TYROPHAGUS TROPICUS Robertson, 1959

MALE (Fig. 65). Length of idiosoma: about 430 μm.

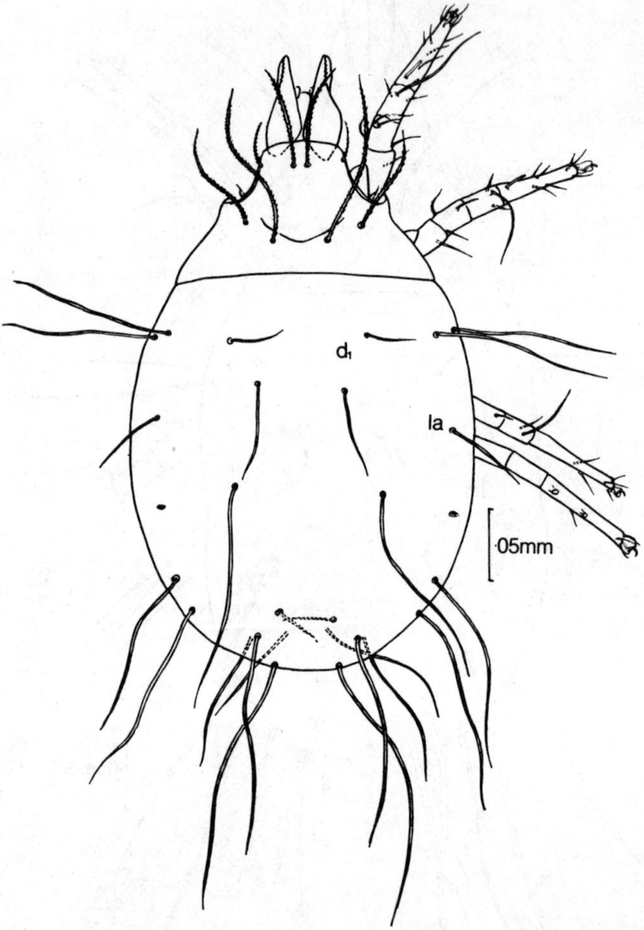

FIG. 65. *Tyrophagus tropicus* Robertson ♂. Dorsal view. Setae of the idiosoma: d_1, *l a*.

This resembles *T. putrescentiae*, but differs in the following characters: *l a* are long, almost twice the length of d_1. The supracoxal seta has a more expanded base than in *T. putrescentiae* and the tip tapers more finely (Fig. 59c). Omega$_1$ on tarsi I and II is slightly expanded at its tip. The penis is short and curved (Fig. 62e).

FEMALE

This closely resembles the male.

BIOLOGY

Robertson (1959) records this mite from tobacco, palm kernel dust, copra and rice.

DISTRIBUTION

England, Ghana, Lagos, Nigeria, New Britain, U.S.A.

TYROPHAGUS BREVICRINATUS Robertson, 1959

MALE (Fig. 66). Length of idiosoma: about 450 µm.

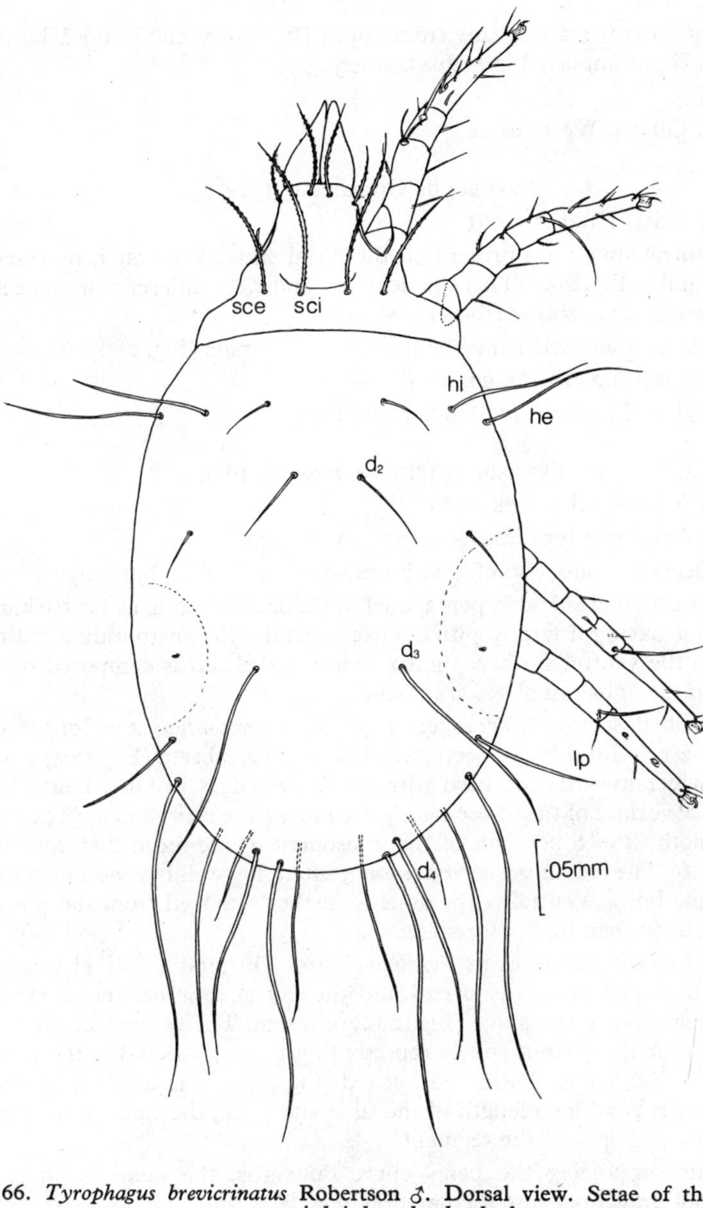

FIG. 66. *Tyrophagus brevicrinatus* Robertson ♂. Dorsal view. Setae of the idiosoma: *sc e*, *sc i*, *h i*, *h e*, d_2, d_3, d_4, *l p*.

This resembles *T. putrescentiae*, but differs in the following characters: the humeral setae, scapulars, d_3, d_4 and *l p* are shorter, d_3, d_4 and *l p* being about twice the length of d_2. The supracoxal seta (Fig. 59d) is short and almost smooth. Omega_1 on tarsi I and II is slightly expanded at the tip. The arms of the penis support are turned outwards and the penis is S-shaped as in *T. putrescentiae* (Fig. 62f).

FEMALE

This closely resembles the male.

BIOLOGY

Robertson (1959) records this from copra [P. Davey coll.] and I have also found it on copra imported into this country.

DISTRIBUTION

England, Ghana, West Africa.

TYROBORUS Oudemans, 1924

Tyrophagus (part) Hughes, 1961.

Three ventral spines are present on the distal ends of the tarsi, representing $p+u$, $q+v$ and s (Fig. 28c). This character is a sufficient difference at the generic level to separate *Tyrophagus* from *Tyroborus*.

The dorsal terminal seta *e* may be thickened as a spine (Fig. 28d).

Other characters as in the genus *Tyrophagus*.

Type species: *Tyroborus lini* Oudemans, 1924.

TYROBORUS LINI Oudemans, 1924

Tyrophagus lini sensu Hughes, 1961.

MALE (Fig. 67). Length of idiosoma: 350 to 470 μm.

The chelicera is stout (Fig. 68), with well-defined teeth on both digits.

The propodosomal shield is pentagonal in shape, extending as far back as the *sc i* setae and having a faintly pitted surface, whilst the surrounding cuticle is smooth. On the ventral surface, the coxalsternal skeleton is composed of thick sclerites and the epimeral plates are distinct (Fig. 69).

The idiosomal setae are arranged as in *T. putrescentiae*, the longer ones, particularly *v e* and *v i*, being pectinate. The supracoxal seta (Fig. 70a) is large, becoming wider towards the base with its lateral edges produced into barbs. On the dorsal surface of the idiosoma, d_1, *l a* and *h v* are short and approximately equal in length (4 to 6 per cent of the idiosoma); d_2 are more than four times longer than d_1. The remaining setae are long, extending well beyond the posterior margin of the body. Ventrally, the anus is further removed from the posterior edge of the body than in *T. putrescentiae*.

The legs are rather stout as in *Tyrolichus casei*. On tarsi I and II omega_1 is slightly club-shaped at its distal end and the dorsal terminal seta *e* varies in width and may resemble a spine (Fig. 28d) or a seta. The ventral terminal ends of the tarsi bear three stout spines representing $q+v$, $p+u$ and *s*, the external $q+v$ and $p+u$ are larger than *s* and hooked (Fig. 28c). Tarsus IV (Fig. 71a) is shorter than the combined length of the tibia and genu, the suckers equidistant from the base and apex of the segment.

The struts supporting the penis curve outwards, the penis is small and S-shaped, not drawn out into a point (Fig. 72a, b).

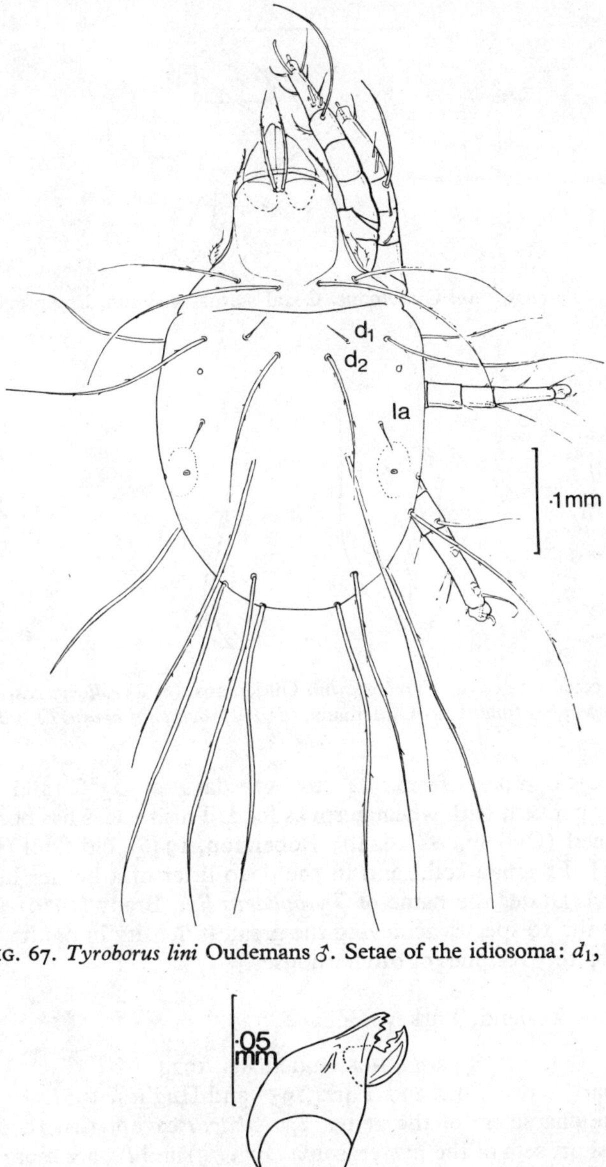

FIG. 67. *Tyroborus lini* Oudemans ♂. Setae of the idiosoma: d_1, d_2, *l a*.

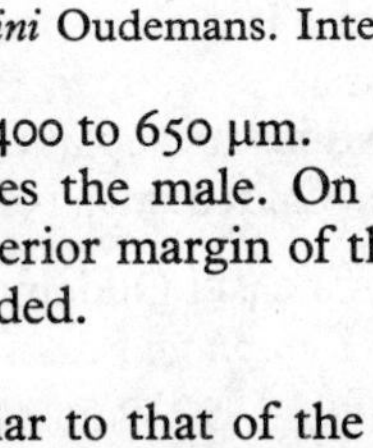

FIG. 68. *Tyroborus lini* Oudemans. Internal view of chelicera.

FEMALE. Length of idiosoma: 400 to 650 µm.

The female closely resembles the male. On the ventral side, the anus (Fig. 73b) is removed from the posterior margin of the body.

No hypopus has been recorded.

LARVA (Fig. 74).

The supracoxal seta is similar to that of the adult. Unlike the adult, *sc i* are considerably shorter than *sc e*. The coxal rods are cylindrical. The sacral setae are long and exceed half the length of the idiosoma.

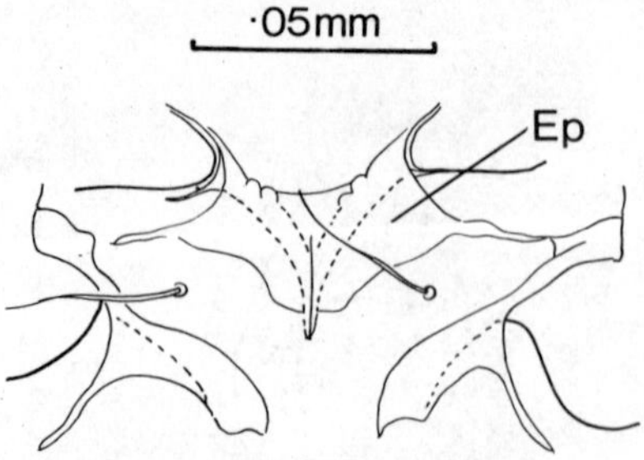

FIG. 69. *Tyroborus lini* Oudemans. Coxal-sternal skeleton. Ep epimeron

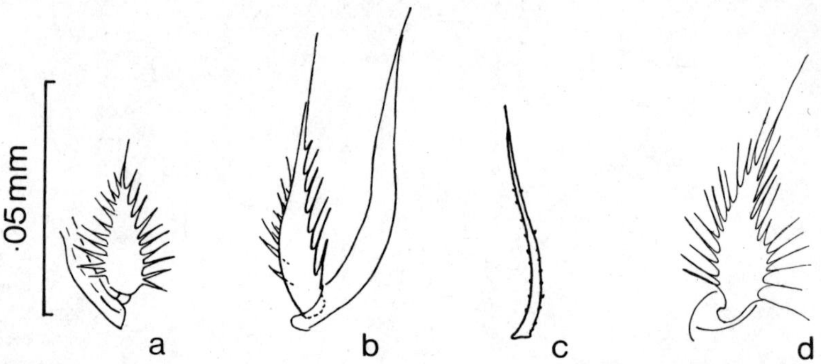

FIG. 70. Supra-coxal seta of (*a*) *Tyroborus lini* Oudemans, (*b*) *Tyrolichus casei* Oudemans, (*c*) *Mycetoglyphus fungivorus* Oudemans, (*d*) *Aleuroglyphus ovatus* (Troupeau).

BIOLOGY

The life cycle takes from 14 to 17 days at 23°C and a relative humidity of 87 per cent with wheat germ as food. This species has been recorded from old linseed (Oudemans, 1924b; Robertson, 1946), old flour (D. Burgess coll.), wheat (J. Freeman coll.) and in the deep litter of a broiler house (J. D. Ethelston coll.). Under the name of *Tyrophagus lini*, Brady (1970) records this mite as one of the 10 species achieving the greatest density in poultry droppings in the pits or battery residues of broiler houses.

DISTRIBUTION

England, New Zealand, Turkey, Holland.

TYROLICHUS Oudemans, 1924

Tyrophagus (part) sensu Türk and Türk, 1957 and Hughes, 1961.

This has the characters of the genus *Tyrophagus* except that the dorsal seta d_1 is the only short seta of the hysterosoma (Fig. 75) and *l a* are more than twice as long as d_1.

The dorsal terminal tarsal seta *e* is a short thick spine (Fig. 76a) and all five ventral terminal setae of the tarsi: *p*, *q*, *s*, *u*, *v* are thickened to form spines (Fig. 77a) of equal size. This justifies its separation from the genus *Tyrophagus*.

Type species: *Tyrolichus casei* Oudemans, 1910.

TYROLICHUS CASEI Oudemans, 1910

Tyroglyphus siro Michael, 1903; *Tyrophagus casei* sensu Hughes, 1961.

MALE (Fig. 75). Length of idiosoma: 450 to 550 µm.

T. casei is more clumsily built than *T. putrescentiae* and has well-tanned legs and gnathosoma.

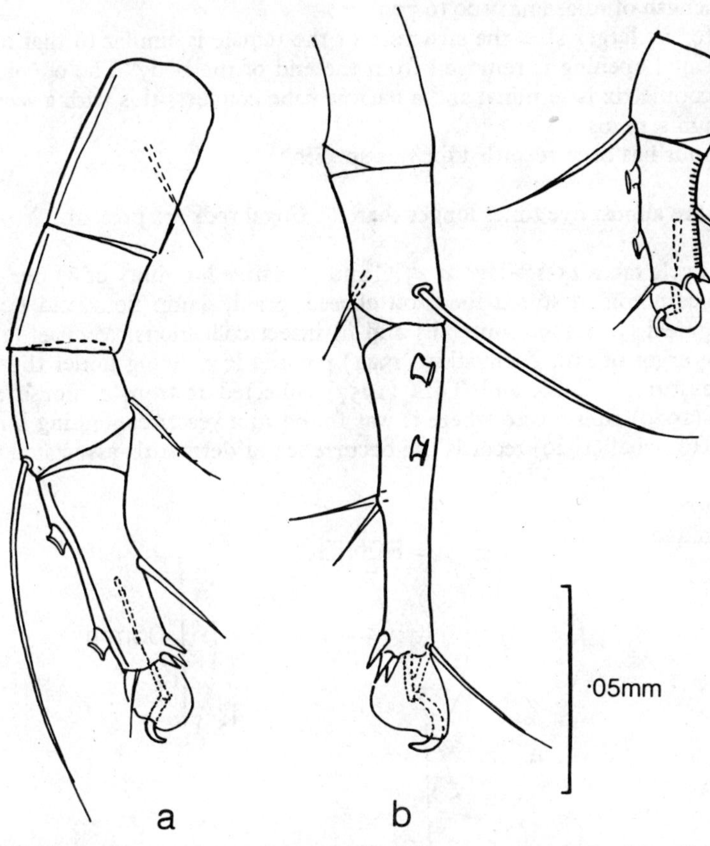

FIG. 71. Lateral view of terminal segments of leg IV of ♂ of (*a*) *Tyroborus lini* Oudemans; (*b*) *Mycetoglyphus fungivorus* Oudemans, (*c*) *Aleuroglyphus ovatus* (Troupeau).

The propodosomal shield is almost square in outline and faintly pitted. The cuticle is smooth. The supracoxal seta (Fig. 70b) is drawn out into a point at its distal extremity, it is expanded at the base and its lateral edges are barbed, the barbs arising at almost an acute angle.

The hysterosomal setae are arranged as in *T. putrescentiae*, the longer ones sparsely pectinate. Setae d_1 are short (5 to 7 per cent of the idiosoma in length); d_2 are two to three times longer than d_1 and *l a* are four to six times longer than d_1. The remaining setae are also long and arranged in the characteristic fan-shaped 'train'.

The legs are stout and ornamented with a faint reticulate pattern. As the tarsi are shorter than those of *T. putrescentiae*, the basal setae and solenidia are closer together. Omega$_1$ (Plate 3, fig. 2) is almost cylindrical, slightly expanded in the middle and arises from the same chitinous depression as the famulus; *e* is a stout conspicuous spine at the distal end of all the tarsi (Fig. 76a) and five spines encircle the base of the claw on the ventral side (Fig. 77a). On tarsus IV, the suckers arise from the median part of the tarsus (Fig. 18).

The support of the penis is curved inwards, the penis itself (Fig. 72c) is almost straight and tapering.

FEMALE. Length of idiosoma: 500 to 700 µm.

Except for its larger size, the structure of the female is similar to that of the male. The anal opening is removed from the end of the body. The opening to the bursa copulatrix is terminal and a narrow tube connects this with a saccular receptaculum seminis.

No hypopus has been recorded for this species.

LARVA

Setae d_2 are almost five times longer than d_1. Coxal rods are present.

BIOLOGY

The life cycle takes 15-18 days at 23°C and a relative humidity of 87 per cent. *T. casei* is common in stored food, on cheese, grain, damp flour, old honeycombs, dogmeal [J. Thompson coll.] and in insect collections. Michael (1903) found it on ergot of rye, Zachvatkin (1941) records it as living under the bark of old tree stumps. Türk and Türk (1957) collected it from a mouse nest. Oudemans (1910) cites a case where it was found in a vessel containing human urine and Henschel (1929) records the occurrence of dermatitis associated with *T. casei*.

DISTRIBUTION

Cosmopolitan.

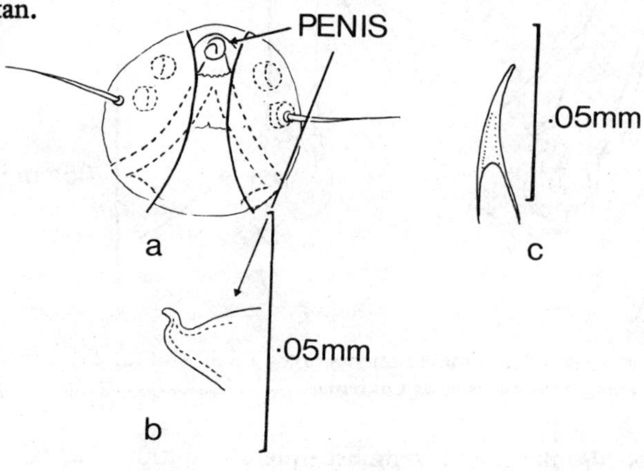

FIG. 72. Penis of (*a*, *b*) *Tyroborus lini* Oudemans, (*c*) *Tyrolichus casei* Oudemans.

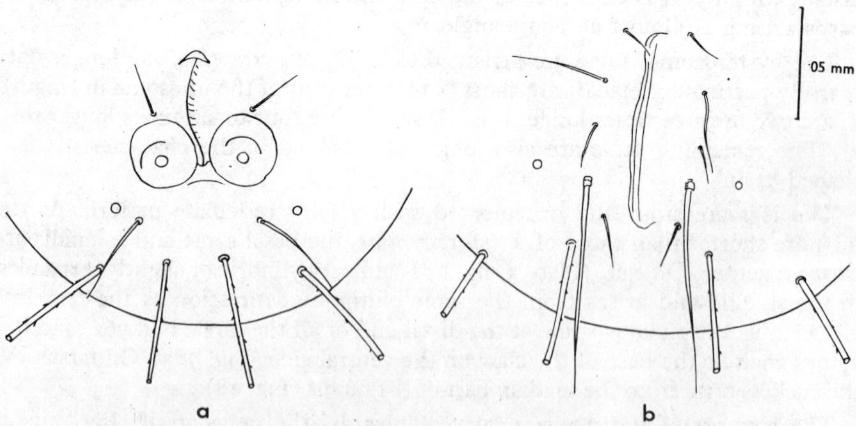

FIG. 73. *Tyroborus lini* Oudemans. Anal region of (*a*) ♂, (*b*) ♀.

MYCETOGLYPHUS Oudemans, 1932

Tyrophagus (part) sensu Türk and Türk, 1957 and Hughes, 1961; *Tyrolichus* sensu Karg, 1971.

Setae *v e* are short and smooth, less than one-quarter the length of *v i*, they arise posterior to *v i* (Fig. 47b).

The dorsal terminal tarsal seta *e* and the five ventral setae *p*, *q*, *s*, *u*, *v* are spine-shaped (Fig. 76b).

Setae *sc i* are longer than *sc e*.

On tibia I, $sigma_1$ is never more than twice as long as $sigma_2$.

Leg I of the male is not enlarged and the femur does not bear a ventral apophysis.

The penis of the male is long (Fig. 78).

Type species: *Mycetoglyphus fungivorus* Oudemans, 1932.

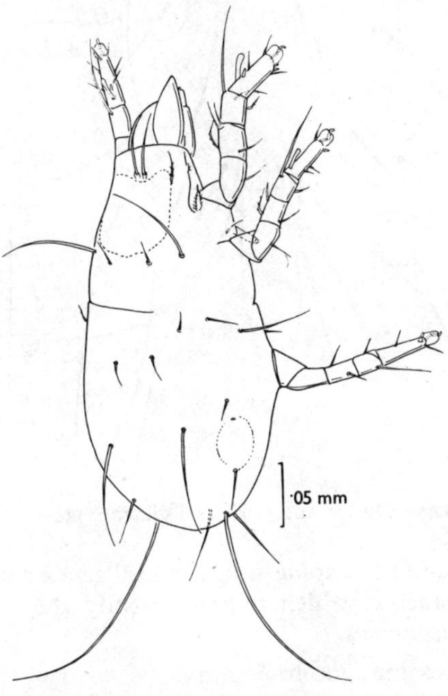

FIG. 74. *Tyroborus lini* Oudemans. Dorso-lateral view of larva.

MYCETOGLYPHUS FUNGIVORUS Oudemans, 1932

Forcellinia fungivora sensu Zachvatkin, 1941; *Tyrophagus fungivorus* sensu Türk and Türk, 1957 and Hughes, 1961; *Tyrolichus fungivorus* sensu Karg, 1971.

Zachvatkin (1941) allocated this species to the genus *Forcellinia*; it differs from the type species *Forcellinia wasmanni* (Moniez, 1892), however, in the relative lengths of the *sc* setae, the shape of the dorsal setae and their relative lengths and the genitalia of the male. Türk and Türk (1957) and Hughes (1961b) included it in *Tyrophagus*, from which it differs, however, in having short *v e* setae arising posterior to *v i* from the lateral margin of the dorsal propodosomal shield, and spine-shaped *e* setae at the distal ends of the tarsi. Karg (1971) regards this last character as sufficient justification for including *M. fungivora* in the genus *Tyrolichus*. In my opinion, the short *v e* setae and

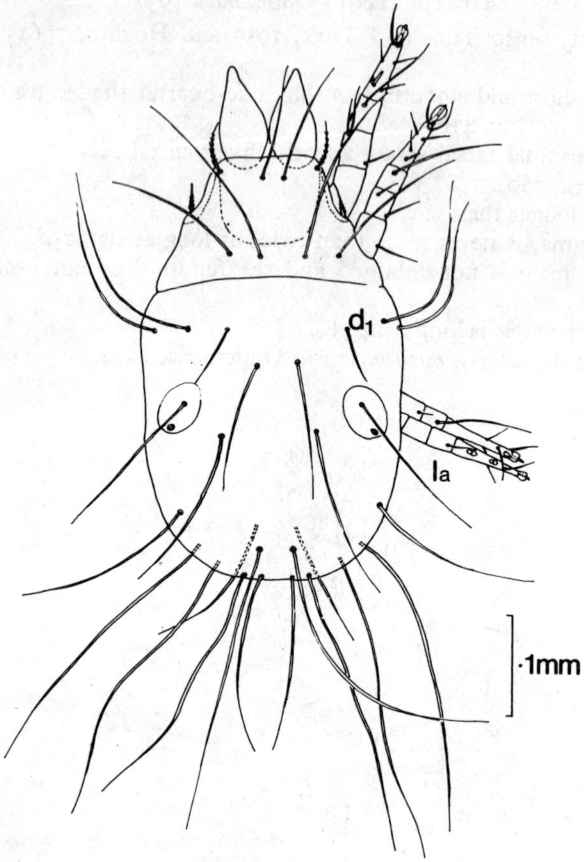

FIG. 75. *Tyrolichus casei* Oudemans ♂. Dorsal view. Setae of the idiosoma: d_1, *l a*.

their position (Fig. 47b), the spine-shaped tarsal seta *e* and the elongated penis of the male are characters which together justify the revival of Oudemans' original genus *Mycetoglyphus*.

MALE. Length of idiosoma: 400 to 600 µm.

This species, with its comparatively well-tanned appendages and large size, has the same general appearance as *Tyrophagus similis*.

The arrangement of setae on the idiosoma and legs also resemble that of *T. similis*, the main differences being: *v e* are very short, less than one-quarter the length of *v i*, they arise posterior to the bases of *v i*, never as far back as the mid point of the lateral edge of the propodosomal shield. Setae *l a* are short (about 6 per cent of the idiosoma), d_1 are 1 to 1.5 times longer than *l a* and d_2 are 1.5 to twice as long as *l a*. The supracoxal seta (Fig. 70c) is a curved structure covered with very small pectinations. On tarsi I and II, omega$_1$ is club-shaped; on tarsi I to III, the distal seta *e* (Fig. 76b) and five ventral spines—slightly unequal in size—are present as in *Tyrolichus casei*; on tarsus IV (Fig. 71b) the suckers are placed in the basal half of the segment, the two setae *r* and *w* being distal to them.

The penis (Fig. 78) is a very long curved tube arising from a basal plate.

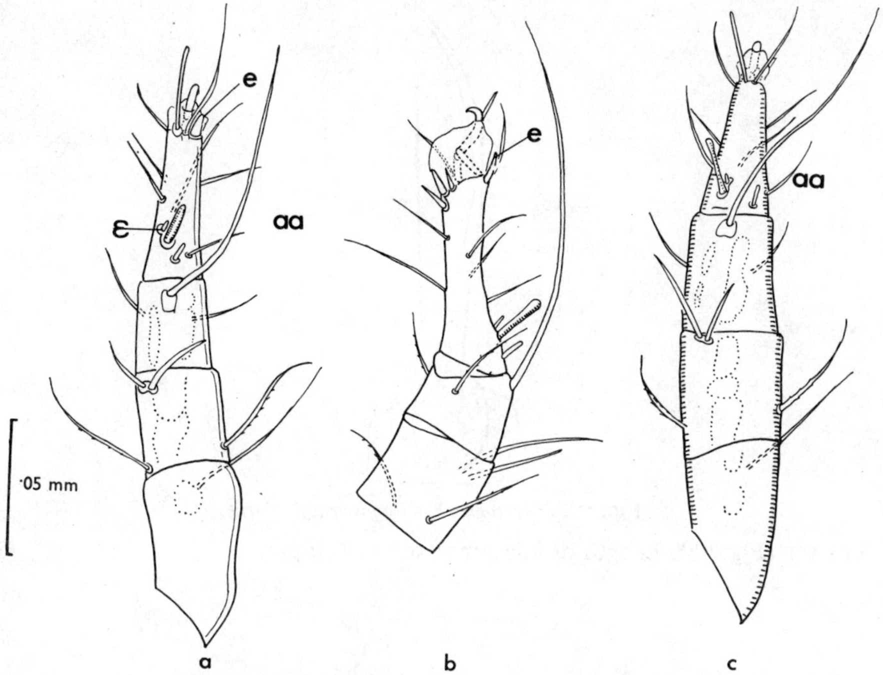

FIG. 76. (*a*) Dorsal view of right leg I of *Tyrolichus casei* Oudemans; (*b*) external view of left leg I of *Mycetoglyphus fungivorus* Oudemans; (*c*) dorsal view of right leg I of *Aleuroglyphus ovatus* (Troupeau). ε (epsilon) famulus; *aa*, *e* seta and spine.

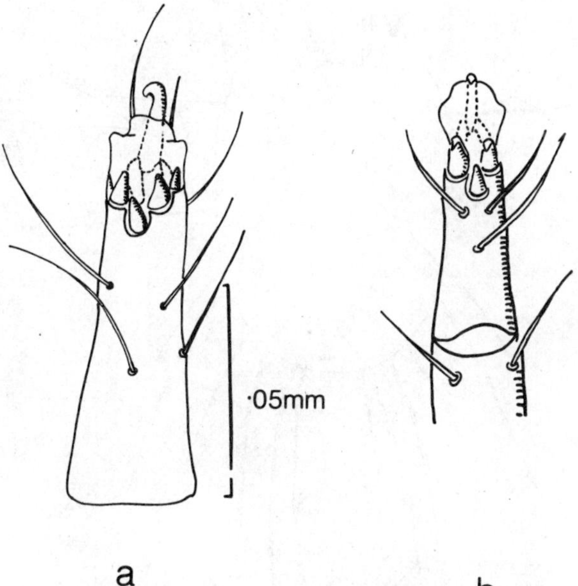

FIG. 77. Ventral view of right tarsus I of (*a*) *Tyrolichus casei* Oudemans ♀, (*b*) *Aleuroglyphus ovatus* (Troupeau) ♂.

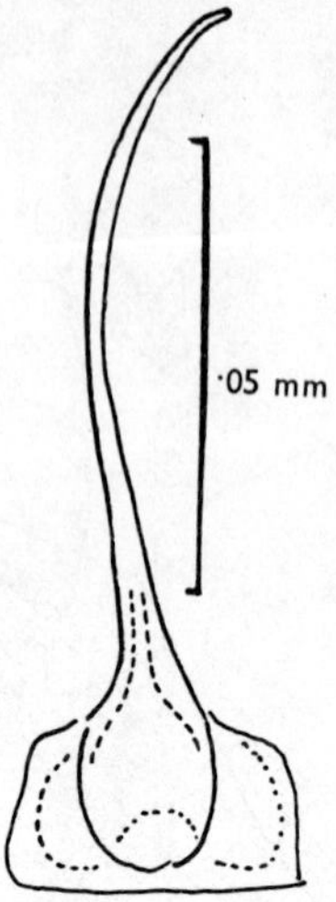

FIG. 78. *Mycetoglyphus fungivorus* ♂. Penis.

FEMALE (Fig. 79). Length of idiosoma: 500 to 600 μm.

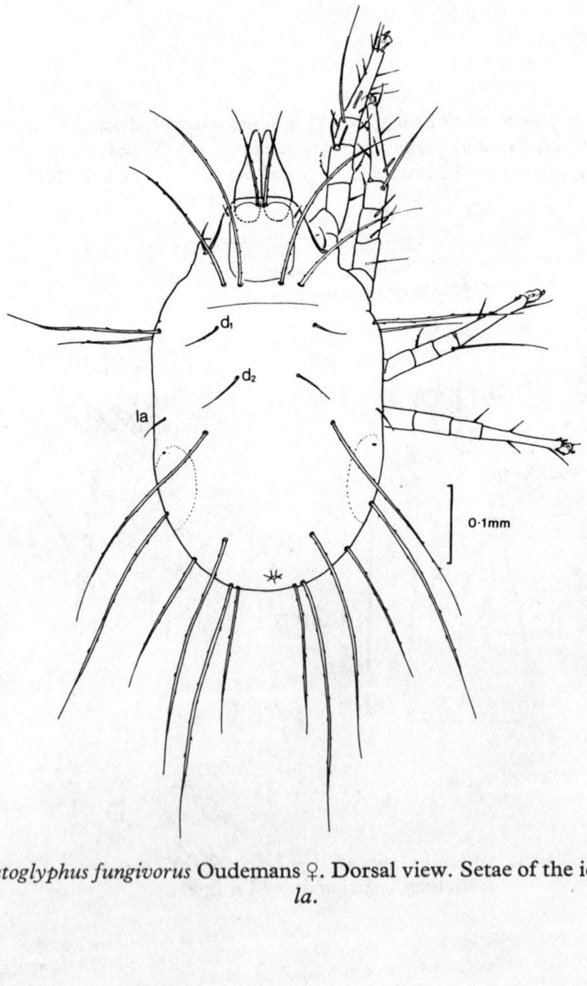

FIG. 79. *Mycetoglyphus fungivorus* Oudemans ♀. Dorsal view. Setae of the idiosoma: d_1, d_2, *la*.

The general appearance of the female is similar to that of the male.

HYPOPUS

According to Zachvatkin, 1941, this is 250 µm long and yellow-brown in colour, with a widely rounded posterior margin. The propodosomal shield has an almost straight anterior edge without any sign of a rostrum; *v i* are absent. The gnathosomal base is completely concealed by the propodosomal shield. Ventrally, the sternum is extended posteriorly, so that coxal shields I and II are completely separate from one another. The sucker plate is well-removed from the posterior edge of the body and is almost circular in outline.

Tarsus I bears a long expanded and three lanceolate setae. Tarsus IV also has two lanceolate setae.

BIOLOGY

Oudemans' original specimens came from mushrooms growing in Sussex, England. It has also been found on celery waste, lettuce, decaying radishes and on wet decaying wood debris (S. Mahunka, 1961). D. A. Griffiths (1960) found that it is common on grassland and beneath stacks of hay and straw. Türk and Türk (1957) record it from moles' nests and nests of *Camponotus ligniperda*, it is possible that ants are used by the hypopus as a means of transport.

DISTRIBUTION

England, U.S.A., S. Africa, Germany, U.S.S.R., Hungary.

ALEUROGLYPHUS Zachvatkin, 1935

Setae *v e* are long and pectinate, more than half the length of *v i* and arising at about the same level (Fig. 80).

Setae *sc i* are shorter than *sc e*.

The supracoxal seta is conspicuous and thickly pectinate (Fig. 70d).

On the tarsus the doral terminal seta *e* is hair-like.

There are three conspicuous ventral spines on the tarsi: $q+v$, $p+u$ and *s* which lie close together (Fig. 77b).

Type species: *Tyroglyphus ovatus* Troupeau, 1878.

ALEUROGLYPHUS OVATUS (Troupeau, 1878)

Tyroglyphus ovatus Troupeau, 1878. Not *Tyroglyphus ovatus* sensu Oudemans, 1912.

MALE (Fig. 80). Length of idiosoma: 480 to 550 µm.

In general build of body and size, *A. ovatus* resembles *Tyroborus lini*. It is distinguished, however, by the more intensely tanned chelicerae and legs, which are dark reddish-brown in colour and contrast with the glistening white appearance of the rest of the body.

The dorsal shield is oblong, with slightly concave lateral margins and a pitted surface. The supracoxal seta (Fig. 70d) is shaped like a leaf, whose lateral edges are produced into long straight pectinations. Setae *sc i* are short, about one-third the length of *sc e*.

Of the dorsal hysterosomal setae, d_1 to d_3, *l a* and *h i* are short, about the same length as *sc i*, whilst d_4 and *l p* are only slightly longer. The only long setae are *sa e*, *sa i* and two pairs of postanals which together form a rather sparse 'train'. All the setae are slightly pectinate and the shorter ones often have bifid ends, the prongs sometimes twisted apart.

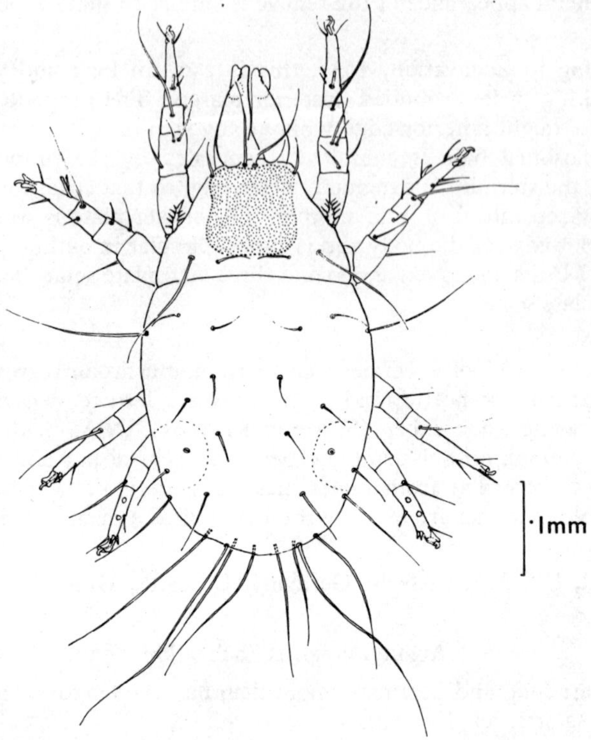

FIG. 80. *Aleuroglyphus ovatus* (Troupeau) ♂. Dorsal view.

The legs are short and stout. On tarsi I and II, omega$_1$ is long and slender, tapering towards a rounded end (Fig. 76c) and arising from the same depression as the very small famulus. At the distal end of the tarsi, the three ventral spines—$p+u$, $q+v$ and s are stout and massive (Fig. 77b), the external ones being hooked at their ends; the dorsal terminal seta e is hair-like. On tarsus IV, the two suckers arise in the median part of the tarsus (Fig. 71c).

The genital folds and genital sense organs are slightly tanned and pale yellow in colour; the penis supports are almost straight and diverge posteriorly, the penis itself is a straight tube. The postanal setae are ventral in position and all three pairs are arranged almost in the same straight line (Fig. 81a).

FEMALE. Length of idiosoma: 580 to 670 µm.

The female closely resembles the male. The anal opening is bordered by four pairs of anal setae (Fig. 81b); of these a_2 are long and extend beyond the posterior edge of the body. The two pairs of postanals are also long and almost in line with one another.

No hypopus has been recorded in this species.

LARVA (Fig. 82).

This closely resembles the adult. The *sc i* setae are much shorter than *sc e* and the coxal rods are blunt tubes. On tarsus I, omega$_1$ almost reaches the end of that segment and expands towards its distal end. A pair of long postanal setae is present.

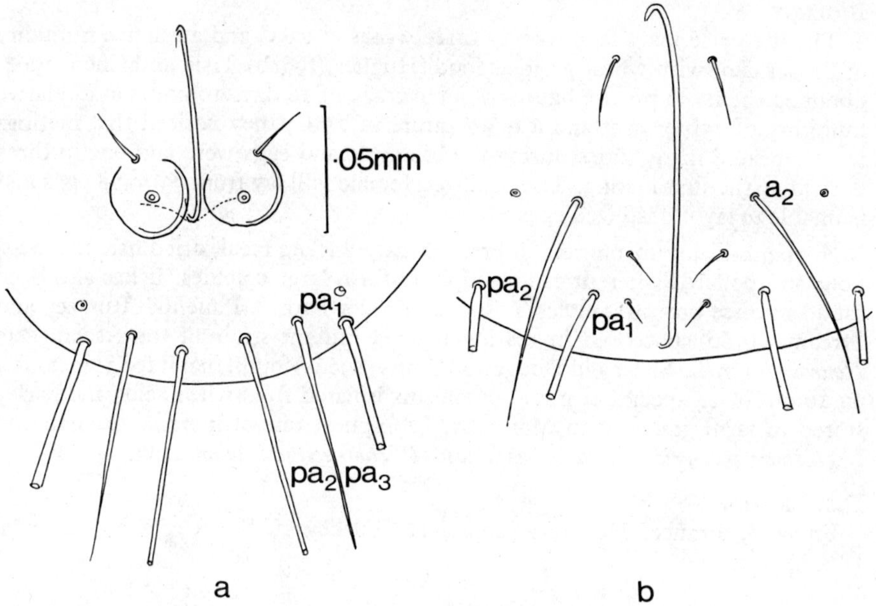

FIG. 81. *Aleuroglyphus ovatus* (Troupeau). Anal region of (*a*) ♂, (*b*) ♀. Setae of the idiosoma: a_2, pa_1, pa_2, pa_3.

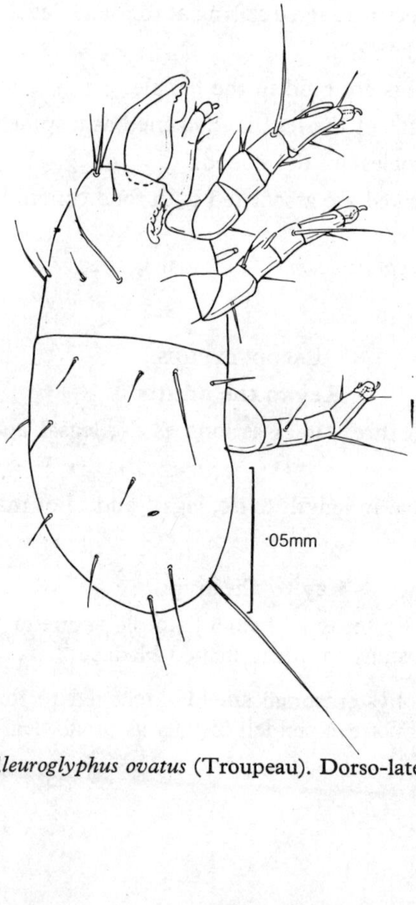

FIG. 82. *Aleuroglyphus ovatus* (Troupeau). Dorso-lateral view of larva.

BIOLOGY

The life cycle takes from two to three weeks at 23°C and a relative humidity of 87 per cent with wheat germ as food (Hughes, 1961b). Hsin and Chen (1964) obtained the more precise figures of an average of 16 days 10 hours at a relative humidity of 75 per cent and a temperature of 25°C; they noticed that matings were repeated many times during the life cycle and eggs were laid one to three days after the first mating. The fertilized female will lay from 33 to 78 eggs and is unable to lay unfertilized eggs.

A. ovatus is not uncommon in bran, wheat, chicken meal, dried fish products, flour and pollards where it can multiply to form large colonies. It has also been found in mice burrows, moles' nests and broiler houses. Pimental, Rumsey and Streams (1960) succeeded in rearing it on a mutant strain of the Ascomycete *Neurospora crassa* Shear and Dodge and Sinha (1966a) found that it fed voraciously on 10 out of 24 species of micro-organisms isolated from wheat, oats and barley stored in farm granaries in Manitoba. Its highest rate of increase occurred on *Nigrospora sphaerica* (Sacc.) Mason and *Trichothecium roseum* Link.

DISTRIBUTION

England, France, Holland, Jugoslavia, Turkey, U.S.S.R., Japan, Canada, U.S.A.

LARDOGLYPHUS Oudemans, 1927

Hoshikadania Sasa and Asanuma, 1951.

Setae *v e* are curved pectinate setae arising at the same level as *v i*.

Setae *sc e* are longer than *sc i*.

The claws of all the legs are bifid in the female.

None of the dorsal setae of the legs is thickened as a spine.

Only heteromorphic males are developed.

The two species described are associated with food containing a large amount of protein.

Type species: *Lardoglyphus zacheri* Oudemans, 1927.

LARDOGLYPHUS

Key to the adults

1. Setae d_4 more than three times as long as d_3; legs I and II of male with bifid claws . . . . *L. zacheri*, p. 81

—. Setae d_4 almost equal in length to d_3; legs I and II of male with undivided claws . . . . *L. konoi*, p. 86

Key to the hypopi

1. Setae arising from hysterosomal shield simple; none of the setae of tarsus IV are expanded distally into leaf-shaped blades . . . . *L. zacheri*, p. 83

—. Setae arising from hysterosomal shield thickened to form spines; two of the setae of tarsus IV are expanded distally as broad, leaf-shaped structures . . . . *L. konoi*, p. 87

LARDOGLYPHUS ZACHERI Oudemans, 1927

HETEROMORPHIC MALE (Fig. 83). Length of idiosoma: 430 to 550 μm.

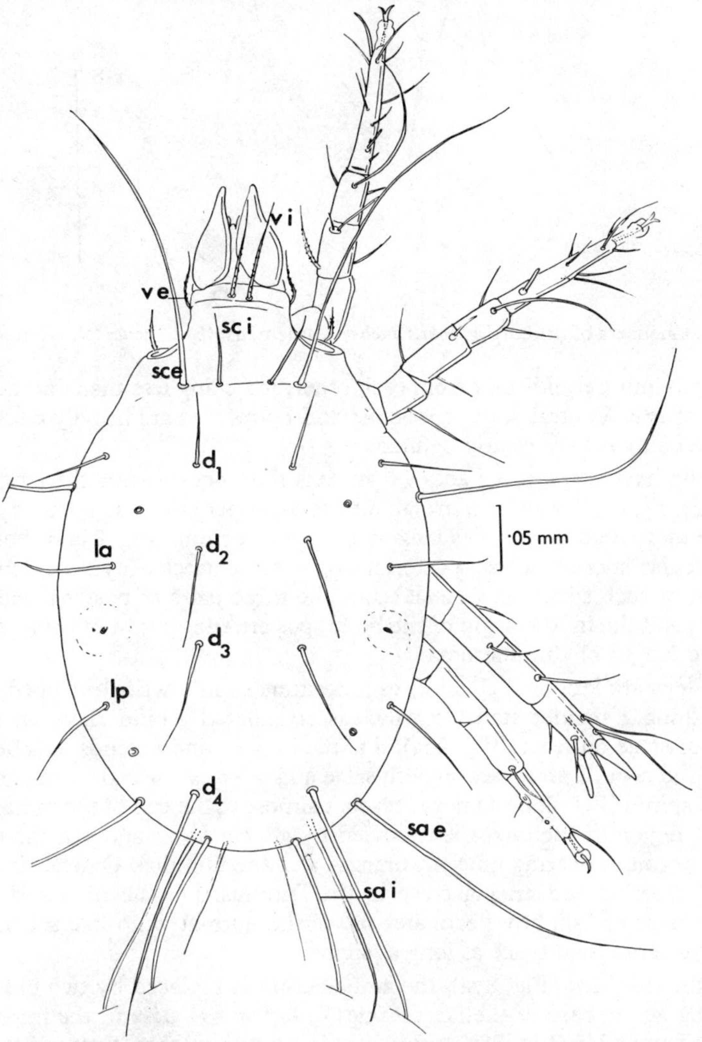

FIG. 83. *Lardoglyphus zacheri* Oudemans ♂. Dorsal view. Setae of the idiosoma: *ve*, *vi*, *sc e*, *sc i*, d_1 to d_4, *la*, *lp*, *sa e*, *sa i*.

The body is widest between the second and third pairs of legs and has a rounded posterior margin. The cuticle is smooth and cream-coloured, the legs, apodemes and chelicerae of a darker shade. The chelicerae are slender (Fig. 84a), with feebly toothed shears. The propodosoma lacks a dorsal shield.

All the setae arising from the idiosoma are smooth or slightly pectinate, many of them becoming noticeably thicker towards the base. Setae *v i* extend forward over the gnathosoma, whilst *v e* embrace its sides; both of these pairs of setae are distinctly pectinate. The supracoxal seta is small, curved and serrated, and Grandjean's organ is an inconspicuous triangular fold of cuticle. The scapular

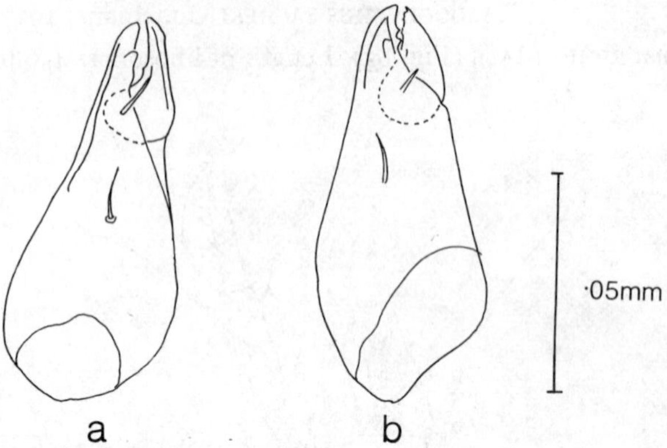

FIG. 84. Chelicera of (*a*) *Lardoglyphus zacheri* Oudemans, (*b*) *L. konoi* (Sasa and Asanuma).

setae are almost equidistant from each other, *sc i* being less than one-quarter the length of *sc e*. Ventrally, the apodemes and epimerites are heavily cuticularized, and the epimeres are clearly defined.

On the hysterosoma, *h i* and *h v* are less than one-quarter the length of *h e*. Setae d_1, d_2, d_3, *l a* and *l p* are about the same length as *sc i*, but d_4, *sa i* and *sa e* are more than three times longer. The anal opening (Fig. 85a) is bordered by two circular suckers whose posterior edges are encircled by a curved sclerite; in front of each sucker is an anal seta. The three pairs of postanal setae extend well beyond the hind margin of the body; pa_3 are the longest of these, exceeding half the length of the idiosoma.

The legs are long and slender; each terminates in a well-developed pretarsus ensheathing a slender stalk, to which is articulated a bifid claw. Of the distal group of setae on leg I (Fig. 86a), *d* is the longest and extends well beyond the end of the claw; *e* and *f* are smooth setae and $q+v$, $p+u$ and *s* are conspicuous ventral spines (Fig. 86b). Omega$_3$ reaches almost to the end of the pretarsus. The median region of the tarsus is encircled by *aa*, *ba*, *la*, *ra* and *wa*, the basal part of this segment bearing omega$_1$, omega$_2$ and the famulus. Omega$_1$ is a slightly curved tapering rod arising close to the famulus. On the tibia and genu the ventral setae are slightly pectinate; phi is the normal whip-like solenidion and sigma$_2$ is more than twice as long as sigma$_1$.

On the third leg (Fig. 87a), the ambulacrum is replaced by two blunt spines, *d* arising at the base of the largest one; *e*, *f*, *r* and *w* arise in the middle of the tarsus. Tarsus IV (Fig. 88a) terminates in an unbranched claw and the copulatory suckers occupy a central position.

FEMALE (Fig. 89). Length of idiosoma: 450 to 600 μm.

The body of the female tapers posteriorly and has a concave posterior margin; the apodemes and epimeres are less well-tanned. The chaetotaxy of the idiosoma is essentially the same as that of the males. The genital opening is a longitudinal slit lying between coxae III and IV. The anus (Fig. 90a) does not reach the posterior margin of the body and is bordered by five pairs of anal seta; two pairs of postanals extend beyond the end of the body, pa_2 being more than half the length of the idiosoma. The bursa copulatrix opens by a small aperture on the posterior end of the body, it communicates with a triangular receptaculum seminis.

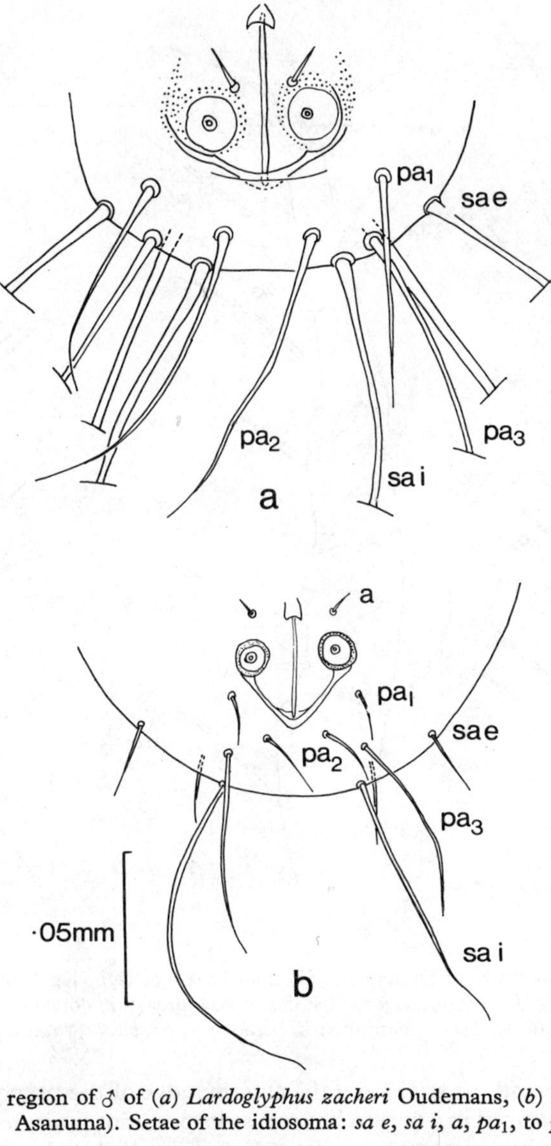

FIG. 85. Anal region of ♂ of (*a*) *Lardoglyphus zacheri* Oudemans, (*b*) *L. konoi* (Sasa and Asanuma). Setae of the idiosoma: *sa e*, *sa i*, *a*, pa_1, to pa_3.

All the legs bear bifid claws (Fig. 86b) and the arrangement of the setae is the same as in the male.

HYPOPUS (Figs. 91, 92). Length of idiosoma: 230 to 300 µm.

The idiosoma is pinkish-brown in colour and pear-shaped in outline. The dorsal surface is arched and covered, anteriorly with a propodosomal shield ornamented with a faint, scale-like pattern, posteriorly with a hysterosomal shield. The anterior margin of the last-mentioned shield is concave and its surface is finely reticulate. Beneath the narrow posterior edge and in the middle of the shield are darkly coloured thickenings of the cuticle. This area of the body is bent ventrally and, with the fourth pair of legs, is dragged along the ground in

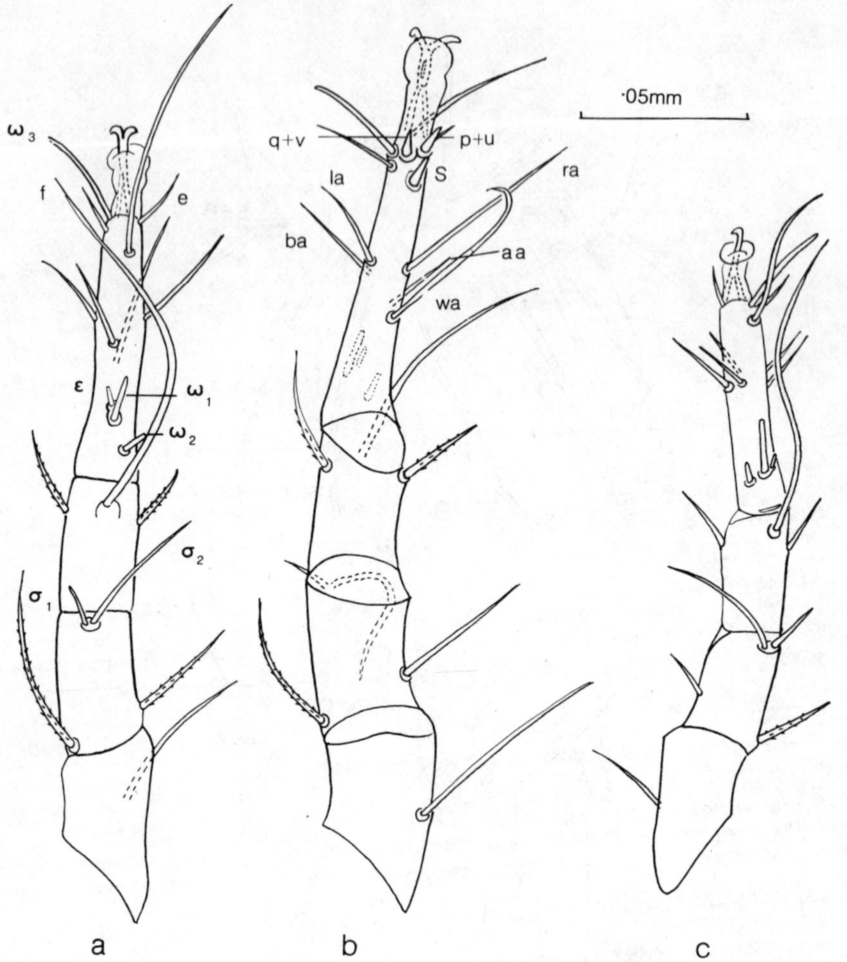

FIG. 86. *Lardoglyphus zacheri* Oudemans: (*a*) dorsal view of right leg I of ♂, (*b*) ventral view of left leg I of ♀. *Lardoglyphus konoi* (Sasa and Asanuma): (*c*) dorsal view of left leg I of ♂. Solenidia: ω_1 to ω_3, σ_1, σ_2. Famulus: ε. Setae: *e*, *f*, *aa*, *ba*, *la*, *ra*, *wa*, *s*, *p*+*u*, *q*+*v*.

walking. Setae *v i* and *v e* arise from the anterior edge of the propodosoma and *sc i* and *sc e* are arranged in a curved line across its posterior limit; *sc i* are shorter than *sc e*. On the hysterosomal shield are four pairs of dorsal setae and three pairs of laterals which are slightly thicker than the more anterior humerals.

The ventral surface of the body is concave and well-cuticularized; apodemes I unite to form a short sternum, but II to IV end freely in the midline. The internal margins of the acetabula are also thickened and those of legs II curve backwards and internally to meet apodemes IV. On coxal shields I and III and also between II and III are three pairs of circular 'pores', similar to those found in the adult. Three pairs of ventral setae are also present between legs II and III, internal to apodemes IV and on either side of the genital opening.

The sucker plate (Fig. 93a) is bounded posteriorly by a colourless margin, the edge of which is visible when the animal is viewed dorsally. It bears two large

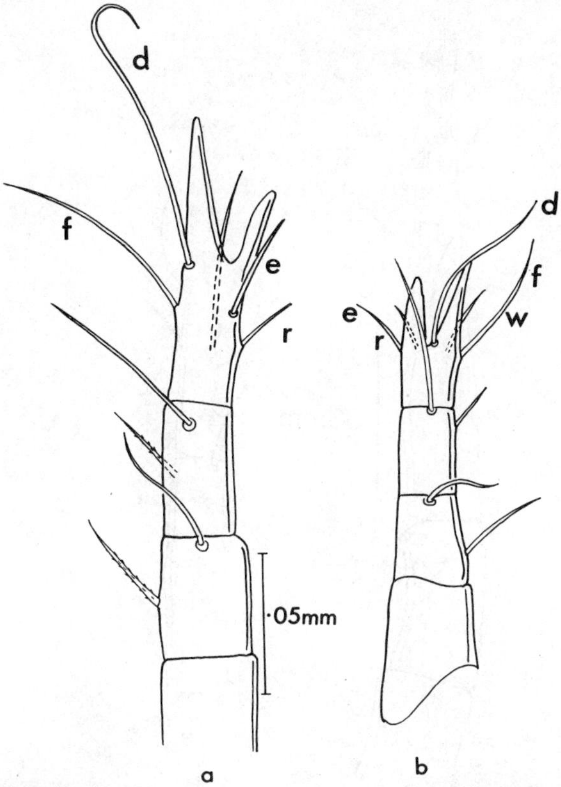

FIG. 87. (*a*) Dorsal view of right leg III of *Lardoglyphus zacheri* Oudemans ♂. (*b*) dorsal view of left leg III of *L. konoi* (Sasa and Asanuma) ♂. Setae of tarsus: *d*, *e*, *f*, *r*, *w*.

central suckers, four posterior (A to D), two anterior (I, K) and ill-defined auxiliary suckers (E to H).

Legs I to III terminate in a membranous pretarsus bearing a uniramous claw. The chaetotaxy of leg I (Fig. 94a) is the same as in the adult, except that on the tarsus, *ba* is missing, and on the genu, only one solenidion is present. Leg IV (Fig. 95a) is considerably shortened and pretarsus and claw are replaced by three dorsal trailing setae (*d*, *e*, *f*) and three shorter ventral ones ($p+u$, $q+v$ and *s*).

LARVA

Coxal rods are present on either side of the coxal setae.

BIOLOGY

At 23°C and 87 per cent relative humidity, the life cycle is completed in 10 to 11 days when dried heart muscle is given as food. Hypopi appear in large numbers when the food becomes short, and these readily attach themselves to larvae of the hide beetle *Dermestes maculatus* De Geer (Hughes, 1956) usually on the smooth surface of the arthrodial membranes. *L. zacheri* has been found on butchers' offal ('gut greaves') by D. A. Griffiths, on bones, hides and sheep-skins; also on the outside of boxes containing calves' stomachs exported as butchers' casings (D. J. Parry coll.).

DISTRIBUTION

England, Germany, S. America, Mexico, Australia.

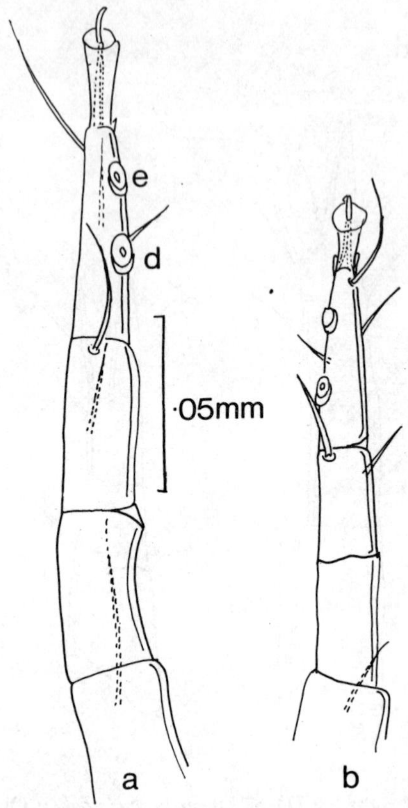

FIG. 88(*a*). Dorsal-lateral view of right leg IV of *Lardoglyphus zacheri* Oudemans ♂; (*b*) dorso-lateral view of left leg IV of *L. konoi* (Sasa and Asanuma) ♂. Tarsal suckers: *d*, *e*.

LARDOGLYPHUS KONOI (Sasa and Asanuma, 1951)

Hoshikadania konoi S. and A., 1951.

MALE (Fig. 96). Length of idiosoma: 300 to 440 µm.

The idiosoma is more rounded than in the preceding species and the animal is more bustling in its movements. The chaetotaxy is the same, but d_4, *sa e*, $p\ a_1$ and $p\ a_2$ are equal in length to d_3 instead of being more than twice as long (Fig. 85). The chelicera (Fig. 84b) has several small teeth on fixed and movable digits. The chitinous sclerite encircling the anal suckers in the male is bent more sharply towards the posterior margin of the body, and the anal setae lie on either side of the anterior end of the anus, instead of midway along its length.

Legs I (Fig. 86c), II and IV bear undivided claws; the tarsus of leg III (Fig. 87b) is shorter than that of *L. zacheri* and the setae are terminal. On leg IV (Fig. 88b) the copulatory suckers are in the middle region of the tarsus.

FEMALE (Fig. 97). Length of idiosoma: 400 to 550 µm.

The chaetotaxy of the idiosoma is similar to that of the male, $p\ a_1$ and *sa e* being relatively short (Fig. 90b). The receptaculum seminis is triangular in shape.

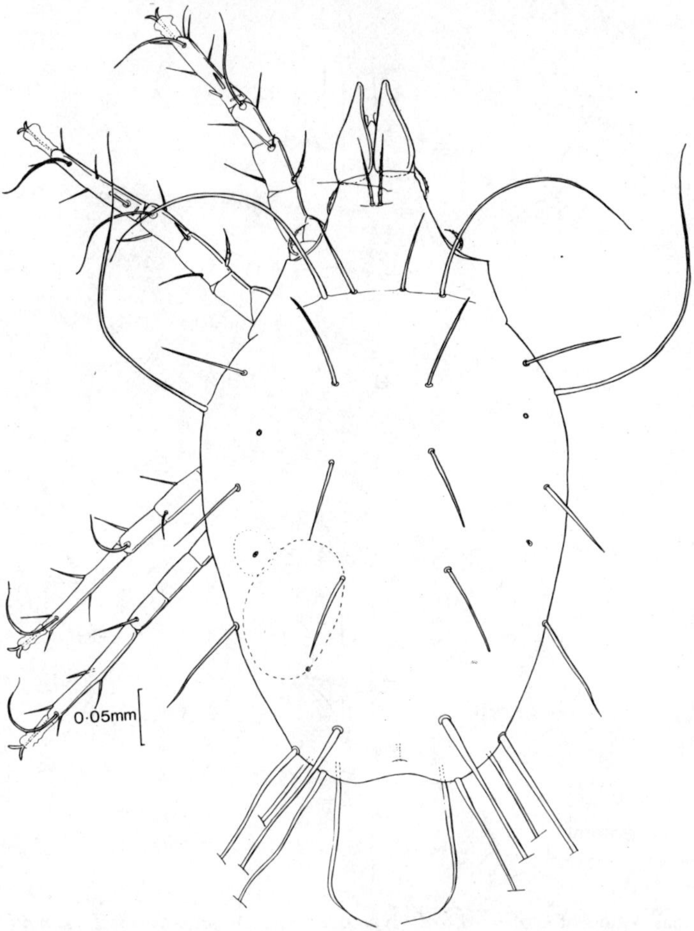

FIG. 89. *Lardoglyphus zacheri* Oudemans ♀. Dorsal view.

HYPOPUS (Figs. 98, 99). Length of idiosoma: 215 to 260 µm.

The setae arising from the hysterosomal shield are thickened to form spines. Ventrally, a posterior projection of apodemes III extends as far back as the setae between apodemes IV. The margin of the sucker plate (Fig. 93b) is wider than in *L. zacheri* and the two central suckers are smaller. Two of the encircling suckers A and D are replaced by horn-like projections and the auxiliary suckers are transparent. The anterior suckers I and K vary in position and are frequently asymmetrical.

The pretarsi of legs I (Fig. 94b) to III are longer and more slender in *L. konoi*. On tarsi I and II the seta *f* is leaf-shaped instead of being simple, and on tarsi III (Fig. 100) all the setae except *d* are expanded distally into transparent laminae. Tarsus IV (Fig. 95b) bears two similar leaf-like structures (*e* and $p+u$), and *rw*, absent in *L. zacheri*, is also present.

BIOLOGY

The life cycle takes nine to eleven days at 23°C and 87 per cent relative humidity, with dried heart muscle as food. Attempts to cross *L. konoi* with

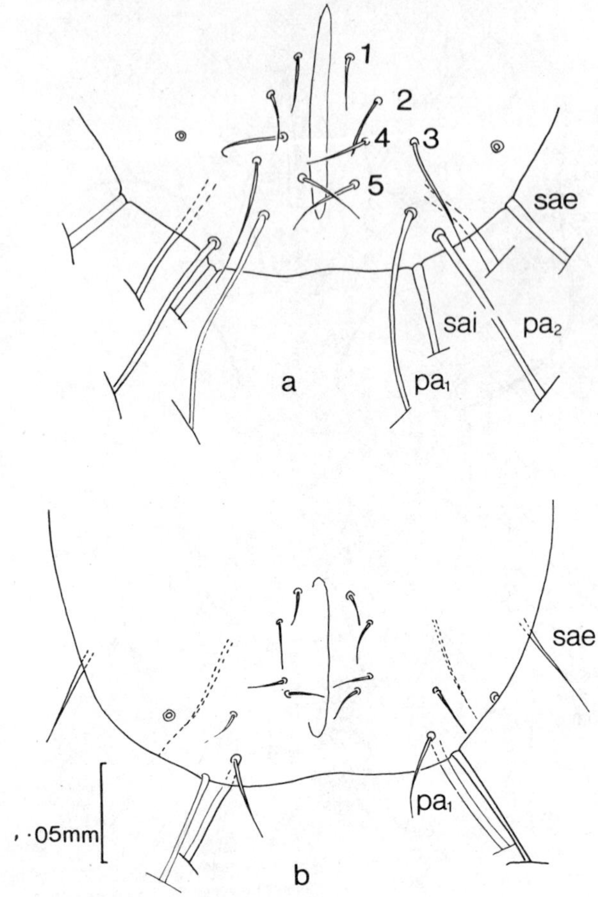

FIG. 90. Anal region of ♀ of (*a*) *Lardoglyphus zacheri* Oudemans, (*b*) *L. konoi* (Sasa and Asanuma). Setae of the idiosoma: 1 to 5 anals, *sa e*, *sa i*, pa_1, pa_2.

L. zacheri were unsuccessful and neither species is parthenogenetic (Hughes, 1956). *L. konoi* is a serious pest of dried fish and salted fish products in Japan (Sasa, 1964) and in Southern India (Pillai, 1957). In this country, it was found on butchers' offal (D. A. Griffiths coll.). The hypopi have been found clinging to *Dermestes carnivorus* (Rack, 1968) and *D. maculatus*.

DISTRIBUTION

England, Germany, Japan, Southern India.

CALOGLYPHUS Berlese, 1923

Caloglyphus sensu Hughes, 1961.

Samšiňák (1960) considers that *Caloglyphus* Berlese, 1923 is a synonym of *Sancassania* Oudemans, 1916. The type species of this is *S. chelone* Oudemans, 1916, known only in the hypopal stage.

The hypopus is a phoretic stage which has evolved to facilitate distribution, and also, in some cases, to tide over adverse conditions. Hence convergent evolution is likely to have taken place and hypopi belonging to different genera may resemble each other. For example, the hypopus of *Lepidoglyphus destructor* is very

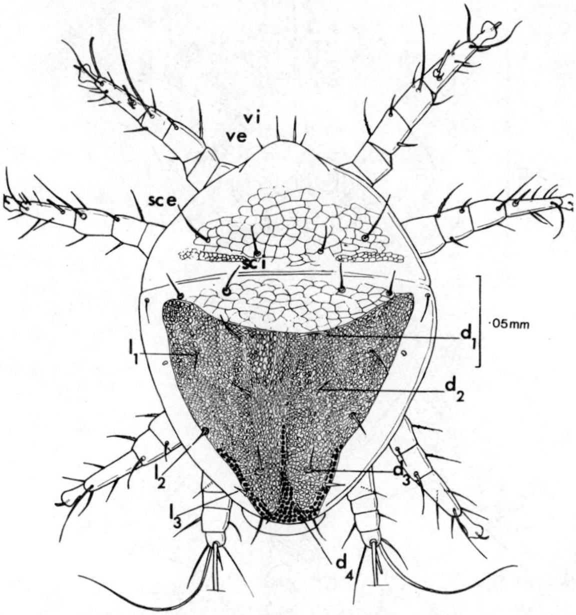

FIG. 91. *Lardoglyphus zacheri* Oudemans. Dorsal view of hypopus. Setae of the idiosoma: *ve*, *vi*, *sc e*, *sc i*, d_1 to d_4, l_1 to l_3.

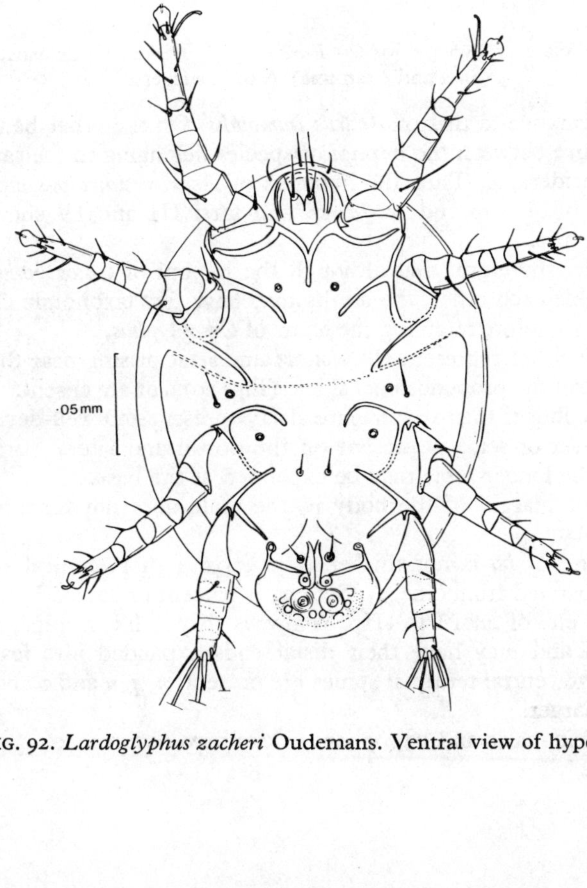

FIG. 92. *Lardoglyphus zacheri* Oudemans. Ventral view of hypopus.

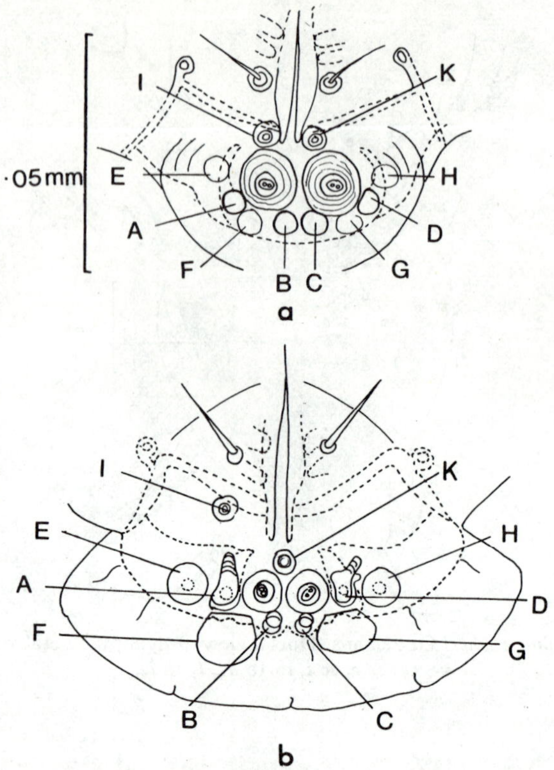

FIG. 93. Sucker plate of hypopus of (*a*) *Lardoglyphus zacheri* Oudemans, (*b*) *L. konoi* (Sasa and Asanuma). A to K suckers.

similar in appearance to that of *Acarus immobilis*. On the other hand, the variation in structure between the hypopi of species belonging to the same genus is sometimes considerable. Thus the hypopus of *Acarus nidicolous* lacks the long scapular setae of *A. siro* and *A. farris*, and tarsi III and IV show a marked reduction in setation.

It is possible, therefore, that although the hypopi of *Caloglyphus* and *Sancassania* resemble each other, the adults may have few taxonomic characters in common. I am therefore retaining the name of *Caloglyphus*.

Setae *v e* are either represented by short fine setae arising near the middle of the lateral edge of the propodosomal shield (Fig. 101), or are absent.

Setae *sc e* are longer than *sc i*; *sc i* are always present and well-developed.

A complete set of setae is present on the dorsal and lateral surfaces of the hysterosoma, the longer setae may be expanded at the base.

The posterior margin of the body in the male does not form a projecting opisthosomal plate.

On legs I and II *ba* is not thickened to form a short conical spine and is always well-separated from omega$_1$, Seta *aa* is present on tarsus I.

At the distal end of tarsi I to III, *e* is always shaped like a spine, *ra* and *f* are usually curved and may have their distal ends expanded into leaflike plates (Fig. 104a). Five ventral terminal spines are present, *p*, *q*, *u* and *v* about equal in size, *s* a little larger.

Heteromorphic males and hypopi are of common occurrence.

Type species: *Tyroglyphus mycophagus* Mégnin, 1874.

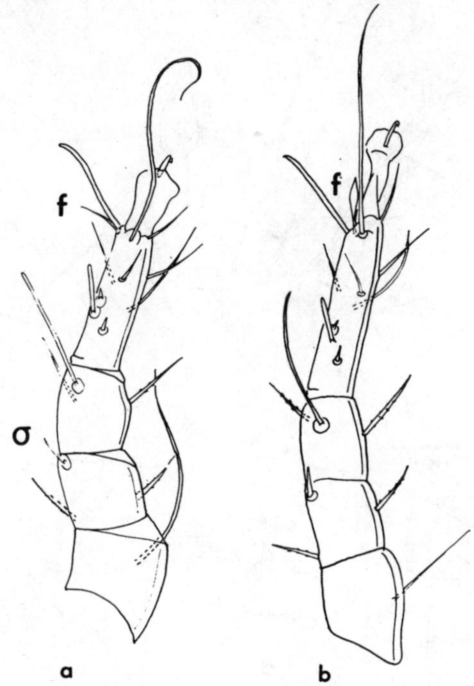

FIG. 94. Dorsal view of right leg I of hypopus of (*a*) *Lardoglyphus zacheri* Oudemans, (*b*) *L. konoi* (Sasa and Asanuma). Solenidion: σ; tarsal seta: *f*.

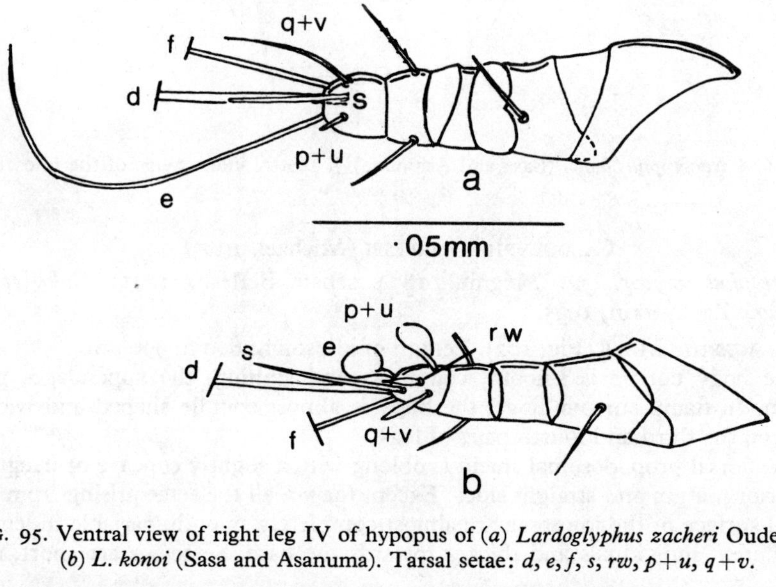

FIG. 95. Ventral view of right leg IV of hypopus of (*a*) *Lardoglyphus zacheri* Oudemans, (*b*) *L. konoi* (Sasa and Asanuma). Tarsal setae: *d*, *e*, *f*, *s*, *rw*, *p*+*u*, *q*+*v*.

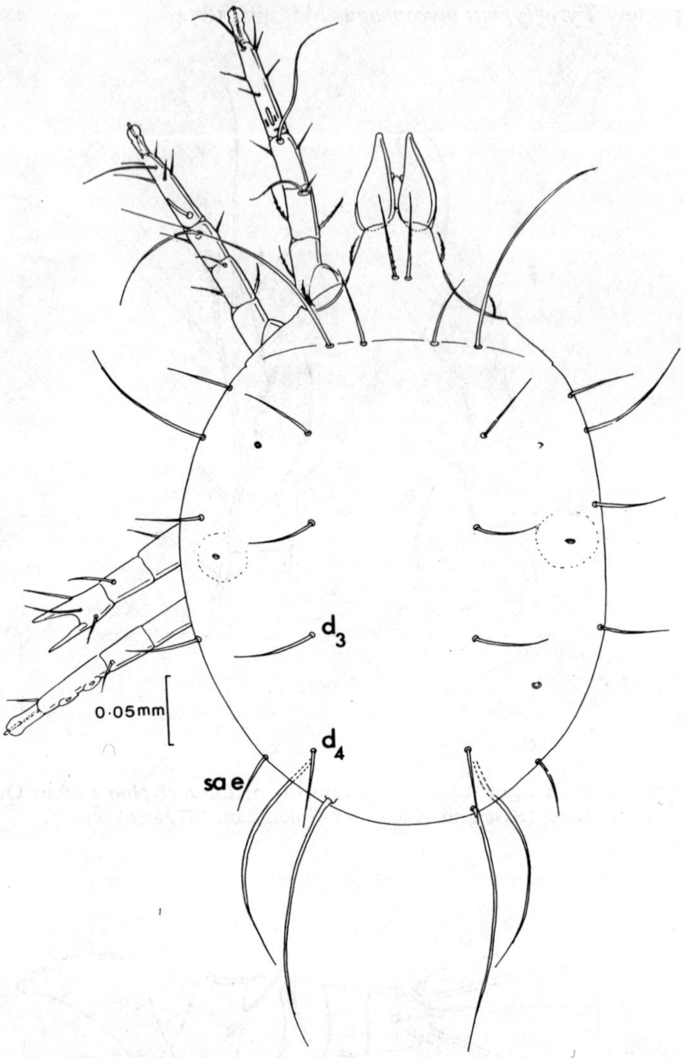

FIG. 96. *Lardoglyphus konoi* (Sasa and Asanuma) ♂. Dorsal view. Setae of the idiosoma: *d*3, *d*4, *sa e*.

CALOGLYPHUS BERLESEI (Michael, 1903)

Tyroglyphus mycophagus Mégnin, 1874 sensu Berlese, 1891; *Caloglyphus rodionovi* Zachvatkin, 1935.

HOMOMORPHIC MALE (Fig. 102). Length of idiosoma: 600 to 900 µm.

The body cuticle is smooth, colourless and shining, the appendages pale brown. In damp surroundings, the body is almost spindle-shaped and widest between the third and fourth pairs of legs.

The dorsal propodosomal shield is oblong with a slightly concave or irregular posterior margin and straight sides. Except for *v i*, all the setae arising from the dorsal surface of the idiosoma are almost completely smooth, variable in length in different individuals and thicker towards the base. Setae *v e* are short, fine setae which arise midway along the lateral edges of the propodosomal shield. The *sc* setae are equidistant from one another, and *sc e* are three to four times longer than *sc i*. The supracoxal seta is well-defined, very slightly pectinate and more than half the length of d_1. Grandjean's organ (Fig. 101) is a truncate spine with small pectinations arising from its surface. On the ventral side epimeral plates are developed, but they are colourless and irregular in outline.

On the dorsal surface of the hysterosoma, d_1 are short (5 to 10 per cent of the idiosoma in length), d_2 are two to three times longer than d_1, *l a* and *h i* are 1.5 to twice as long as d_1; d_3, d_4 and *l p* are long, d_4 extending well beyond the end of the body. Ventrally (Fig. 103), *p* a_2 are three to five times longer than *p* a_1, *p* a_3 are longer than *p* a_2 and like them, stretch beyond the posterior margin. Conspicuous round anal suckers are present as in the previous genus.

The gnathosoma is narrow and tapering, the chelicerae with toothed shears and a conspicuous mandibular spine.

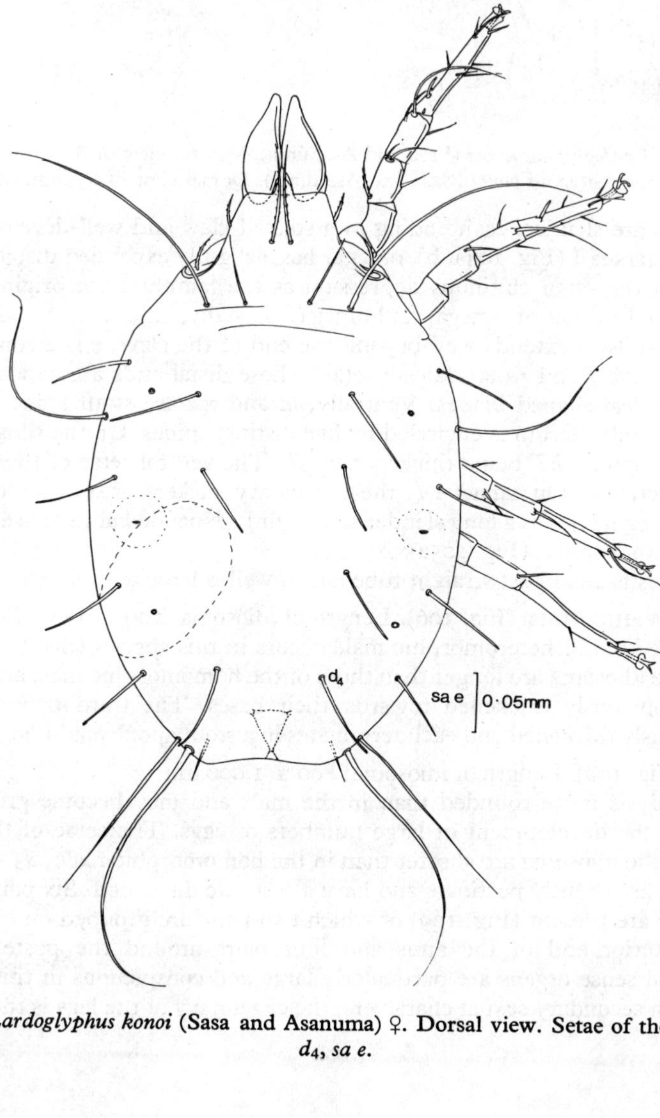

FIG. 97. *Lardoglyphus konoi* (Sasa and Asanuma) ♀. Dorsal view. Setae of the idiosoma: d_4, *sa e*.

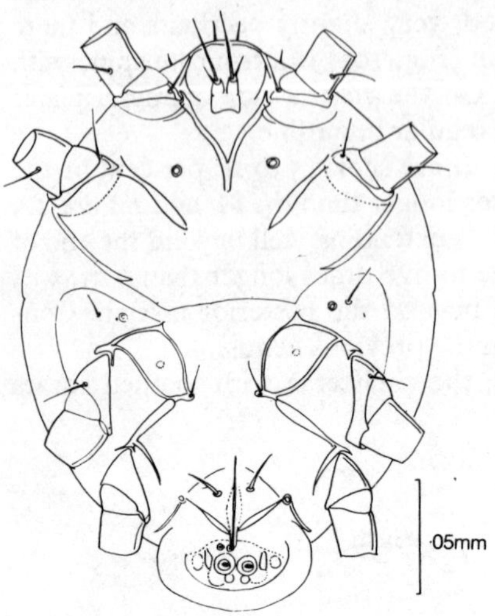

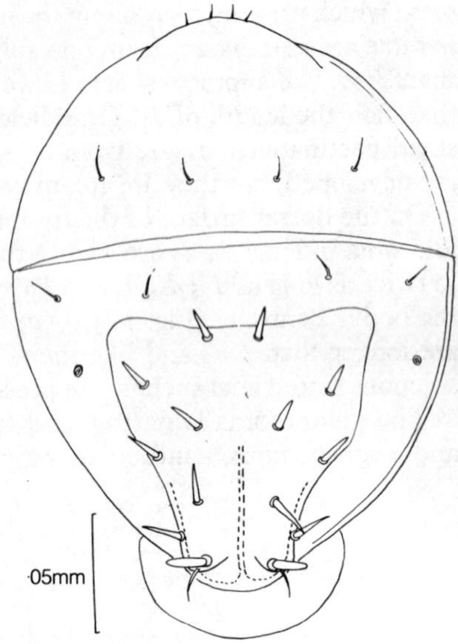

FIG. 98. *Lardoglyphus konoi* (Sasa and Asanuma). Ventral view of hypopus (left).
FIG. 99. *Lardoglyphus konoi* (Sasa and Asanuma). Dorsal view of hypopus (right).

The legs are slender, each ending in a stalked claw and well-developed pretarsus. On tarsus I (Fig. 104a, b), omega$_1$ has a slightly expanded distal end and arises from the same chitinous depression as the famulus; the origin of *aa* is slightly distal to that of omega$_1$ and omega$_2$. Distally, omega$_3$ is a rod of even diameter, whilst *d* extends well-beyond the end of the claw; *e* is a conspicuous stout spine and *f* and *ra* are falcate setae, whose distal ends are expanded into transparent leaf-shaped blades. Ventrally, *la* and *wa* are stout spines and the base of the ambulacrum is encircled by five distinct spines. On the tibia, *gT* and *hT* are also spines, *hT* being thicker than *gT*. The ventral setae of the genu are slightly pectinate. On tarsus IV, the copulatory suckers arise from the distal half of the segment; *f* is a long slender seta slightly expanded at its free end, and *r* and *w* are stout spines (Fig. 105a).

The penis is an almost straight tube and is well-sclerotized.

HETEROMORPHIC MALE (Fig. 106). Length of idiosoma: 700 to 1,000 μm.

Only one kind of heteromorphic male occurs in this species (See p. 17). The setae of the idiosoma are longer than those of the homomorphic male and tend to be more obviously thickened towards their bases. The third pair of legs is conspicuously thickened and each terminates in a stout apophysis (Fig. 107a, b).

FEMALE (Fig. 108). Length of idiosoma: 800 to 1,000 μm.

The body is more rounded than in the male and may become greatly distended by the development of large numbers of eggs. The setae of the dorsal surface of the idiosoma are shorter than in the homomorphic male; d_4 are not as long as d_3, are slightly pectinate and have a truncate distal end. Six pairs of anal microsetae are present (Fig. 109) of which two pairs are grouped on either side of the anterior end of the anus and four pairs around the posterior end. The genital sense organs are particularly large and conspicuous in this species. Apart from secondary sexual characters, the chaetotaxy of the legs is the same as in the male.

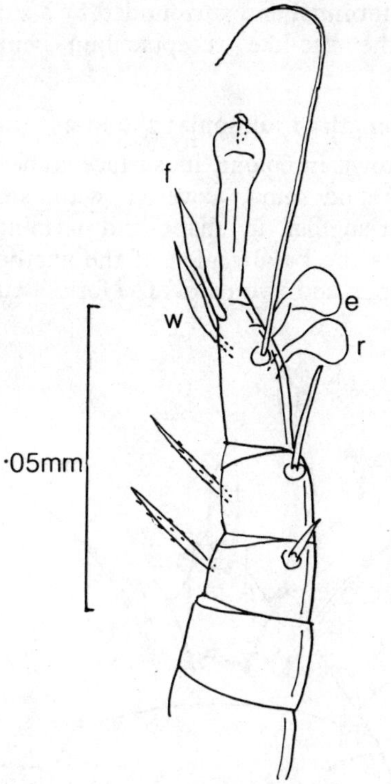

FIG. 100. *Lardoglyphus k onoi* (Sasa and Asanuma). Dorsal view of right leg III of hypopus. Tarsal setae: *e*, *f*, *r*, *w*.

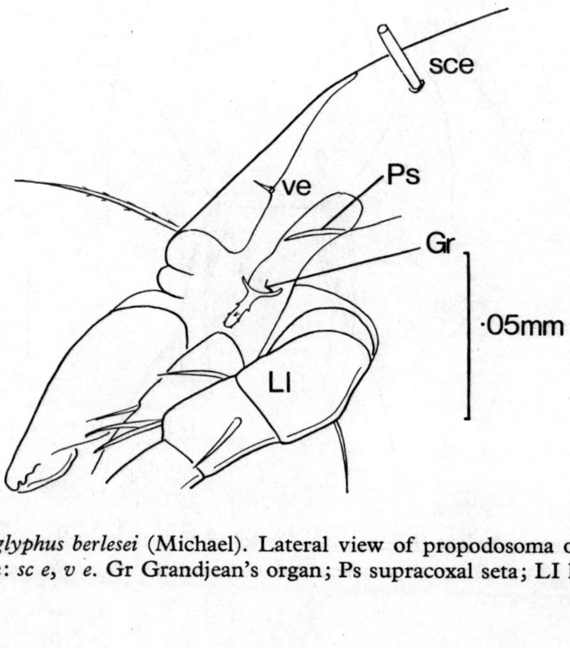

FIG. 101. *Caloglyphus berlesei* (Michael). Lateral view of propodosoma of protonymph. Setae: *sc e*, *v e*. Gr Grandjean's organ; Ps supracoxal seta; LI leg I.

The bursa copulatrix is terminal and surrounded by a small sclerotized plate. It is connected with the sac-like receptaculum seminis by a narrow tube (Fig. 110).

HYPOPUS (Fig. 111a, b). Length of idiosoma: 250 to 350 μm.

The idiosoma is dark brown in colour, its surface arched and except for the anterior part of the propodosoma, covered with smooth cuticle. The propodosoma is almost triangular in shape and terminates anteriorly in a rounded apex which covers the basal region of the gnathosoma; the *v i* setae arise from the tip of this apex and two pairs of *sc* form a curved transverse row

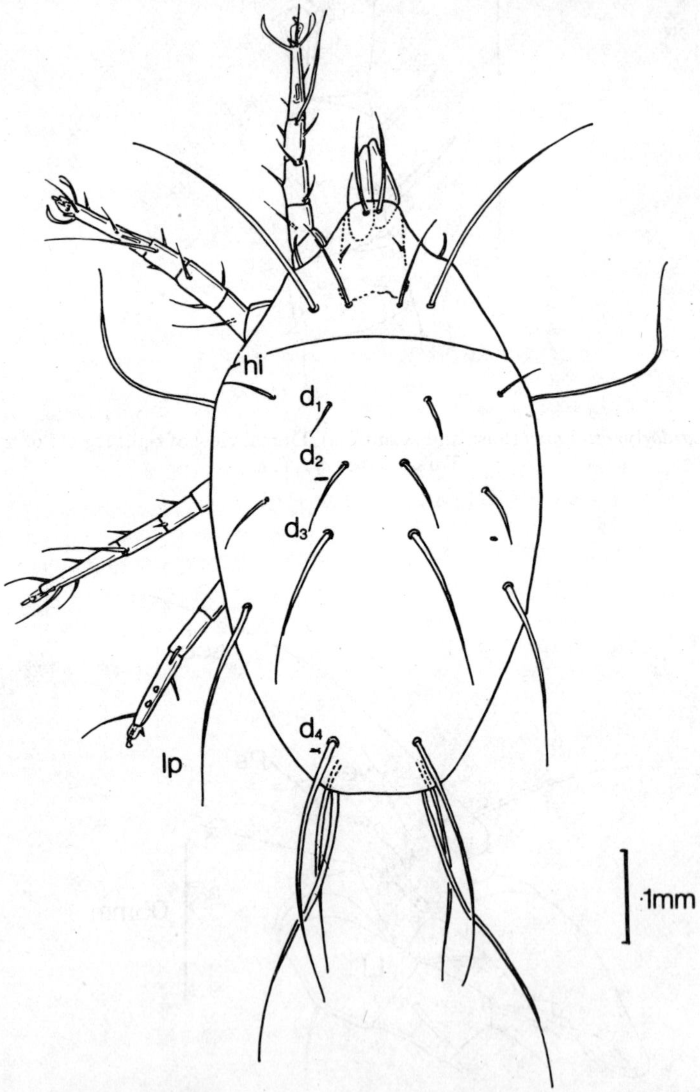

FIG. 102. *Caloglyphus berlesei* (Michael) ♂. Dorsal view. Setae of the idiosoma: d_1 to d_4, *h i*, *l p*.

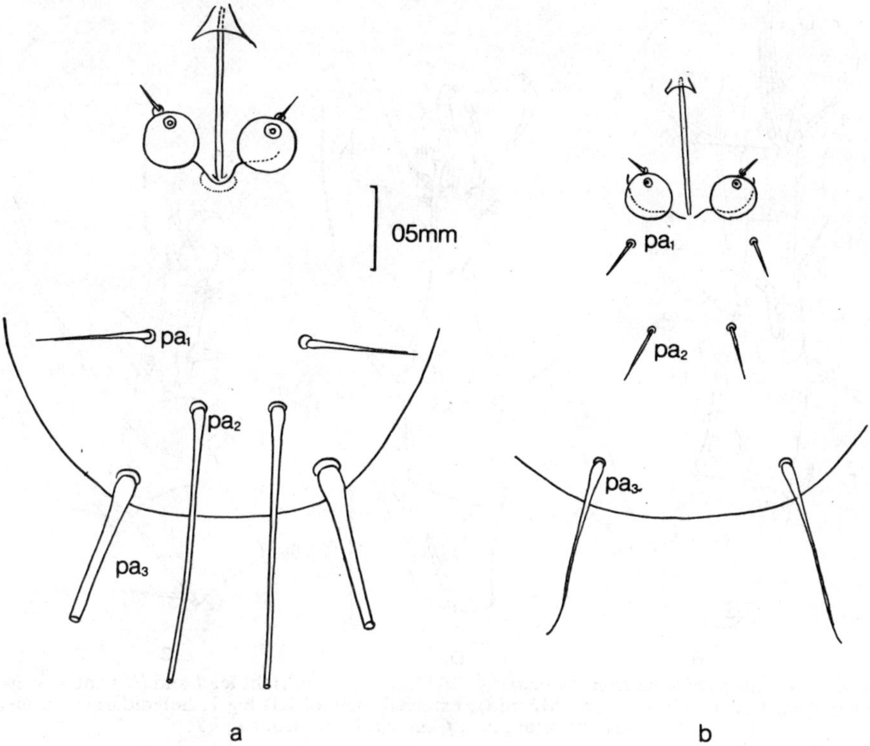

FIG. 103. Anal region of (*a*) *Caloglyphus berlesei* (Michael)♂, (*b*) *C. mycophagus* (Mégnin) ♂. Post-anal setae: $p\ a_1$ to $p\ a_3$.

across the propodosoma. The hysterosoma is four to five times longer than the propodosoma and bears fine inconspicuous setae arranged as shown in Fig. 111b.

Ventrally, the sternal shield is well-defined laterally by epimerites II which have a slightly curved outline; coxal shields II are virtually closed, although their internal margins are very indistinct. Coxal shields III and IV are completely enclosed and separated from one another along the midline. All the coxal shields have thickened edges. The genital shield and sucker disc are particularly well-sclerotized. Coxal suckers are present on coxal shields I and III, and a pair of suckers and setae are placed on either side of the genital slit. The sucker plate (Fig. 112) is perforated by the usual eight suckers, the central and anterior ones being of almost equal diameter.

The greater part of legs I and II is visible when the animal is viewed from above and all four pairs of legs have well-developed claws and pretarsi. On tarsi I and II, the claw is encircled by five curved leaf-shaped setae (Fig. 113a), of which two are smaller than the remaining three. An additional seta *e* has its distal extremity expanded into a cup-shaped sucker; $omega_1$ is longer than the width of the tarsus at its base, but shorter than $omega_1$ of tarsus II. Seta *ba* is smooth. On tibiae and genua I and II *hT*, *gT* and *mG* are spines, shorter in length than $omega_1$. On tarsus IV, *r* (Fig. 111b) is a long curved seta with a serrated edge extending almost to the end of the tarsus.

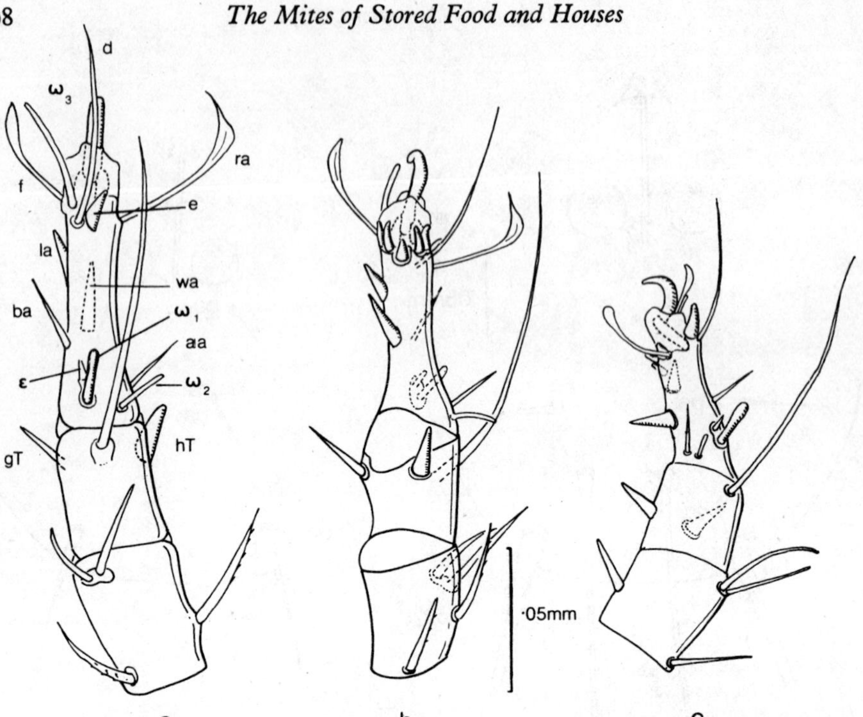

FIG. 104. *Caloglyphus berlesei* (Michael)♂. (*a*) Dorsal view of right leg I and (*b*) ventral view of left leg I (*c*) *C. mycophagus* (Mégnin): external view of left leg I. Solenidia: ω_1 to ω_3; famulus: ε: setae: *d, e, f, aa, ba, la, ra, wa, gT, hT*.

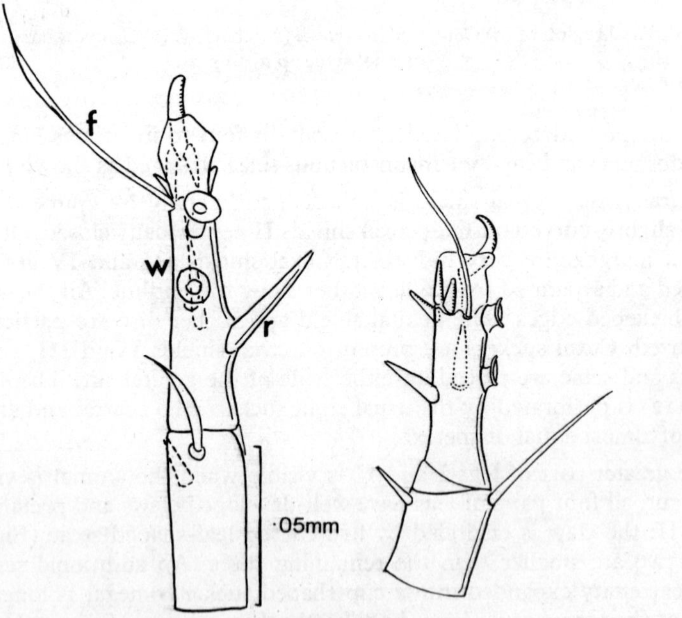

FIG. 105. (*a*) Dorsal view of distal segments of right leg IV of *Caloglyphus berlesei* (Michael) ♂; (*b*) external view of distal segments of right leg IV of *C. mycophagus* (Mégnin) ♂. Tarsal setae: *f, r, w*.

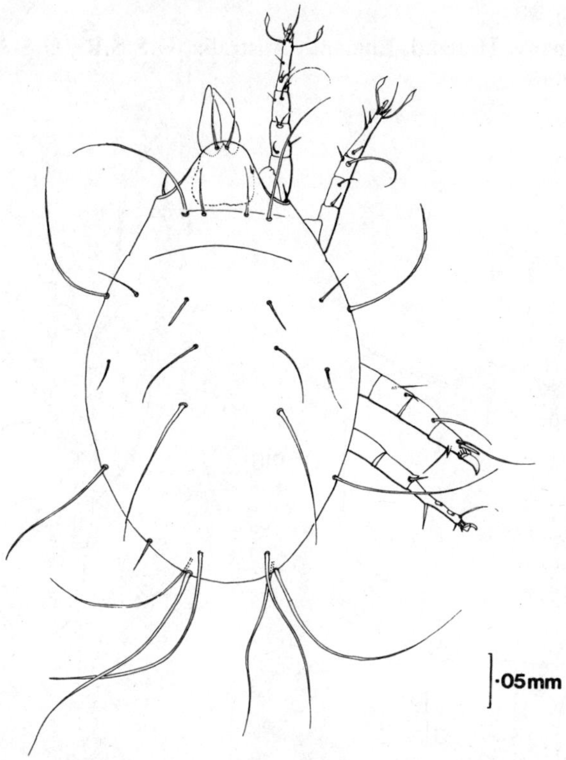

FIG. 106. *Caloglyphus berlesei* (Michael). Dorsal view of heteromorphic male.

LARVA (Fig. 114)

This lacks the leaf-shaped setae on the legs and has well-developed coxal rods.

BIOLOGY

The life cycle takes from eight to nine days at 22°C under saturated conditions and with wheat germ as food. With such a high humidity, the rate of breeding is very rapid and, in one example, one female laid 1,034 eggs in 39 days, as many as 213 eggs being produced in 24 hours. A small number of larvae which emerged from these eggs eventually passed through the hypopal stage even though they were supplied with abundant food and water (Hughes, 1961b). Pimental, Rumsey and Streams (1960) reared *C. berlesei* on the Ascomycete *Neurospora crassa* Shear and Dodge; the cultures were kept at 27°C and a relative humidity of 73 per cent and a weighed amount of the fungus was introduced into the containers every day. The mites consumed between 0.01 mg and 0.05 mg per mite per day, the eggs hatched in 24 to 48 hours and the life cycle was completed in 9 days.

In food stores, *C. berlesei* is occasionally found on damp, mouldy wheat and on copra, linseed and groundnuts which have been kept under exceptionally damp conditions and are covered with a thin film of water. It is sometimes found in insect cultures and will attack the eggs and larvae, causing considerable damage. The hypopi are carried by stored product insects, particularly *Tenebrio molitor* L. In recent years *C. berlesei* has often been found in large numbers in the deep litter of broiler houses in company with *Androlaelaps casalis*.

DISTRIBUTION

Italy, Germany, Holland, England, Australia, U.S.S.R., U.S.A. It is probably cosmopolitan.

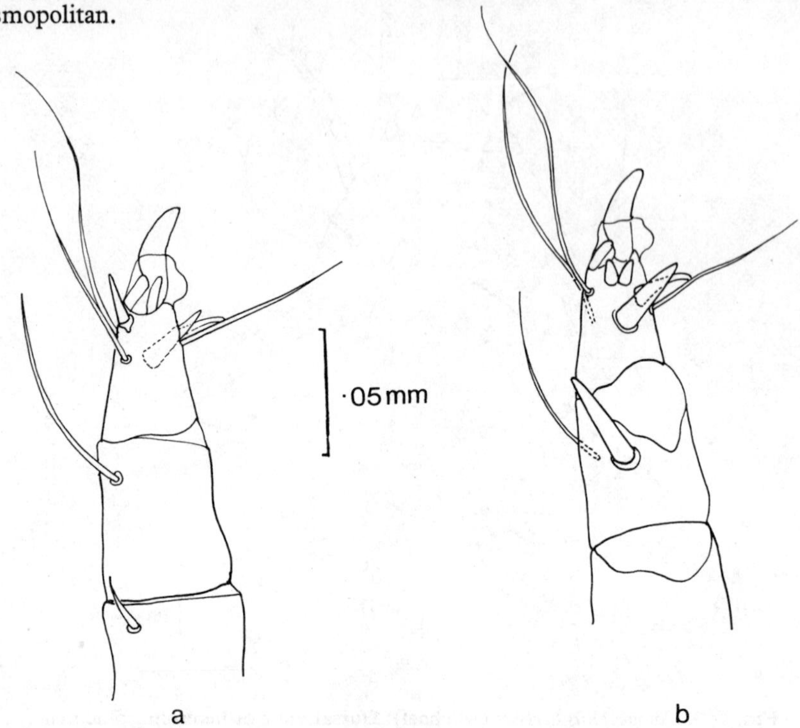

FIG. 107. *Caloglyphus berlesei* (Michael). Heteromorphic male. Extremity of leg III in (*a*) dorsal, (*b*) ventral view.

CALOGLYPHUS MYCOPHAGUS (Mégnin, 1874)

MALE (Fig. 115). Length of idiosoma: about 640 μm.

The body is more rounded than in the previous species. The dorsal propodosomal shield has an almost straight posterior margin. The arrangement of the dorsal setae is similar to that of *C. berlesei*. Setae *v i* and *sc i* are more distinctly pectinate and the supracoxal seta is considerably shorter, less than half the length of d_1. On the hysterosoma, d_1, d_2 and *l a* are almost equal in length and d_3 and *l p*, although variable in length, are not as long as in *C. berlesei* (cf. Figs. 102 and 115. Ventrally (Fig. 103b), the postanal setae are more widely spaced and $p\ a_2$ are less than twice as long as $p\ a_1$.

The chaetotaxy of tarsus I (Fig. 104c) is similar to that of *C. berlesei*, although the tarsus itself is shorter. On tarsus IV (Fig. 105b), the two suckers are in the distal half of this segment and the long terminal seta *f* may be slightly expanded.

I have not found any heteromorphic males.

FEMALE (Fig. 116). Length of idiosoma: about 780 μm.

The idiosoma is almost globular in shape. Of the dorsal setae, d_4 are equal to or longer than d_3 and extend beyond the posterior edge of the body; the arrangement of the setae is the same as in *C. berlesei*. Six pairs of anal setae (Fig. 109b) are present, the posterior group of setae arising in front of the posterior end of the anus. The bursa copulatrix is terminal and opens into a globular receptaculum seminis.

R. L Heinemann and R. D. Hughes (1970) have described the structure of the reproductive organs in males and females, and also the process of meiosis in virgin females, inseminated females and males. This species is not parthenogenetic.

Hypopi have not been bred or described for this species.

BIOLOGY

H. C. Gough and E. B. Brown found *C. mycophagus* on *Asparagus plumosus* seedlings grown in boxes. Cavities in shoots and roots were full of the mites and there appeared to be no evidence of preceding decay. Mégnin found his specimens on mushrooms.

DISTRIBUTION

England, France.

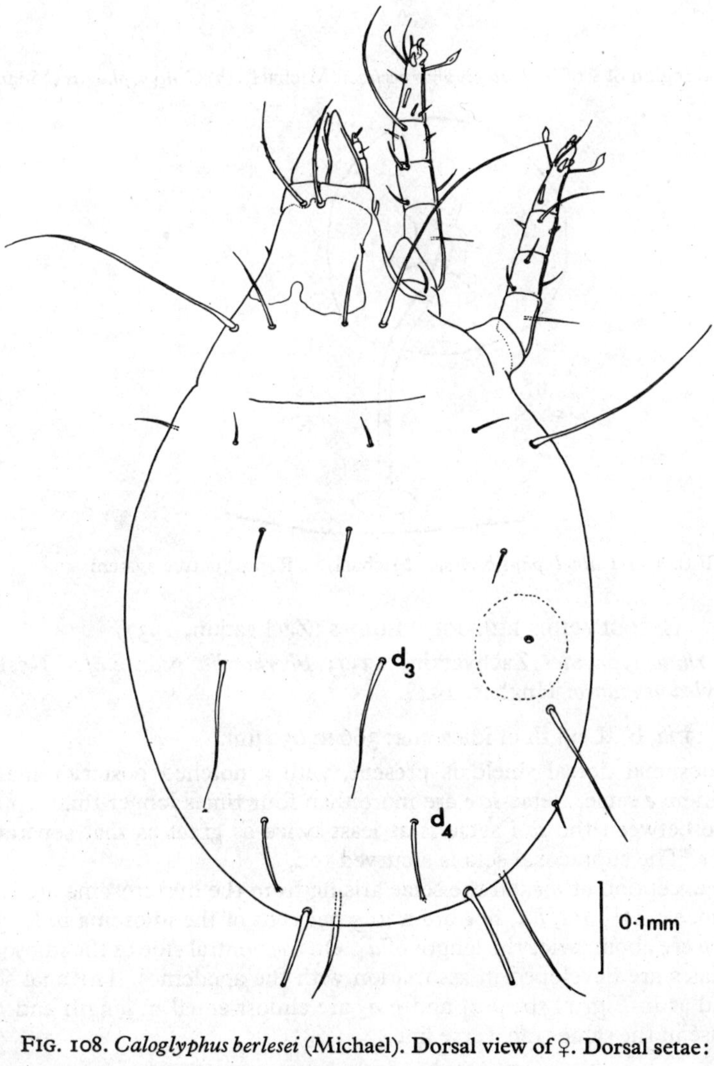

FIG. 108. *Caloglyphus berlesei* (Michael). Dorsal view of ♀. Dorsal setae: d_3, d_4.

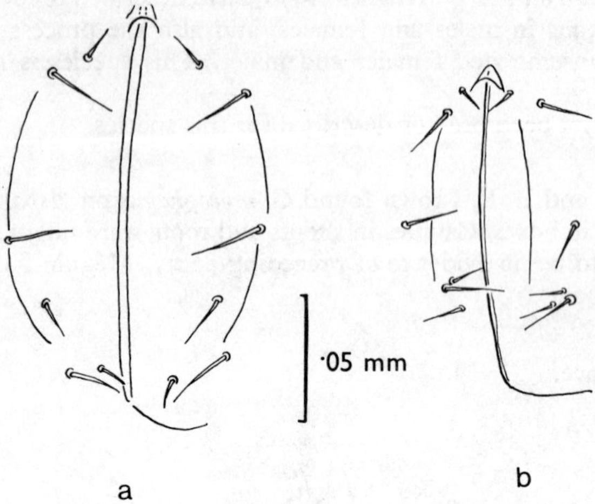

FIG. 109. Anal region of ♀ of (*a*) *Caloglyphus berlesei* (Michael), (*b*) *C. mycophagus* (Mégnin).

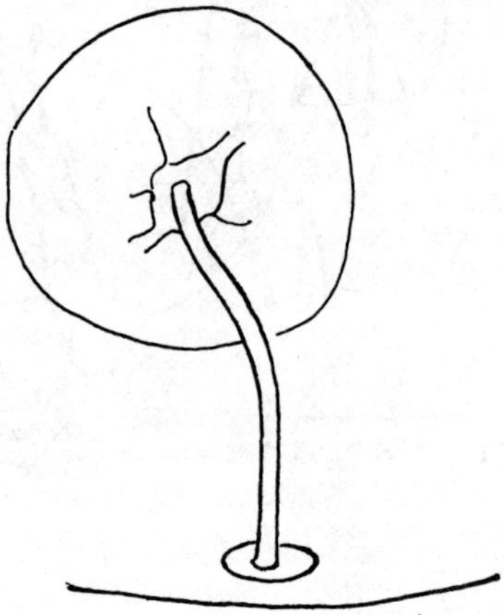

FIG. 110. *Caloglyphus berlesei* (Michael) ♀. Reproductive system.

CALOGLYPHUS RHIZOGLYPHOIDES (Zachvatkin, 1937)

Acotyledon rhizoglyphoides Zachvatkin, 1937; *Eberhardia pedispinifer* Nesbitt, 1945; *Acotyledon munroi* Hughes, 1948.

MALE (Fig. 117a, b). Length of idiosoma: 360 to 650 μm.

A propodosomal dorsal shield is present, with a notched posterior margin and short fine *v e* setae. Setae *sc e* are more than four times longer than *sc i* and the distance between the *sc i* setae is at least twice as great as that separating *sc i* from *sc e*. The supracoxal seta is a curved rod.

With the exception of *h e*, all the setae arising from the hysterosoma are short and fine. Setae d_1, d_2, *h i*, *l a*, *sa e* are 2 to 5 per cent of the idiosoma in length; d_3, d_4 and *l p* are about twice the length of d_1. On the ventral side of the idiosoma, epimeral plates are developed in association with the apodemes. The anal setae are arranged as in Fig. 117b; $p\ a_1$ and $p\ a_2$ are almost equal in length and $p\ a_2$ and $p\ a_3$ arise in the same transverse line.

On tarsus I (Fig. 118d), *f* is a curved seta, not flattened and leaflike at its distal end, and *l a* and *w a* are slender setae, not shaped like spines; *ra* arises at the same level as *ba*. On the tibia *hT* is also slender and on the genu, *mG* is smooth. The copulatory suckers on tarsus IV (Fig. 119b) are equidistant from the base and apex of the segment.

No heteromorphic males have been recorded for this species.

FEMALE (Fig. 120). Length of idiosoma: 530 to 700 µm.

This closely resembles the male. Around the anal opening (Fig. 128b) are arranged six pairs of anal setae.

HYPOPUS (Figs. 121a, b).

This is an ethereal creature, very pale in colour with ventrally curved body edges. The propodosomal shield is triangular in shape, with a blunt apex concealing the base of the gnathosoma. A strip of thin cuticle joins the propodosomal and hysterosomal shields. The arrangement of the dorsal setae is the same as in *C. berlesei*. The basal segment of the gnathosoma is short, its apex only slightly bilobed.

Ventrally, the sternum is short, about half the length of the propodosoma; coxal shields II are closed, apodemes II and epimerites II having a distinctly curved outline. No transverse line separates the sternal and ventral shields. The anterior edges of coxae III and IV are clearly outlined, and the coxae separated from their counterparts by a distinct space as in *C. berlesei*. The posterior

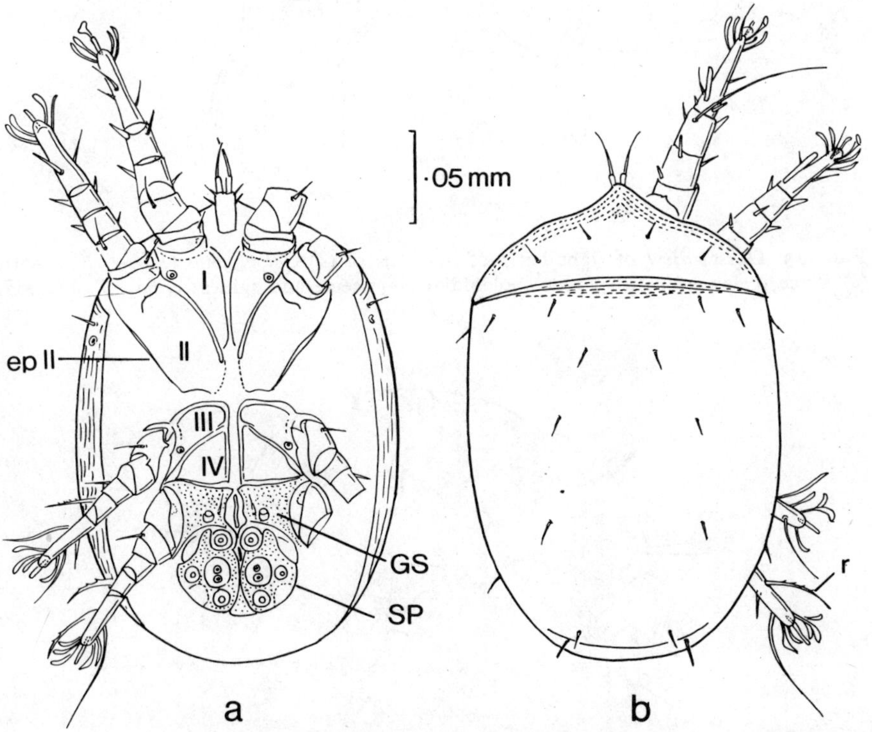

FIG. 111. *Caloglyphus berlesei* (Michael). Dorsal (*a*) and ventral (*b*) views of hypopus. I to IV coxal shields; ep II epimerite II; GS genital shield; SP sucker plate; *r* serrated seta.

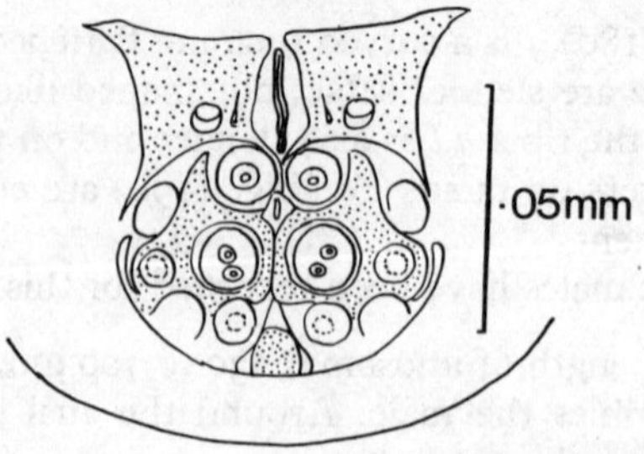

FIG. 112. *Caloglyphus berlesei* (Michael). Sucker plate of hypopus.

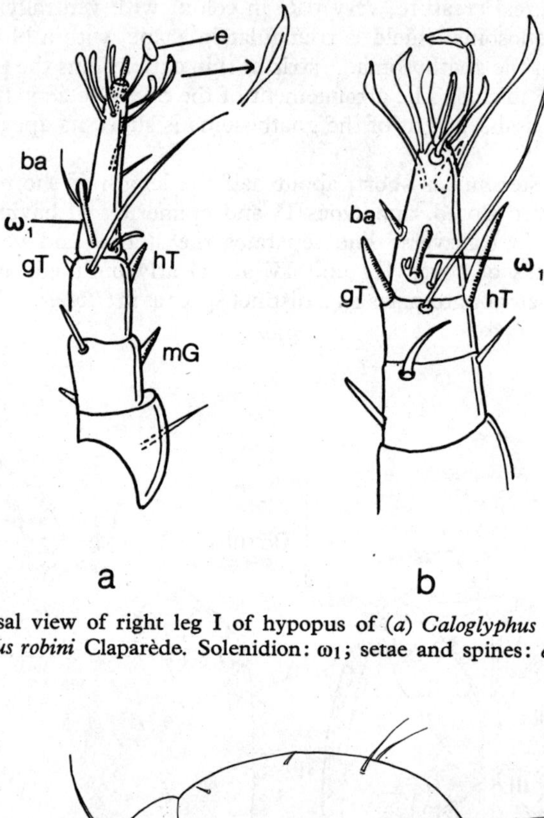

FIG. 113. Dorsal view of right leg I of hypopus of (*a*) *Caloglyphus berlesei* (Michael), (*b*) *Rhizoglyphus robini* Claparède. Solenidion: ω_1; setae and spines: *e, ba, gT, hT, mG.*

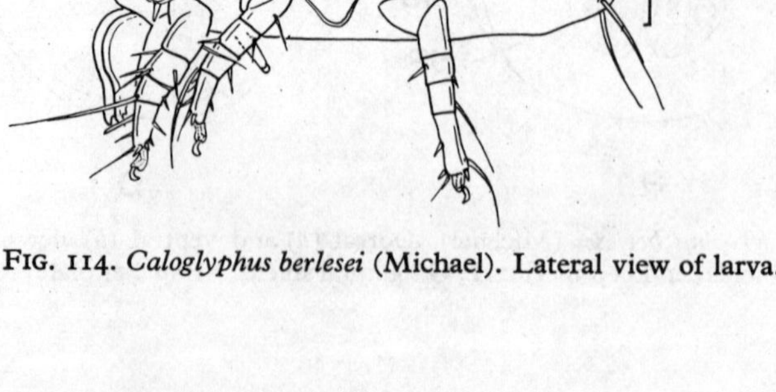

FIG. 114. *Caloglyphus berlesei* (Michael). Lateral view of larva.

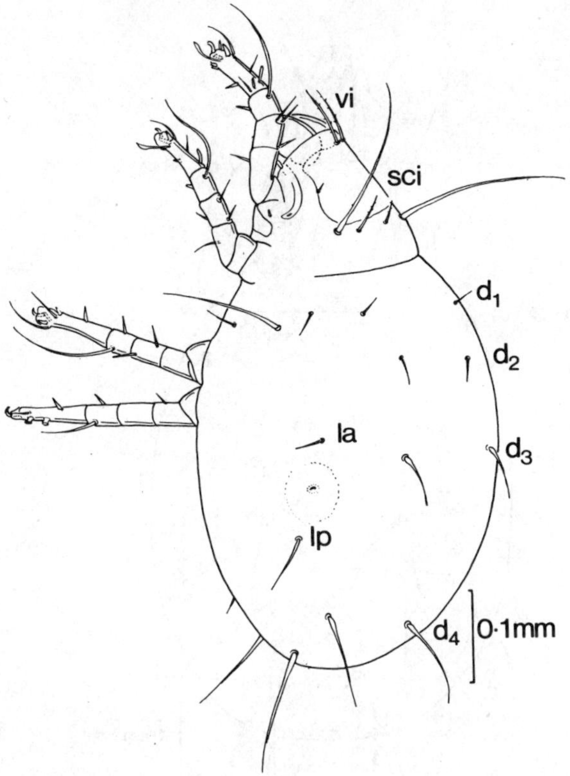

FIG. 115. *Caloglyphus mycophagus* (Mégnin). Dorso-lateral view of ♂. Setae of the idiosoma: *v i*, *sc i*, d_1 to d_4, *l a*, *l p*.

margin of the ventral shield is not clearly outlined. Vestigial suckers are present on coxae I and III.

The sucker plate (Fig. 122) is oval in shape, the suckers feebly developed except for the posterior ones which are strongly birefringent. These probably correspond to A, B, C and D (see Fig. 23). The two large central suckers barely project from the surface of the plate and the two anterior suckers are also poorly developed. Both anal and genital openings are distinct. The sucker plate has a flat border and the posterior margin of the body is also flattened.

On tarsus I (Fig. 123b) an expanded seta is not developed, but one extending well beyond the end of the claw, arises from the same position. The setae enveloping the claw are curved and only slightly widened distally. Omega_1 is long and slender, tapering in the middle region and expanding towards either end. The ventral setae of the genu and tibia (*hT* and *mG*) are not enlarged to form spines and are less conspicuous than in *C. berlesei*.

BIOLOGY

This species has been found in damp chaff, under haystacks (D. A. Griffiths), also on mildewed grain, wheat and flax (A. A. Zachvatkin, 1941). E. and F. Türk (1957) found it in moles' nests and the nests of *Lasius brunneus*; they also obtained the hypopi from *Sorex araneus*, *Crocidura leucodon* and *Microtus arvalis*.

DISTRIBUTION

England, N. Ireland, Germany, Canada, Portugal, Angola and the U.S.S.R.

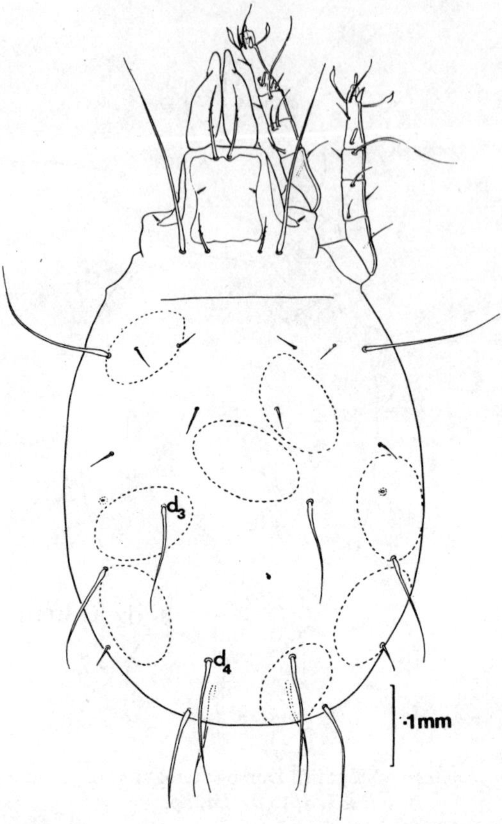

FIG. 116. *Caloglyphus mycophagus* (Mégnin). Dorsal view of ♀. Dorsal setae: d_3, d_4.

CALOGLYPHUS OUDEMANSI (Zachvatkin, 1937)

Caloglyphus krameri Berlese 1881 sensu Hughes, 1961 is considered by Samšiňák, 1966 to represent the taxon *Cosmoglyphus oudemansi* (Zachvatkin, 1937).

The genus *Cosmoglyphus* type species *Tyroglyphus krameri* Berlese, 1881 was erected by Oudemans (1932), to include those reddish-brown hypopi which possess a propodosoma with wing-shaped anterior angles and a small sucker plate without a membranous edge. Zachvatkin 1937 regarded *Cosmoglyphus* as a subgenus of *Eberhardia* Oudemans, 1924 and in 1941 included both *Eberhardia* and *Cosmoglyphus* in the genus *Acotyledon* Oudemans, 1903. In the genus *Acotyledon* Oudemans had placed those hypopi with a feebly sclerotized cuticle, an oval idiosoma and a reduced sucker plate. Adults with hypopi of the *Cosmoglyphus* or *Acotyledon* type are difficult to separate at the generic level from *Caloglyphus* and in 1961 I merged *Acotyledon* with *Caloglyphus*.

Samšiňák 1966 considers that the genus *Cosmoglyphus* comprises a sharply defined group of species characterised by their relatively small size (♂ 250 to 600 μm) (♀ 450 to 750 μm), by having a well-developed flattened supracoxal seta with barbed lateral edges and one seta at the distal end of the tarsus in the male sometimes widened into the shape of a leaf. I do not consider that these characters are sufficient to justify the retention of the genus *Cosmoglyphus* and I am therefore including *Cosmoglyphus* in the genus *Caloglyphus*.

HOMOMORPHIC MALE (Fig. 124). Length of idiosoma: 430 to 500 μm.

This species resembles *C. berlesei* in the colour and texture of the cuticle, but the body is more elongated and less bulbous in shape. A dorsal propodosomal shield is present with an almost straight posterior margin, the *v e* setae arising midway along its lateral edges. Setae *sc e* are two to three times longer than *sc i*, the *sc i* setae being the same distance from each other as from the *sc e* setae. The supracoxal seta (Fig. 125a, b) is curved and flattened, with four to eight lateral projections and sometimes with a few barbs arising from the flattened surface. The number and position of the barbs varies from one individual to another.

The dorsal setae vary in length and appearance. The shorter ones, such as d_1, d_2, *l a*, *h i* and *h v* are smooth, stiff, sometimes with a rounded distal end; the longer setae, i.e., d_3, d_4, *l p*, *sa e* and *sa i* have pointed ends, *sa e* being at least three times longer than d_1. At the ventral posterior end of the body (Fig. 126a), $p\ a_1$ are almost the same distance apart as $p\ a_3$, and $p\ a_2$ arise almost midway between $p\ a_3$ and anterior to them.

The general appearance of the legs and the arrangement of the setae closely resemble those of *C. berlesei*. On leg I (Fig. 118a, b), omega$_1$ is slightly expanded distally, *ra* is a slender seta, not enlarged distally as in *C. berlesei*, and the distal end of *f* expands abruptly into a transparent leaf-shaped plate. On tarsus IV (Fig. 119a) the tarsal suckers are equidistant from the base and apex of the segment.

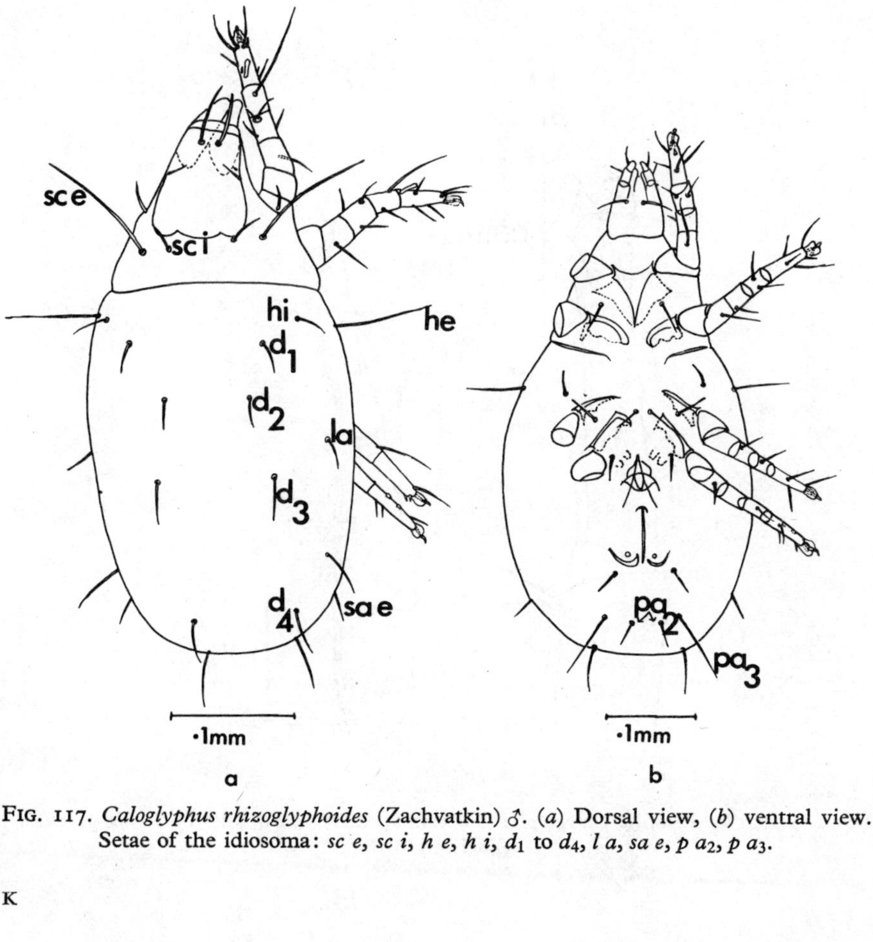

FIG. 117. *Caloglyphus rhizoglyphoides* (Zachvatkin) ♂. (*a*) Dorsal view, (*b*) ventral view. Setae of the idiosoma: *sc e*, *sc i*, *h e*, *h i*, d_1 to d_4, *l a*, *sa e*, $p\ a_2$, $p\ a_3$.

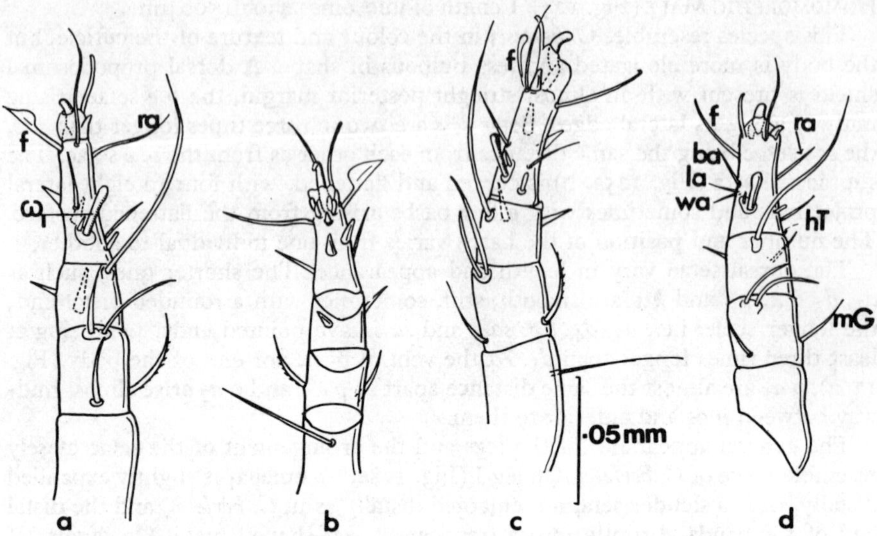

FIG. 118. *Caloglyphus oudemansi* (Zachvatkin) ♂. Dorsal (*a*) and ventral (*b*) views of right leg I. (*c*) *C. hughesi* (Samšiňák) ♂. Dorso-lateral view of right leg I. (*d*) *C. rhizoglyphoides* (Zachvatkin) ♂. Dorsal view of right leg I. Solenidion: ω_1; setae: *f*, *ba*, *la*, *ra*, *wa*, *hT*, *mG*.

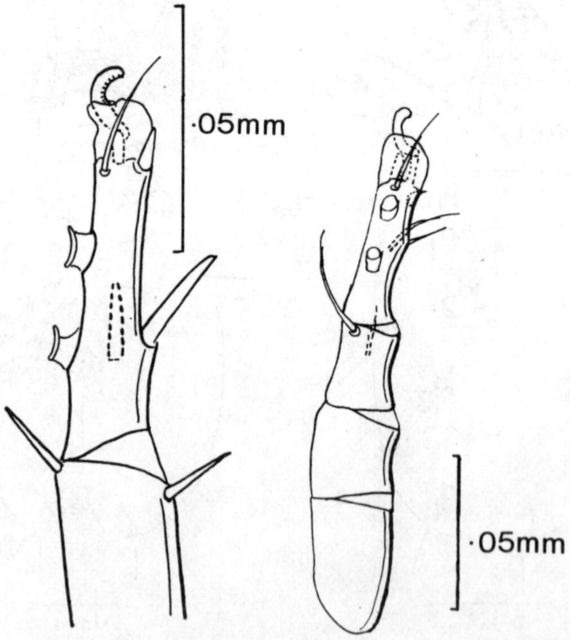

FIG. 119. Left leg IV of (*a*) *Caloglyphus oudemansi* (Zachvatkin) ♂ and (*b*) *C. rhizoglyphoides* (Zachvatkin) ♂.

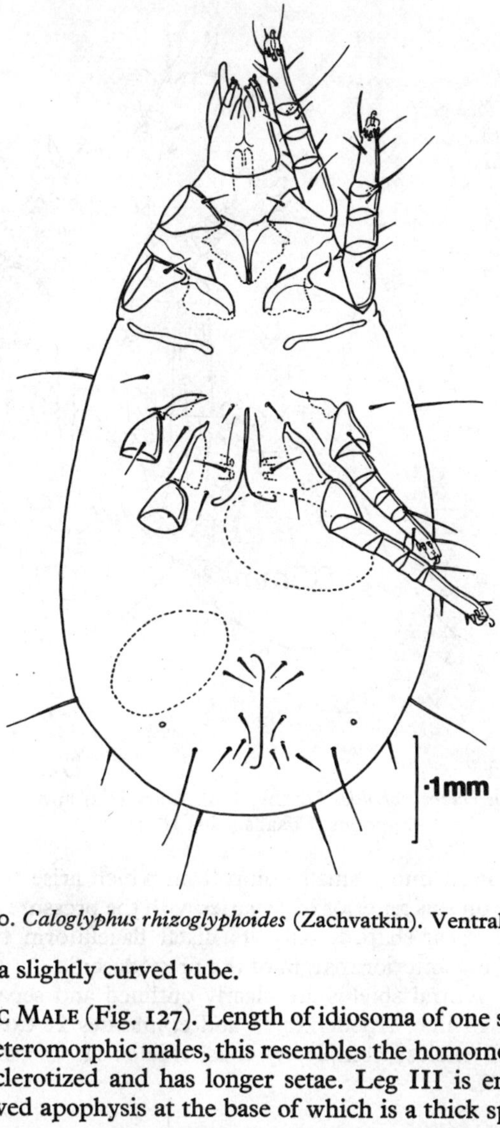

FIG. 120. *Caloglyphus rhizoglyphoides* (Zachvatkin). Ventral view of ♀.

The penis is a slightly curved tube.

HETEROMORPHIC MALE (Fig. 127). Length of idiosoma of one specimen: 450 µm.

As in other heteromorphic males, this resembles the homomorphic form, but is more heavily sclerotized and has longer setae. Leg III is enlarged, the tarsus ending in a curved apophysis at the base of which is a thick spine.

FEMALE. Length of idiosoma: 530 to 775 µm.

The female closely resembles the male although the dorsal setae are shorter in proportion to the length of the body. Setae d_3 and d_4 have pointed ends. Ventrally (Fig. 128a) six pairs of anal microsetae are present and the anal opening is removed from the posterior margin of the body. On legs I and II, the seta *f* is not expanded distally into a leaf-shaped plate.

HYPOPUS (Fig. 129a, b). Length of idiosoma: 250 to 300 µm.

The idiosoma is almost circular in outline and reddish-brown in colour. The dorsal surface is arched and the lateral edges extended as a thin, rather transparent flange; the setae arising from the idiosoma are all fine and bent. The anterior edge of the idiosoma is indented on either side of the midline, and the

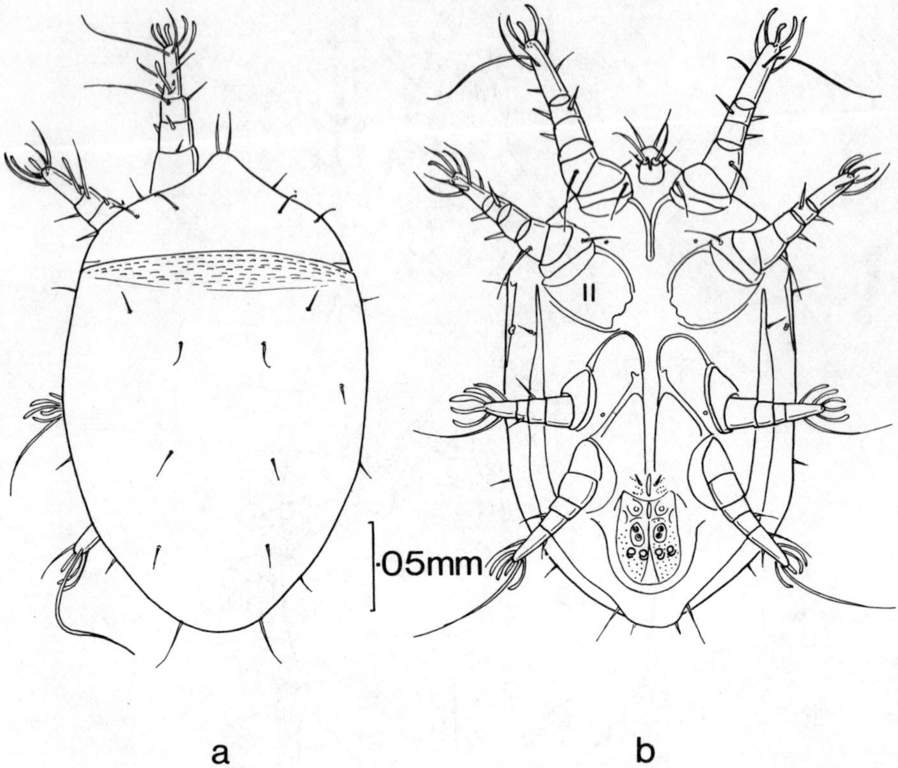

FIG. 121. *Caloglyphus rhizoglyphoides* (Zachvatkin). Dorsal (*a*) and ventral (*b*) views of hypopus. Coxal shields II: II.

median region is raised into a small hump from which arise the *v i* setae; the degree of curvature on this anterior edge varies with the pressure of the coverslip. The gnathosoma is pear-shaped, only its distal flagelliform bristles (aristae) protruding beyond the anterior margin of the propodosoma.

The sternal and ventral shields are clearly outlined and separated from one another by an arched line. Apodemes II and epimerites II extend back as far as this line so as to enclose completely coxal shields II. Coxal shields III are

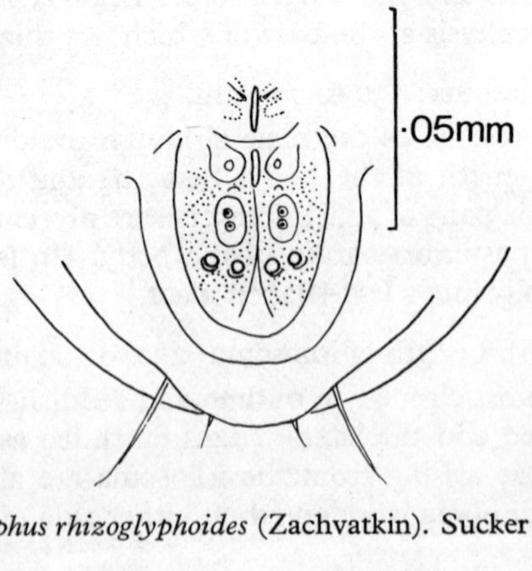

FIG. 122. *Caloglyphus rhizoglyphoides* (Zachvatkin). Sucker plate of hypopus.

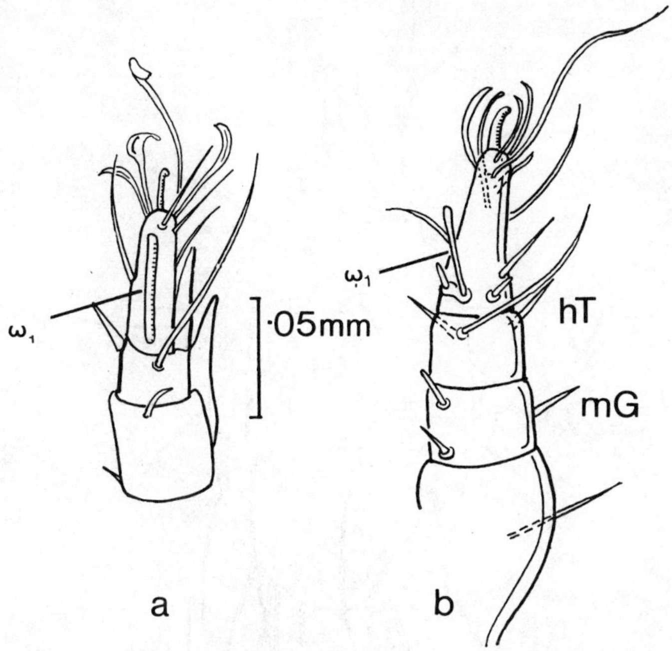

FIG. 123. Dorsal view of right leg I of hypopus of (*a*) *Caloglyphus oudemansi* (Zachvatkin and (*b*) *C. rhizoglyphoides* (Zachvatkin). Setae and spines: *hT*, *mG*. Solenidion: ω_1.

open, a pair of setae marking the anterior ends of apodemes IV. The posterior edge of coxal shields IV is marked by an anteriorly convex line.

The sucker plate is smaller than that of *C. berlesei* and more anterior in position; it lies midway between the posterior margin of the sternal shield and the posterior edge of the body and bears the normal complement of suckers. Suckers are also present on either side of the genital opening and on coxal shields I and III.

The distal three segments of legs I and the distal two of legs II only are visible when the animal is viewed from above. Legs III and IV are bent outwards at the tarso-tibial joint. On leg I (Fig. 123a) omega$_1$ is very long and extends forward almost to the base of the claw; *ba* is also very long, almost as long as the tarsus itself. Three of the apical setae are sickle-shaped with transparent leaf-shaped ends; *e* is a long curved seta with a cup-shaped expanded end. On the tibia and genu, *hT* and *mG* are enlarged as colourless flat spines which lie close against the side of the leg. The tarsi of the remaining legs possess the same leaf-shaped setae clustered round the base of the claw.

LARVA

This is similar to that of *C. berlesei*. The setae have the same rounded ends found in the other developmental stages.

BIOLOGY

This species was originally found in the nests of the ant *Plagiolepis longipes*. It also occurred on damp nuts (Brazil nuts, groundnuts) which had been in

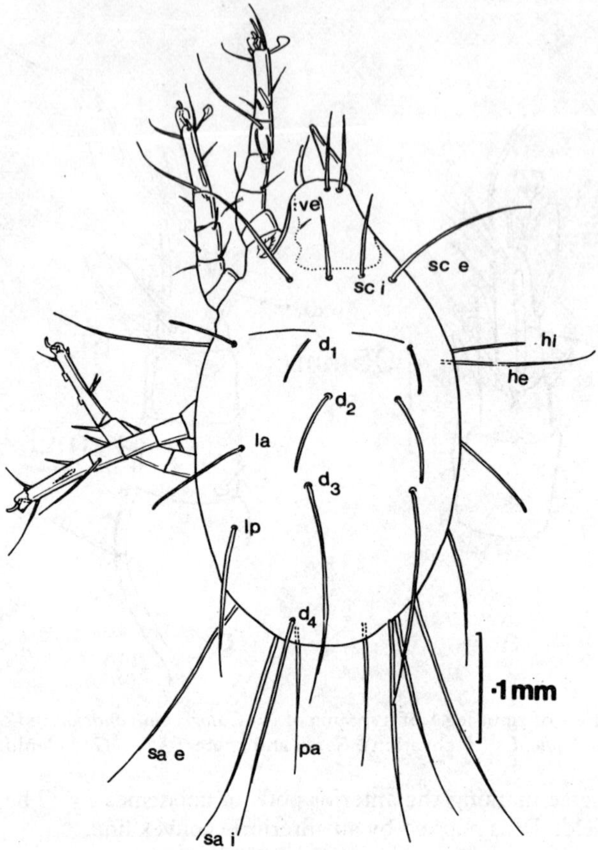

FIG. 124. *Caloglyphus oudemansi* (Zachvatkin). Dorsal view of ♂. Setae of the idiosoma: *v e, sc e, sc i, h e, h i,* d_1 to d_4, *l a, l p, sa e, sa i, p a.*

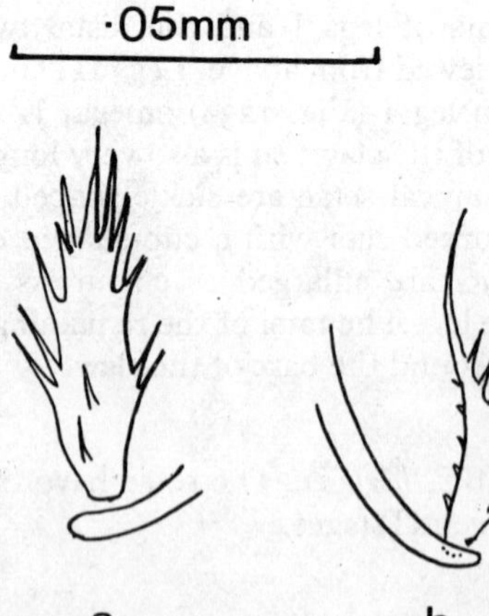

FIG. 125. (*a, b*). Supracoxal seta of *Caloglyphus oudemansi* (Zachvatkin).

storage for some time (K. G. Smith coll.), on damp grass and decaying plants, beneath haystacks, and in the deep litter of broiler houses. It requires a moist environment, similar to that needed by *C. berlesei* and, in my cultures, hypopi became more numerous when the culture became short of food.

DISTRIBUTION
England, Italy, U.S.S.R., Australia, Java, India.

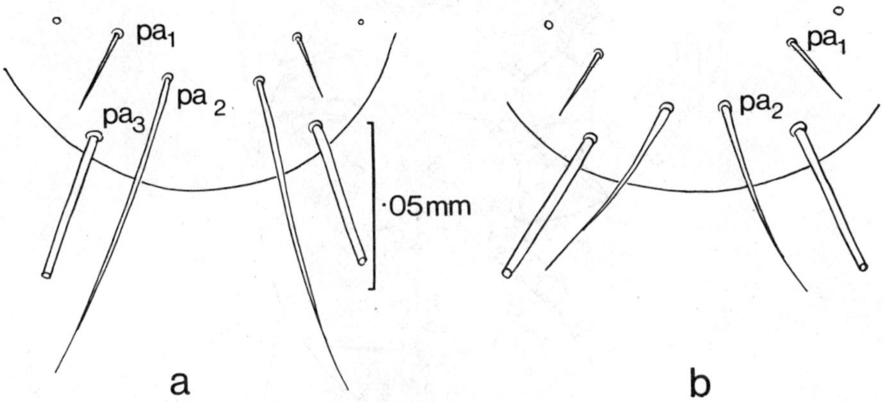

FIG. 126. Post-anal region of (*a*) *Caloglyphus oudemansi* (Zachvatkin) and (*b*) *C. hughesi* Samšiňák. Post-anal setae: pa_1 to pa_3.

CALOGLYPHUS HUGHESI (Samšiňák, 1966)

Caloglyphus redikorzevi sensu Hughes, 1961.

Eberhardia redikorzevi was first described by Zachvatkin in 1937, who in 1941 united the genera *Eberhardia* and *Cosmoglyphus* as the genus *Acotyledon*. In 1961 I incorporated the genus *Acotyledon* into *Caloglyphus* and described *Caloglyphus redikorzevi* (Zachvatkin). After examining my material, Samšiňák (1966) considered that my *Caloglyphus redikorzevi* is a different taxon from that described by Zachvatkin. He therefore re-named it *Cosmoglyphus hughesi*. As previously stated, (p. 106), I consider the genus *Cosmoglyphus* forms part of *Caloglyphus*.

MALE (Fig. 130). Length of idiosoma: 400 to 500 µm.
This resembles *C. oudemansi*. All the dorsal setae, with the exception of *h e* and *sc e*, terminate in rounded spatulate ends (Fig. 130b). Setae d_3, d_4, *l p* and *sa e* are shorter than those of the previously described species, *sa e* not exceeding twice the length of d_1. On leg I, *f* is a curved seta whose distal end is only slightly enlarged (Fig. 118c). At the ventral posterior margin of the body (Fig. 126b), the arrangement of the post anal setae is the same as that of *C. oudemansi*, except that $p\,a_2$ are shorter and more slender, less than three times longer than $p\,a_1$.

FEMALE (Fig. 131). Length of idiosoma: 500 to 700 µm.
This differs from *C. oudemansi* in having shorter setae with spatualate ends.

BIOLOGY
This species was found by G. C. Williams on rice bran imported from Burma.

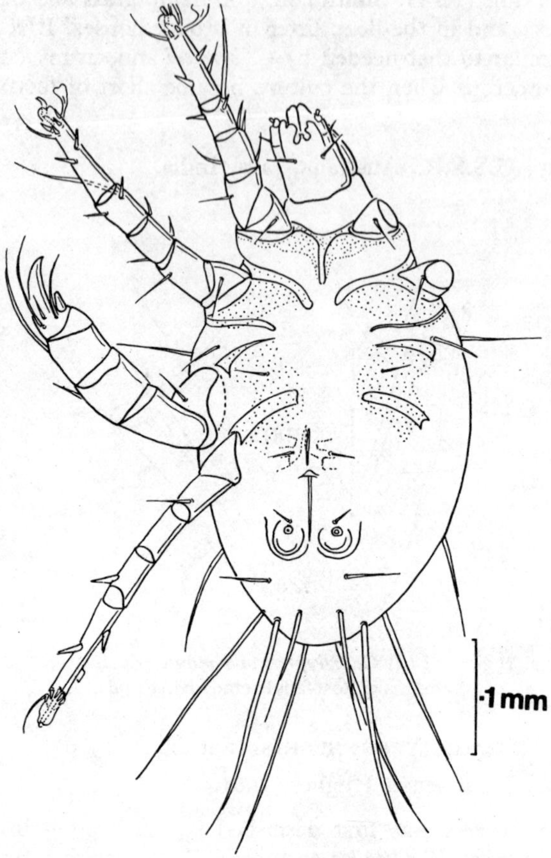

FIG. 127. *Caloglyphus oudemansi* (Zachvatkin). Ventral view of heteromorphic ♂.

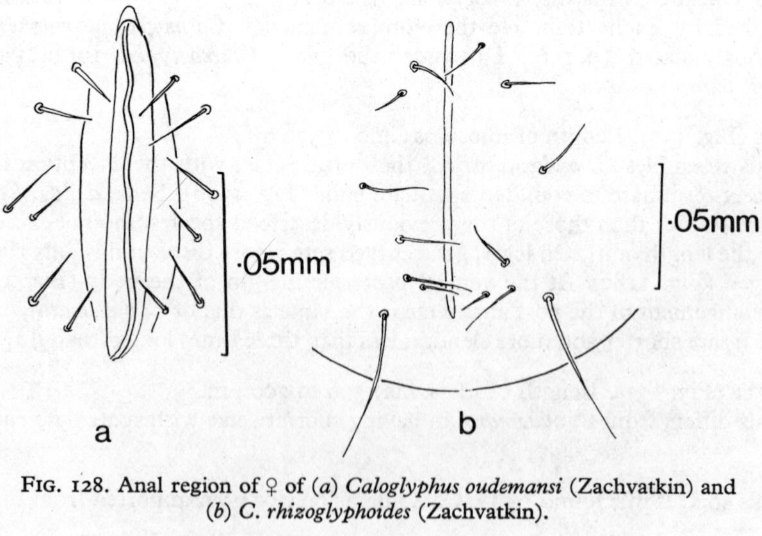

FIG. 128. Anal region of ♀ of (*a*) *Caloglyphus oudemansi* (Zachvatkin) and (*b*) *C. rhizoglyphoides* (Zachvatkin).

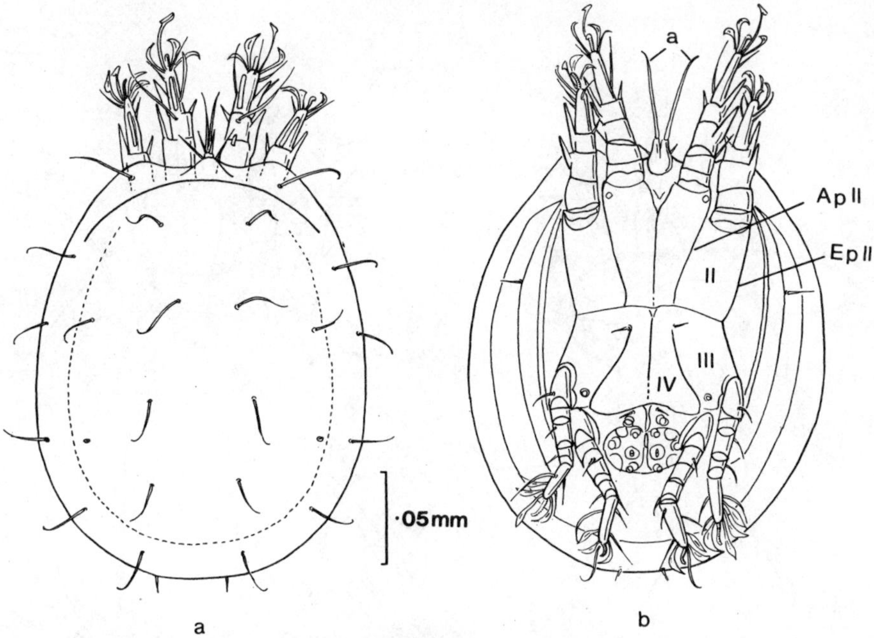

FIG. 129. *Caloglyphus oudemansi* (Zachvatkin). Dorsal (*a*) and ventral (*b*) views of hypopus. Ap II apodeme II, Ep II epimerite II, a aristae of gnathosoma, II to IVcoxal shields II to IV.

RHIZOGLYPHUS CLAPARÈDE, 1869

Setae *v e* are either represented by short fine setae arising near the middle of the lateral edge of the dorsal propodosomal shield, or are absent.

Setae *sc e* are longer than *sc i*; *sc i* may be absent.

A supracoxal seta is present.

The posterior margin of the body in the male does not form a projecting opisthosomal plate.

On legs I and II, *ba* (Fig. 135b) is shaped like a cone and is inserted close to $omega_1$. Seta *aa* is missing from tarsus I. Some of the terminal tarsal setae may have their distal ends slightly expanded.

Heteromorphic males and hypopi are of common occurrence.

Type species: *Rhizoglyphus robini* Claparède, 1869.

Vitzthum, 1929, Zachvatkin, 1941, Baker and Wharton, 1952, Türk and Türk, 1957 and Hughes, 1961b, designate *Tyroglyphus echinopus* Fumouze and Robin, 1868 as the type species of *Rhizoglyphus*, rather than *R. robini*, Clarparède,1869.

Claparède studied a culture of *Rhizoglyphus* obtained from potatoes, dahlia tubers and cabbage stalks and reared on hyacinth bulbs (quoted from Michael, 1901). At that time, the significance of the hypopus and heteromorphic male was not understood and Claparède believed that the hypopus was a male and, with the female, described and figured them as *Hypopus dujardinii*. The two types of male he regarded as the male and female of a separate species belonging to another genus. He called this *Rhizoglyphus robini* (See Michael, 1901, Van Eyndhoven, 1963 and 1968). Owing to this confusion, Michael, 1901 regarded the species *Tyroglyphus echinopus* F. and R. rather than *Rhizoglyphus robini* Claparède as type species of *Rhizoglyphus* and, until Van Eyndhoven drew attention to the mistake, subsequent authors have copied him.

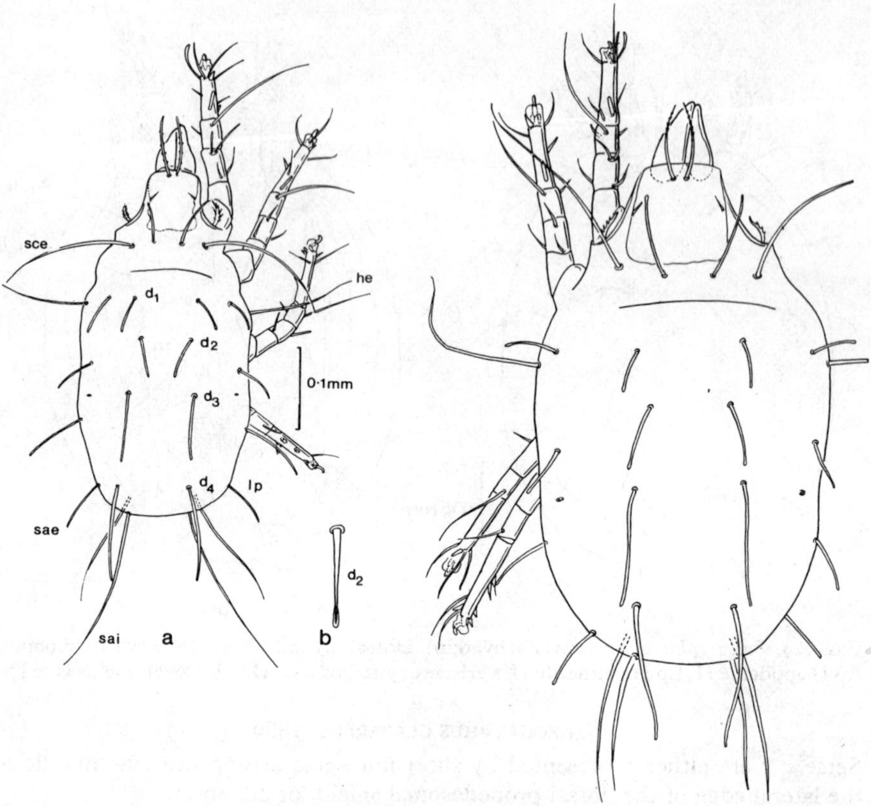

FIG. 130. *Caloglyphus hughesi* (Samšiňák) ♂. (*a*) Dorsal view, (*b*) seta d_2. Setae of the idiosoma: *sc e*, *h e*, d_1 to d_4, *l p*, *sa e*, *sa i* (left).
FIG. 131. *Caloglyphus hughesi* (Samšiňák). Dorsal view of ♀ (right).

RHIZOGLYPHUS

Key to the adults

1. Setae *sc i* represented by microsetae, shorter than the supracoxal setae .... *R. robini*, p. 116

—. Setae *sc i* more than twice as long as the supracoxal seta.... *R. callae*, p. 122

RHIZOGLYPHUS ROBINI Claparède, 1869

Rhizoglyphus echinopus (Fumouze and Robin, 1868) sensu Hughes, 1961.

In 1868, Fumouze and Robin described *Tyroglyphus* (=*Rhizoglyphus*) *echinopus* which was found by M. Kunckel living on the flowers of a dried-up hyacinth. This is figured as lacking setae *sc i*, d_1 and d_2. In their Plate 20, fig. 2 the penis is drawn with an oval base and in Plate 21, fig. 6 it is depicted as having a round base. Fumouze and Robin are careful to state that this is how it appears after being subjected to pressure. 'Le penis et les ventouses du mâle telles qu'elles se presentent après qu'ils ont été soumis à un certain degré de pression (p. 304)'. It can be assumed, therefore, that the uncrushed specimen had an oval base to the penis.

In 1961 I published a description of two species of *Rhizoglyphus*. To the one with very short *sc i* setae I gave the name of *R. echinopus* (Fumouze and Robin) and to the other, distinguished by having much longer *sc i* setae, the name of *R. callae* Oudemans. I did not mention the shape of the penis base.

Van Eyndhoven (1960, 1963 and 1968) published the results of studies on the morphology and host plants of species of *Rhizogyphus*. He concludes that *Rhizoglyphus* species having long *sc i*, and a rounded penis base as depicted by Fumouze and Robin Plate 21, fig. 6 are found on hyacinth and he considers these to be *R. echinopus*, whereas those with short *sc i*, and a conical penis base as in Fumouze and Robin's drawing on Plate 20, fig. 2 are associated with potatoes and *Narcissus*. These he called *R. robini* Claparède. Van Eyndhoven failed to notice Fumouze and Robin's comments on the effect of pressure on the specimen and therefore assumed that *R. echinopus* F. and R. had a rounded, rather than a conical, penis base. He solved the difficulty of the long *sc i* setae by assuming that they (and other setae as well) had become detached and the French authors had failed to see the bases. Robin is recognized as being an outstandingly accurate observer, and it is unlikely that he would not have noticed that the *sc i* setae, as well as d_1 and d_2, were missing from all his specimens.

I am therefore assuming that the true *R. echinopus* F. and R. lacks these dorsal setae and has an oval base to the penis. I am accepting Van Eyndhoven's name

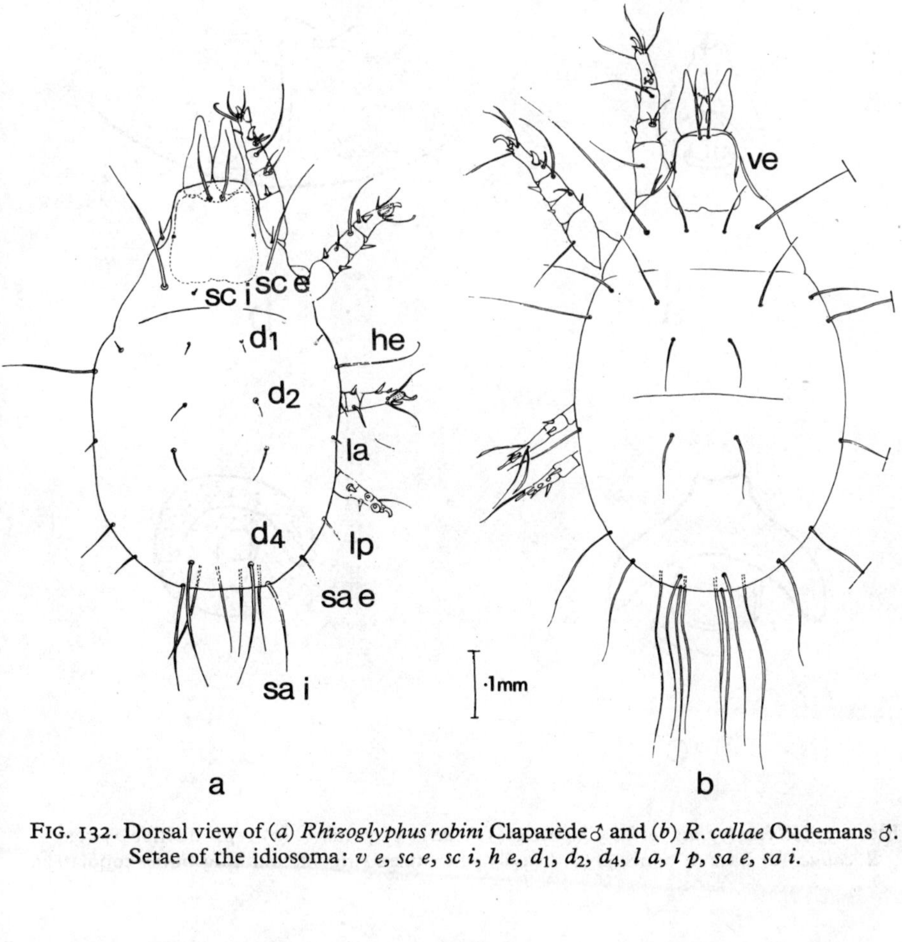

FIG. 132. Dorsal view of (*a*) *Rhizoglyphus robini* Claparède ♂ and (*b*) *R. callae* Oudemans ♂. Setae of the idiosoma: *v e, sc e, sc i, h e*, d_1, d_2, d_4, *l a, l p, sa e, sa i*.

robini Claparède for the species described by me in 1961 as *echinopus*. I do not accept Van Eyndhoven's contention that *R. echinopus* (Fumouze and Robin) has conspicuously long *sc i* setae.

HOMOMORPHIC MALE (Fig. 132a). Length of idiosoma: 450 to 720 µm.

The body is smooth, colourless and glistening, the appendages reddish-brown in colour. The propodosomal shield is oblong, its posterior margin slightly irregular in outline. Setae *v e* may be microsetae or absent, although their position is always indicated. The setae arising from the dorsal surface of the idiosoma are smooth. Setae *sc e*, *h e*, d_4 and *sa i* exceed one quarter the length of the idiosoma and, of the remaining setae, *sc i*, d_1, d_2, *h i* and *l a* are less than 10 per cent of the idiosoma in length. Setae d_4, *l p* and *sa e* are longer than d_1 and always present. The supracoxal seta resembles a bristle and is longer than d_1. Ventrally, the apodemes are darkly coloured and attached to distinct plates. The genital opening lies between coxae IV with paired folds concealing a short penis whose supporting struts are almost conical (Fig. 133a). The anal opening is relatively short, its posterior end flanked by anal suckers (Fig. 134) which do not possess a well-cuticularized ring. Three pairs of postanal setae are present, $p\ a_1$ being much shorter than the more posterior $p\ a_2$ and $p\ a_3$ which arise in a

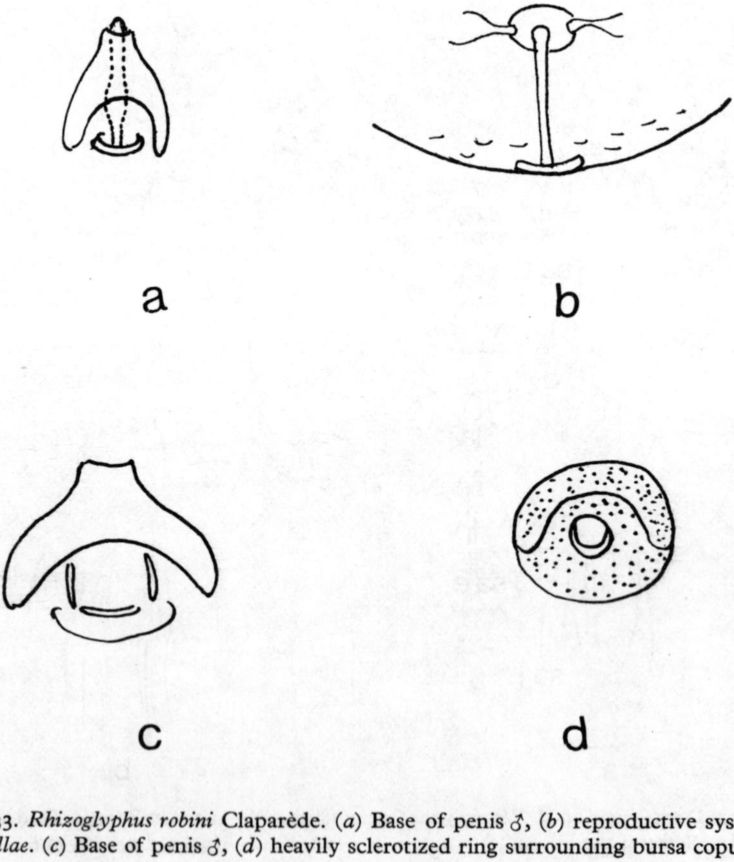

FIG. 133. *Rhizoglyphus robini* Claparède. (*a*) Base of penis ♂, (*b*) reproductive system ♀. *R. callae*. (*c*) Base of penis ♂, (*d*) heavily sclerotized ring surrounding bursa copulatrix.

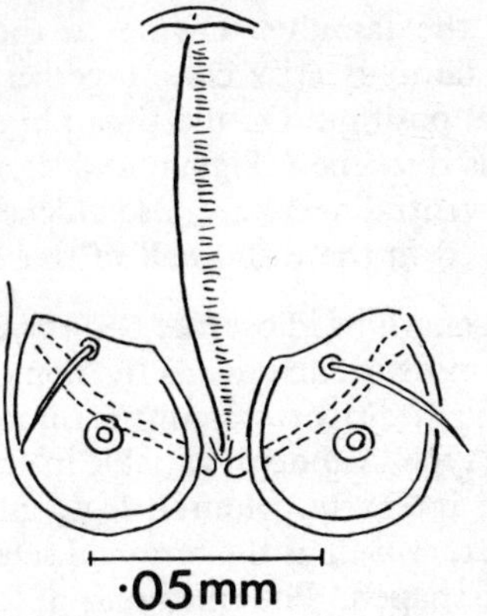

FIG. 134. *Rhizoglyphus robini* Claparède ♂. Anal region.

transverse row across the ventral end of the hysterosoma and extend well beyond its posterior edge.

The gnathosoma is normal in structure, the chelicerae distinctly toothed. The legs are short and stout, each terminating in a stout claw and stalk whose base is enveloped in a reduced pretarsus (Figs. 29, 135a) and encircled ventrally by five well-defined spines. On tarsus I, setae *d*, *f* and *ra* are curved and may be slightly expanded distally, whilst *e* and *wa* are spines; *ba* is a stout spine arising

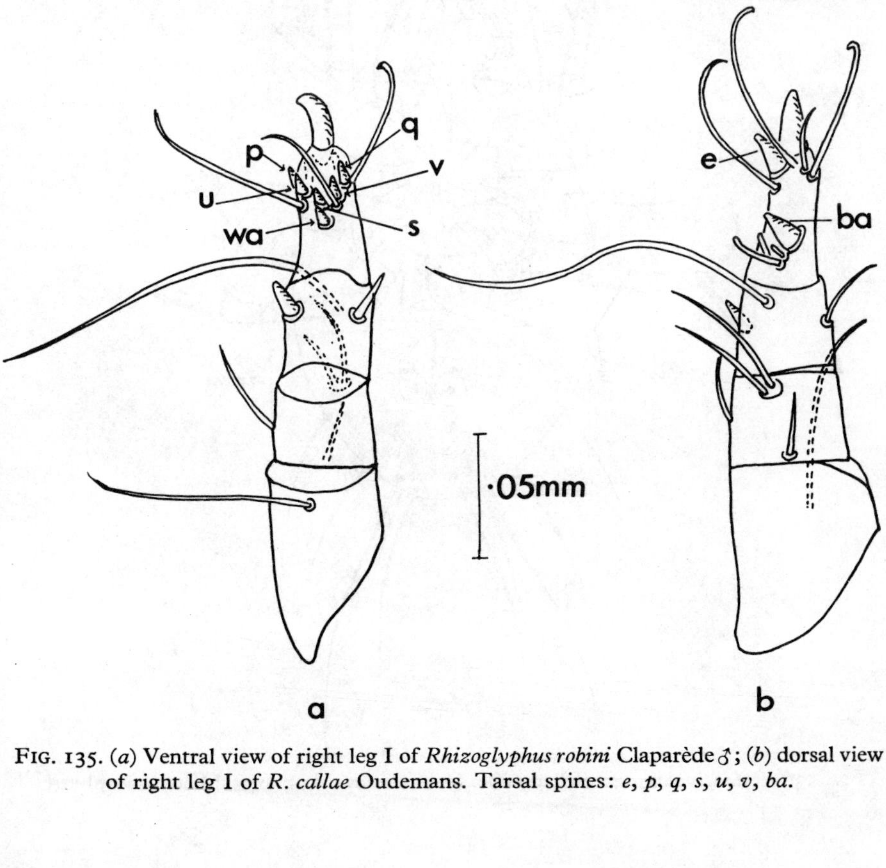

FIG. 135. (*a*) Ventral view of right leg I of *Rhizoglyphus robini* Claparède ♂; (*b*) dorsal view of right leg I of *R. callae* Oudemans. Tarsal spines: *e*, *p*, *q*, *s*, *u*, *v*, *ba*.

immediately in front of the famulus. Owing to the shortness of the tarsus, omega$_1$, omega$_2$ and the famulus arise close together at the base of the tarsus. Omega$_3$ occupies its usual position. On the tibia phi extends beyond the end of the claw and the seta *gT* is thickened. Sigma$_1$ and sigma$_2$ on the genu are almost equal in length and the ventral setae are also thickened as spines. Tarsus IV has a pair of suckers placed in the distal half of this segment.

HETEROMORPHIC MALE. Length of idiosoma: 600 to 780 μm.

This is not so frequent in its occurrence as the homomorphic male and Woodring (1969b) found that in a healthy stock culture, more than 99 per cent of males were of the homomorphic type. Although variable in size and general appearance, it is larger and has more intensely coloured legs, gnathosoma and apodemes. The dorsal setae are longer, as well as the terminal setae (*f*, *ra*, *d*) of legs I to III whose distal ends are leaf-shaped. The third pair of legs terminates in a curved process and occasionally this modification occurs on one side of the body only (Fig. 136).

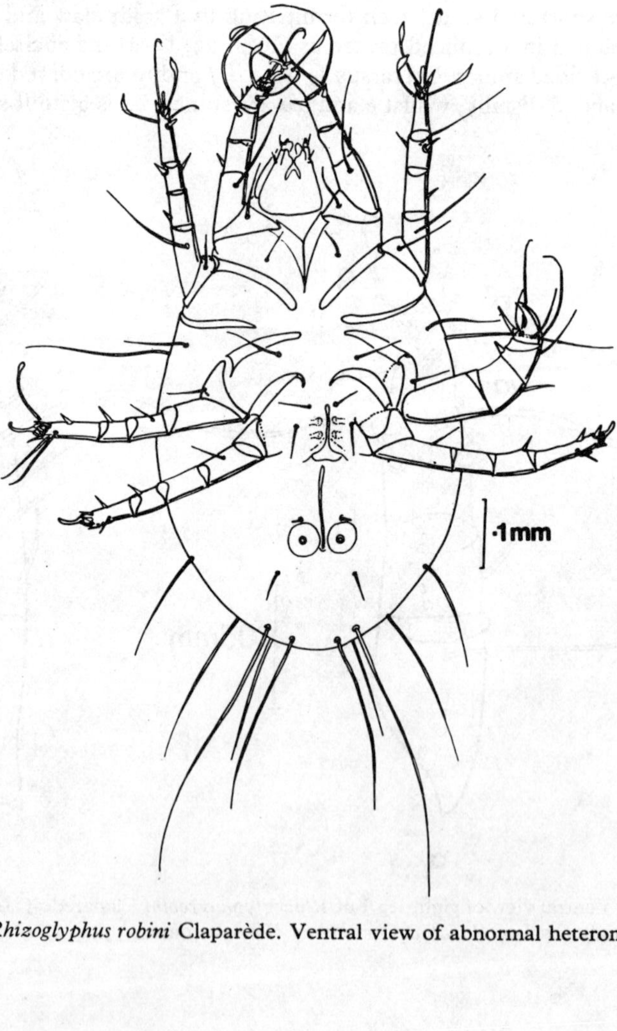

FIG. 136. *Rhizoglyphus robini* Claparède. Ventral view of abnormal heteromorphic ♂.

FEMALE. Length of idiosoma: 500 to 1,100 µm.

This closely resembles the male. The genital opening is between legs III and IV. The anal opening is surrounded by six pairs of anal setae of which the external posterior pair is much longer than the remaining five. The opening to the bursa copulatrix is terminal and surrounded by a feebly sclerotized plate. A duct leads from it to a receptaculum seminis which is connected to the ovaries by a pair of ducts (Fig. 133b).

HYPOPUS (Fig. 137a, b). Length of idiosoma: 250 to 350 µm.

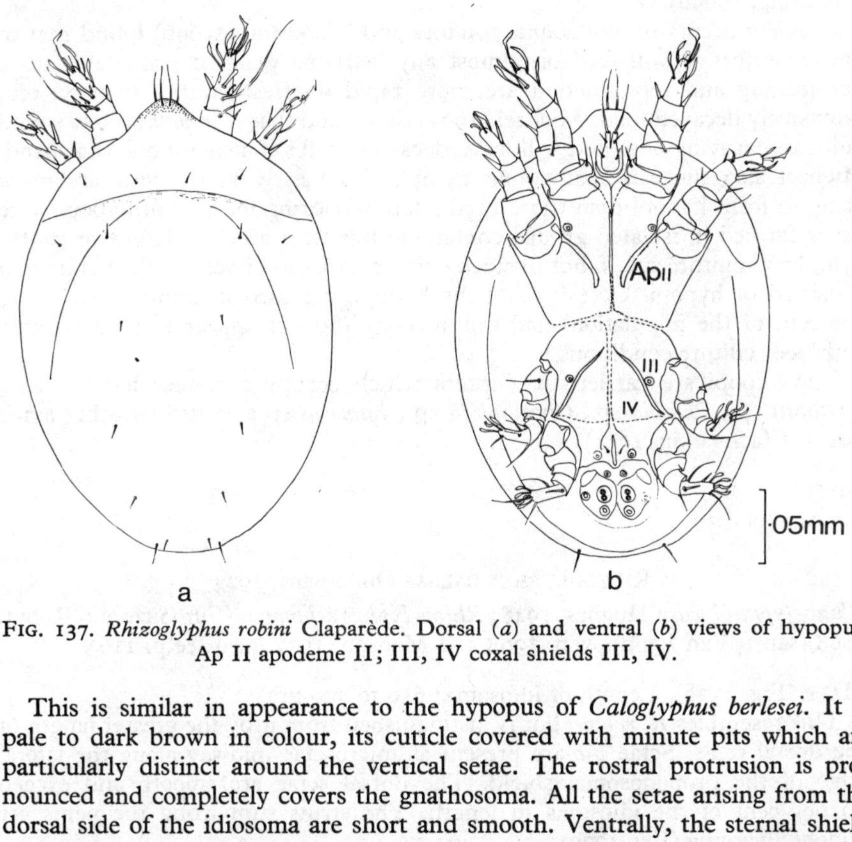

FIG. 137. *Rhizoglyphus robini* Claparède. Dorsal (*a*) and ventral (*b*) views of hypopus. Ap II apodeme II; III, IV coxal shields III, IV.

This is similar in appearance to the hypopus of *Caloglyphus berlesei*. It is pale to dark brown in colour, its cuticle covered with minute pits which are particularly distinct around the vertical setae. The rostral protrusion is pronounced and completely covers the gnathosoma. All the setae arising from the dorsal side of the idiosoma are short and smooth. Ventrally, the sternal shield is well-defined and neither apodemes II nor the sternum reach its posterior edge, although a shallow groove extends from the end of the sternum to the posterior margin of the sternal plate and gives the appearance of a junction. Coxal shields III and IV are outlined and separated from the genital plate. Coxal suckers are present on coxae I and III, and genital suckers and setae on either side of the genital opening. The sucker disc has two large central suckers and six peripheral ones of equal size.

The legs are shorter and more massive than in *Caloglyphus berlesei* and only the terminal three segments of legs I are visible when the animal is viewed from above. On tarsus I (Fig. 113b) the claw is surrounded by a terminally expanded seta and five leaf-shaped setae. Omega$_1$ is shorter than the base of the tarsus and *ba* is shaped like a spine. On genu I, the ventral spines *gT* and *hT* exceed the length of omega$_1$. On tarsus IV, *d* extends only slightly beyond the end of the claw.

LARVA

The setae d_3 and *l a* are longer in proportion to the body than in the other stages. Coxal rods are present which have smoothly rounded ends.

BIOLOGY

The life cycle, excluding the hypopal stage, can take from 7 to 27 days at temperatures between 13 and 26°C (Garman, 1937, Zachvatkin, 1941 and Woodring, 1969b).

R. robini occurs on bulbs and potatoes and Woodring (1969b) found that in the laboratory it will feed on almost any decaying plant and animal matter, but feeding and reproduction are more rapid on fresh, rather than on old, extensively decayed food. Michael (1903) also found that *R. robini* prefers sound bulbs to decaying ones and will leave decaying bulbs if a sound one is at hand. Michael also suggests that young nymphs bred early in the year are more likely to form hypopi than those bred later. Woodring found that hypopi were never formed in isolated groups containing less than 20 eggs, larvae or protonymphs in culture tubes, but in stock cultures, 1 to 2 per cent of the population consisted of hypopi. Occasionally, the hypopi increased in number (up to 25 per cent of the population) and this increase did not appear to be correlated with poor culture conditions.

The hypopi are carried by Diptera which occupy a similar habitat, e.g., *Scatopsis* sp., *Phorbia* sp., *Chortophila* sp., *Eumerus* sp and also by other small insects (Zachvatkin, 1941).

DISTRIBUTION

Cosmopolitan.

RHIZOGLYPHUS CALLAE Oudemans, 1924

Rhizoglyphus lucasii Hughes, 1948; *Rhizoglyphus echinopus* (Fumouze and Robin, 1868) sensu Van Eyndhoven, 1968 and Manson, 1972 (also see p. 116).

MALE (Fig. 132b). Length of idiosoma: 650 to 700 µm.

This resembles *R. robini*, but is distinguished from it by the greater length of the dorsal setae. Setae *v e* are present as microsetae midway along the lateral edge of the propodosomal shield. The dorsal setae are smooth and exceed 10 per cent of the idiosoma in length. The struts supporting the penis are widely diverging (Fig. 133c).

FEMALE. Length of idiosoma: 680 to 720 µm.

This resembles the male. The bursa copulatrix is surrounded by a heavily sclerotized ring (Fig. 133d) and communicates directly with a large and irregularly shaped receptaculum seminis (Van Eyndhoven, 1968).

BIOLOGY

R. callae occurs on *Narcissus* and *Freesia* bulbs (Rack, 1968), tulip bulbs (Mahunka, 1961) and *Gladiolus* corms. Specimens found on tulip bulbs were attacking the moist inner leaves of the bulb. Van Eyndhoven (1961) has obtained it from hyacinth, lily and onion bulbs and Oudemans' specimen came from *Calla* bulbs imported into Holland from Java. It has also been recorded from wet and decaying wheat spillage and from the fatty debris in a lard factory (G. Hosie coll.)

R. callae is reproductively isolated from *R. robini* (personal observation).

DISTRIBUTION

England, Holland, Hungary, Germany, Egypt, Japan, New Zealand.

THYREOPHAGUS RONDANI, 1874

Setae *v e*, *sc i*, *h i*, d_1, d_2 and *l a* are absent in the adult.

In the male, the posterior margin of the body is prolonged into an opisthosomal lobe strengthened with a dorsal plate and lying in the same plane as the ventral surface of the body.

On tarsus I, *ba* and *la* are absent. Five small ventral spines are present on the distal ends of the tarsi.

Hypopi and heteromorphic males have not been found.

Type species: *Acarus entomophagus* Laboulbène, 1852.

THYREOPHAGUS ENTOMOPHAGUS (Laboulbène, 1852)

Acarus entomophagus Laboulbène, 1852; *Tyroglypus entomophagus* Laboulbène and Robin, 1862. Also see Michael, 1903.

Not *Thyreophagus entomophagus* sensu Türk and Türk, 1957.

MALE. Length of idiosoma: 300 to 450 µm.

The idiosoma is oval and elongated with a colourless shining cuticle and pale brown stumpy legs. The opisthosomal lobe (Figs. 138, 139) is flattened and concave ventrally, so that the anal suckers lie in a depression. The dorsal opisthosomal plate has an irregular anterior margin and a thickened posterior one.

The propodosomal dorsal shield extends as far back as the scapular setae, it is rounded posteriorly and its lateral edges are incised at the level where *v e* setae arise in the genus *Rhizoglyphus*. Setae *v i* project forward over the gnathosoma and the supracoxal seta is a curved rod; *sc e* are the longest setae arising from the body. Setae *h e* and *l p* project from the sides of the hysterosoma and d_4 are displaced to the base of the opisthosomal lobe. Setae *sc i*, *v e*, d_1, d_2 and *l a* are missing. On the ventral side of the opisthosomal lobe, $p\ a_1$ and $p\ a_2$ are microsetae and $p\ a_3$ are long and extend well beyond the posterior end of the body; *sa e* are external to $p\ a_2$. Feebly sclerotized epimeral plates are developed in association with the apodemes. The genital opening lies between coxae IV.

The chelicerae have toothed shears and the small sensory rod on the distal segment of the pedipalp is unusually long.

All the legs terminate in stalked claws whose bases are enveloped in a well-developed pretarsus, similar to that found in the genus *Tyrophagus*. On tarsus I (Fig. 140a, b) omega_1 tapers towards its free end, and omega_2 is a blunt rod which arises just in front of it. At the distal end of the tarsus, *d* extends beyond the end of the claw and *f*, *wa* and *ra* are slender setae; *e* is a small spine and *p*, *q*, *u*, *v* and *s* are also small spines lying at the base of the claw, *p* and *q* being smaller than *u*, *v* and *s*. Omega_3 occupies its normal position. The chaetotaxy of the remaining segments of the legs is normal, and none of the setae is thickened to form a spine. Owing to the shortness of tarsus IV, the copulatory suckers are close together. A spine is present in the position occupied by phi on tibia IV.

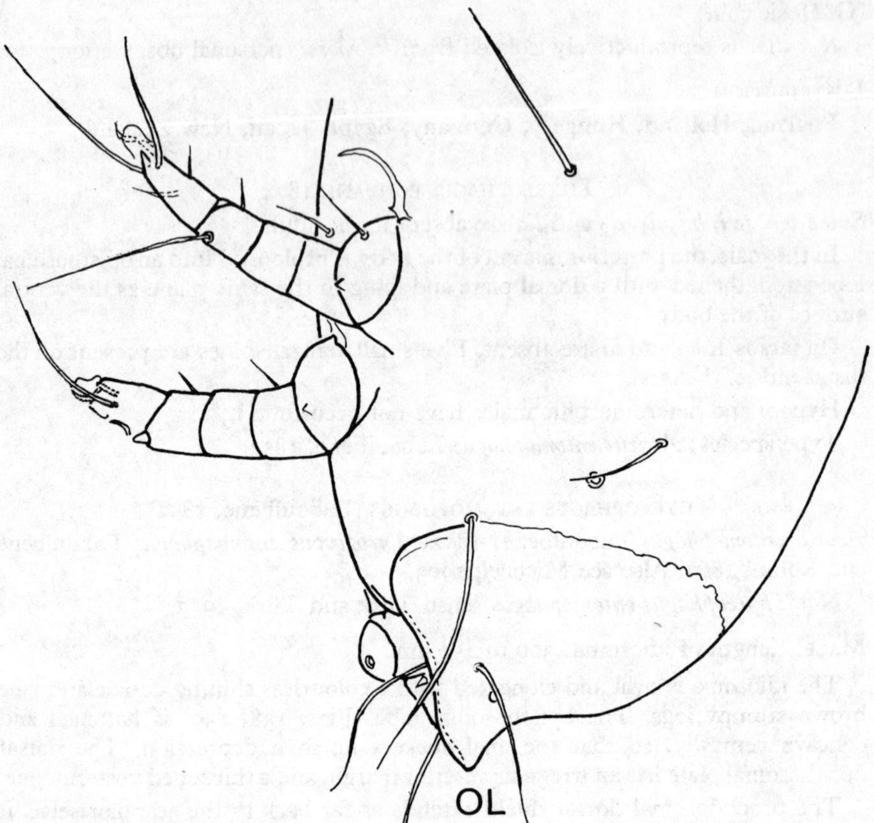

FIG. 138. *Thyreophagus entomophagus* (Laboulbène) ♂. Lateral view of terminal half of idiosoma. OL opisthosomal lobe.

FEMALE (Fig. 141). Length of idiosoma: 450 to 600 μm.

The body is longer and more slender than that of the male, the posterior margin being slightly pointed and not prolonged into an opisthosomal lobe. The genital opening lies between coxae III and IV, it is well separated from the anus which extends back as far as the posterior edge of the body; two pairs of long anal setae arise on either side of it. The opening of the bursa copulatrix is terminal, at the tip of a small papilla, and a narrow looped tube connects it with the receptaculum seminis (Fig. 142).

Except for that of the posterior end of the body, the chaetotaxy of body and legs is the same as in the male, I have never found a hypopal nymph, but one is described by Lignières (1893) and a translation of this description is given by Michael, 1903.

LARVA (Fig. 143).

No coxal rods are present and, in addition to those setae present in the adult, *l a* are also developed as short, fine setae. The pretarsi of all the legs are unusually well developed. A pair of long setae arises from the posterior margin of the body.

BIOLOGY

The life cycle takes from 28 to 38 days at a relative humidity of 75 per cent, 26 to 32 days at 85 per cent, the temperature in both cases being 18°C and the culture medium, wheat germ and yeast (G. A. Brett and R. K. Thornback,

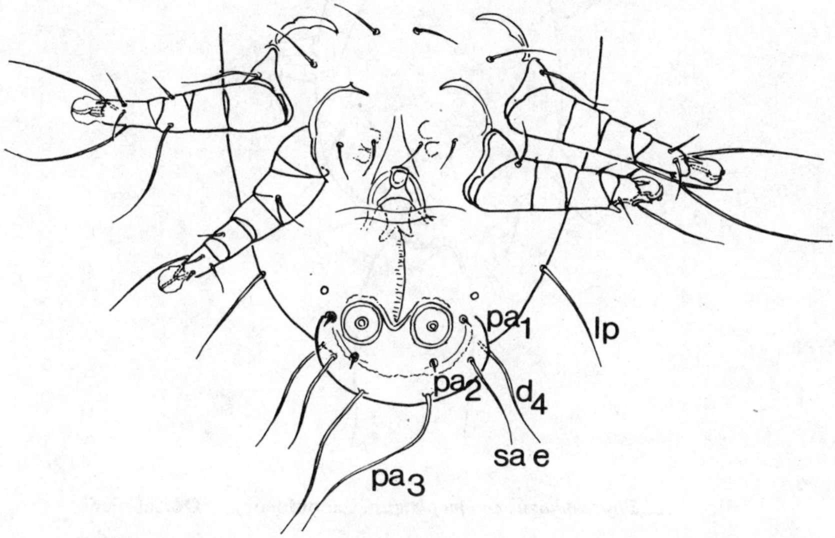

FIG. 139. *Thyreophagus entomophagus* (Laboulbène) ♂. Ventral view of terminal half of idiosoma. Setae of the idiosoma: *p a*$_1$ to *p a*$_3$, *d*$_4$, *l p*, *sa e*.

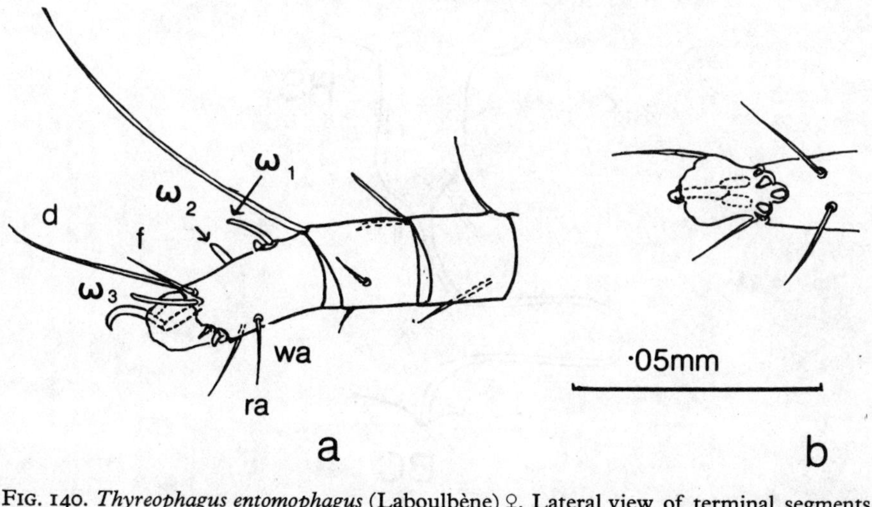

FIG. 140. *Thyreophagus entomophagus* (Laboulbène) ♀. Lateral view of terminal segments of right leg I, (*b*) ventral view of tarsus I to show the five ventral spines. Solenidia ω$_1$ to ω$_3$; setae: *d*, *f*, *ra*, *wa*.

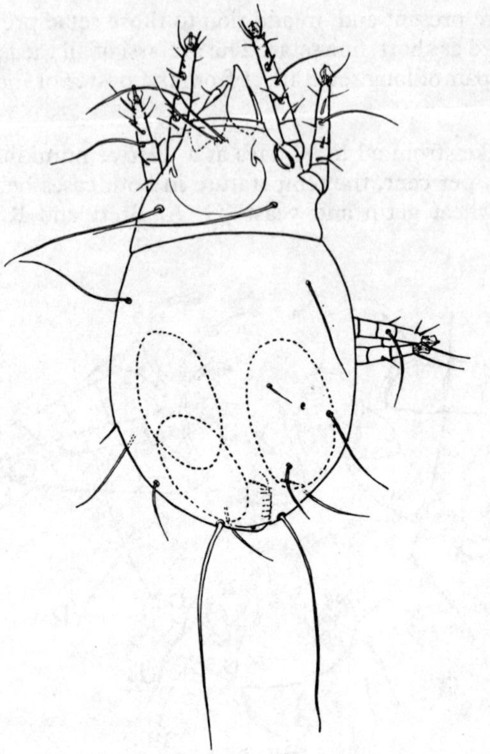

FIG. 141. *Thyreophagus entomophagus* (Laboulbène) ♀. Dorsal view.

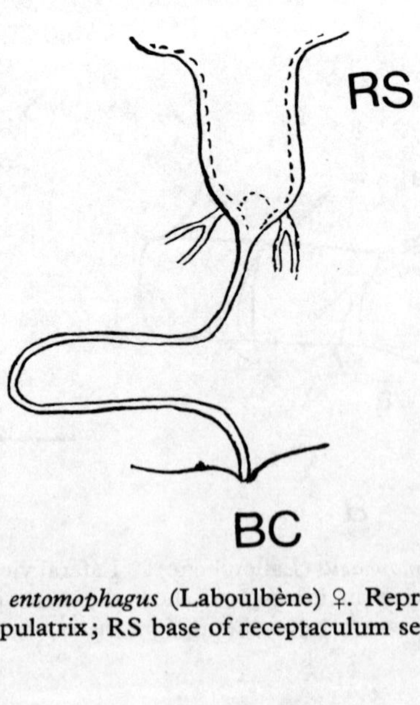

FIG. 142. *Thyreophagus entomophagus* (Laboulbène) ♀. Reproductive system. BC bursa copulatrix; RS base of receptaculum seminis.

pers. comm.). They consider that *Thyreophagus* can penetrate into compressed flour more successfully than *A. siro*.

This species is fairly widely distributed in small numbers in flour, sometimes increasing in numbers sufficient to cause serious damage; in such cases, K. G. Smith (1969) considers that infestation probably originates from the milling machinery. Laboulbène found *T. entomophagus* in collections of dried insects, Michael (1903) collected it from Russian ergot of rye and Zachvatkin (1941) associates it with old flour. Wasylik (1959) found *Thyreophagus* in a sparrow's nest.

DISTRIBUTION

England, Italy, France, U.S.S.R., Germany, U.S.A., Poland.

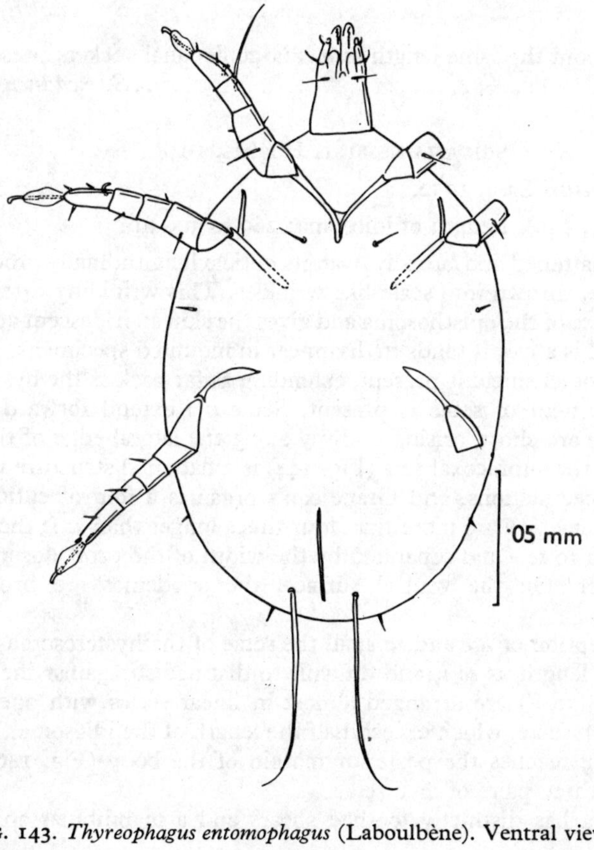

FIG. 143. *Thyreophagus entomophagus* (Laboulbène). Ventral view of larva.

SUIDASIA Oudemans, 1905

Aphelenia Oudemans, 1924; *Chibidania* Sasa, 1952.

The cuticle is finely wrinkled or covered with a scale-like pattern.

Setae *v e* are fine and arise posterior to *v i*.

Setae *sc e* are more than four times longer than *sc i* which are always present and considerably nearer to *sc e* than to each other.

A complete set of smooth short setae is present on the dorsal and lateral surfaces of the hysterosoma.

The posterior margin of the body in the male does not form a projecting opisthosomal plate.

Anal copulatory suckers may be missing in the male.

On the apex of tarsus I no dorsal spines are present, but three conspicuous ventral spines are developed. Omega_1 of tarsus I differs in shape from that of tarsus II.

Type species: *Suidasia pontifica* Oudemans, 1905.

SUIDASIA

Key to the adults

1. Setae *h e* considerably longer than *h i* (Fig. 144); no anal suckers in the male .... *S. nesbitti*, p. 128

—. Setae *h e* about the same length as *h i*; large flat anal suckers present in the male.. *S. medanensis*, p. 131

SUIDASIA NESBITTI Hughes, 1948

Chibidania tokyoensis Sasa, 1952.

MALE (Figs. 144, 145). Length of idiosoma: 269 to 300 µm.

The body is flattened and broadly oval, its cuticle longitudinally grooved and, at the same time, thrown into scale-like wrinkles. This wrinkling extends on to the ventral surface of the opisthosoma and gives the skin an iridescent appearance when the animal is alive; it tends to disappear in mounted specimens. A smooth dorsal propodosomal shield is present, extending as far back as the hysterosoma.

A full complement of setae is present. Setae *v i* extend forward over the gnathosoma; *v e* are short, arising midway along the lateral edge of the propodosomal shield; the supracoxal seta (Fig. 144) is a flattened structure with well-marked lateral pectinations, and Grandjean's organ is a fold of cuticle with a serrated edge. Setae *sc e* are more than four times longer than *sc i*; the *sc i* setae are inserted near to *sc e* and separated by the width of the propodosomal shield from each other. On the ventral surface, the apodemes are broad, stout structures.

With the exception of *h e* and *sa e*, all the setae of the hysterosoma are short, about the same length as *sc i*, and difficult to distinguish against the wrinkled cuticle. Setae d_1 to d_4 are arranged almost in linear series with one another; *sa e* are long limp setae, which exceed half the length of the idiosoma. Ventrally, the anal opening reaches the posterior margin of the body (Fig. 146a) and is surrounded by three pairs of anal setae.

Each chelicera has distinctly toothed shears and a mandibular spine on its ventral surface.

The legs are rather short and stumpy with the same chaetotaxy in both sexes. On leg I (Fig. 147a, b), *d* extends beyond the end of the claw, *e* and *f* are shorter setae and omega_1 is a curved rod; ventrally, *p*, *q* and *s* are conspicuous bent spines, *s* arising in the middle of the tarsus; *u* and *v* are long and slender and pressed tightly against the pretarsus, so that only the base is clearly visible (D. A. Griffiths, pers. comm.). The setae and solenidia arising from the base of the tarsus tend to be close together, owing to the shortness of the segment;

on tarsus I $omega_1$ is a slender curved rod which extends forward overlapping the base of *ba*; on tarsus II, it is shorter and thicker (Fig. 148a, b). The famulus is directed toward the tibia and is often concealed by $omega_1$; *aa, ba, la, ra* and *wa* are all fine setae; $omega_2$ arises close to *ba*. On the genu, $sigma_2$ is more than three times longer than $sigma_1$ and the setae of this and the remaining segments of leg I are not thickened to form spines. On tarsus IV (Fig. 149) the copulatory suckers are nearer to the base and apex of the tarsus than to each other.

The penis (Fig. 150) lies between coxae IV, it is a rather large curved tube.

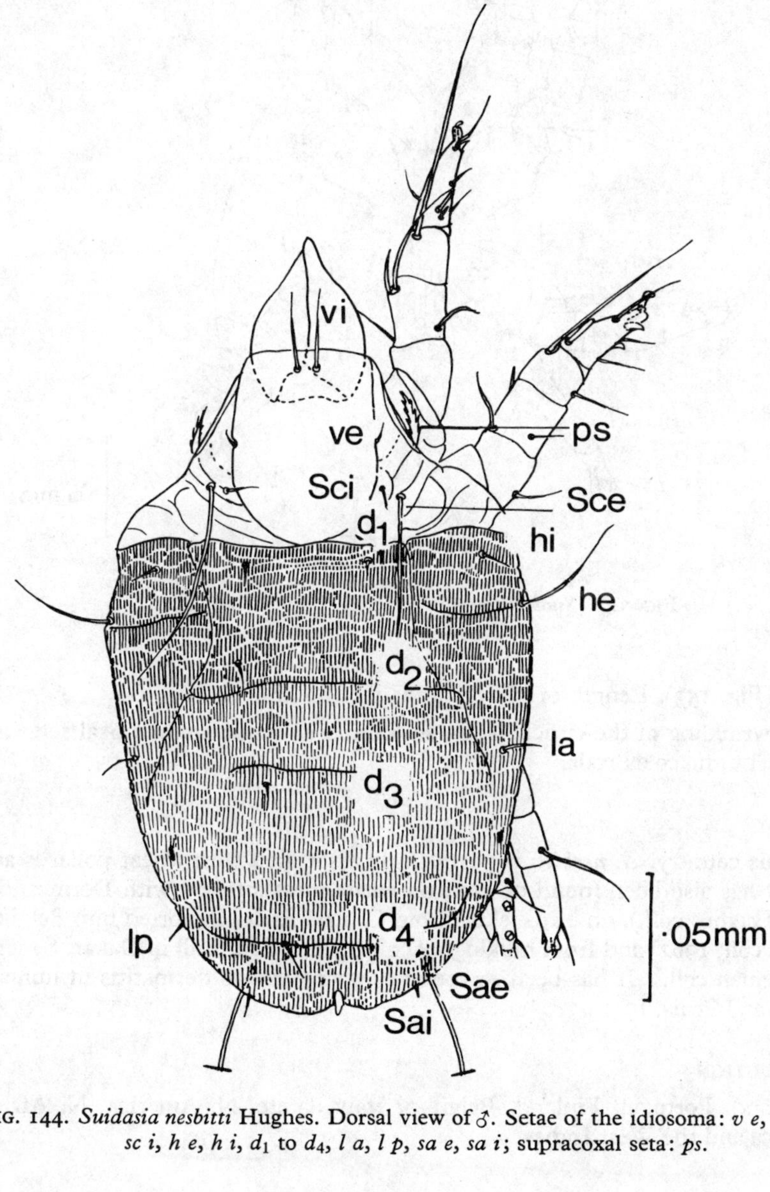

FIG. 144. *Suidasia nesbitti* Hughes. Dorsal view of ♂. Setae of the idiosoma: *v e, v i, sc e, sc i, h e, h i*, d_1 to d_4, *l a, l p, sa e, sa i*; supracoxal seta: *ps*.

FEMALE (Fig. 151). Length of idiosoma: 300 to 340 µm.

The female closely resembles the male. The anal opening (Fig. 152a) reaches the posterior margin of the body and is surrounded by five pairs of anal setae, the third pair of setae being further from the anus than the others.

The genital opening extends between coxae III and IV.

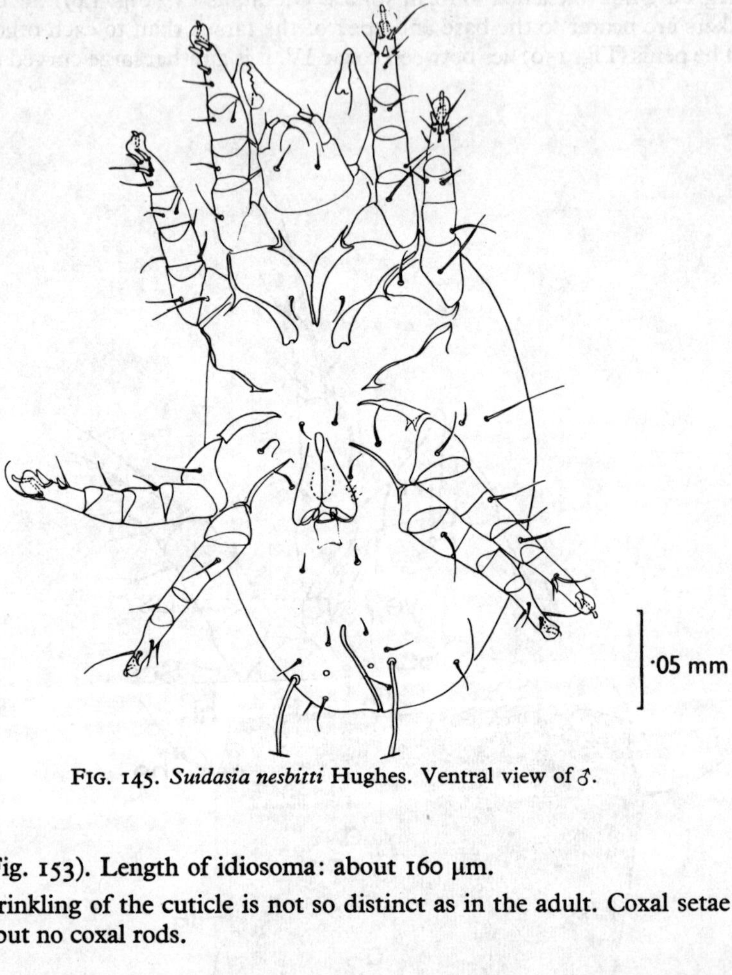

FIG. 145. *Suidasia nesbitti* Hughes. Ventral view of ♂.

LARVA (Fig. 153). Length of idiosoma: about 160 µm.

The wrinkling of the cuticle is not so distinct as in the adult. Coxal setae are present, but no coxal rods.

BIOLOGY

In this country, *S. nesbitti* is particularly associated with wheat pollards and bran; it has also been found on rice, on whale meat infested with Dermestidae (T. E. Hughes coll.), on birds' skins from Central Africa imported into Belgium (Benoit coll, 1962) and from milking machinery in a flour mill in Dakar, Senegal (J. Freeman coll.). It has been recorded as giving rise to dermatitis in humans (Kilpio and Pirila, 1952).

DISTRIBUTION

England, Portugal, Finland, Belgium, Italy, Crete, N. America, N. Africa, S. Africa and the West Indies.

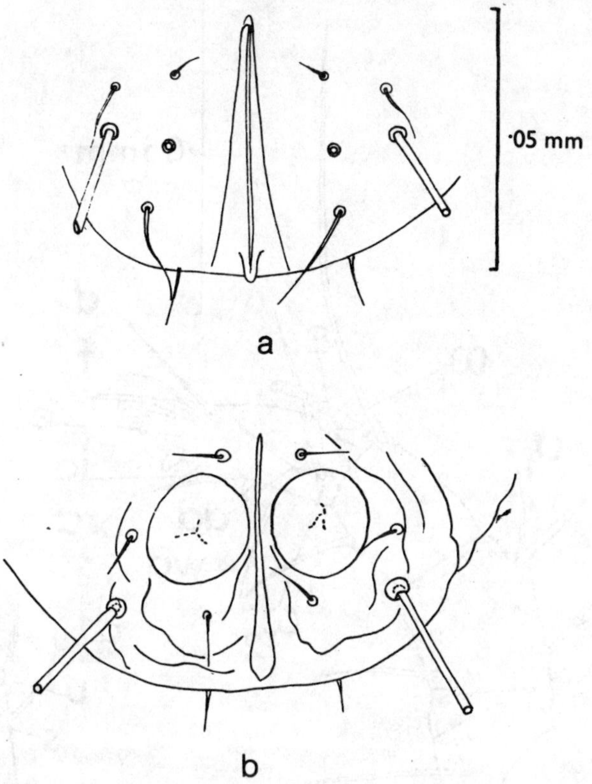

FIG. 146. Anal region of ♂ of (*a*) *Suidasia nesbitti* Hughes, (*b*) *S. medanensis* Oudemans.

SUIDASIA MEDANENSIS Oudemans, 1924

Suidasia insectorum Fox, 1950.

MALE (Fig. 154). Length of idiosoma: 300 to 320 μm.

The shape of the body is the same as in the previous species, but the wrinkling of the cuticle (Fig. 155) is more scale-like in appearance and lacking in longitudinal folding. The wrinkling can only be seen in freshly mounted specimens.

The arrangement and length of the setae resemble those of *S. nesbitti*, the main differences being that *v e* are more anterior in position, intermediate between *v i* and the supracoxal seta, and *h e* are equal in length to *h i*.

The anal opening (Fig. 146b) is near the end of the body and is flanked by suckers; three pairs of anal setae encircle these suckers.

The chaetotaxy of leg I (Fig. 156a, b) differs from that of *S. nesbitti* in that the ventral spines *u* and *v*, and the famulus are not visible.

FEMALE (Fig. 157). Length of idiosoma: 290 to 360 μm.

This closely resembles the male. The anal setae tend to be arranged in linear series on either side of the anus (Fig. 152b).

LARVA (Fig. 19). Length of idiosoma: about 160 μm.

The larva differs from that of *S. nesbitti* in having coxal rods as well as coxal setae. Each rod is shaped like a wine bottle, the 'neck' ending in a transparent bulb-like structure.

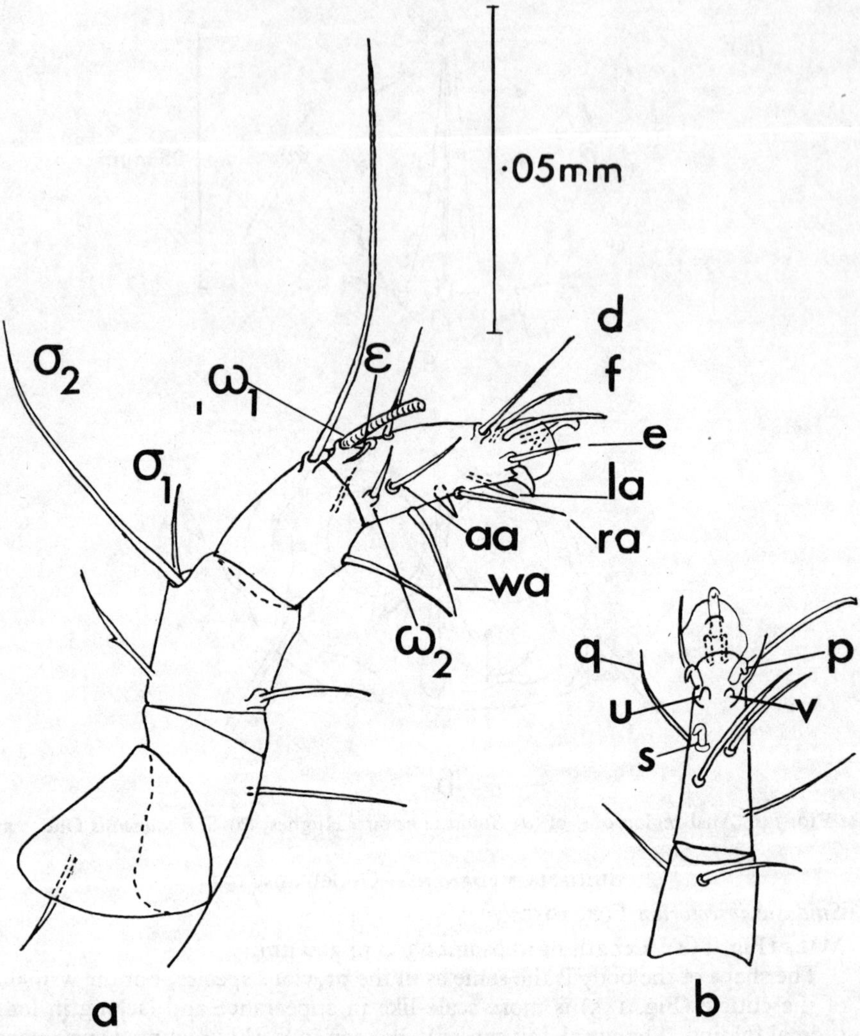

FIG. 147. *Suidasia nesbitti* Hughes ♀. (*a*) External view of right leg I, (*b*) ventral view of tibia and tarus of left leg I. Solenidia: ω_1 to ω_2, σ_1, σ_2; famulus: ε; setae and spines: *d, e, f, aa, ba, la, ra, wa, u, v, s p, q*.

BIOLOGY

The life history takes from 16 to 18 days at 23°C and a relative humidity of 87 per cent, wheat germ being used as food. This species will not breed with *S. nesbitti* (A. M. Hughes, 1954). It has been recorded from a bee's nest (*Xylocopa* sp.) by Oudemans, 1924b, from rice bran, groundnuts, cowpeas, dead mosquitoes (I. Fox, 1950) and from Pacific island copra (W. Horrigan coll.). Possibly its natural habitat is a bee's nest.

DISTRIBUTION

England, Germany, N. Africa, Sumatra, Puerto Rico. In Angola it is more frequently found than *S. nesbitti* (J. A. M. Guimaraes, pers. comm.).

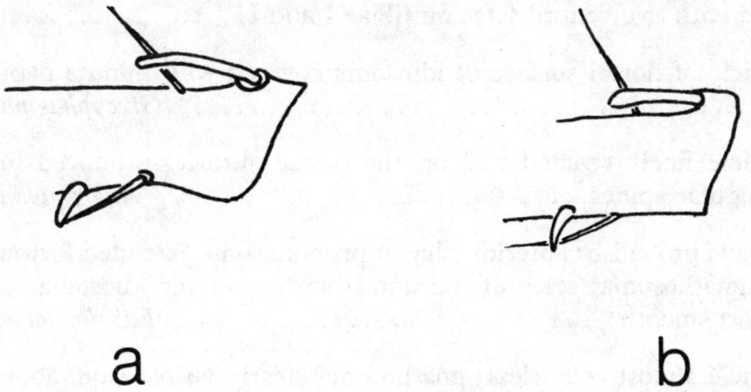

FIG. 148. *Suidasia nesbitti* Hughes ♂. Base of right tarsus I (*a*) and tarsus II (*b*).

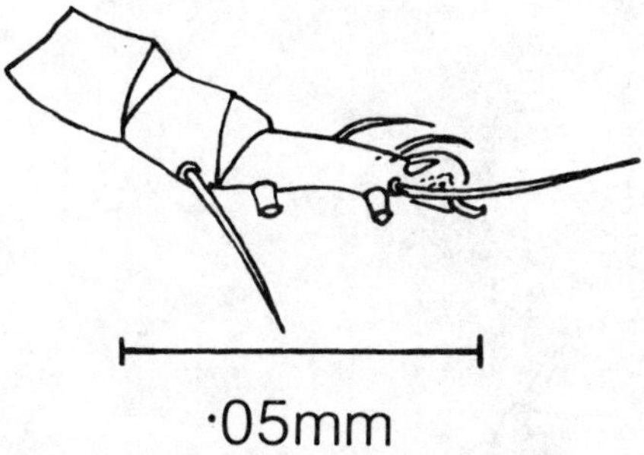

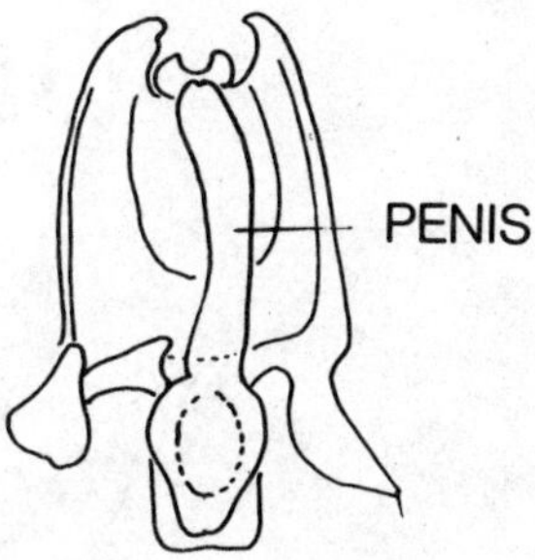

FIG. 149. *Suidasia nesbitti* Hughes ♀. External view of right leg IV (left).
FIG. 150. *Suidasia nesbitti* Hughes ♂. Penis and associated sclerites (right).

Family GLYCYPHAGIDAE Berlese, 1887

The species are free-living, often associated with insects or small mammals, or found in the nests of mammals.

The dorsal side of the idiosoma is usually not divided by a transverse constriction into propodosoma and hysterosoma.

The dorsal propodosomal shield may be reduced or absent.

The cuticle is rarely smooth, usually rough or covered with small papillae.

The claw is usually inserted at the distal end of a fleshy pretarsus, only joined to the end of the tarsus by two thin 'tendons'. It is sometimes absent.

Anal and tarsal suckers are usually absent in the male.

GLYCYPHAGIDAE

Key to the subfamilies

1. Majority of idiosomal setae long, densely pectinate, bipectinate or leaf-shaped 2

—. Idiosomal setae mainly short 4

2. Peripheral idiosomal setae flattened and pectinate, bipectinate or foliate (Fig. 190), often forming a fringe encircling the idiosoma; tarsi short and stout, often with a dorsal keel; only one ventral seta on tibiae I and II . . . . *Ctenoglyphinae*, p. 165

—. Idiosomal setae densely pectinate; tarsi long and slender, without a dorsal keel; with two ventral setae on tibiae I and II........................3

3. Cuticle of dorsal surface of idiosoma covered with minute papillae...... ....Glycyphaginae, p. 135

—. Cuticle finely striated and on the dorsal surface, produced into minute triangular spines..............................*Aëroglyphinae*, p. 177

4. Cuticle brownish; anterior edge of propodosoma extended forward to cover the gnathosoma; setae of the dorsal surface of the idiosoma (except *v i*), almost smooth.............................. *Labidophorinae*, p. 178

—. Cuticle almost colourless; gnathosoma clearly visible from above; setae of the dorsal surface of the idiosoma flattened, with fringed edges. Associated with bat roosts..............................*Nycteriglyphinae*, p. 174

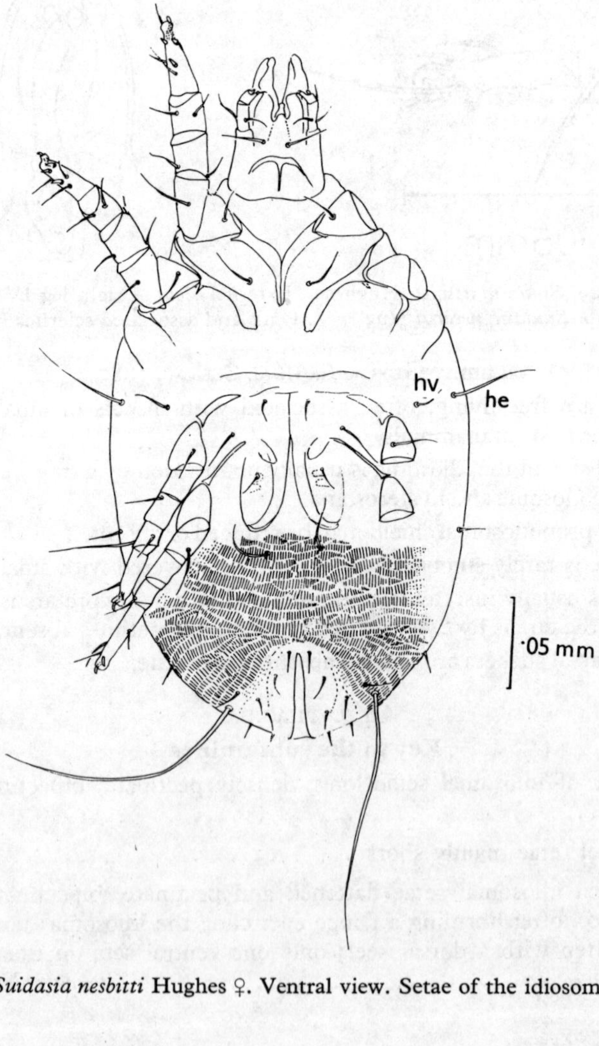

FIG. 151. *Suidasia nesbitti* Hughes ♀. Ventral view. Setae of the idiosoma: *h v*, *h e*.

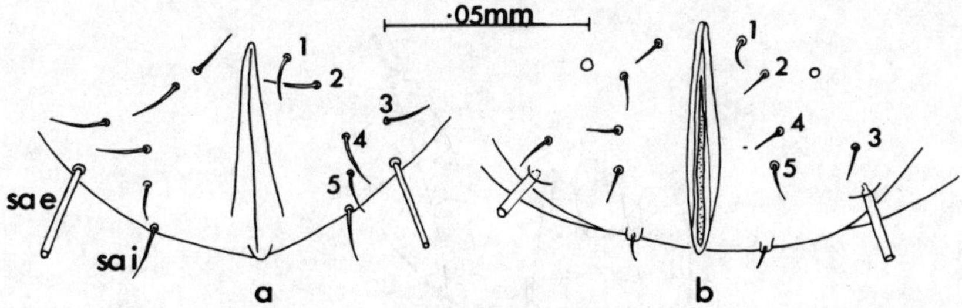

FIG. 152. Anal region of ♀ of (*a*) *Suidasia nesbitti* Hughes, (*b*) *S. medanensis* Oudemans. Setae of the idiosoma: *sa e*, *sa i*; anal setae: 1 to 5.

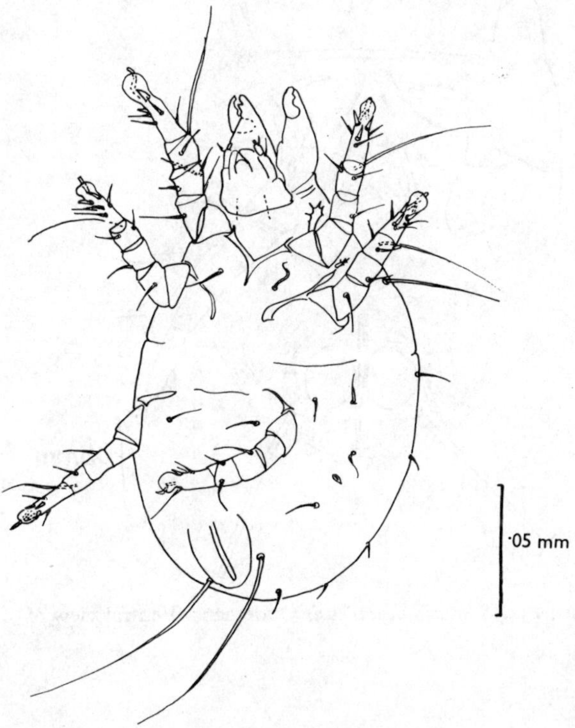

FIG. 153. *Suidasia nesbitti* Hughes. Ventro-lateral view of larva.

GLYCYPHAGINAE Zachvatkin, 1941

The idiosomal setae are always long and densely pectinate.

The tarsi are long and slender, without a keel.

One or two ventral setae are inserted on tibiae I and II. The setae of tibia and genu are generally pectinate.

No anal or tarsal suckers are present in the male.

The penis in the male is usually inconspicuous.

The cuticle is usually covered with small papillae.

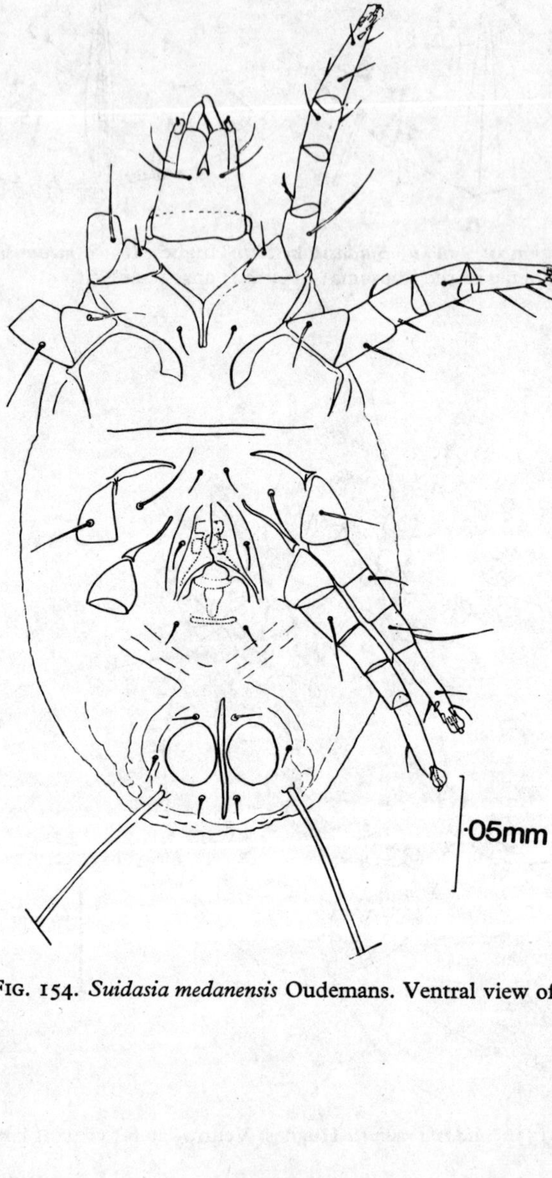

FIG. 154. *Suidasia medanensis* Oudemans. Ventral view of ♂.

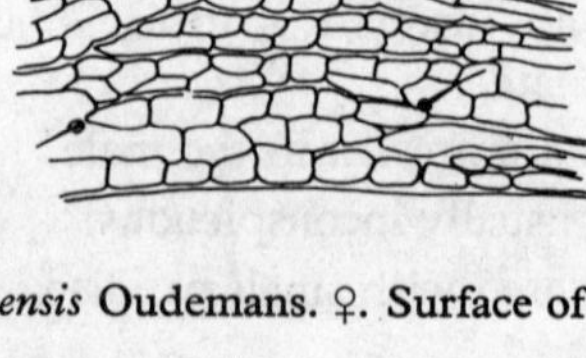

FIG. 155. *Suidasia medanensis* Oudemans. ♀. Surface of cuticle around the setae d_2.

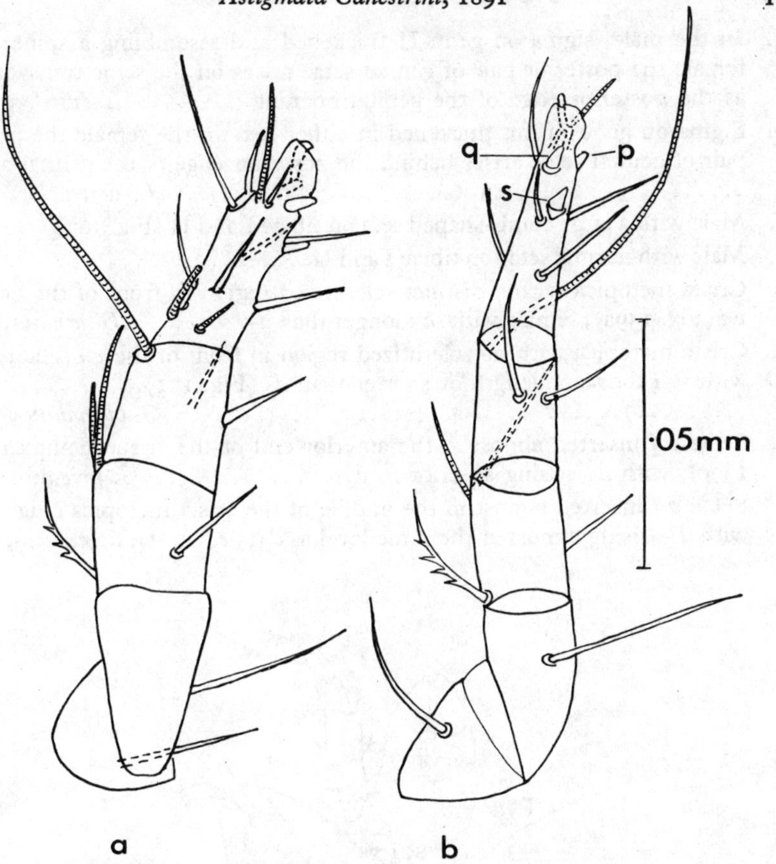

FIG. 156. *Suidasia medanensis* Oudemans ♀. (*a*) External view of right leg I, (*b*) ventral view of left leg I. Ventral spines: *p*, *q*, *s*.

GLYCYPHAGINAE

Key to the adults

1. With setae *v i* and *v e* arising close together (Fig. 186)................ ....*Blomia freemani*, p. 161

—. With *v e* well-separated from *v i*..2

2. With a subtarsal scale (Fig. 173) and without a crista metopica..........3

—. Without a subtarsal scale, usually with a crista metopica (Fig. 158)..... ....*Glycyphagus* 6

3. On genu I, solenidion $sigma_2$ more than three times longer than $sigma_1$; setae *la*, *ra*, *ba* on tarsus I arise in the distal third of the tarsus......... ....*Lepidoglyphus* 4

—. On genu I, $sigma_1$ and $sigma_2$ are almost the same length; *la*, *ra* and *ba* on tarsus I arise from the basal half of the segment.................. ....*Austroglycyphagus geniculatus*, p. 158

4. Ventral seta *nG* of genua III widened to form a pectinate scale (Fig. 179) ....*L. michaeli*, p. 157

—. *nG* not widened to form a pectinate scale.............................5

5. In the male, sigma on genu II thickened and resembling a spine; in the female the posterior pair of genital setae arises on the same transverse line as the posterior edge of the genital opening.......... *L. fustifer*, p. 156
—. Sigma on genu II not thickened in either sex. In the female the posterior pair of genital setae arises behind the posterior edge of the genital opening .... *L. destructor*, p. 150
6. Male with a large comb-shaped seta on tibiae I and II (Fig. 167b)........7
—. Male with normal setae on tibiae I and II...............................8
7. Crista metopica with a distinct sclerotized region in front of the origin of *v i* (Fig. 159a); female with *sa i* longer than d_2.......... *G. ornatus*, p. 144
—. Crista metopica with no sclerotized region in front of the *v i* setae; female with *sa i* the same length or shorter than d_2 (Fig. 171)................ .... *G. bicaudatus*, p. 147
8. Setae *v i* inserted almost at the anterior end of the crista metopica; (Fig. 159c); with d_2 arising anterior to d_3................ *G. privatus*. p. 144
—. Setae *v i* inserted almost in the middle of the crista metopica (Fig. 159b); with d_2 arising almost at the same level as d_3....... *G. domesticus*, p. 140

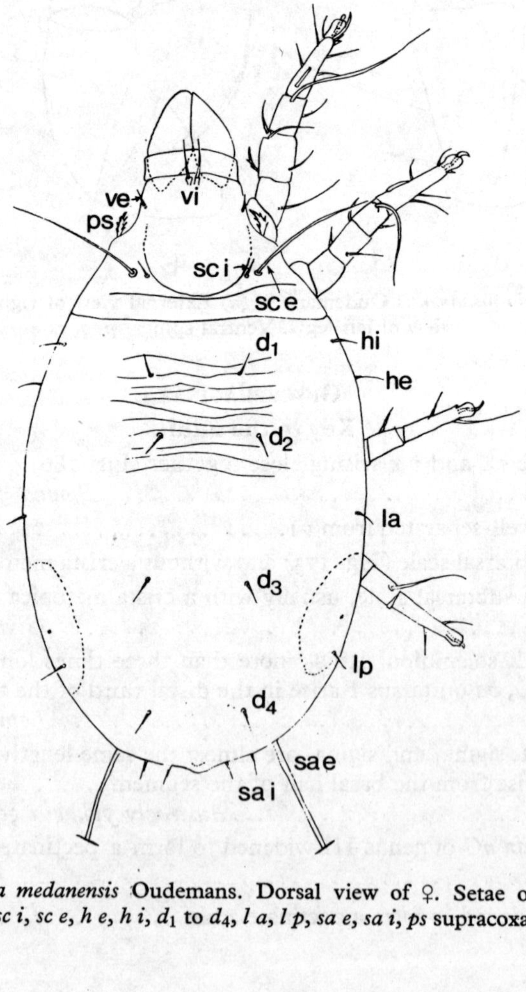

FIG. 157. *Suidasia medanensis* Oudemans. Dorsal view of ♀. Setae of the idiosoma: *v e, v i, sc i, sc e, h e, h i*, d_1 to d_4, *l a, l p, sa e, sa i, ps* supracoxal seta.

GLYCYPHAGUS Hering, 1938 s. lat.

The genus *Lepidoglyphus* was created by Zachvatkin (1936) to include those species of *Glycyphagus* possessing a pectinate subtarsal scale and lacking a crista metopica. In 1941 he changed the status of *Lepidoglyphus* so that it became a subgenus of *Glycyphagus*. Subsequent to 1941, *Lepidoglyphus* continued to be regarded as a genus (Cooreman, 1942, Türk and Türk, 1957, Sellnick, 1958). In 1961, I included *Lepidoglyphus* in the genus *Glycyphagus*, but in this work, I am accepting Zachvatkin's original conception and regarding *Lepidoglyphus* as a distinct genus.

GLYCYPHAGUS Hering, 1938 s. str.

A narrow elongated dorsal propodosomal shield or crista metopica is present (Fig. 158).

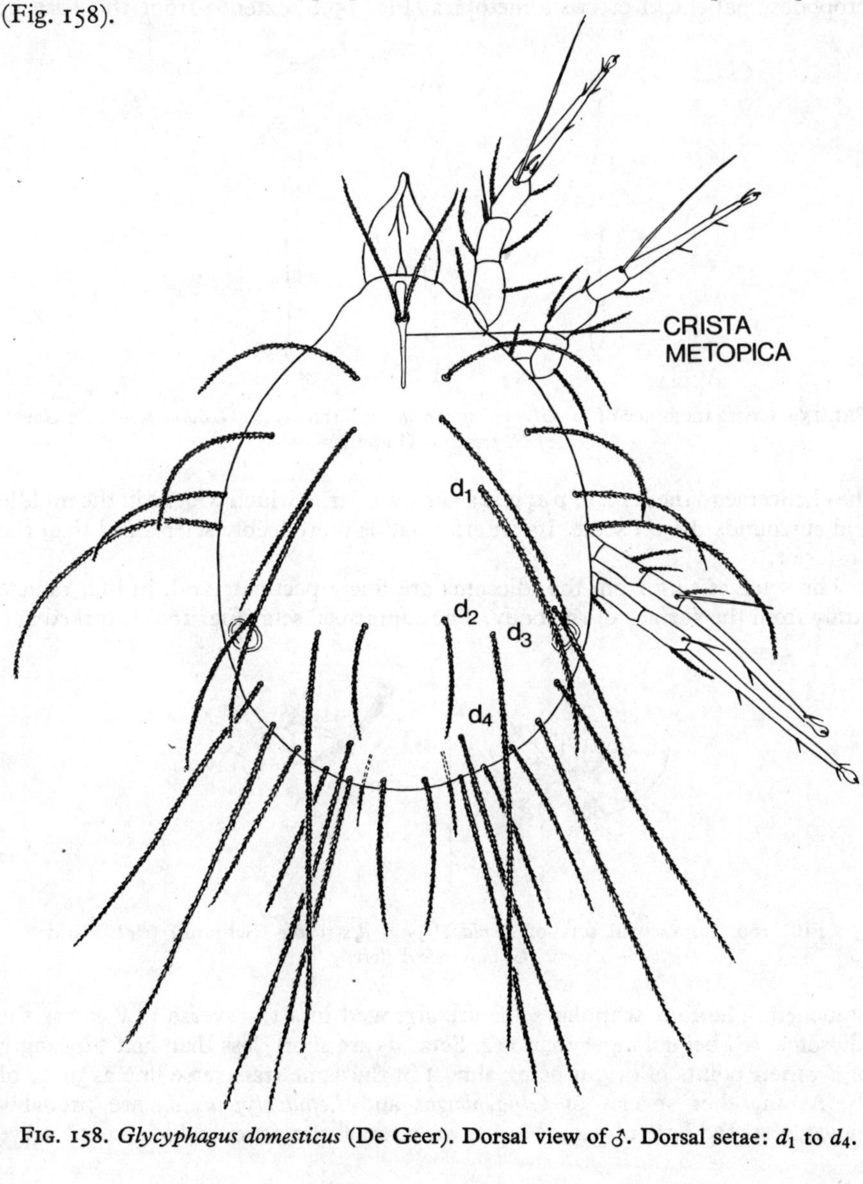

FIG. 158. *Glycyphagus domesticus* (De Geer). Dorsal view of ♂. Dorsal setae: d_1 to d_4.

The tarsi are not ensheathed in a pectinate subtarsal scale.
On genu I, $sigma_2$ is more than twice as long as $sigma_1$ (Fig. 161).
Two ventral setae are present on tibiae I and II.
Male and female genital openings lie between coxae II and III.
Type species: *Acarus prunorum* Hering, 1838.

GLYCYPHAGUS DOMESTICUS (De Geer, 1778)

Acarus domesticus De Geer, 1778; *Oudemansium domesticum* Zachvatkin, 1936.

MALE (Fig. 158). Length of idiosoma: 320 to 400 μm.

The idiosoma is rounded in shape. The cuticle is dull, owing to the fact that it is covered with minute papillae (see frontispiece). The narrow elongated dorsal propodosomal shield or crista metopica (Fig. 159b) extends from the base of

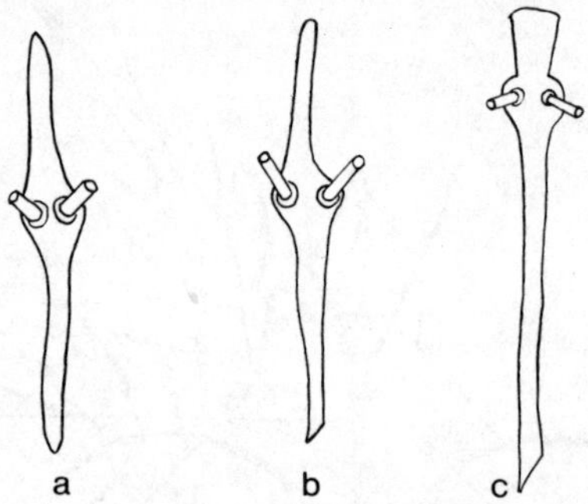

FIG. 159. Crista metopica of (*a*) *Glycyphagus ornatus* Kramer, (*b*) *G. domesticus* (De Geer), (*c*) *G. privatus* Oudemans.

the chelicerae to the level of *v e*; it is a narrow sclerite which widens in the middle and surrounds the *v i* setae. Its anterior half is more feebly sclerotized than the posterior.

The setae arising from the idiosoma are finely pectinate and, in life, radiate stiffly from the surface of the body. The supracoxal seta (Fig. 160) is forked and

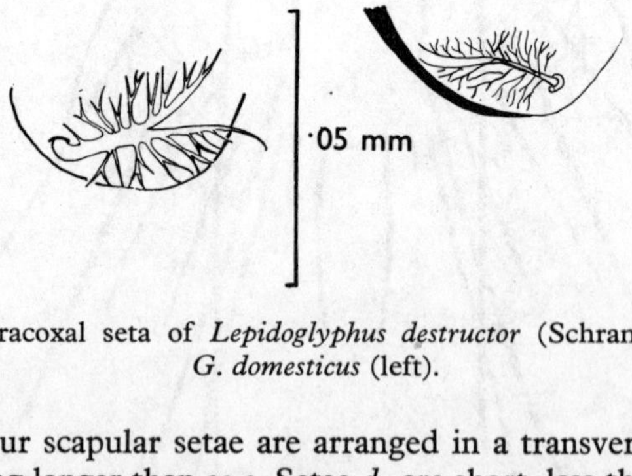

FIG. 160. Supracoxal seta of *Lepidoglyphus destructor* (Schrank) (right) and *G. domesticus* (left).

branched. The four scapular setae are arranged in a transverse row across the idiosoma, *sc i* being longer than *sc e*. Setae d_2 are short, less than half the length of d_1, their points of origin being almost in the same transverse line as those of d_3. As in other species of *Glycyphagus* and *Lepidoglyphus* d_3 are probably movable, as the base of each one has an internal projection which could serve as a muscle attachment. Three pairs of lateral setae are present—l_1 to l_3. From the posterior margin of the body project three pairs of postanals and *sa i* and *sa e*.

Ventrally, the apodemes of legs I meet to form a short sternum, apodemes II are also well-developed, apodemes III and IV are slender and apodemes III each have an anteriorly-directed process. The genital opening lies between coxae II and III.

The legs are long, with tapering segments and each terminates in a pretarsus and small claw. No subtarsal scale is present, but it is replaced by a pectinate seta—*wa*—which is present on all the legs and arises from the middle of the tarsus (Fig. 161a); the setae *la*, *ba* and *ra* arise between the base of *wa* and the

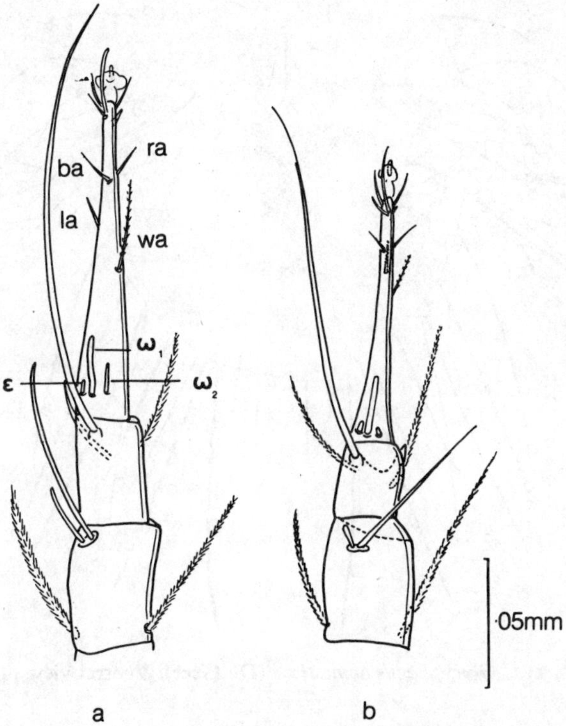

FIG. 161. Dorsal view of right leg I of (*a*) *Glycyphagus domesticus* (De Geer) ♂, (*b*) *G. privatus* Oudemans ♂. Solenidia: ω_1, ω_2; famulus: ε; setae: *ba*, *la*, *ra*, *wa*.

apex of the tarsus. On tarsus I, omega$_1$ is a slender rod longer than that on tarsus II and omega$_2$ is about half its length, much longer than the famulus. On genu I, sigma$_1$ is less than half the length of simga$_2$, about the same length as omega$_1$. On tibiae III and IV, *kT* is well-removed from the distal edge of the segment.

FEMALE (Fig. 162). Length of idiosoma: 400 to 750 μm.

The female closely resembles the male. The genital opening extends to the posterior edge of acetabula III, it is shorter than the distance separating it from the anterior end of the anus; a small crescentic epigynium covers the anterior end of the genital folds. The posterior pair of genital setae is inserted just behind

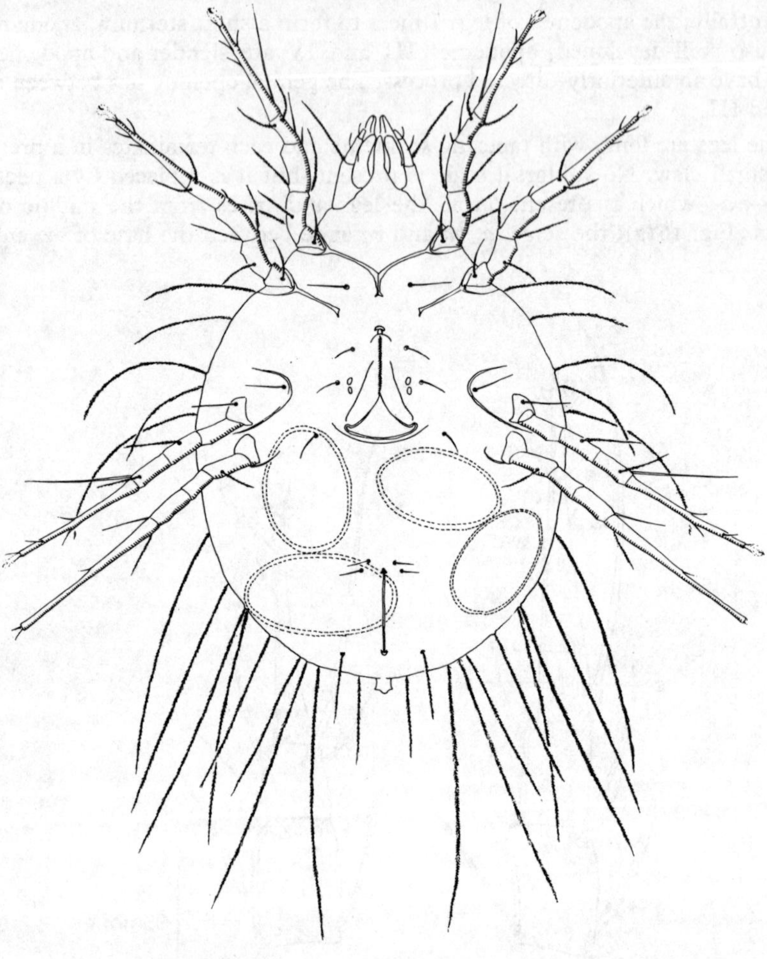

FIG. 162. *Glycyphagus domesticus* (De Geer). Ventral view of ♀.

the hind end of the genital opening. Two pairs of setae lie on either side of the anterior end of the anus. A tubular bursa copulatrix projects from the hind margin of the body.

HYPOPUS (Fig. 163). Length of hypopus and case: about 330 μm.

The hypopus is an oval white body with small bud-like appendages; the protonymphal cuticle which surrounds it, is covered with reticulations.

LARVA

The crista metopica is similar in structure to that of the adult, but is more feebly sclerotized. Prominent coxal rods are present.

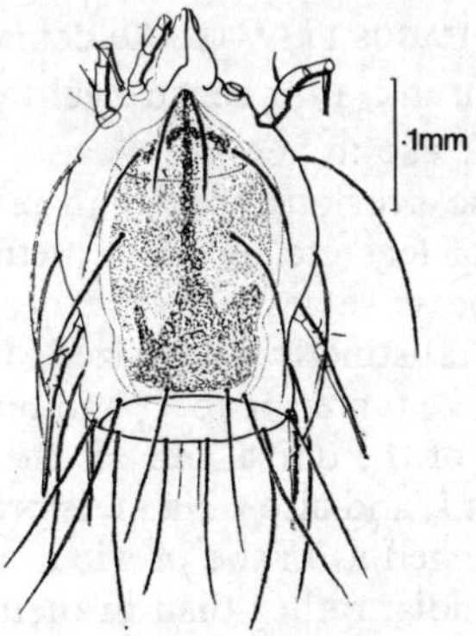

FIG. 163. *Glycyphagus domesticus* (De Geer). Dorsal view of hypopus enclosed in protonymphal cuticle.

BIOLOGY

The life cycle is completed in 22 days at a temperature of 23 to 25°C and a relative humidity of 80 to 90 per cent. The hypopus appears to be formed independently of external conditions and as many as 50 per cent of the protonymphs may pass through this stage (A. M. Hora, 1934). The hypopus, possibly owing to the fact that much of its internal structure dedifferentiates (A. M. and T. E. Hughes, 1939), is very resistant to drought and may remain encysted from five days to six months or even a few years. (D. A. Griffiths, pers. comm.).

This is a widely distributed species, often occurring in large numbers on dried plant and animal remains in houses and stables, but not in grassland (Griffiths, 1960). It has been found in flour, wheat, hay, linseed, tobacco, cheese, ham (Chabasse and Genthorn, 1966), sugarbeet seed (W. Chmielewski, 1969) dried calves stomach (J. S. Burton coll.), and in bees' and birds' nests, on damp mouldy wallpaper and in the sweepings from barns. It is the most common pest of stockfish in Iceland (G. Gigja, 1963) and will increase to enormous numbers on furniture covered with rush or sea grass fibre, or upholstered with green Algerian fibre; here it feeds on moulds growing on the fibre. Sinha (1968a) has shown, however, that it is more restricted in its fungal diet than *Lepidoglyphus destructor*.

On one occasion, it was found in the cold room of the Pathology Department of a Hospital where it was found crawling over the agar plates stored there. The mites were feeding on the culture medium, that of the cholera bacillus being particularly acceptable. They were probably introduced as a result of the room being used as a temporary storage for food. (L. M. Stables pers. comm.).

Robertson (1946) states that a number of cases of dermatitis have been reported in New Zealand, where workers were handling infested cheese. Joyeux and Baer, 1945 have identified *G. domesticus* as the vector of the tapeworm *Catenotaenia pusilla*, the adult of which is found in rodents. Davies (1926) records it as being responsible for ear canker in rabbits. *G. domesticus*, among other species, has on several occasions been associated with symptoms of asthma in human beings (Deschiens, 1951).

DISTRIBUTION

Cosmopolitan, but more frequently found in Europe, than in Canada, Japan and Australia (Sinha, 1968a).

GLYCYPHAGUS PRIVATUS Oudemans, 1903

Glycyphagus cadaverum Schrank, 1781 sensu Zachvatkin, 1941.

MALE. Length of idiosoma: 280 to 360 μm.

G. privatus resembles the previous species in texture of cuticle, length and thickness of setae, length of legs etc. A crista metopica (Fig. 159c) is present, extending backwards to the *sc* setae; *v i* arise near its anterior edge and the region in front of this is indistinctly sclerotized. Of the dorsal setae, d_2 are inserted as in *L. destructor*, i.e. anterior to d_3 and on the same level as d_1. Apart from this, the arrangement of the dorsal setae is the same as in *G. domesticus*.

The distal margin of tibia I, and also tibia II, is produced as a thin rim and the setae and solenidia are arranged as in the previous species, any difference being in length of setae and solenidia, rather than in their position. On tarsus I (Fig. 161b) omega$_2$ is short, equal in length to the famulus, and on genu I, sigma$_1$ is shorter, i.e., much less than half the length of omega$_1$.

FEMALE. Length of idiosoma: 370 to 450 μm.

The genital opening is longer than in *G. domesticus*, extending as far back as the posterior edge of acetabula IV, its length exceeds the distance between it and the anal slit.

The hypopus has not been described for this species.

LARVA

In an unpublished drawing, Oudemans shows this as having a crista metopica similar to that of the adult, and also small button-shaped coxal rods. On genu I, sigma$_1$ and sigma$_2$ have the same proportions as in the adult.

BIOLOGY

G. privatus has been found in Manitoba wheat, barley, grass seed and in the dust collected from silos and warehouses (O'Farrell and Butler, 1948). Zachvatkin (1941) records it from grass, clover and flax seed, he also found it in fields and in the piles of debris left after winnowing. Wasylik (1959) found *G. privatus* in a sparrow's nest and Haarlov and Alani (1970) record it from house dust.

GLYCYPHAGUS ORNATUS Kramer, 1881

MALE (Fig. 164). Length of idiosoma: 430 to 500 μm.

The idiosoma is oval, widest between the second and third pairs of legs and slightly constricted behind the fourth pair. The crista metopica (Fig. 159a) resembles that of *G. domesticus*, the *v i* setae arising from the widened region in the middle of the sclerite.

The setae of the idiosoma are long and densely pectinate and arise from distinct sclerotized bases. Setae d_2 are short, their terminal ends barely reaching the bases of d_4, they vary in position in different individuals and may arise anterior or posterior to d_3. Setae d_3 exceed the idiosoma in length and may possibly be movable, as each has a small internal projection at its base. The remaining posterior setae are also extremely long. The supracoxal seta (Fig. 165a) is forked and branched.

Tarsi I and II are curved, particularly tarsi II (Fig. 166) and the tibiae and genua are expanded distally. The ventral setae of all the legs are comparatively long and pectinate. On tibia I and II, *hT* is modified into a triangular hairy structure whose internal edge bears nine to ten teeth on leg I, four to five on leg II. On genu I (Fig. 167a, b) sigma$_1$ is shorter than sigma$_2$. On tibiae III and IV, the arthrodial membrane extends to the base of *kT*.

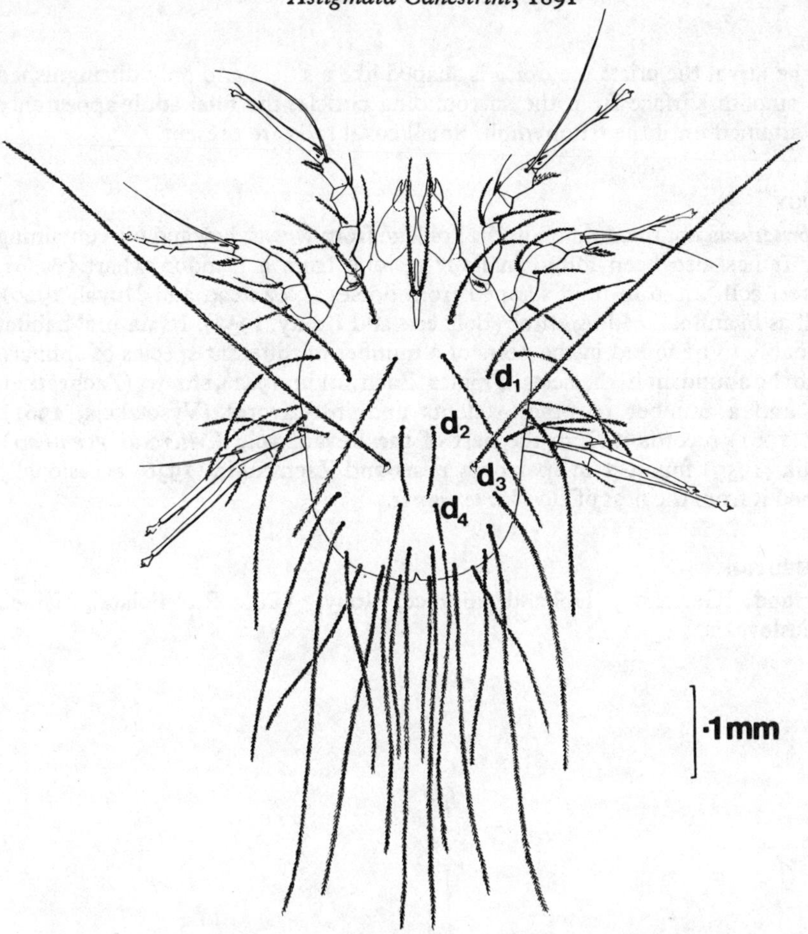

FIG. 164. *Glycyphagus ornatus* Kramer. Dorsal view of ♂. Dorsal setae: d_1 to d_4.

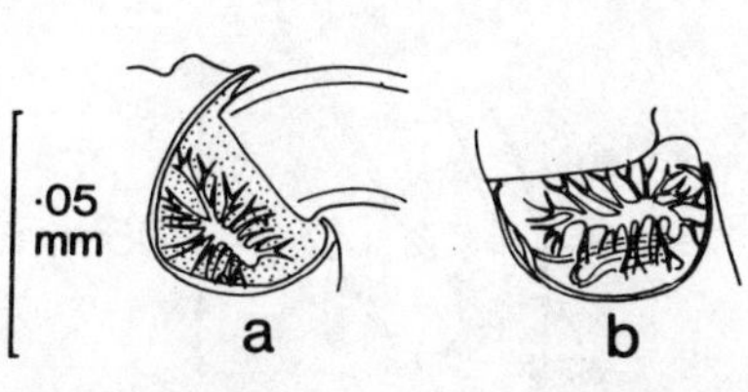

FIG. 165. Supracoxal seta of (*a*) *Glycyphagus ornatus* Kramer, (*b*) *G. bicaudatus* Hughes.

FEMALE (Fig. 168). Length of idiosoma: 540 to 600 µm.

The posterior margin of the genital opening is at the same level as apodemes III and separated by a space from the posterior pair of genital setae. The bursa copulatrix opens at the tip of a small mound projecting from the end of the body.

The tarsi of all the legs are almost straight and *hT* on tibiae I and II are normal setae (Fig. 167a). On tarsus I, the distal setae *la*, *ra*, *ba* and *wa* are close together; in *G. domesticus* (Fig. 161a) these setae are more widely spaced.

No hypopus has been recorded for this species.

Larva

In the larva, the crista metopica is shaped like a shield and only distinguished by its smooth surface from the surrounding cuticle, the final adult appearance is not attained until the tritonymph. Small coxal rods are present.

Biology

G. ornatus is not uncommon in the spillage from wheat, hay and oil-containing seeds. It has also been found in flour residue from a London wharf (A. M. Shenker, coll.), also in mud scraped from horses (Newstead and Duval, 1920), as well as in animal feeding stuffs (Bollaerts and Breny, 1951). Its natural habitat is probably to be found in the nests of a number of different species of animals, as it can be abundant in the nests of moles (Zachvatkin, 1941), shrews (Zachvatkin, 1941) and a number of other rodents and insectivores (Vysotzkaja, 1961). Costa (1961) records it from the nest of the Levant vole (*Microtus guentheri*), Wasylik (1959) found it in sparrow's nests and Zachvatkin (1936) occasionally obtained it from the nest of *Bombus terrestris*.

Distribution

England, Germany, Holland, France, Italy, U.S.S.R., Poland, Israel, Czechoslovakia.

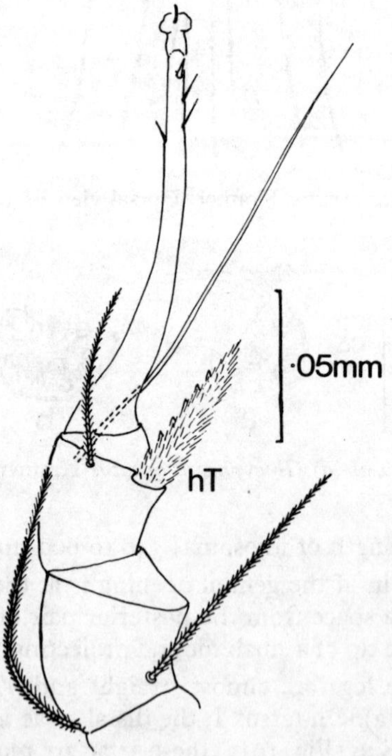

FIG. 166. *Glycyphagus ornatus* Kramer ♂. Ventral view of left leg II. Modified seta: *hT*.

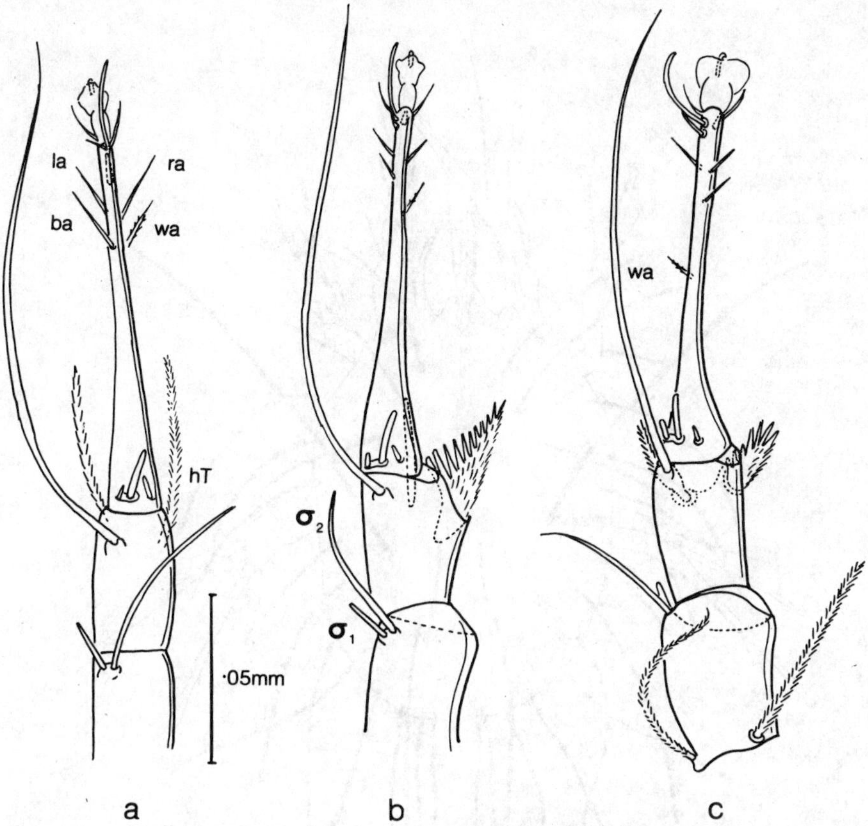

FIG. 167. *Glycyphagus ornatus* Kramer. Dorsal view of right leg I of (*a*) ♀, (*b*) ♂. *G. bicaudatus* (*c*). Dorsal view of right leg I of ♂. Solenidia: σ_1, σ_2; setae: *ba*, *la*, *ra*, *wa*, *hT*.

GLYCYPHAGUS BICAUDATUS Hughes 1961

MALE (Fig. 169). Length of idiosoma of two specimens: 390 to 433 μm.

This is closely related to *G. ornatus*. The crista metopica (Fig. 170b) is not so well developed and is narrower; the region anterior to the *vi* setae may or may not be developed. The supracoxal seta (Fig. 165b) has a much branched single main stem.

The setae arising from the dorsal side of the idiosoma are arranged as in *G. ornatus*, they are strongly pectinate and are inserted deeply into the surface cuticle. Setae d_3 and d_4 differ from the corresponding setae in *G. ornatus* in being flattened and swollen towards the base.

The tarsi of legs I and II are curved, the seta *wa* (Fig. 167c) arises near the middle of the tarsus, not in its anterior half as in *G. ornatus*. On tibiae I and II, *hT* is modified into a triangular scale, with its anterior edge produced into seven or eight teeth on legs I and five to six on legs II.

FEMALE. Length of idiosoma of four specimens: 433 to 635 μm.

The region of the crista metopica (Fig. 170a) anterior to the *v i* setae is visible as a faint line beneath the cuticle. The body setae resemble those of the male, except that *sa i* are quite short—about the same length as d_2—and slightly swollen in the middle (Fig. 171).

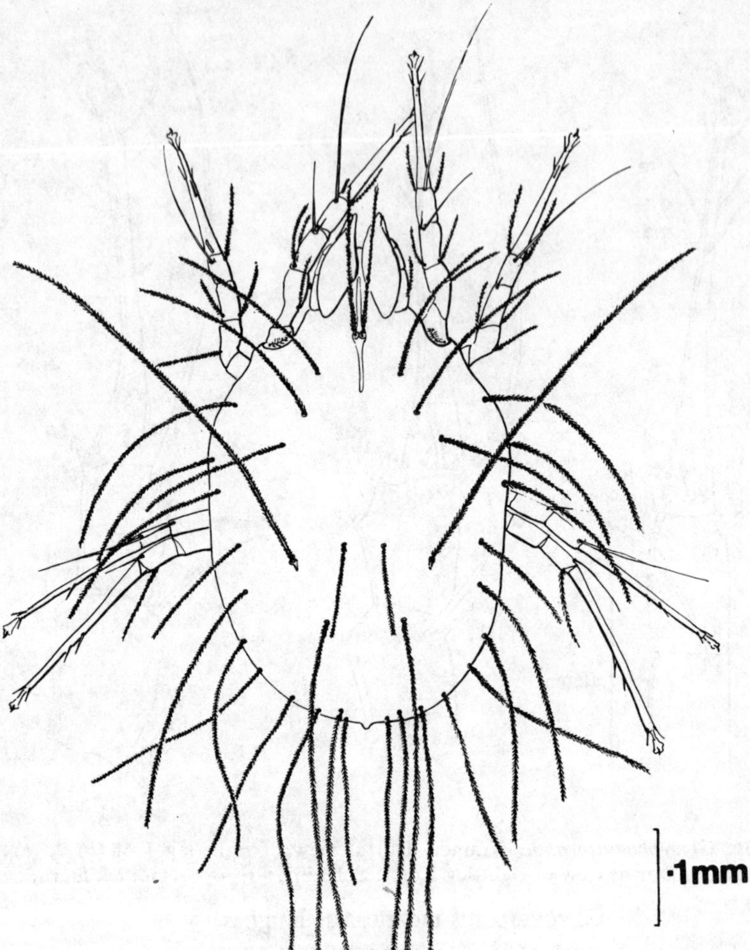

FIG. 168. *Glycyphagus ornatus* Kramer. Dorsal view of ♀.

As in the female of *G. ornatus*, the setae arising from tarsus I and II are longer than in the male and *hT* on tibiae II and I are unmodified.

No external tubular bursa copulatrix is present, its opening being a round hole which leads into a receptaculum seminis, the walls of this internal sac are visible, its base supported by a pair of curved sclerites.

DISTRIBUTION

G. bicaudatus was found by G. E. Woodroffe in a mouse nest situated in the grounds of the Pest Infestation Laboratory, Slough, Berks, England. It is allied to *G. zachvatkini* Volgin 1961 which was found in the nest of *Picus canus* L.

and on *Microtus fertis* Büchn, Khabarosk, U.S.S.R. The two species differ in the structure of the crista metopica and the chaetotaxy of the legs.

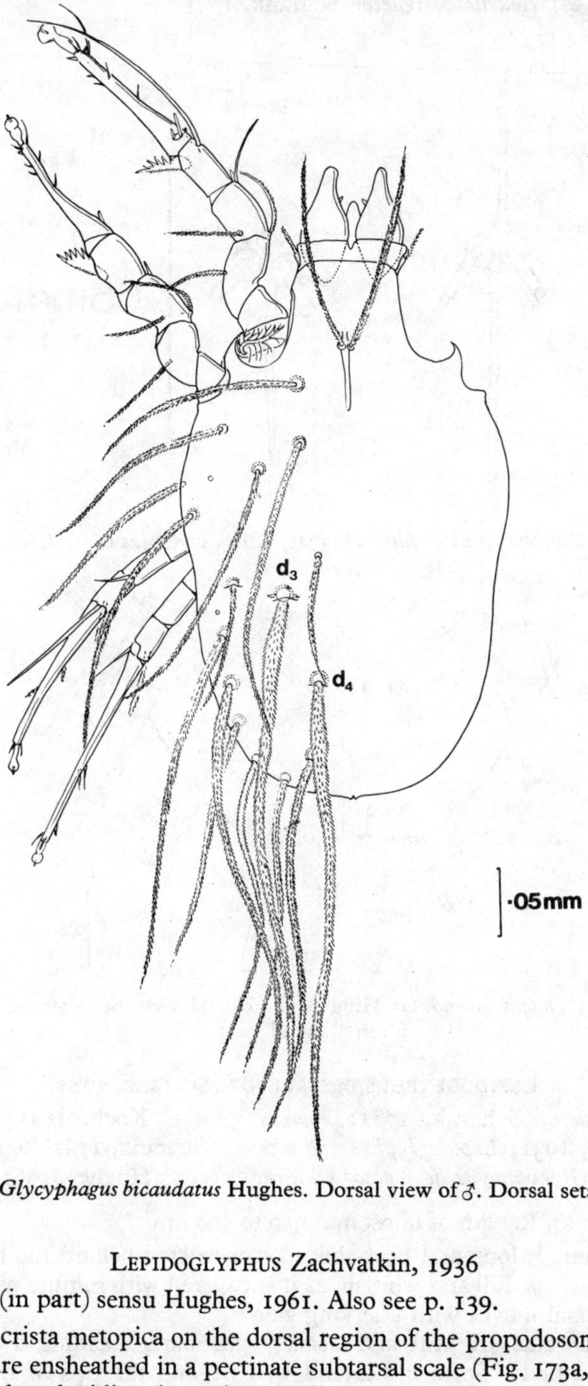

FIG. 169. *Glycyphagus bicaudatus* Hughes. Dorsal view of ♂. Dorsal setae: d_3, d_4.

LEPIDOGLYPHUS Zachvatkin, 1936

Glycyphagus (in part) sensu Hughes, 1961. Also see p. 139.

Without a crista metopica on the dorsal region of the propodosoma.

The tarsi are ensheathed in a pectinate subtarsal scale (Fig. 173a, b).

On genu I, the solenidion $sigma_2$ is more than twice as long as $sigma_1$ (Fig. 173a).

Two ventral setae are present on tibiae I and II.
Male and female genital openings lie between coxae II and III.
Type species: *Acarus destructor* Schrank, 1781.

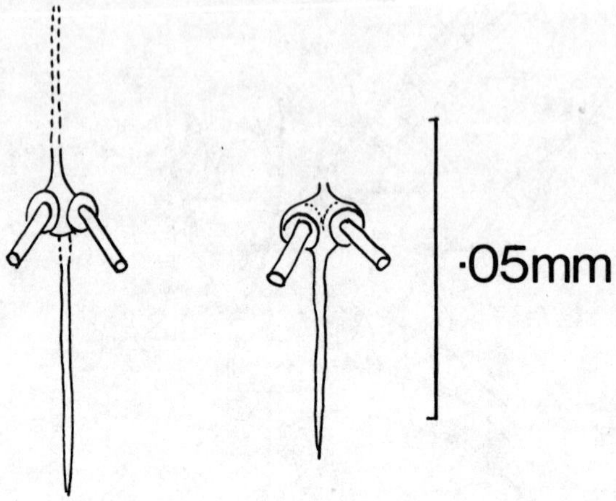

FIG. 170. *Glycphagus bicaudatus* Hughes. Crista metopica of ♀ (left), ♂ (right).

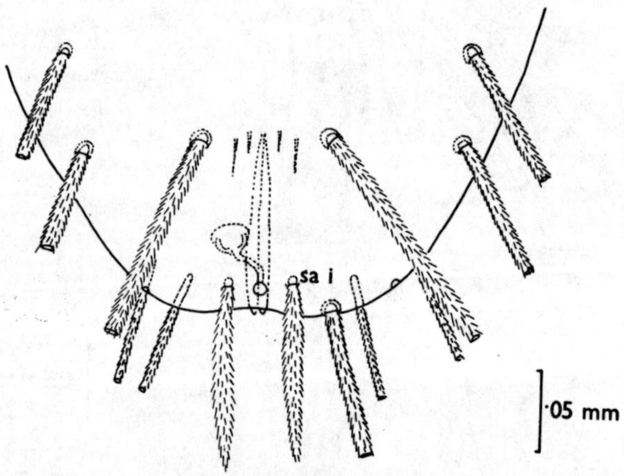

FIG. 171. *Glycyphagus bicaudatus* Hughes. ♀. Dorsal view of posterior end of body. Internal sacral seta: *sa i*.

LEPIDOGLYPHUS DESTRUCTOR (Schrank, 1781)

Acarus destructor Schrank, 1781; *Acarus spinipes* Koch, 1841; *Glycyphagus anglicus* Hull, 1931; *Lepidoglyphus cadaverum* (Schrank, 1781) sensu Türk and Türk, 1957, *Glycyphagus destructor* (Schrank) sensu Hughes, 1961.

MALE (Fig. 172). Length of idiosoma: 350 to 500 µm.

The idiosoma is long and pear-shaped, constricted behind the fourth pair of legs. The cuticle is dull and whitish, as it is covered with minute papillae. When alive, the animal moves with a jerking gait.

The dorsal setae are stiff and densely pectinate, standing away from the surface of the body. Setae *v i* extend well beyond the tips of the chelicerae, *v e* are more posterior in position and the same distance apart as *sc i*. The scapular seta are arranged in a transverse row across the idiosoma behind legs II, *sc i* being almost the same length as *v i*. The supracoxal seta (Fig. 160) is a much branched bifid rod. Two pairs of humerals are present; d_1 are longer than d_2 which barely reach the posterior margin of the body; d_3 are posterior to d_2 and almost in linear series with d_1 and d_4. Three pairs of laterals are present, increasing in length towards the posterior end of the body. From the hind margin of the body project *sa i*, *sa e* and three pairs of postanals, one pair of which is short

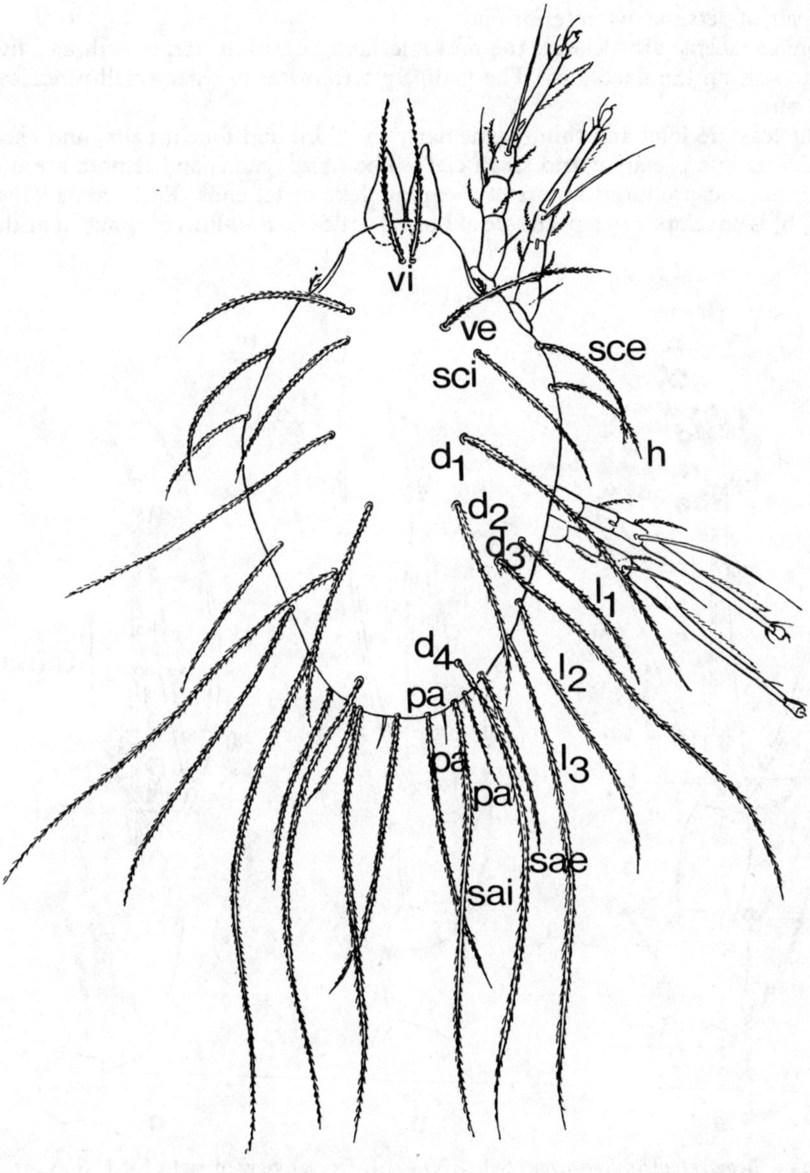

FIG. 172. *Lepidoglyphus destructor* (Schrank) ♂. Dorsal view. Setae of the idiosoma: *v e*, *v i*, *sc e*, *sc i*, *h*, d_1 to d_4, l_1 to l_3, *sa e*, *sa i*, *p a*.

and smooth. Setae d_3, d_4, l_3 and *sa i* are the longest setae arising from the idiosoma.

Ventrally, the apodemes of legs I meet to form a short sternum and apodemes II are also well-developed; apodemes III and IV are much reduced, their function of muscle attachment being taken over by epimerites II, each of which has a strong anteriorly-directed process. The genital opening lies between coxae III, its anterior end marked by a triangular plate; two pairs of genital setae lie on either side and a third pair arises behind the posterior edge of the genital opening. The anal opening extends to the posterior margin of the body, it has one pair of setae at its anterior end.

The chelicerae are slender, the movable limb bears four large teeth and five are present on the fixed arm. The pedipalp terminates in three small processes (Fig. 7b).

The legs are long and thin, particularly the third and fourth pairs, and each terminates in a pretarsus and small claw. The tibiae, genua and femora are not expanded and produced into a thin rim at their distal ends. Each tarsus (Fig. 173a, b) is enveloped in a pectinate subtarsal scale—*wa*—which is attached at the

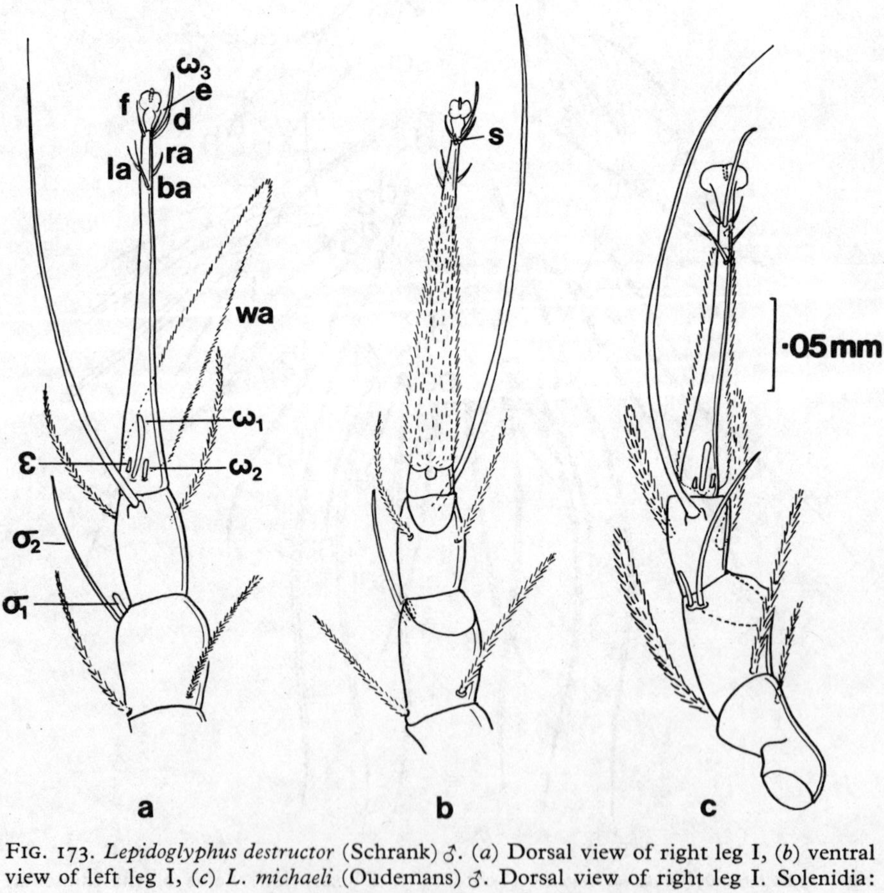

FIG. 173. *Lepidoglyphus destructor* (Schrank) ♂. (*a*) Dorsal view of right leg I, (*b*) ventral view of left leg I, (*c*) *L. michaeli* (Oudemans) ♂. Dorsal view of right leg I. Solenidia: ω_1 to ω_3, σ_1, σ_2; famulus: ε; setae: *d*, *e*, *s*, *ba*, *la*, *ra*; tarsal scale: *wa*.

base. When in its natural position, this scale gives the tarsus a ciliated appearance, but in mounted specimens it tends to get shifted to one side. At the distal end of the tarsus, the pretarsus is encircled by three setae (*d, e, f*), three small spines and a solenidion omega$_3$, and close behind these are *la, ba* and *ra*. At the base of the tarsus, omega$_1$, omega$_2$ and the famulus arise close together, omega$_1$ is a curved rod and omega$_2$ is less than half the length of omega$_1$. On genu I sigma$_1$ is expanded distally, sigma$_2$ is more than four times longer than sigma$_1$ and the ventral setae of the genu and tibia are pectinate. On legs III and IV (Fig. 174a), the ventral seta *kT* does not arise from the edge of the arthrodial membrane.

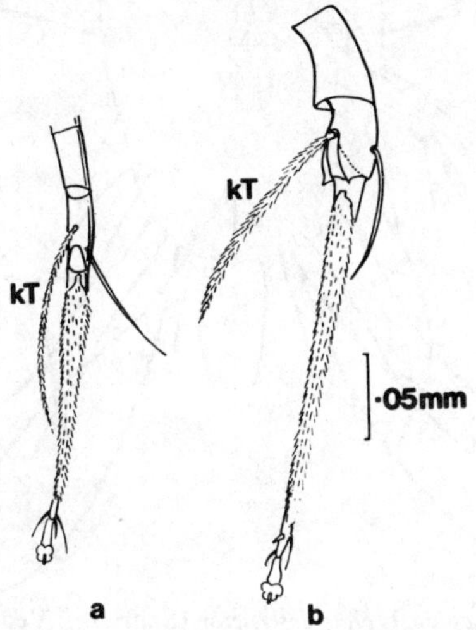

FIG. 174. Ventral view of right leg IV of (*a*) *Lepidoglyphus destructor* (Schrank) and (*b*) *L. michaeli* (Oudemans) ♀. Tibial seta: *kT*.

FEMALE (Fig. 175). Length of idiosoma: 400 to 560 μm.

This resembles the male in shape and arrangement of setae. The genital folds meet for the greater part of their length and a crescentic epigynium covers their anterior end; the posterior pair of genital setae is on a level with the hind end of the genital opening, between apodemes III and IV. The bursa copulatrix is a short tube with a partially lobed edge (Fig. 177). The anus extends to the posterior margin of the body and two pairs of setae—the outer longer than the inner—are inserted on either side of its anterior end.

HYPOPUS (Fig. 176, 25). Length of hypopus and case: about 350 μm.

The hypopus is inactive and remains enclosed within the protonymphal cuticle. It has an oval, colourless body with reduced legs (Fig. 25) and a distinct transverse suture across the dorsal surface, dividing the body into propodosoma and hysterosoma. The apodemes of legs I and II are feebly sclerotized and the rudiment of the genital opening is between legs IV.

On legs I to III, the claw is as long as the tarsus, on leg IV it is much shorter. A long solenidion arises from the base of tarsus I and may represent omega$_1$; this is represented by a short solenidion on tarsus II.

LARVA (Fig. 178).

Each tarsus is enveloped in a scale as in the adult. The coxal rods are small.

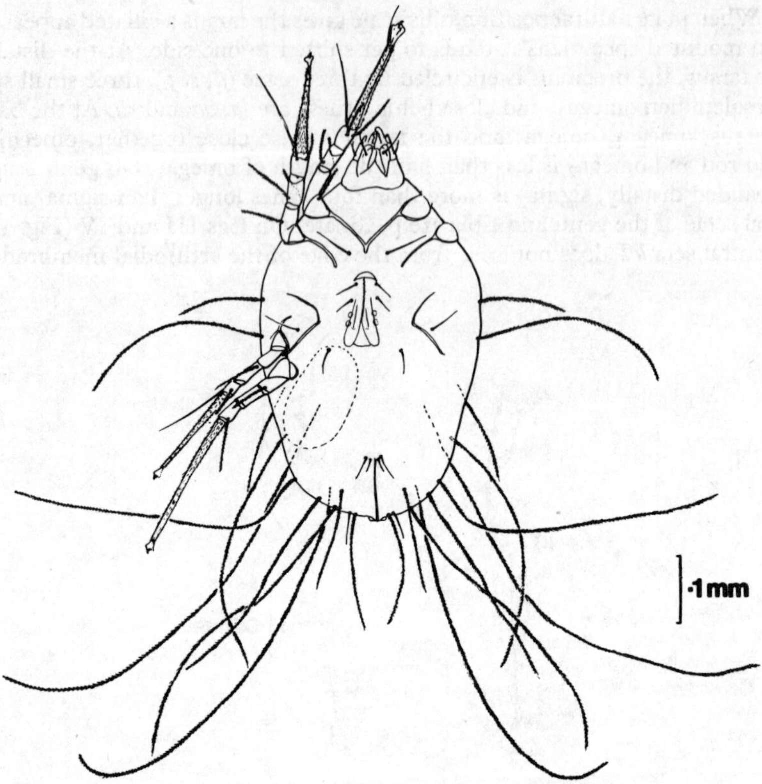

FIG. 175. *Lepidoglyphus destructor* (Schrank). Ventral view of ♀.

BIOLOGY

This is one of the commonest species of stored product mite and is frequently found in association with *Acarus siro* and *Cheyletus eruditus* or *C. malaccensis*. Sheals (1956) found it in soil samples taken from old grassland at Bellahouston Park, Glasgow and it has also been recorded from soil in the agricultural zones of Canada, Japan and the U.S.S.R. (Zachvatkin 1941, Sinha 1963, Sinha and Wallace 1966). In the midland counties of England, Griffiths (1960) did not find it in grassland, but it was always present in the stacks of grain, straw and hay standing in the open field or in a permanent stackyard. Sinha (1968b) is of the opinion that, in Japan, infestation of granaries originates from field straw stacks and, as in this country, the mite is more common in samples taken from floor debris than from the stored grain. In Manitoba and Saskatchewan, temperatures inside granaries may fall to about −18°C and *L. destructor* with certain other stored product mites, will recover after exposure to this temperature (Sinha 1964b). The hypopal stage is less susceptible to such low temperatures than the adults, nymphs and larvae (Ushatinskaya 1954).

In addition to oats, rye, wheat and barley, *L. destructor* has also been found in linseed, rice, dried fruit, sugar beet seed in Poland (W. Chmielewski 1969), on dried calves stomachs (J. S. Burton coll.), in dead insects and on dried mammal skins, in rodent and bumble bee nests. Rack (1968) records it from mattress stuffing taken from a damp house and Brady (1970) found that it could increase to large populations on poultry droppings in broiler houses.

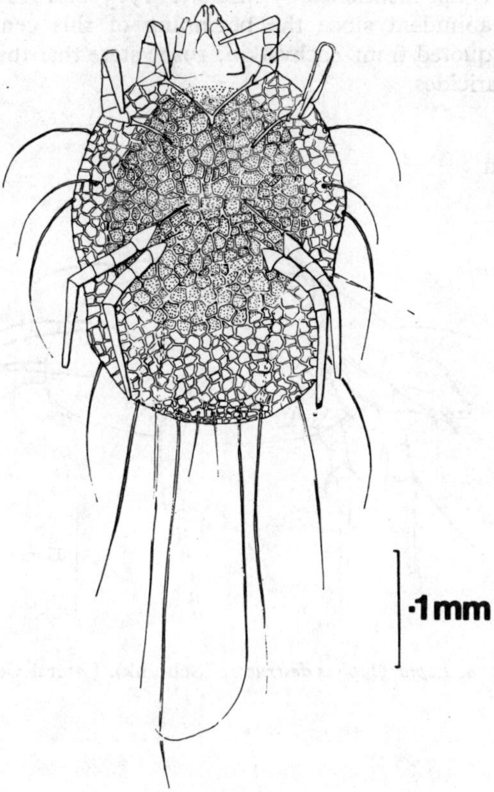

FIG. 176. *Lepidoglyphus destructor* (Schrank). Hypopus enclosed in protonymphal cuticle. Ventral view.

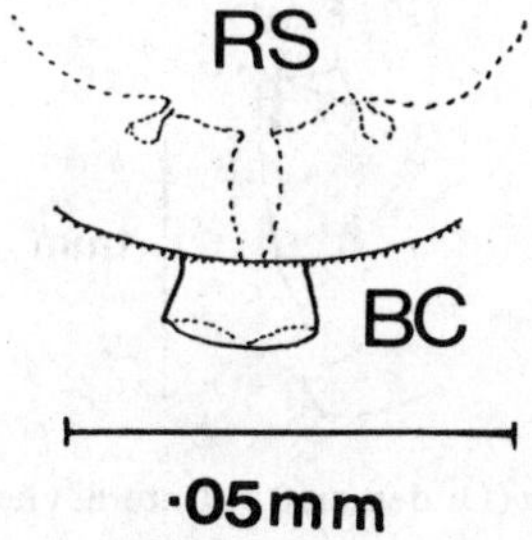

FIG. 177. *Lepidoglyphus destructor* (Schrank). Genitalia of ♀. RS receptaculum seminis, BC bursa copulatrix.

According to Bollaerts and Breny (1951), *L. destructor* is strictly mycophagous and Sinha (1968a) found that it was able to survive when kept with cultures of 12 of the 25 species of stored product fungi that he used in his experiments; *Acarus siro*, however, could survive on 17 species. Analysis of the gut contents showed that different fungi are favoured by the two species, so that it is doubtful whether competition for food occurs in stored grain.

L. destructor is not mentioned by Michael (1903) and it is possible that it has become more abundant since the beginning of this century. Smirnov and Polejaev 1936 (quoted from Zachvatkin, 1941) state that this species has a high resistance to acaricides.

DISTRIBUTION

Cosmopolitan.

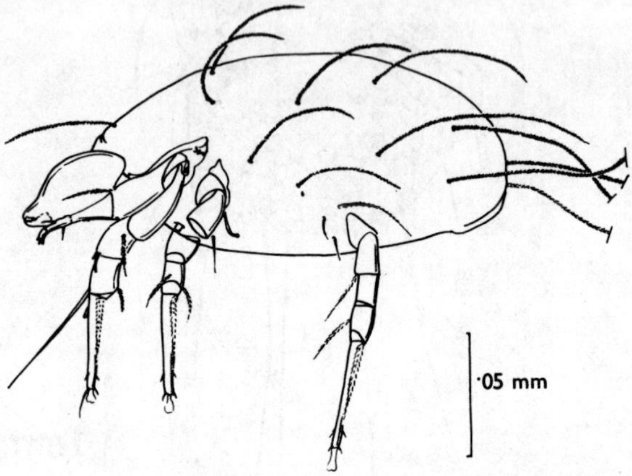

FIG. 178. *Lepidoglyphus destructor* (Schrank). Lateral view of larva.

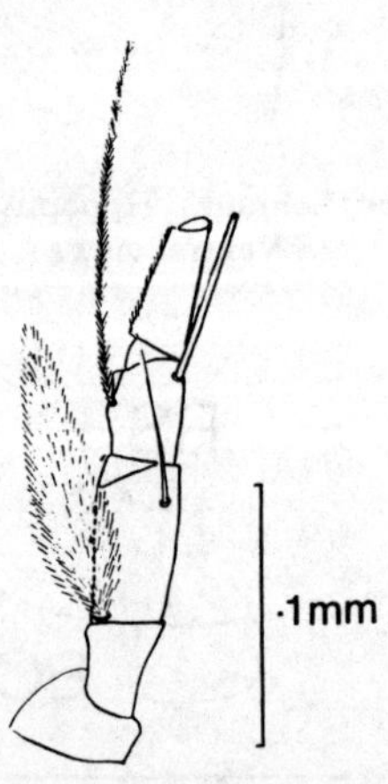

FIG. 179. *Lepidoglyphus michaeli* (Oudemans) ♀. Lateral view of basal region of right leg III.

LEPIDOGLYPHUS FUSTIFER (Oudemans, 1903)

Glycyphagus fustifer Oudemans, 1903

This was found originally on furniture in a house. Zachvatkin (1941) also found it in large numbers on rye in the store of a collective farm. Žďárková (1967) collected a few individuals from cotton, poppy and turnip seed.

DISTRIBUTION

Germany, Czechoslovakia, U.S.S.R.

LEPIDOGLYPHUS MICHAELI (Oudemans, 1903)

Glycyphagus michaeli Oudemans, 1903.

MALE. Length of idiosoma: 450 to 550 µm.

This resembles *L. destructor* in general appearance, but can be distinguished by its larger size and extreme rapidity of movement. The chaetotaxy of the idiosoma is the same as in the previous species, but the setae themselves are more densely pectinate. In setal length, the most obvious difference is that *sc i* are much longer than *v i* instead of being of about the same length.

The leg segments, particularly the tibia and genu of legs IV (Fig. 174b), are expanded distally into a thin transparent flange enveloping the base of the succeeding segment. The ventral setae of the tibiae are more 'bushy' than in *L. destructor* (Fig. 173c) and, in both sexes, the ventral seta *nG* of legs III (Fig. 179) is widened to form a 'furry' scale. On tibiae III and IV, the distal arthrodial membrane extends as far back as the base of *kT* as in *G. ornatus* and the cuticle on either side is produced into a thin plate, so that *kT* arises at the base of a deep cleft (Fig. 174b).

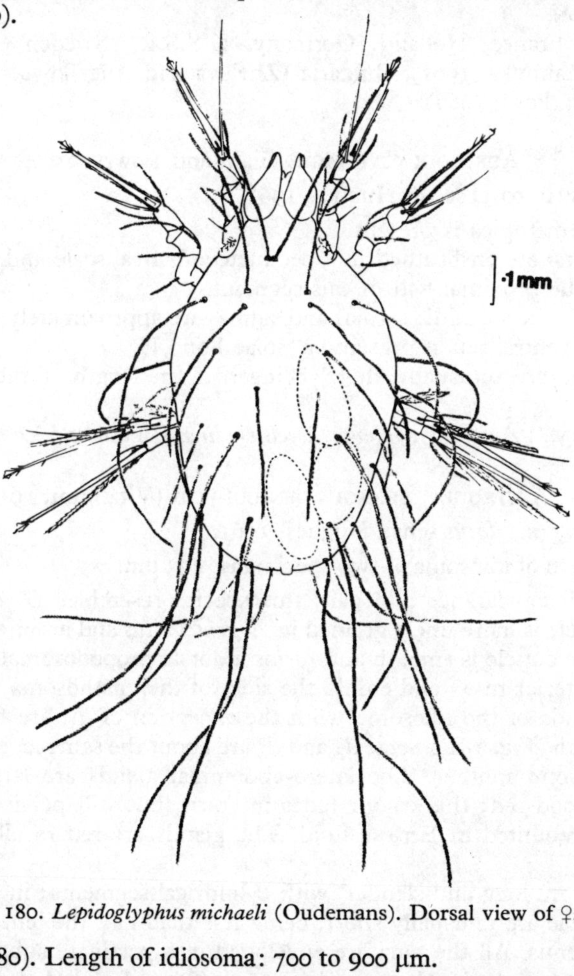

FIG. 180. *Lepidoglyphus michaeli* (Oudemans). Dorsal view of ♀.

FEMALE (Fig. 180). Length of idiosoma: 700 to 900 µm.

The genital opening is further forward than in *L. destructor*, its posterior edge on a level with the anterior ends of apodemes III, and the posterior pair of genital setae far behind it; the anterior end of the opening is covered by a crescentic epigynium. The bursa copulatrix is a short inconspicuous tube.

HYPOPUS. Length of idiosoma: about 260 µm.

According to Michael (1901) the hypopus remains enclosed within the protonymphal cuticle, which becomes dried and covered with a reticulate pattern as in *L. destructor*. The hypopus itself has a pear-shaped body with rudimentary limbs; it is without a sucker plate and is only capable of slight movement.

BIOLOGY

This species is not so common in stored food as the previous one, but under natural conditions it is more widely distributed. It is abundant in dried vegetable matter, on hay and fodder in stables (Michael 1901) and in rodent and insectivore nests (Vysotzkaya, 1961). It has also been found in brewer's yeast and on moist grain. Zachvatkin (1941) states that it is more sensitive to desiccation than is *L. destructor*, but in granaries in the U.S.S.R., it may be the dominant species.

DISTRIBUTION

England, France, Holland, Germany, U.S.S.R., Sweden (Sellnick 1958), Hungary (Mahunka, 1961), Bulgaria (Zheleva and Angelkova, 1964), Czechoslovakia (Ždárková, 1967).

AUSTROGLYCYPHAGUS (Fain and Lowry, 1974)

Glycyphagus (in part) sensu Hughes, 1961.

No crista metopica is present.

All the tarsi are ensheathed in a pectinate subtarsal scale and setae *ba*, *la* and *ra* arise in the proximal half of this segment.

On genu I, the solenidia $sigma_1$ and $sigma_2$ are approximately equal in length.

Only one ventral seta is present on tibiae I and II.

The tibiae are unusually short, being half the length of the corresponding genua.

Type species: *Austroglycyphagus weelawadjiensis* Fain and Lowry, 1974.

AUSTROGLYCYPHAGUS GENICULATUS (Vitzthum, 1919)

Glycyphagus geniculatus sensu Hughes, 1961.

MALE. Length of idiosoma of two specimens: 433 µm.

In general appearance and gait, this species resembles *G. domesticus*. The surface cuticle is more finely grained in *A. geniculatus* and around the base of the *v i* setae, the cuticle is smooth and forms a dorsal propodosomal shield. The *v e* setae are anterior to *v i* and enfold the sides of the gnathosoma. All the setae of the dorsal side of the idiosoma, with the exception of d_1, are finely pectinate; d_1 are smooth (Fig. 181). Setae d_2 and d_3 are about the same length and in linear series with one another. The latero-abdominal glands are large and, in life, coloured blood-red; this colour fades in spirit, but still persists for a time in specimens mounted in Berlese fluid. The glands are red in all developmental stages.

The legs are long and slender, with cylindrical segments; in all four pairs of legs the tibiae are unusually short, being less than half the length of the corresponding genua. All the tarsi are enveloped in a pectinate subtarsal scale as in the genus *Lepidoglyphus*, but the chaetotaxy of tarsi I and II shows considerable differences. On tarsus I (Fig. 182) $omega_1$ is unusually long and curved and is

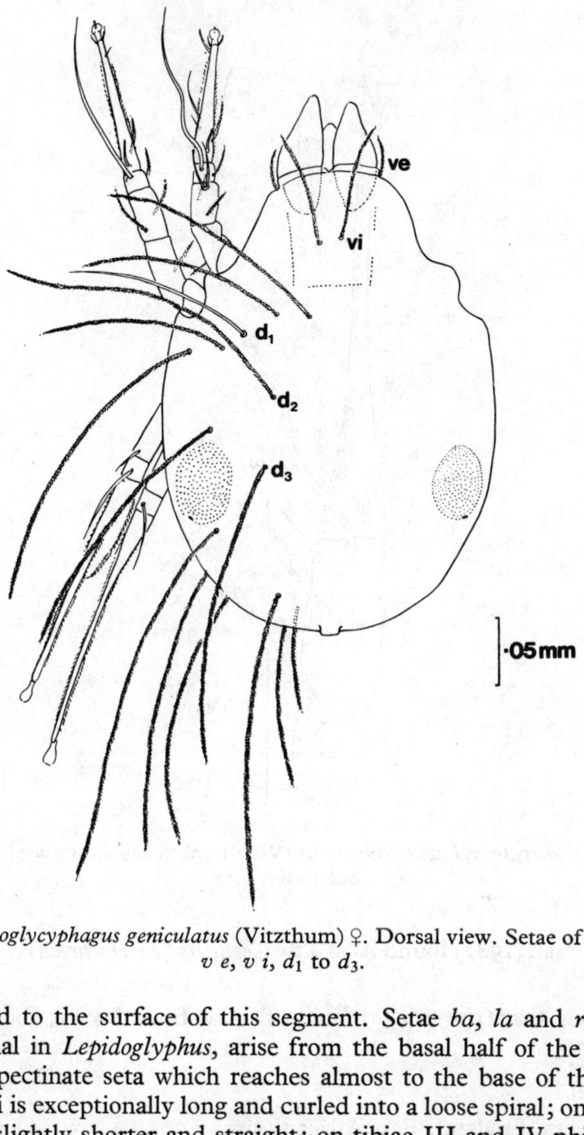

FIG. 181. *Austroglycyphagus geniculatus* (Vitzthum) ♀. Dorsal view. Setae of the idiosoma: *v e*, *v i*, d_1 to d_3.

closely applied to the surface of this segment. Setae *ba*, *la* and *ra* which are almost terminal in *Lepidoglyphus*, arise from the basal half of the segment, *la* being a long pectinate seta which reaches almost to the base of the pretarsus. On tibia I, phi is exceptionally long and curled into a loose spiral; on tibia II this solenidion is slightly shorter and straight; on tibiae III and IV phi is less than one half the length of the corresponding tarsus. As in the genus *Ctenoglyphus*, seta *hT* is missing from tibiae I and II. On genu I, $sigma_1$ and $sigma_2$ are subequal in length.

FEMALE (Fig. 181). Length of idiosoma of four specimens: 430 to 500 μm.

This closely resembles the male. The genital opening is placed well in front of the posterior genital setae. The bursa copulatrix is a short wide tube.

BIOLOGY

A. geniculatus was found by G. E. Woodroffe (1954) in birds' nests collected near Slough, Berks, and also in grain bin residues. Vitzthum's specimens were found in the nest of a bee *Xylocopa* (*Koptorthosoma*) *nigrita* F. d'Amani in East

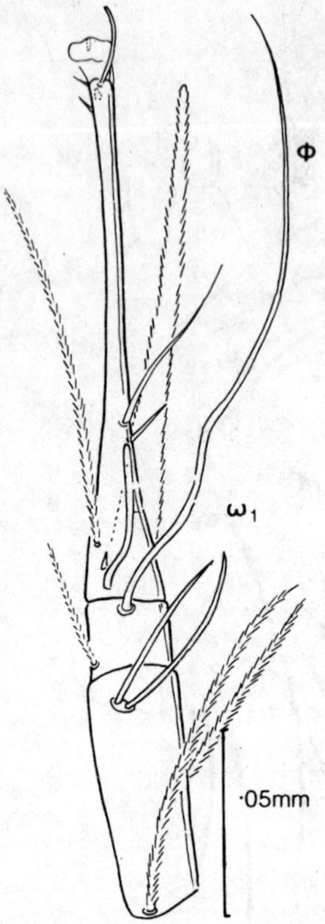

FIG. 182. *Austroglycyphagus geniculatus* (Vitzthum) ♀. Dorsal view of right leg I. Solenidia: ω, Φ.

Africa. Cooreman (1942) found it on a fly *Ceratitis* (*Trirhithrum*) *coffeae* collected in Zaire.

It is possible that *Glycyphagus latipes* Berlese from Java is the same species as *A. geniculatus*.

DISTRIBUTION

Slough, Berks, England; East Africa, Zaire.

BLOMIA Oudemans, 1928

Without a dorsal shield or crista metopica.

Setae *v e* and *v i* arise close together on the vertex.

Without a pectinate subtarsal scale.

No claws are present.

Only one solenidion is present on genu I.

Male and female genital openings lie between coxae IV.

Type species: *Glycyphagus tjibodas* Oudemans, 1911.

BLOMIA FREEMANI Hughes, 1948

MALE (Fig. 183). Length of idiosoma: 320 to 350 μm.

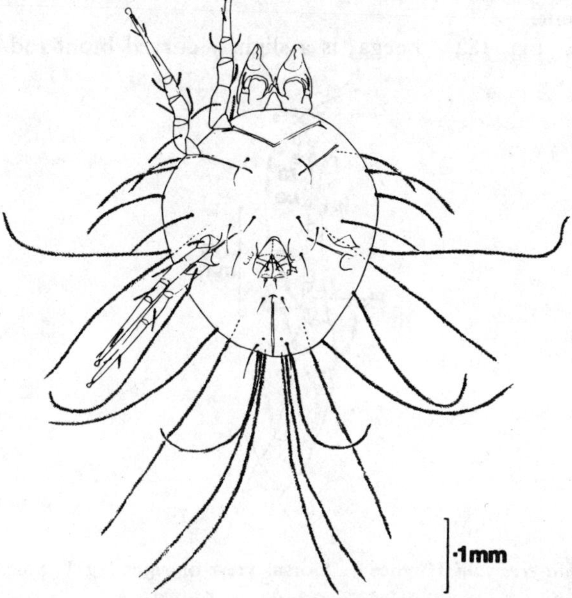

FIG. 183. *Blomia freemani* Hughes. Ventral view of ♂.

The body is almost globular in shape, widest between the second and third pairs of legs and slightly tapering towards the posterior end. The cuticle is colourless and rough, covered with innumerable fine projections. When alive, the animal moves with quick, bustling haste, resembling superficially the protonymph of *Glycyphagus domesticus*. No dorsal propodosomal shield or crista metopica is present. Ventrally, the apodemes are slender sclerites, placed obliquely in the body, those of legs I only just meeting in the midline.

Standing away from the body are the densely pectinate setae, the anterior ones gently curved. The two pairs of verticals arise close together and extend forward almost to the tips of the chelicerae, *v i* being posterior to *v e*. The supracoxal seta is a richly branched structure. Setae *sc i*, *sc e* and *h i* arise almost in the same transverse line; *h e* and d_1 also originate in line with one another and are of about the same length. Setae d_2 are close together, arising equidistant from d_1 and d_3, they are much shorter than any of the other dorsal setae and are only sparsely pectinate; d_3, d_4, l_1 to l_3, *sa i* and *sa e* are longer setae, the posterior ones exceeding the idiosoma in length.

The genital opening is between coxae IV and concealed by obliquely placed genital folds, on the internal side of which are the genital sense organs. It is encircled by three pairs of genital setae, the members of the third pair being close together immediately behind the opening. The penis is a curved tube supported by two basal sclerites. The anus extends to the posterior margin of the body; two pairs of smooth anal setae arise on either side of its anterior and posterior end and a long pair of pectinate postanal setae projects from the hind end of the body.

The chelicerae are large and well-sclerotized, with two teeth on the movable limb and two large and two small on the fixed.

The tarsi of all the legs are long and slender, exceeding the combined length of the two preceding segments; each terminates in a small lobed pretarsus. Claws are absent.

On tarsus I (Fig. 184), omega$_3$ is a slightly curved blunt rod exceeding the

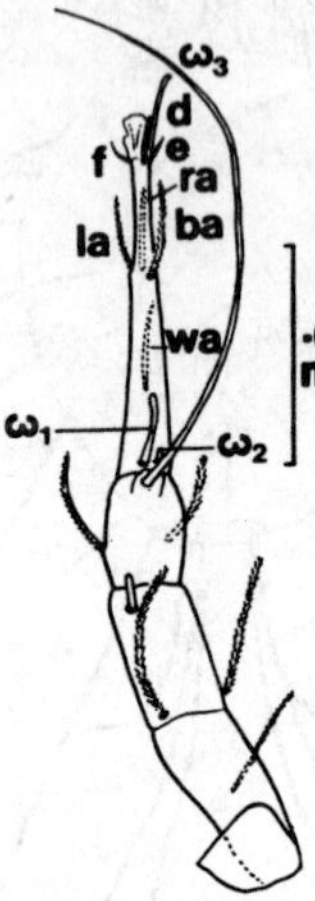

FIG. 184. *Blomia freemani* Hughes ♂. Dorsal view of right leg I. Solenidia: ω_1 to ω_3; setae: *d, e, f, ba, la, ra, wa*.

pretarsus in length, *d*, *e* and *f* are short setae; on the ventral side of the distal end of the tarsus are three small spines. The seta *ba* is nearer to the apex than to the base of the tarsus, and arises at the same level as *la* and *ra*; all three setae, together with *wa*, are pectinate. Omega$_1$ is a rod with a slightly expanded end, whose base is on the same level as a short omega$_2$. The famulus is not visible. On tarsus II, omega$_1$ is shorter and the base of *ba* is nearer to it. The ventral setae of genua and tibiae I and II are all pectinate. Phi is extremely long and slender on all the legs, extending beyond the end of the pretarsus; on leg IV (Fig. 185) it arises from the middle of the tibia. Only one solenidion arises from genu I and none from genua II and III. Tarsus IV is narrow and hinged at an angle to the apex of the tibia by a large arthrodial membrane.

FEMALE (Fig. 186). Length of idiosoma: 440 to 520 µm.

Except for its larger size, the female closely resembles the male. The genital opening is covered by obliquely-placed folds (Fig. 187) on the underside of which are two pairs of genital sense organs; three pairs of genital setae project from the cuticle on either side of it. The anus is near the posterior margin of the body and is associated with six pairs of anal setae, of these two pairs lie on either side of its anterior end and four pairs at the posterior (Fig. 188). The two external posterior pairs are longer than the others and distinctly pectinate. The bursa copulatrix is a long thin tube with a cleft end.

BIOLOGY

This mite was found in large numbers in a dock silo, on Manitoba wheat in association with *Acarus siro* and *Lepidoglyphus destructor*. Smaller numbers were present in bagged Manitoba wheat in a warehouse, and among split wheat in three sections of one large flour mill, where a small permanent population seemed to be established (O'Farrell and Butler 1948). No hypopus has been found.

DISTRIBUTION

Northern Ireland; Buckinghamshire, England.

Zachvatkin (1936) has also described *B. thori* from the Smolensk region of the U.S.S.R where it was found on flax seed, and *B. kulagini* from wheat stored in Moscow granaries. *B. khalilovae* has also been found in granaries and in millet and barley (Zachvatkin, 1949). An undescribed species of *Blomia* has also been found associated with floor dust in Barbados, W.I.

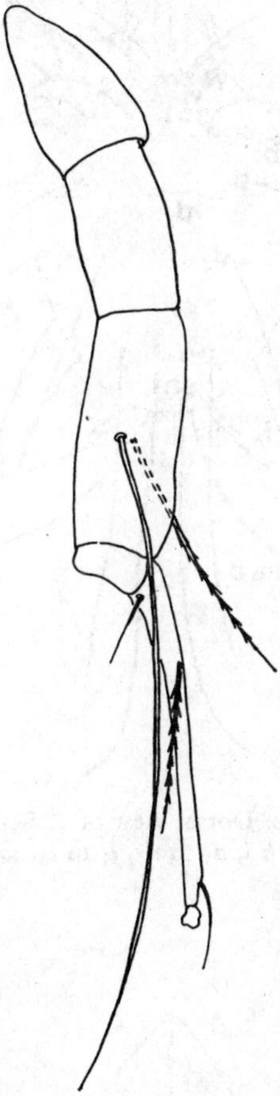

FIG. 185. *Blomia freemani* Hughes ♂. Dorsal view of leg IV.

CTENOGLYPHINAE Zachvatkin, 1941

The peripheral idiosomal setae are widely pectinate, bi-pectinate or like leaves forming a fringe encircling the idiosoma.

The tarsi are not slender, often with a dorsal keel.

Only one ventral seta (*gT*) is present on tibiae I and II.

No anal or tarsal suckers are present in the male.

The penis in the male is often very long.
The cuticle is usually rough or covered with fine warts.
No hypopus has been recorded.

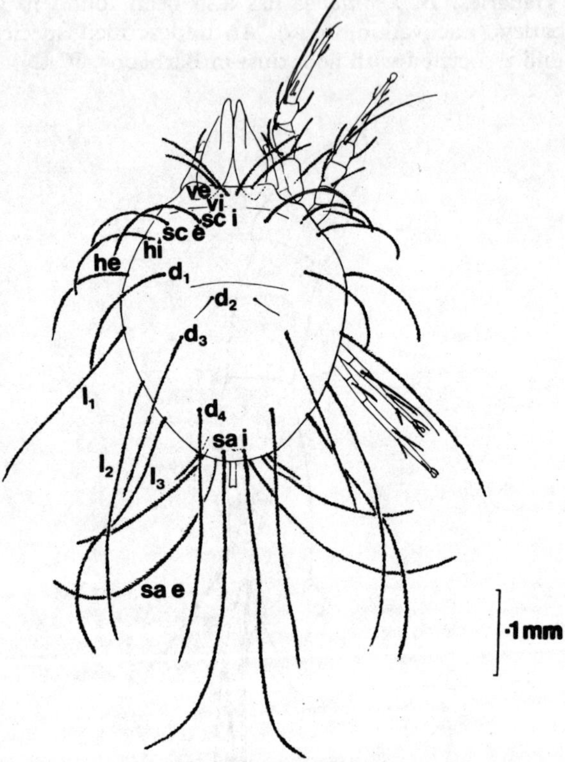

FIG. 186. *Blomia freemani* Hughes. Dorsal view of ♀. Setae of the idiosoma: *v e*, *v i*, *sc e*, *sc i*, *h e*, *h i*, d_1 to d_4, l_1 to l_3, *sa e*, *sa i*.

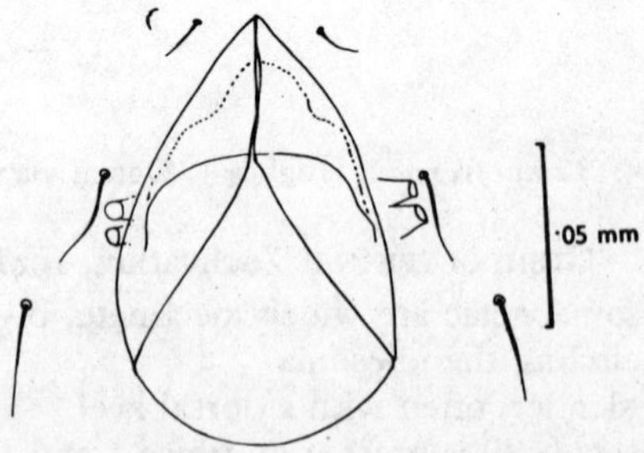

FIG. 187. *Blomia freemani* Hughes ♀. Genital opening.

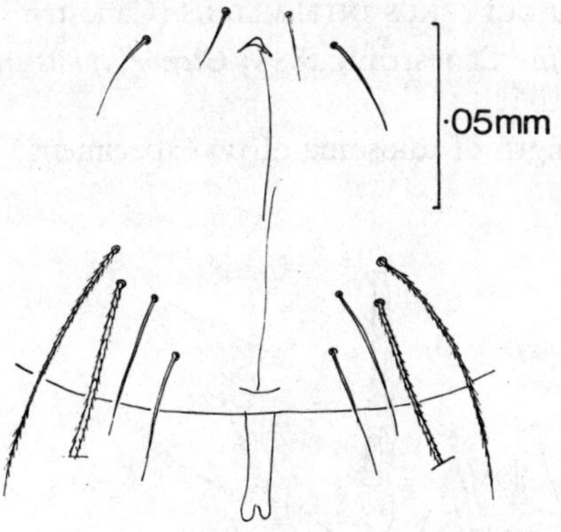

FIG. 188. *Blomia freemani* Hughes ♀. Anus and bursa copulatrix.

CTENOGLYPHINAE

Key to the adults

1. Males and females alike. Body setae pectinate and ribbon-shaped (Fig. 190d). Penis of male short............*Diamesoglyphus intermedius*, p. 165

—. Sexually dimorphic. Body setae distinctly bipectinate (Fig. 190a). Penis of male long..*Ctenoglyphus* 2

2. Body setae leaf-shaped with barbs joined by a transparent membrane thickened at the periphery........................*C. palmifer*, p. 173

—. Body setae narrower, barbs of setae free............................3

3. Barbs of setae straight, each arising at an acute angle to the main axis in the female. Setae d_1 almost equal to d_2 in the male.....*C. plumiger*, p. 169

—. Barbs of setae arising at right angles to the main axis of the seta and bent in the female. Seta d_1 twice as long as d_2 in the male....*C. canestrinii*, p. 171

DIAMESOGLYPHUS Zachvatkin, 1941

Ctenoglyphus (part) sensu Hughes 1961.

The genus *Diamesoglyphus* was included by me (1961) in the genus *Ctenoglyphus*. In this work, I am recognizing it as a distinct genus as I consider that the lack of sexual dimorphism and uniform appearance of the idiosomal setae are characters of generic importance.

No sexual dimorphism occurs and the idiosoma is rounded in both sexes.

The cuticle is finely roughened.

The setae arising from the dorsal surface of the idiosoma are all alike: narrow, flat and of varying widths, finely pectinate along the edges.

Only one solenidion arises from genu I.

In the male, the penis is short.

Type species: *Glycyphagus intermedius* Canestrini, 1888.

DIAMESOGLYPHUS INTERMEDIUS (Canestrini, 1888)

Glycyphagus intermedius Canestrini, 1888; *Ctenoglyphus intermedius* (Canestrini, 1888) sensu Hughes, 1961.

MALE (Fig. 189). Length of idiosoma of two specimens: about 400 µm.

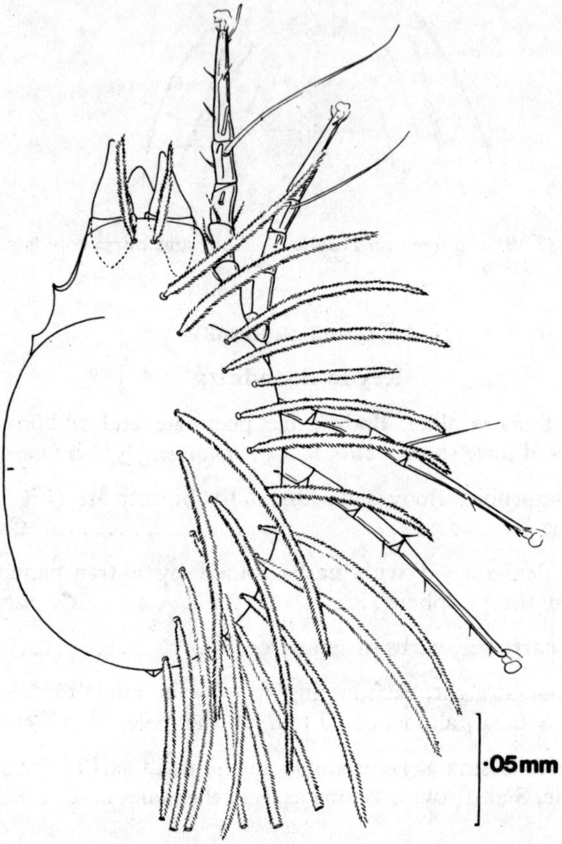

FIG. 189. *Diamesoglyphus intermedius* (Canestrini). Dorsal view of ♂.

The body is pale brown in colour and shaped like that of a *Tyrophagus*, the chelicerae clearly visible from above and a transverse groove extending across the dorsal side of the body behind the second pair of legs. The cuticle is rough and covered on the dorsal surface with fine projections which are more widely spaced than in *Glycyphagus*. Ventrally, the cuticle is smooth. The apodemes of legs I meet to form a short sternum, those of legs II to IV are free. The penis is a short tube between coxae IV and genital folds and sense organs are not visible.

The setae arising from the dorsal surface are flattened like ribbons and bipectinate, with additional barbs arising from the main axis near its base (Fig. 190d). The peripheral setae encircle the body like a halo, whilst the central ones are almost in linear series and project at right angles from the surface; the anterior are slightly wider than the more posterior ones.

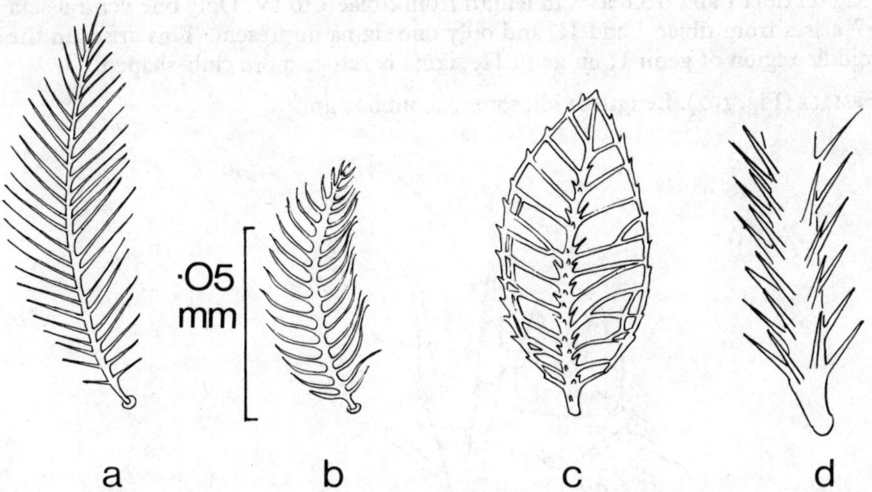

FIG. 190. Setae of (*a*) *Ctenoglyphus plumiger* (Koch), (*b*) *C. canestrinii* (Armanelli), (*c*) *C. palmifer* (Fumouze and Robin), (*d*) *Diamesoglyphus intermedius* (Canestrini).

The legs are relatively long and slender, each terminates in a pretarsus which bears a vestigial claw. On tarsi and tibiae I and II a longitudinal ridge is present on the dorsal surface (Fig. 191c). On tarsus I omega$_1$ is a long straight rod

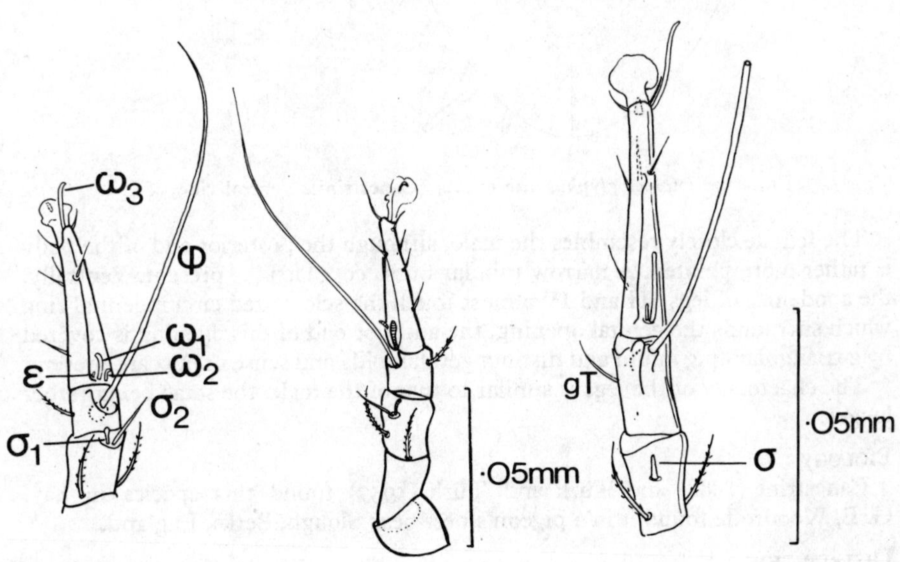

FIG. 191. (*a*) *Ctenoglyphus plumiger* (Koch) ♀. Dorsal view of right leg I. (*b*) *C. canestrinii* (Armanelli) ♀. External view of left leg I. (*c*) *Diamesolygphus intermedius* (Canestrini). Dorsal view of right leg I. Solenidia: ω_1, ω_2, ω_3, Φ, σ; famulus: ε; tibial seta: *gT*.

lying close to $omega_2$, $omega_3$ extends beyond the end of the tarsus. Phi is very long on tibia I and decreases in length from tibiae I to IV. Only one ventral seta *gT* arises from tibiae I and II, and only one sigma in present. This arises in the middle region of genu I; on genu II, sigma is rather more club-shaped.

FEMALE (Fig. 192). Length of idiosoma: about 600 μm.

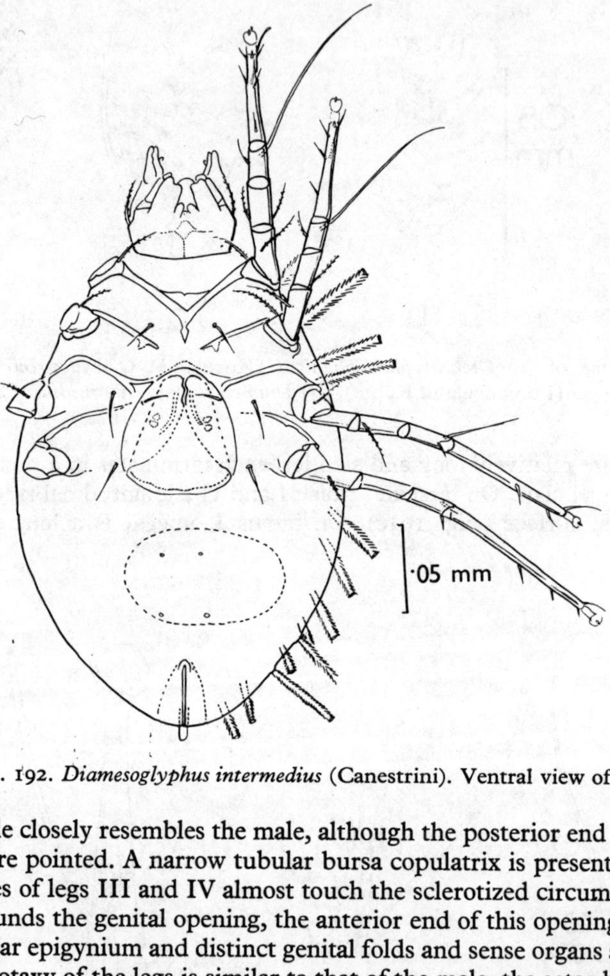

FIG. 192. *Diamesoglyphus intermedius* (Canestrini). Ventral view of ♀.

The female closely resembles the male, although the posterior end of the body is rather more pointed. A narrow tubular bursa copulatrix is present. Ventrally, the apodemes of legs III and IV almost touch the sclerotized circumgenital ring which surrounds the genital opening, the anterior end of this opening is covered by a triangular epigynium and distinct genital folds and sense organs are present.

The chaetotaxy of the legs is similar to that of the male, the setae being rather longer.

BIOLOGY

Canestrini (1888) and Türk and Türk (1957) found this species in hay. G. E. Woodroffe found it in a pigeon's nest near Slough, Berks, England.

DISTRIBUTION

Italy, England, Germany.

Ctenoglyphus hughesi Attiah, 1969 which was found in flour in Cairo, should also be included in this genus as it has little sexual dimorphism and the idiosomal setae are similar in length to those of *D. intermedius*.

CTENOGLYPHUS Berlese, 1884

Glycyphagus (in part) sensu Michael, 1901.

Sexual dimorphism is usually pronounced, the male being smaller than the female, with a rounded body and less strongly pectinate setae. The idiosoma is flattened in the female and overhangs the gnathosoma. In some, but not all species, a transverse groove separates the propodosoma from the hysterosoma.

The cuticle is rough and, in the female, covered with irregular projections.

The setae arising from the margin of the idiosoma are bipectinate, sometimes leaflike in shape.

Both $sigma_1$ and $sigma_2$ are present on genu I.

In the male, the penis is exceptionally long.

Type species: *Acarus plumiger* Koch, 1835.

CTENOGLYPHUS PLUMIGER (Koch, 1835)

Acarus plumiger Koch, 1835.

MALE (Fig. 193). Length of idiosoma: 190 to 200 μm.

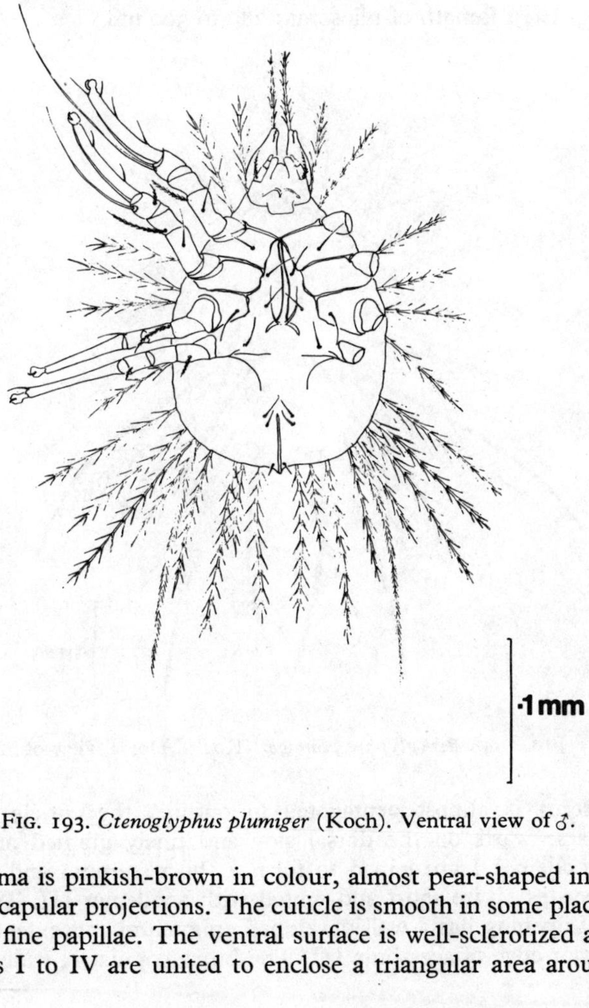

FIG. 193. *Ctenoglyphus plumiger* (Koch). Ventral view of ♂.

The idiosoma is pinkish-brown in colour, almost pear-shaped in outline and without any scapular projections. The cuticle is smooth in some places, in others covered with fine papillae. The ventral surface is well-sclerotized and the apodemes of legs I to IV are united to enclose a triangular area around the long curved penis. The penis is supported on a triangular base lying between apodemes IV.

All the setae arising from the dorsal side of the body are bipectinate and inserted deeply into the body; in addition, the main axis of each seta is also covered with minute projections. The setae arising from the central region of the idiosoma, i.e., d_3 and d_4 are particularly long and pectinate. Setae d_1 are almost equal in length to d_2. In life, these dorsal setae radiate from the body.

The legs are long and thick, tarsi I and II being distinctly keeled on their dorsal sides (Fig. 191a). Each terminates in a pretarsus and claw, the pretarsus arising from a ventral indentation at the end of the tarsus. Omega$_1$ arises from a groove at the base of the keel and is flanked by omega$_2$ and a very small famulus; omega$_3$ enfolds the pretarsus and the remaining tarsal setae are short and fine. On tibia I, phi is long and thick and arises from a deep cleft in the side of the tibia; it becomes progressively shorter on tibiae II and III and is a small rod-shaped structure on tibia IV. On genu I, sigma$_2$ is much longer than sigma$_1$ and sigma$_1$ is expanded distally. Only one ventral seta arises from tibiae I and II, but both are present on genua I and II.

FEMALE (Fig. 194). Length of idiosoma: 280 to 300 μm.

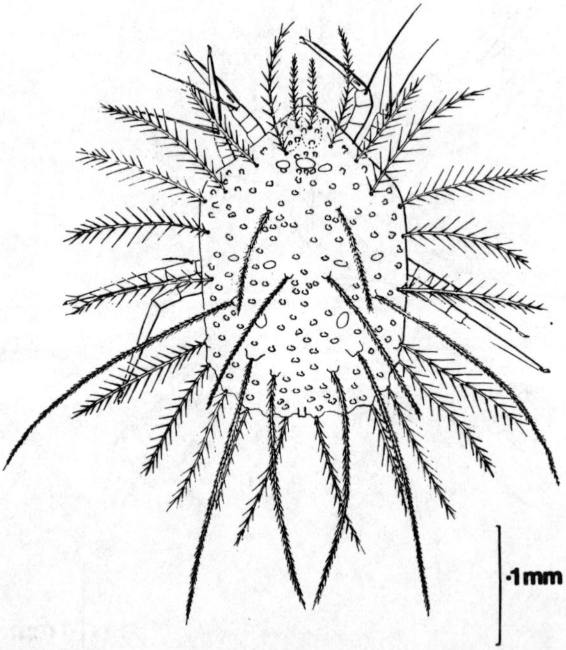

FIG. 194. *Ctenoglyphus plumiger* (Koch). Dorsal view of ♀.

The idiosoma is almost pentagonal in outline, the cuticle covered with irregular coarse warts on the dorsal side and finely grained on the ventral. Ventrally, apodemes I are joined to form a short sternum and II to IV end freely; epimerites II are short and united with apodemes III. The large elongated genital opening has a well-developed epigynium and extends as far back as the posterior edge of acetabula III. The bursa copulatrix widens towards its base and is covered with fine warts. The anal slit extends to the posterior edge of the body; two pairs of anal setae arise on either side of its anterior end.

The idiosomal setae (Fig. 190a) are longer than in the male, the peripheral ones having more pronounced straight barbs which arise at an acute angle to the ventral shaft; in life, these setae are smoothly curved, so that the body appears to be outlined by a feathery calyx. The setae arising from the centre of the dorsal surface, i.e., d_1 to d_4 and *sa i*, project almost at right angles to the body and are narrow and thickly pectinate.

The legs are more slender than those of the male and phi on tibia I is not so well-developed.

LARVA

The idiosomal setae are only feebly pectinate and they become increasingly complex with each succeeding moult.

BIOLOGY

C. plumiger is widespread in distribution and is frequently found in haystacks (Griffiths, 1960). It occurs sparingly in damp oats, wheat, barley and grass seed and sometimes increases to large numbers in hay. It has also been recorded from spillage of fishmeal (O'Farrell and Butler, 1948) and in honeycombs (Michael, 1901).

DISTRIBUTION

England, Northern Ireland, Germany, Italy, France, Hungary, Holland, U.S.S.R., Australia.

CTENOGLYPHUS CANESTRINII (Armanelli, 1887)

Glycyphagus canestrinii Armanelli, 1887.

MALE. Length of idiosoma: 180 to 200 µm.

The shape of the idiosoma is similar to that of *C. plumiger* and the cuticle has a finely granular texture. *C. canestrinii* is yellowish in colour, the gnathosoma and legs tinged with pink.

Except for d_3, all the setae of the idiosoma are bipectinate, the barbs being straight, almost parallel with one another and diminishing in length towards the base of the seta. The main axis of each has a slightly rough surface and the whole seta is similar in structure to those found in the female of *C. plumiger*. Seta d_3 are almost as long as the idiosoma, narrow and finely pectinate; d_1 are twice as long as d_2.

The legs are longer in proportion to the length of the body than in the female and similar in structure to those of *C. plumiger*.

FEMALE (Fig. 195). Length of idiosoma: 300 to 320 µm.

The idiosoma is almost square in outline. The surface of the cuticle is covered with large irregular warts which stick out of the body and have unevenly lobed ends. A distinct transverse groove separates the propodosoma from the hysterosoma. A long bursa copulatrix shaped as in Fig. 196, projects from the end of the body.

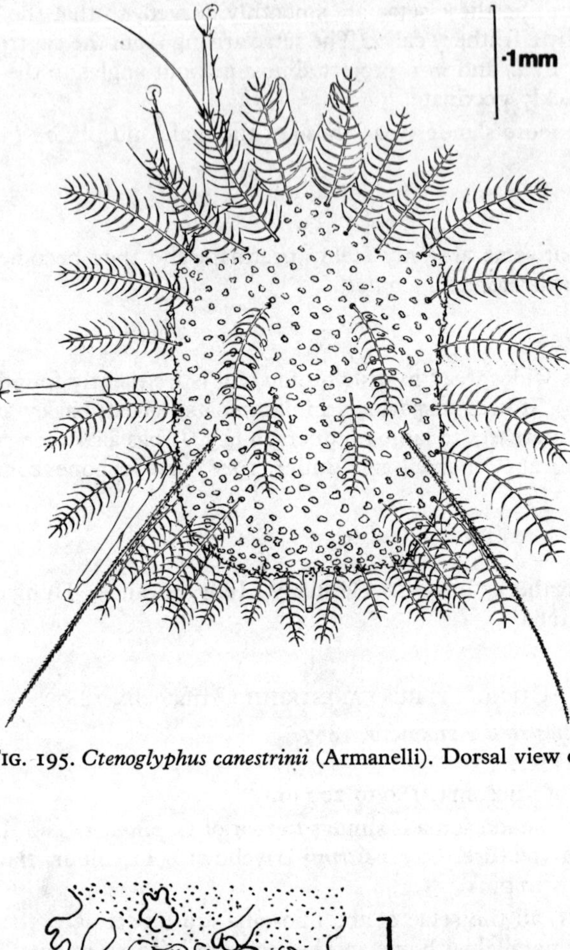

FIG. 195. *Ctenoglyphus canestrinii* (Armanelli). Dorsal view of ♀.

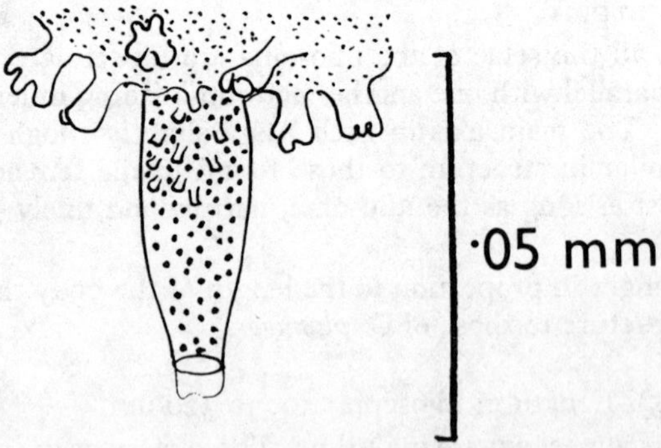

FIG. 196. *Ctenoglyphus canestrinii* (Armanelli). ♀. Bursa copulatrix.

The legs (Fig. 191b) are similar in structure to those of the previous species. The nymphs and larvae resemble the male in structure rather than the female.

BIOLOGY

This species is found in the dust of stables and hay lofts, often associated with *C. plumiger*. It is also found under hay and straw stacks, but less frequently than the previous species. Mahunka (1961) records it from soil obtained from a stable and also in a manure heap. O'Farrell and Butler, (1948) found it in spillage from a country oat mill.

DISTRIBUTION

England, Italy, Hungary, U.S.S.R., N. Ireland.

CTENOGLYPHUS PALMIFER (Fumouze and Robin, 1868)

Glycyphagus palmifer Fumouze and Robin, 1868.

MALE. Length of idiosoma: 180 to 200 μm.

The idiosoma is square, with almost parallel margins and a distinct transverse groove behind legs II. The cuticle has a granular texture and is yellowish in colour.

The setae of the idiosoma are mainly peripheral. Setae d_3 and l_3 to l_5 are narrowly and thickly pectinate, d_3 being about the same length as the idiosoma. Setae d_4 and both pairs of *sa*, are large and leaf-shaped, each has a central roughened axis from which projects a series of barbs (Fig. 190c); these barbs are sometimes branched, or they may anastomose, the two halves of the 'leaf' often being asymmetrical. These barbs are connected together by a thin transparent membrane, thickened at the edges and produced at intervals into small spines. The more anterior setae are narrow and lanceolate.

The legs lack the keel that is present on tarsi I and II of *C. canestrinii*, and on genu I, $sigma_1$ and $sigma_2$ are almost equal in length.

FEMALE (Fig. 197). Length of idiosoma: about 260 μm.

The idiosoma resembles that of the male in shape and colour. The cuticle covering the hysterosomal region is thickened to form a series of irregular shallow ridges.

The idiosoma is outlined by a halo of thirteen pairs of setae, the most anterior pair being bipectinate, the remaining leaflike (Fig. 190c) and similar to those found in the male. The setae all differ from one another in detailed structure, particularly one pair in the sacral region which is much narrower and more pointed than the rest. Four pairs of dorsal setae are present, d_1, d_2 and d_4 being much shorter than d_3.

The legs have no keels on tarsi and tibiae I and II.

According to Michael (1901), the larva has bipectinate setae.

BIOLOGY

This mite is not uncommon in wall dust and sawdust of cellars (Michael, 1901). It was also found in chaff, in stable dust and in the spillage from an oat mill (O'Farrell and Butler, 1948).

DISTRIBUTION

England, France, Italy, Germany, N. Ireland.

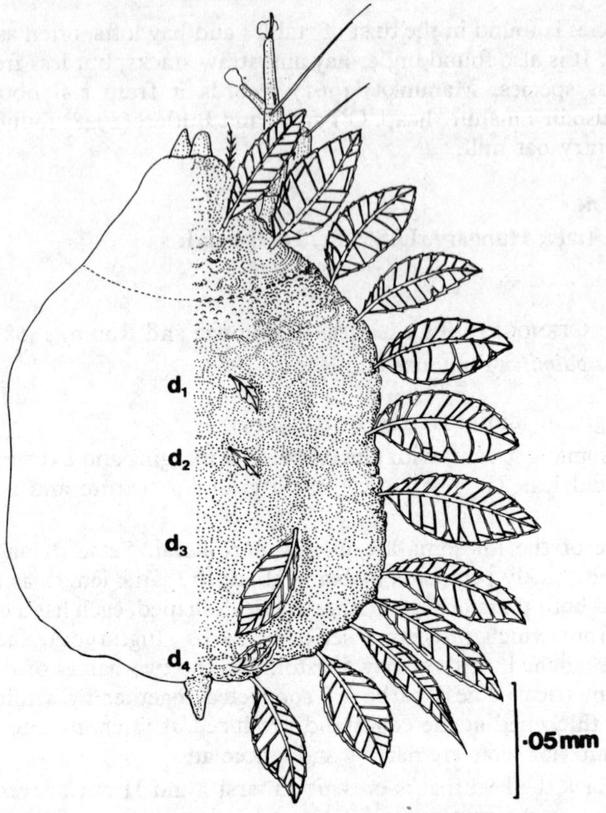

FIG. 197. *Ctenoglyphus palmifer* (Fumouze and Robin). Dorsal view of ♀. Dorsal setae: d_1 to d_4.

NYCTERIGLYPHINAE Fain, 1963

Small mites with a flattened body and without a groove between the propodosoma and hysterosoma.

The cuticle is finely striated or has a scale-like pattern.

The dorsal idiosomal setae are short and finely pectinate.

The legs are relatively short and all end in a globular pretarsus and well-developed claw. Tarsus I has two or three solenidia and a famulus.

Little sexual dimorphism occurs. The epigynium in the female is fused with apodemes I. The opening to the bursa copulatrix is at the end of a dorsal tube.

No anal or tarsal suckers are present in the male.

No hypopus has been found.

The members of the subfamily are associated with bats.

COPROGLYPHUS Türk and Türk, 1957

With the characters of the subfamily.

The seta *d* on tarsi I and II is of about the same length as omega$_3$ (Fig. 200).

With one solenidion arising from the dorsal side of genu I. [In *Nycteriglyphus*, two solenidia arise in this position].

Type species: *Coproglyphus stammeri* Türk and Türk, 1957.

COPROGLYPHUS STAMMERI Türk and Türk, 1957

Nycteriglyphus stammeri sensu Hughes, 1961.

MALE (Fig. 198). Length of idiosoma: about 230 μm.

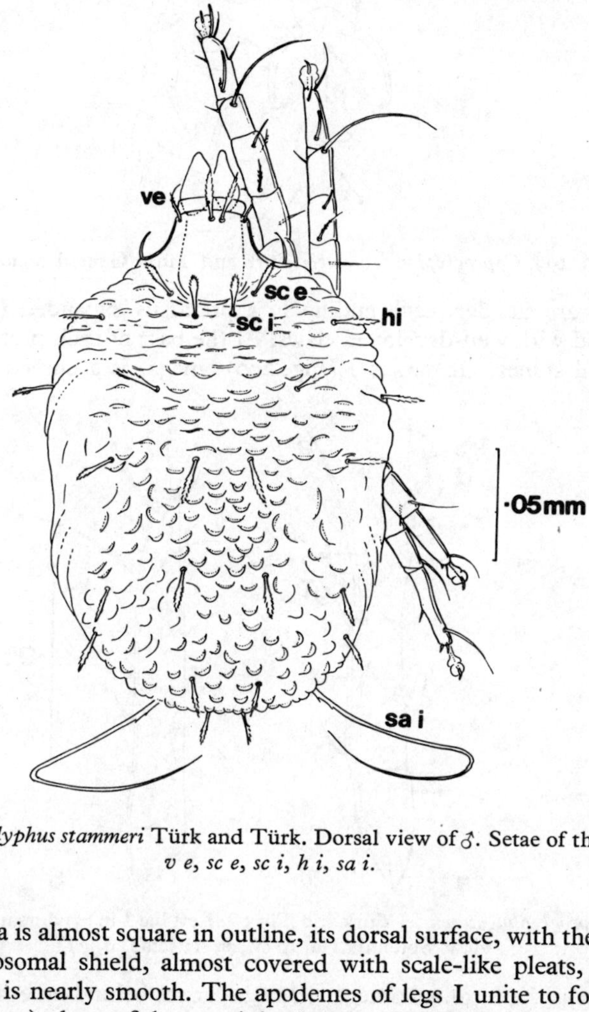

FIG. 198. *Coproglyphus stammeri* Türk and Türk. Dorsal view of ♂. Setae of the idiosoma: *v e*, *sc e*, *sc i*, *h i*, *sa i*.

The idiosoma is almost square in outline, its dorsal surface, with the exception of the propodosomal shield, almost covered with scale-like pleats, whilst the ventral surface is nearly smooth. The apodemes of legs I unite to form a short sternum (Fig. 199), those of the remaining legs are free. The epimerite of legs II is well-developed. The genital opening lies between coxae III, it is encircled almost completely by the epiandium. The penis is long and slender, supported by a complex series of struts.

All the dorsal setae are flattened, with fringed edges. The *v e* setae enfold the sides of the gnathosoma. Setae *sc e* are anterior to *sc i*, and *h i* originate on the same transverse line as *sc i* . The *sa i* setae are long and smooth, except for a few pectinations arising from the slightly thickened base. The supracoxal seta is curved, about the same length as omega$_1$.

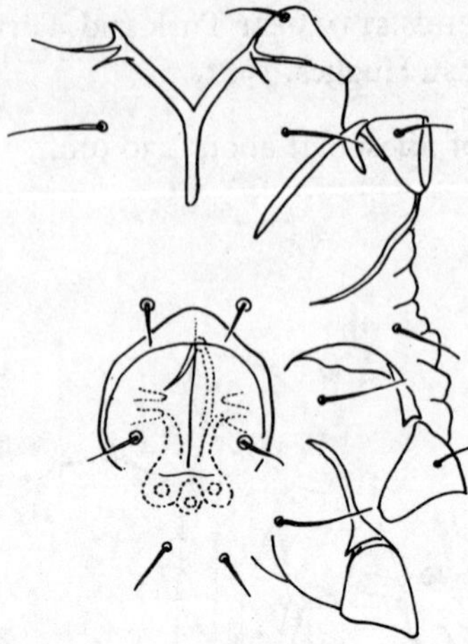

FIG. 199. *Coproglyphus stammeri* Türk and Türk. Genital region of ♂.

The legs are slender, each ending in a pretarsus expanded into a globular pulvillus and with well-developed claws. At the base of each pretarsus are three stout ventral spines. On tarsus I (Fig. 200) omega$_1$ is a curved rod about the

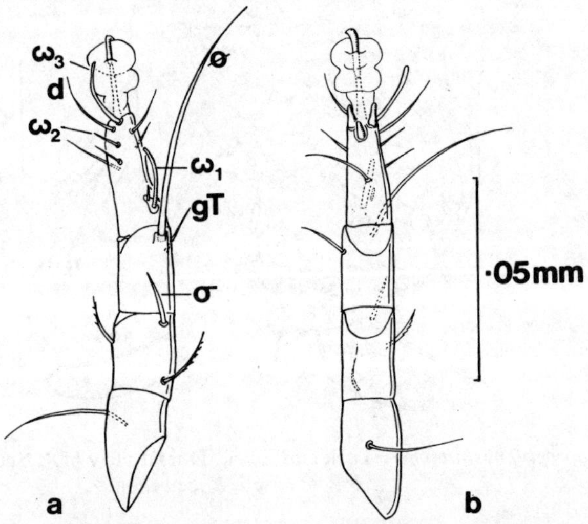

FIG. 200. *Coproglyphus stammeri* Türk and Türk ♀. Left leg I in (*a*) dorsal view, (*b*) ventral view. Solenidia: ω_1 to ω_3, Φ, σ; setae; *d*, *gT*.

same length as the terminal omega$_3$. Omega$_2$ arises behind omega$_3$ and has been called omega$_4$ by Fain (1963). On tibia I and II, phi exceeds the length of the tarsus and only one ventral seta *gT* is present. Only one sigma arises from genu I. Phi on tibia III is almost equal in length to that arising from tibia IV. No tarsal suckers are present on legs IV.

FEMALE. Length of idiosoma: about 230 µm.

The female closely resembles the male. Apodemes I touch one another without forming a sternum, their internal ends just touching the transverse epigynium (Fig. 201). A tubular bursa copulatrix arises from the end of the body (Fig. 202).

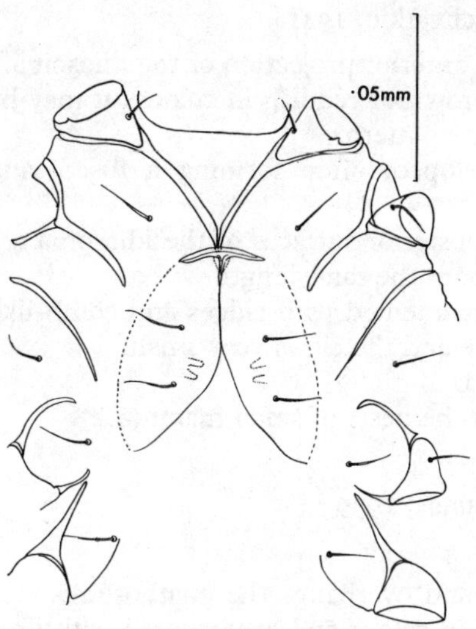

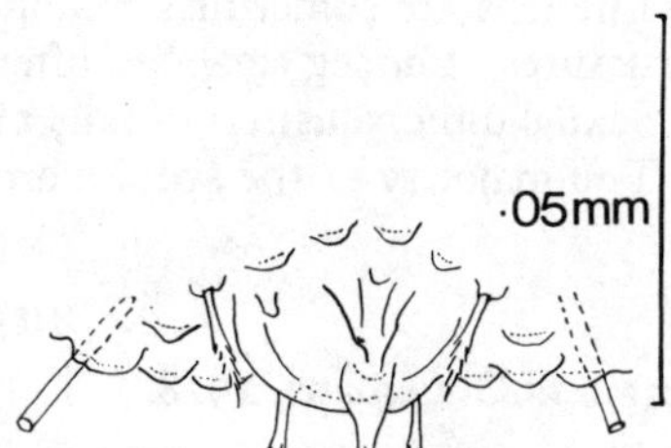

FIG. 201. *Coproglyphus stammeri* Türk and Türk. Genital region of ♀ (left).
FIG. 202. *Coproglyphus stammeri* Türk and Türk. Bursa copulatrix of ♀ (right).

DISTRIBUTION

A number of specimens was found in bat roosts near Slough, Berks, by G. E. Woodroffe. Türk and Türk found *Coproglyphus stammeri* on bat droppings near Erlangen, W. Germany.

AËROGLYPHINAE Zachvatkin, 1941

The idiosoma is flattened, without a transverse constriction, between propodosoma and hysterosoma.

The cuticle is densely striated, except in the region of the dorsal propodosomal shield, and the cuticle of the dorsal surface has numerous triangular spines embedded in it.

The dorsal idiosomal setae are slightly flattened and densely pectinate, especially along the edges. In length, they vary from 20 to 50 per cent of that of the idiosoma.

This subfamily contains only one genus *Aëroglyphus* Zachvatkin, 1941, type species: *Glycyphagus peregrinans* Berlese, 1892 which is associated with a number of species of Hymenoptera. N. Banks (1906) described *Glycyphagus robustus* which also belongs to the genus *Aëroglyphus* and which was found on a seed lot in Ohio.

A. robustus, which has been redescribed more fully by Cooreman 1959, was found by Sinha *et al.*, (1962) in stored grain in many granaries in Canada. Further investigation showed that this species is resistant to low temperatures and can overwinter as the hypopus at temperatures of —39°C. The effect of humidity and temperature on its biology have been studied by Sinha (1966b) and Barker (1967c).

A. robustus has not been recorded in this country.

LABIDOPHORINAE Zachvatkin, 1941

The gnathosoma is usually overhung by an anterior projection of the idiosoma.

The cuticle is usually strongly tanned, brown or reddish in colour; it may be smooth, granular or covered with a reticulate pattern.

The coxal-sternal skeleton is well-developed, often forming a fused ring round the female genital opening.

The setae arising from the dorsal and posterior surfaces of the idiosoma are short, smooth or feebly pectinate, and of about the same length.

The legs are sometimes modified and ornamented with ridges and comb-like structures. The leg setae are often pectinate and the claws very small.

Sexual dimorphism is usually pronounced.

The majority of the species are found in the nests of small mammals.

GOHIERIA Oudemans, 1939

Ferminia Oudemans, 1928.

The propodosoma is extended forward and overhangs the gnathosoma.

The cuticle is evenly sclerotized, brown in colour and ornamented with fine, almost smooth setae. No distinct propodosomal shield or crista metopica is present.

The apodemes of all the legs are slender and unite to surround the genital opening.

The genua and tibiae of the legs are distinctly ridged, particularly in the female. The distal edges of the femora and genua are expanded.

Sexual dimorphism is slight.

Tracheal tubes are present in the female.

Type species: *Glycypgagus fuscus* Oudemans, 1902.

GOHIERIA FUSCA (Oudemans, 1902)

Glycyphagus fuscus Oudemans, 1902; *Ferminia fusca* (Oudemans, 1902).

MALE (Fig. 203). Length of idiosoma: 300 to 320 μm.

Both dorsal and ventral surfaces of the body are well tanned and covered with pinkish-brown pitted cuticle. A hood-shaped extension of the dorsal side of the idiosoma overlies the gnathosoma. The anterior edge of the hysterosoma is marked by a transverse pleat, so that, in the living animal, the hysterosoma appears to be covered by a separate shield. The apodemes of all the legs are slender struts, those of legs I joining to form a short sternum; this unites with the internal ends of apodemes II to IV to enclose a colourless area of cuticle in front of the genital opening. The cuticle of the ventral surface is more coarsely pitted than that of the dorsal, and untanned cuticle joins dorsal and ventral surfaces together. The genital opening is between coxae IV, the penis being a straight tube whose apex is directed posteriorly. The anal opening extends to the posterior margin of the body and one pair of anal setae arises at its anterior end.

With the exception of *v i* which are pectinate, all the setae of the idiosoma are only slightly serrated, those arising from the propodosoma being directed forwards, those from the hysterosoma laterally and backwards. The *v e* setae are almost in the same transverse line as the pectinate supracoxal setae, each of which lies in a groove; *sc i*, *sc e* and *h i* are on a level with one another and four pairs of dorsals are arranged in linear series.

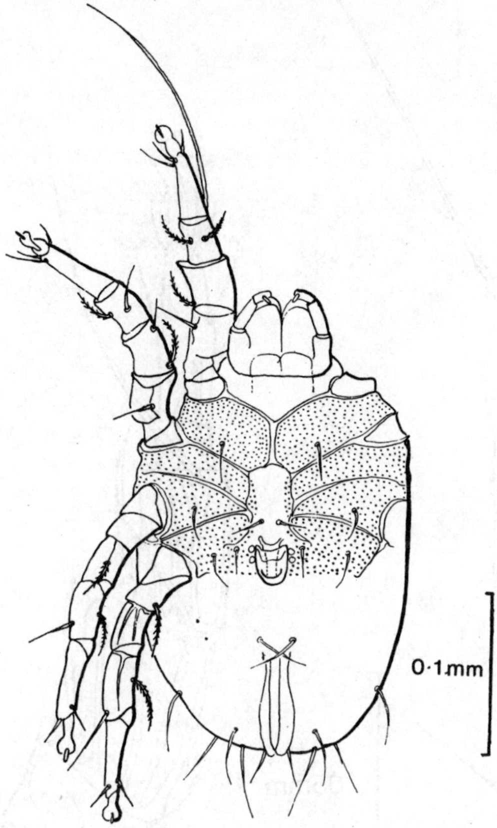

FIG. 203. *Gohieria fusca* (Oudemans). Ventral view of ♂.

The legs are short and massive, the pretarsus arising from the ventral side of the apex of each tarsus. The distal ends of the genua and femora are expanded, so as to partially envelope the succeeding segment. The dorsal surfaces of genua and tibiae are distinctly ridged. Tarsus I (Fig. 204a) is foreshortened and those setae which normally occupy the middle of the segment (*la*, *ra*, *wa*) arise close to *s* at the base of the pretarsus, whilst $omega_1$, $omega_2$, the famulus and *ba* occupy their normal positions. On tibia I, phi is unusually long, but decreases in length from tibiae I to IV. On genu I, $sigma_1$ is much longer than $sigma_2$. Both ventral setae are present on tibia I. Legs III and IV (Fig. 205) are noticeably bent, with unusually long pretarsi.

FEMALE (Figs. 206, 207). Length of idiosoma: 380 to 420 μm.

The idiosoma is paler in colour than in the male, more square in outline and with finer setae. Ventrally, almost all that part of the body between coxae I and IV is occupied by the large genital folds. Apodemes I are united with the epigynium: a transverse plate at the anterior end of the genital opening; apodemes II nearly reach the circumgenital ring, and III and IV touch at their internal ends. Small genital sense organs project from the posterior end of the genital folds.

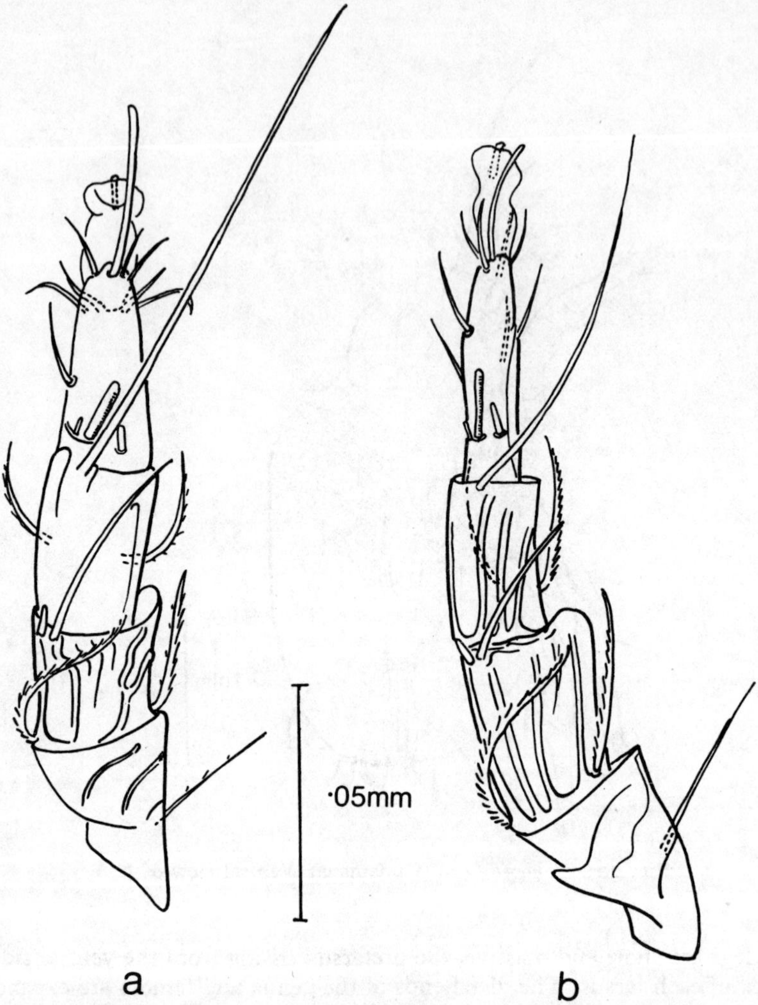

FIG. 204. *Gohieria fusca* (Oudemans). Dorsal view of right leg I of (*a*) ♂, (*b*) ♀.

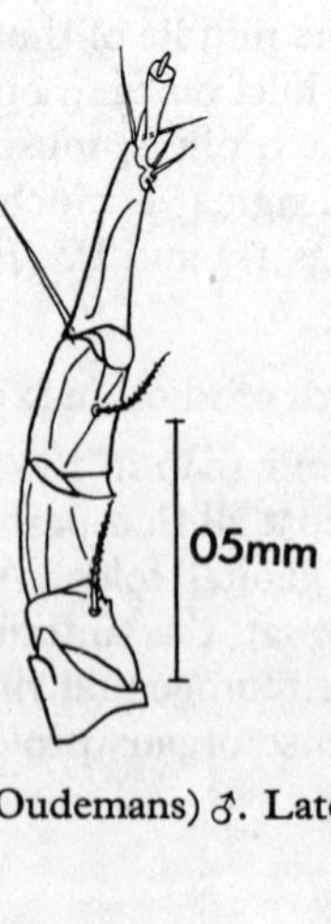

FIG. 205. *Gohieria fusca* (Oudemans) ♂. Lateral view of left leg IV.

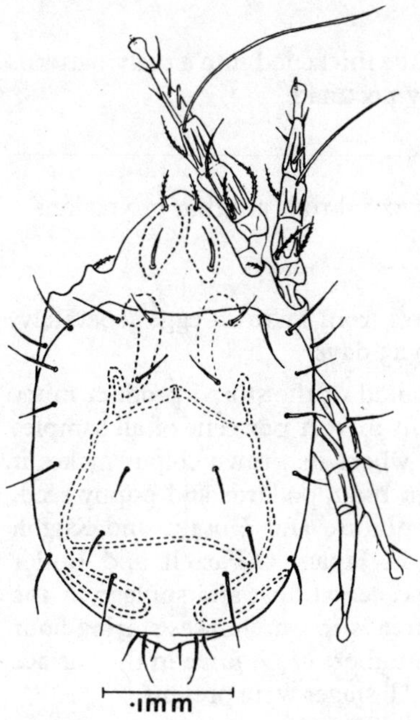

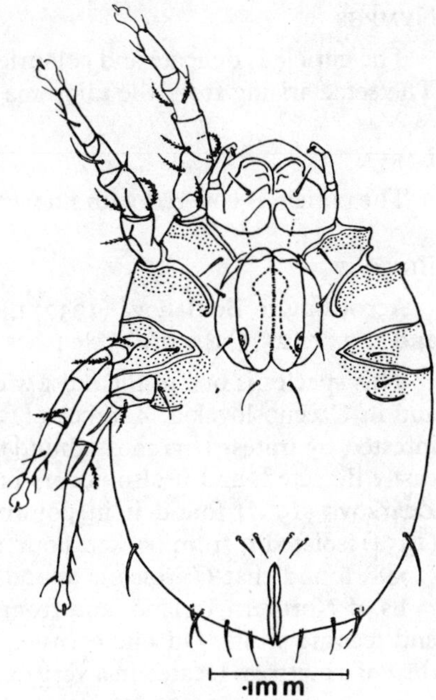

FIG. 206. *Gohieria fusca* (Oudemans). Dorsal view of ♀ (left).
FIG. 207. *Gohieria fusca* (Oudemans). Ventral view of ♀ (right).

A pair of well-developed tracheae is present which in life, are full of air. Each trachea branches into an anterior region which expands into an air sac, and a long winding posterior part which according to van der Hammen 1968, crosses the one from the other side, but does not join it.

The folds on either side of the anal opening project beyond the posterior margin of the body. Two pairs of anal setae arise on either side of the anterior end of the opening. The opening of the bursa copulatrix is overhung by a small projection; it is connected with a receptaculum seminis by a tube (Fig. 208).

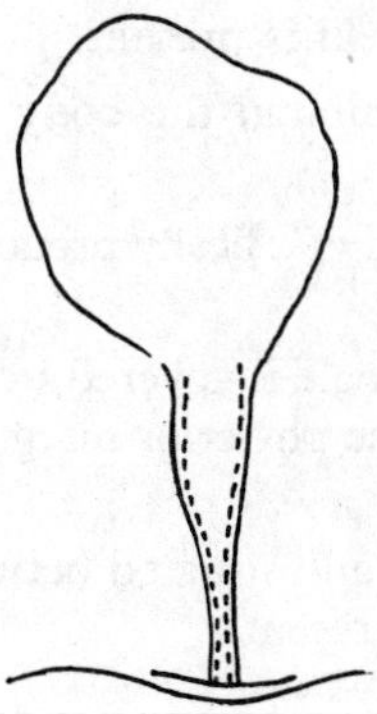

FIG. 208. *Gohieria fusca* (Oudemans). Genitalia of ♀.

The legs are more slender than those of the male, the longitudinal ridges better developed. On tarsus I (Fig. 204b) the setae are spaced out along the tarsus instead of being concentrated at the distal end.

NYMPHS

The cuticle is delicate and colourless, its surface thickened into a scaly pattern. The setae arising from the idiosoma are slightly pectinate.

LARVA

The cuticle is covered with fine warts and the coxal rods are thin projections.

BIOLOGY

According to Boulanova (1937) the female lays from 11 to 24 eggs separately, and at 24 to 25°C the life cycle takes from 11 to 23 days.

This species is one of the most widely distributed of the stored product mites and in Czechoslovakia, Žďárková (1967) found it in 10.1 per cent of all samples infested by mites. It is most abundant in flour where its brown colour makes it easily detected, and it also occurs on rice, corn, bran, pollards and poppy seed. Žďárková (1967) found it in poultry protein mixture and Hosaya and Kugoh (1954) isolated it from packed flour and sugar in Japan. O'Farrell and Butler (1948) found that *G. fusca* is abundant in mixed cereal dust and spillage in the mills of Northern Ireland, and from there it spreads to warehouses storing flour and feeding stuffs. On one occasion, I found numbers of *G. fusca* in the surface dust of a mattress located in a very damp house. All stages were present.

DISTRIBUTION

England, N. Ireland, France, Germany, Holland, Belgium, Czechoslovakia, U.S.S.R., Turkestan, New Zealand, Japan, Egypt.

CHORTOGLYPHIDAE Berlese, 1897

Free-living mites, often found in food stores and granaries, and in the dust of houses.

The idiosoma has a firm, oval outline and is not divided into propodosoma and hysterosoma.

No dorsal propodosomsal shield is present.

The cuticle is smooth and shining, the body setae short and nearly all are smooth.

The claw is inserted at the end of a fleshy pretarsus. Only one sigma is present on genu I.

The genital opening of the female is covered by two sclerotized plates situated between coxae III and IV, whose posterior margins form a smooth bow-shaped curve.

The penis of the male is long and situated between coxae I and II; tarsal and conspicuous anal suckers are present.

CHORTOGLYPHUS Berlese, 1884

The characters of the family are those of the genus.

Type species: *Tyroglyphus arcuatus* Troupeau, 1879.

CHORTOGLYPHUS ARCUATUS (Troupeau, 1879)

Tyroglyphus arcuatus Troupeau, 1879. *Chortoglyphus nudus* Berlese, 1884.

MALE (Fig. 209). Length of idiosoma: 250 to 300 μm.

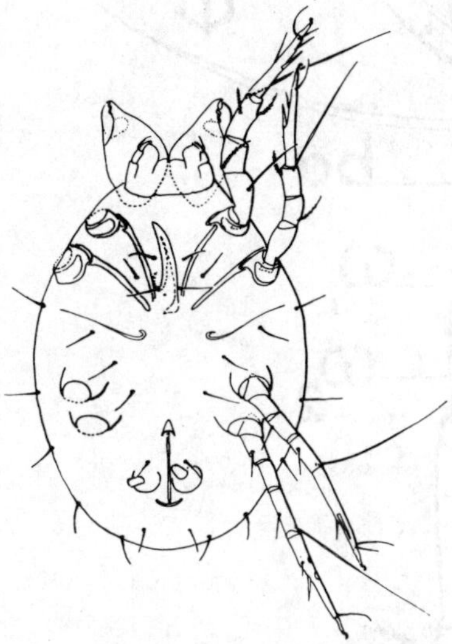

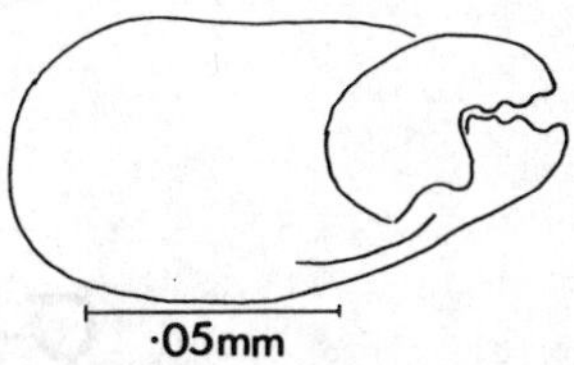

FIG. 209. *Chortoglyphus arcuatus* (Troupeau). Ventral view of ♂ (left).
FIG. 210. *Chortoglyphus arcuatus* (Troupeau). Chelicera of ♀ (right).

The body is a regular oval, strongly arched and variable in colour. By Michael and Zachvatkin it is said to be greenish, with pink legs and gnathosoma; all the specimens I have seen are distinctly pink, the appendages being rather darker in colour. In life, the anterior edge of the idiosoma overhangs the gnathosoma, so that only the tips of the chelicerae are visible from above. The chelicerae (Fig. 210) are unusually large, with distinctly toothed shears. No propodosomal dorsal shield is present.

The body setae are all short and fine, about 3 to 5 per cent of the idiosoma in length. Setae *v i* project forward over the gnathosoma, *v e* are slightly longer and clearly pectinate, arising at the same level as *v i*. The scapulars are arranged in a transverse row and are almost equidistant from one another. The supracoxal seta is a thin pectinate rod. The four pairs of doral setae are almost in linear series with one another, and only two pairs of laterals are present. Ventrally, the genital opening is between coxae I and II, so that apodemes I and II are forced apart and form part of the transparent genital folds. No sternum is present. The anal slit is some distance from the posterior edge of the body and is flanked by a pair of conspicuous anal suckers. Each consists of a circular sclerotized base to which is attached a membranous funnel. A pair of preanal setae lies in front of the suckers and a pair of postanals is behind.

The legs are long and slender and each terminates in a prominent pretarsus to which is attached a small claw. On tarsus I (Fig. 211), omega$_1$ is a long curved rod arising close to a small omega$_2$. Seta *wa* is enlarged to form a stout spine and *ba* is a thin seta. The solenidion phi is unusually long and extends well beyond the end of the tarsus on all the legs. Only one solenidion sigma arises

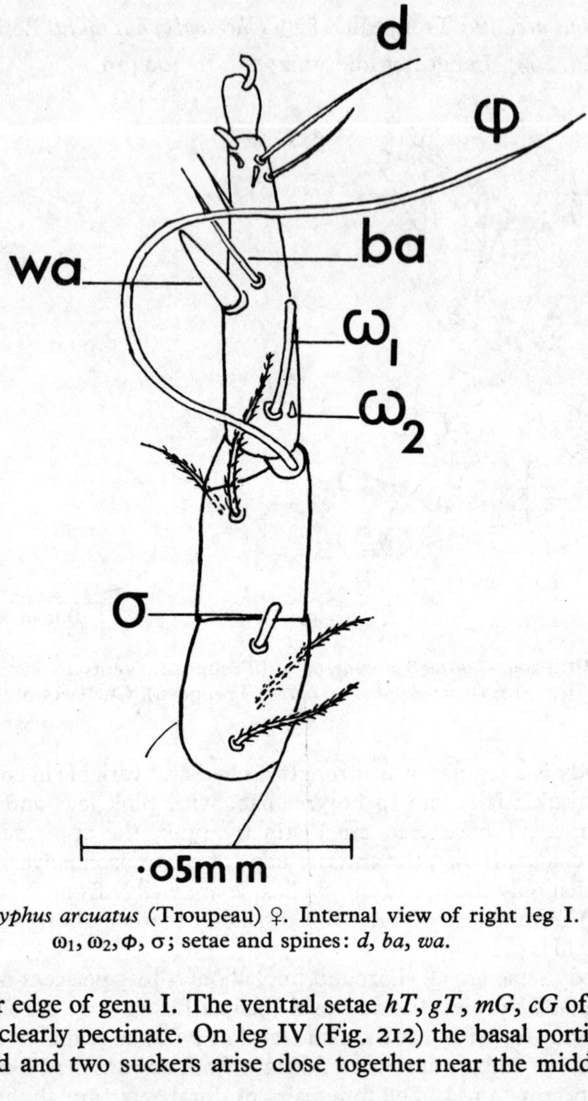

FIG. 211. *Chortoglyphus arcuatus* (Troupeau) ♀. Internal view of right leg I. Solenidia: $\omega_1, \omega_2, \Phi, \sigma$; setae and spines: *d*, *ba*, *wa*.

from the anterior edge of genu I. The ventral setae *hT*, *gT*, *mG*, *cG* of the genu and tibia are all clearly pectinate. On leg IV (Fig. 212) the basal portion of the tarsus is enlarged and two suckers arise close together near the middle of the tarsus.

The penis is a very large curved tube with a bifurcate base.

FEMALE (Figs. 213, 214). Length of idiosoma: 350 to 400 µm.

The female closely resembles the male in arrangement and length of the dorsal setae. Ventrally, apodemes I unite to form a short sternum. Apodemes II are long and placed transversely across the body, parallel to an elongated sclerite lying between coxae II and III. Apodemes III and IV are not developed. The genital folds are broad plates, with strongly sclerotized and curved posterior borders, without genital sense organs on their internal surfaces. The anus is near the posterior margin of the body and five pairs of anal setae are associated with it. The bursa copulatrix is a small round opening on the dorsal side of the posterior end of the body.

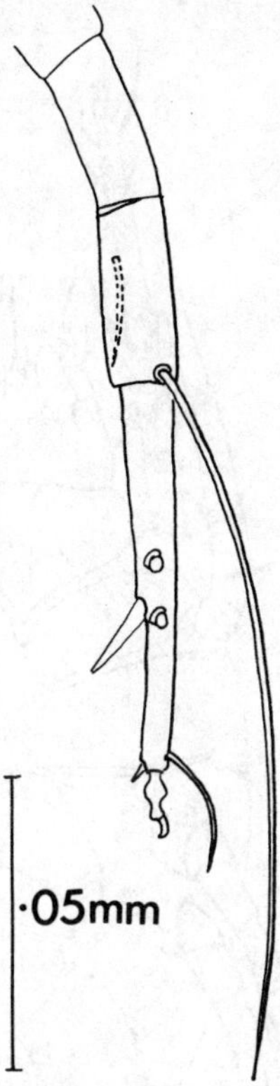

FIG. 212. *Chortoglyphus arcuatus* (Troupeau) ♂. Dorso-lateral view of right leg IV.

Legs I and II are shorter than those of the male, whilst legs IV are longer, tarsus IV being considerably longer than the two preceding segments taken together.

No hypopus has been recorded for this species.

LARVA

Small nipple-shaped coxal rods are present. Rudiments of genital sense organs are developed in the protonymph and tritonymph.

BIOLOGY

This species is common in the floor dust of barns, mills, stables and granaries, and is frequently found in flour and heaps of old straw. Zachvatkin (1941) also found it in wheat, rye, oats and grass seed. Robertson (1946) records a heavy infection on red clover seed and Ždárková (1967) often found it on cereals and stored poultry mixtures. Attiah (1969) and O'Farrell and Butler (1948) both isolated it from rice. Bollaerts and Breny (1951) always found it in association with *Gohieria fusca* and *Acarus siro*. Wasylik (1959) collected it from a sparrow's nest and Brady (1970) found it in the litter of broiler houses. I found it in large numbers in the floor dust (collected by R. Bruce Pearson) of a house in the Barbados and, on one occasion, in the surface dust of a mattress located in a farmhouse.

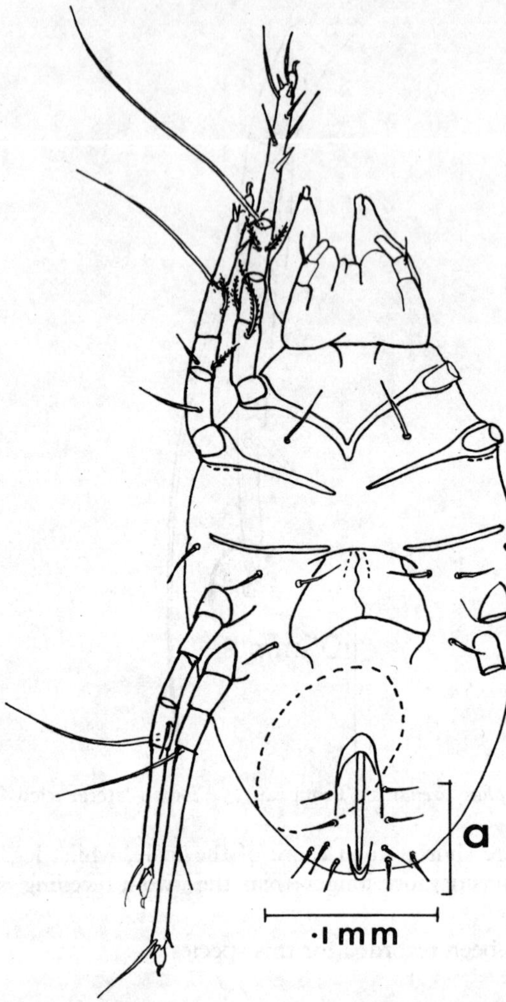

FIG. 213. *Chortoglyphus arcuatus* (Troupeau). Ventral view of ♀. Anal setae: *a*.

DISTRIBUTION

England, France, Belgium, Italy, Germany, Holland, Poland, Czechoslovakia, U.S.S.R., U.A.R., New Zealand, Barbados.

CARPOGLYPHIDAE Oudemans, 1923

Free-living mites not known to be associated with any particular habitat.

The cuticle is smooth (*Carpoglyphus*) or covered with a number of strongly sclerotized plates (*Hericia*).

The apodemes of legs I and II are united with the sternum in both sexes (*Carpoglyphus*) or in the male only (*Hericia*).

The claws are large and joined to the end of the tarsus by a well-developed pretarsus.

Except for the posterior setae, most of the idiosomal setae are smooth.

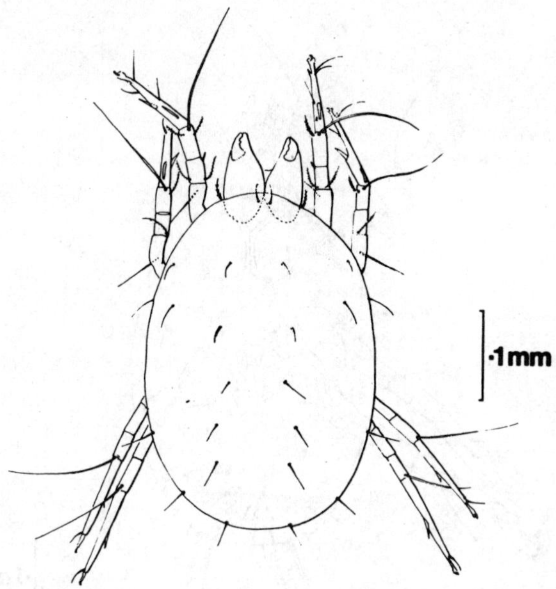

FIG. 214. *Chortoglyphus arcuatus* (Troupeau). Dorsal view of ♀.

CARPOGLYPHUS Robin, 1869

The idiosoma is slightly flattened, oval and covered with a shining smooth cuticle.

No dorsal propodosomal shield is present and no transverse groove separates the propodosoma from the hysterosoma.

The apodemes of legs I and II are united with the sternum in both sexes.

The body setae are smooth and *v e* are almost in the same transverse line as coxae II. Three pairs of lateral setae l_1 to l_3 are present.

On tibia I, phi arises from the middle region of this segment.

No coxal rods are present in the larva.

Hypopi are sometimes formed.

Type species: *Acarus lactis* L., 1758.

CARPOGLYPHUS

Key to the adults

1. Distance between the d_1 setae greater than that between the d_2 setae (Fig. 221). Most of the dorsal setae are pointed............*C. munroi*, p. 191

—. Distance between the d_1 setae about the same as that between the d_2 setae (Fig. 219). Most of the dorsal setae have rounded ends....*C. lactis*, p. 187

CARPOGLYPHUS LACTIS (Linné, 1758)

Acarus lactis L, 1758; *Carpoglyphus passularum* Robin, 1869; *Glycyphagus anonymus* Haller, 1882.

MALE (Fig. 215). Length of idiosoma: 380 to 400 μm.

The idiosoma is oval and slightly flattened, only coloured by the contents of the alimentary canal showing through the translucent cuticle. The legs and

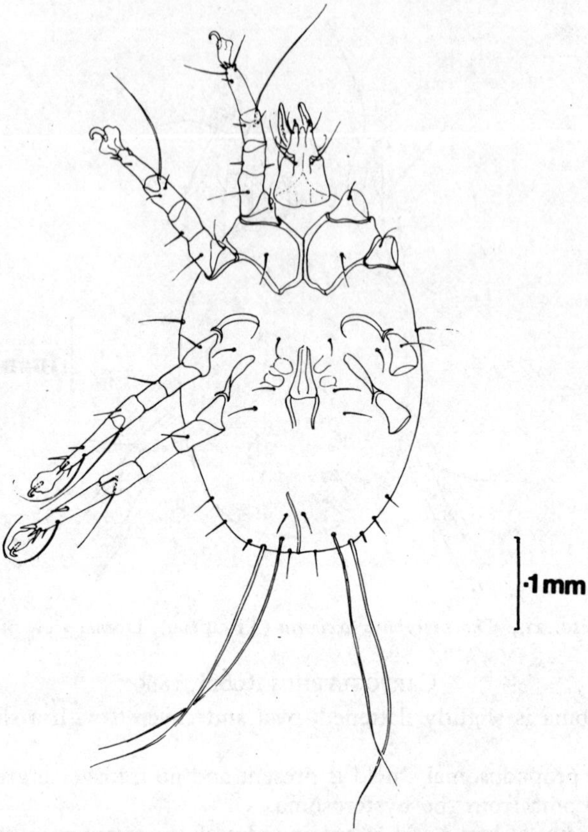

FIG. 215. *Carpoglyphus lactis* (L.). Ventral view of ♂.

gnathosoma are slightly pinkish. The shoulder region is well-defined and the posterior margin truncate or slightly concave. No dorsal propodosomal shield is present. The latero-abdominal glands are displaced to the posterior corners of the idiosoma and contain a coloured substance. A slightly raised cornea is found on either side of the base of the gnathosoma, but there is no pigmented retina. Ventrally, the apodemes are well-sclerotized, and apodemes I join in the mid-line to form a sternum which bifurcates at its posterior end to articulate with the apodemes of legs II (Fig. 216b). The genital opening is between coxae III and

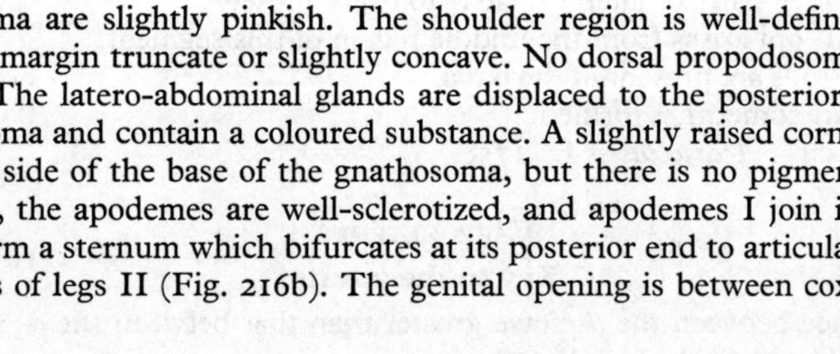

FIG. 216. Coxal-sternal skeleton of ♂ of (*a*) *Carpoglyphus munroi* Hughes and (*b*) *C. lactis* (L.).

IV. The penis is a blunt tube with its apex directed anteriorly, and the genital sense organs are unusually long. Two pairs of genital setae are present, about equal in length. The anus extends as far as the posterior margin of the body.

With the exception of *v e* and two pairs of setae arising posteriorly, all the setae are short (7 to 12 per cent of the length of the idiosoma), with rounded ends. Setae *v i* do not extend as far as the tips of the chelicerae; *v e* are more posterior and intermediate in position between *v i* and *sc i*; d_1 to d_4 and *sa i* are arranged in two longitudinal rows down the middle of the idiosoma. The supracoxal seta is a blunt rod. From the posterior end of the body project two pairs of long setae: *p* a_1 and *sa e* (Fig. 215).

The gnathosoma is shaped like a cone and very mobile. The chelicera (Fig. 217) has thin narrow shears.

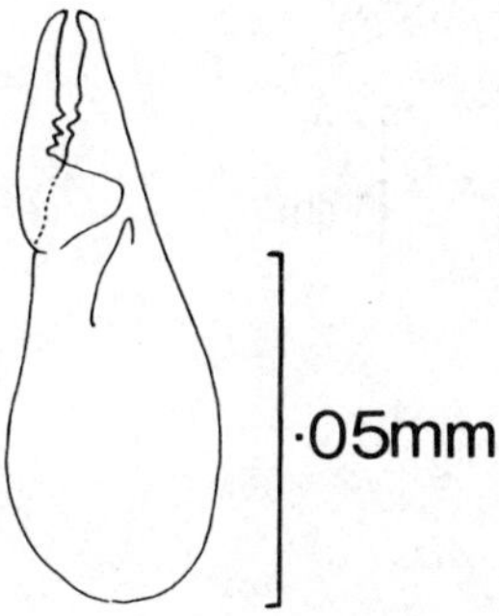

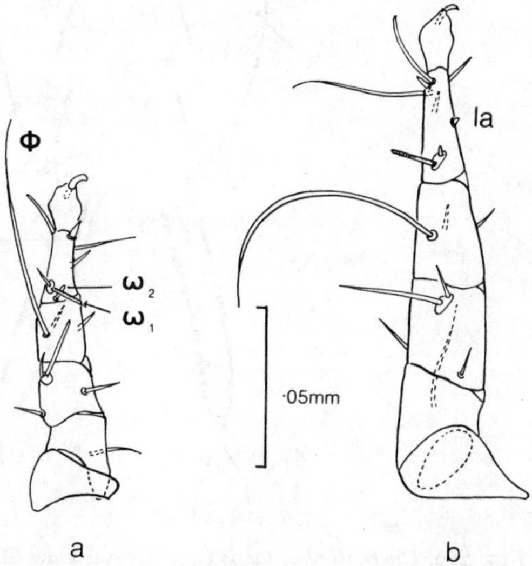

FIG. 217. *Carpoglyphus lactis* (L.). Chelicera of ♀ (left).

FIG. 218. Dorsal view of left leg I of ♂ of (*a*) *Carpoglyphus lactis* (L.) and (*b*) *C. munroi* Hughes. Solenidia: ω_1, ω_2, Φ; tarsal seta or ω_2: *la*.

All the legs terminate in well-developed pretarsi strengthened by two thin 'tendons' which extend from the end of the tarsus nearly to the sickle-shaped claw. On tarsus I (Fig. 218a) some of the median and distal setae are spines. Omega$_1$ is a cylindrical rod frequently bent externally, so as to cover the base of omega$_2$. On tibia I and II, phi arises from the middle region and both ventral setae are present.

FEMALE (Fig. 219). Length of idiosoma: 380 to 420 μm.

The female closely resembles the male. On the ventral side of the idiosoma, the sternum and apodemes II are fused to form an epigynium covering the anterior end of the genital opening. The genital folds are not well-sclerotized and lie between coxae II and III. The anal opening extends almost to the posterior margin of the body (Fig. 220b) and there is only one pair of anal setae. The bursa copulatrix is a round opening on the posterior dorsal end of the body.

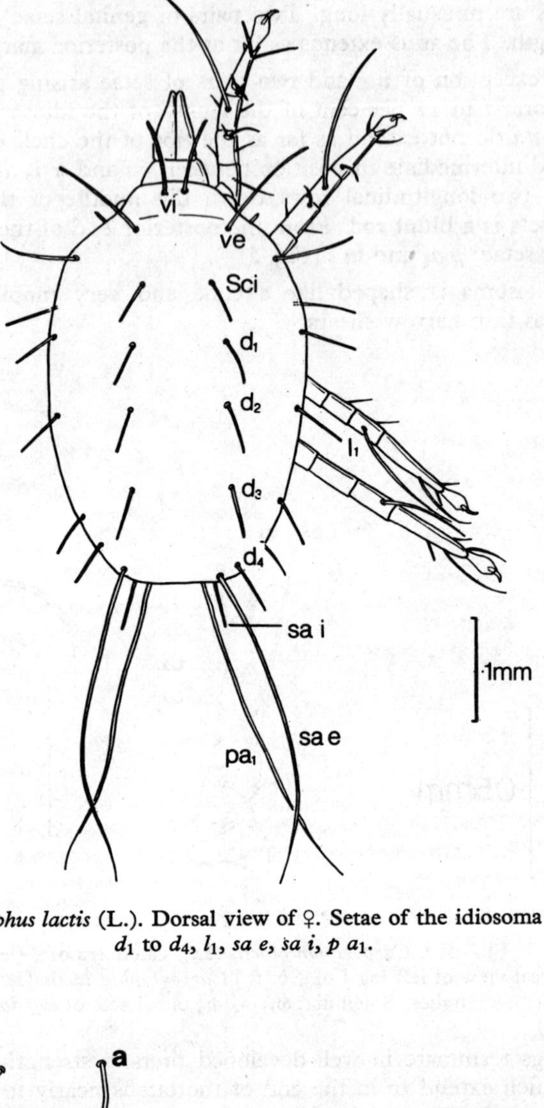

FIG. 219. *Carpoglyphus lactis* (L.). Dorsal view of ♀. Setae of the idiosoma: *v i, v e, sc i,* d_1 to d_4, l_1, *sa e, sa i, p* a_1.

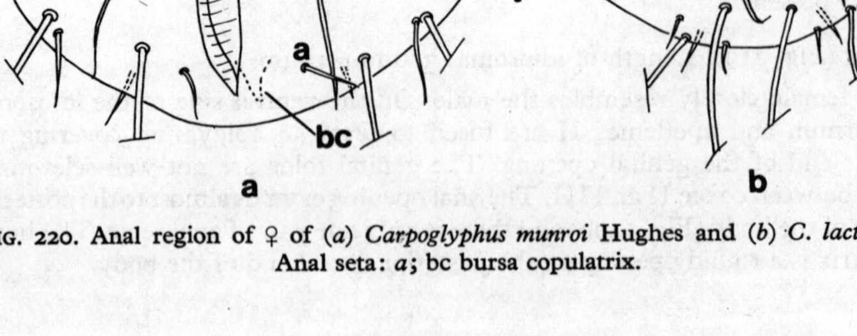

FIG. 220. Anal region of ♀ of (*a*) *Carpoglyphus munroi* Hughes and (*b*) *C. lactis* (L.). Anal seta: *a*; bc bursa copulatrix.

The legs are more slender than those of the male, with less well-developed pretarsi.

HYPOPUS

This is rarely found, but it has been described by Vitzthum 1940 and bred in the laboratory by W. Chmielewski 1967b. I have found it on one occasion only and it is possible that only certain strains of *C. lactis* can give rise to the hypopus.

LARVA

This has no coxal rods.

BIOLOGY

C. lactis is found almost exclusively on food substances containing sugars from which bacterial activity gives rise to the lower fatty acids such as lactic, acetic and succinic acids (Vitzthum, 1943). It has been found on all kinds of dried fruit, honeycombs, pollen in bee hives (D. C. Lee coll.), on fruit drink residue, confection of senna, wine where it clings to pieces of cork, rotting potatoes, cheese, old flour, cocoa beans and groundnuts (Žďárková, 1967). In one instance it was breeding in the caramel used in sweet manufacture, and stored Christmas puddings are often attacked. Taboda (1954) attributes a case of pulmonary acariasis in Spain to a species of *Carpoglyphus*.

Oboussier (1939) says that *C. lactis* feeds on the sugary secretion covering figs which is scraped away by the chelicerae, the growth of mould being prevented by a secretion from the salivary glands which are very large in this species. The eggs are attached to the substrate by a rapidly hardening stalk, some 25 to 30, more rarely 72, eggs being laid by one female in a week. The mites can survive on an average from 40 to 50 days.

Davies (1926) found *C. lactis* breeding in scabs on a domestic rabbit.

DISTRIBUTION

Europe, North America, Argentina.

CARPOGLYPHUS MUNROI Hughes, 1952

MALE (Fig. 221). Length of idiosoma: 320 to 500 μm.

In colour and general appearance, *C. munroi* resembles *C. lactis*. It differs in having a more rounded dorsal surface and more pointed posterior margin. Ventrally, the coxal setae between apodemes I and II (Fig. 216a) are replaced by small chitinous rings resembling the bases of setae; also, the posterior pair of genital setae is more than twice as long as the anterior.

The dorsal setae are longer than those of *C. lactis* and pointed. Setae *v e* are more than twice as long as *v i* and the remaining dorsal setae are 10 to 16 per cent of the idiosoma in length. Setae *sc i* and d_1 to d_4 are not so regularly arranged in linear series, the d_1 setae being further apart than *sc i* and d_2.

The chaetotaxy of the legs differs in that $omega_2$ is missing and may be represented by a small process sunk into the cuticle (Fig. 218b).

FEMALE (Fig. 222). Length of idiosoma: 450 to 600 μm.

This resembles the male, but the dorsal setae are considerably longer; thus d_1 are 28 to 30 per cent of the idiosoma in length and l_1 38 to 45 per cent. Ventrally, the coxal seta is present between apodemes I and II and the genital opening extends between coxae I and II as in *C. lactis*. The genital setae are of

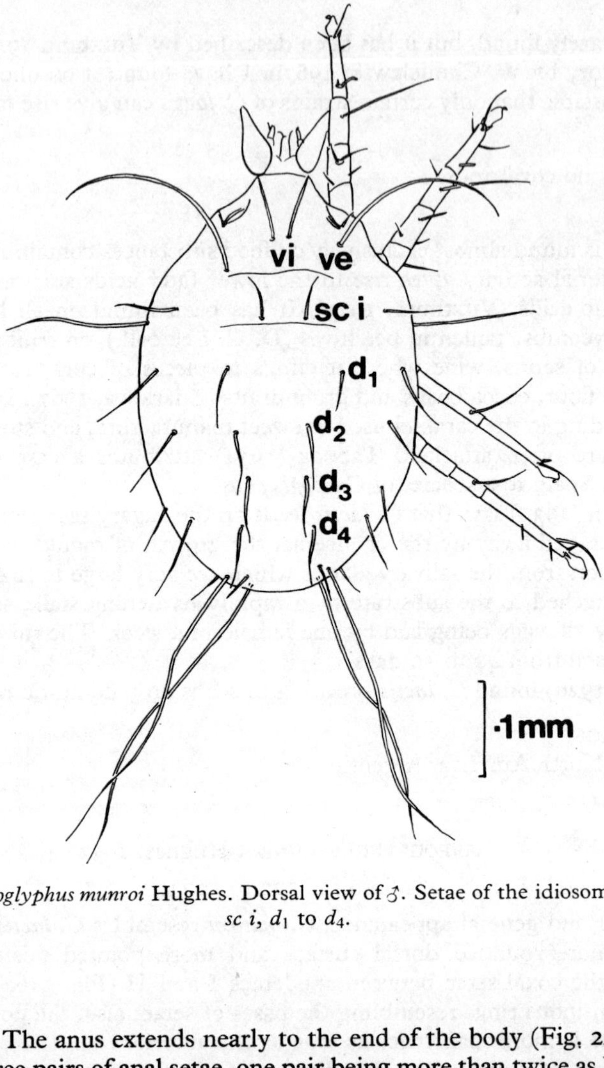

FIG. 221. *Carpoglyphus munroi* Hughes. Dorsal view of ♂. Setae of the idiosoma: *v i*, *v e*, *sc i*, d_1 to d_4.

equal length. The anus extends nearly to the end of the body (Fig. 220a) and is flanked by three pairs of anal setae, one pair being more than twice as long as the other two. A dorsal tubular bursa copulatrix is present.

The pretarsi and claws are less well-developed than in the male. The chaetotaxy of leg I is the same, except that the distal spines are longer and omega$_2$ is shaped like a seta.

Biology and distribution

C. munroi has only been recorded from England. It was first found in a disused bell-tower near Ascot, Surrey, in a mass of cobwebs containing dead insects. Presumably it feeds on the bodies of this insect. G. E. Woodroffe also collected it from bat roosts near Slough, Berks.

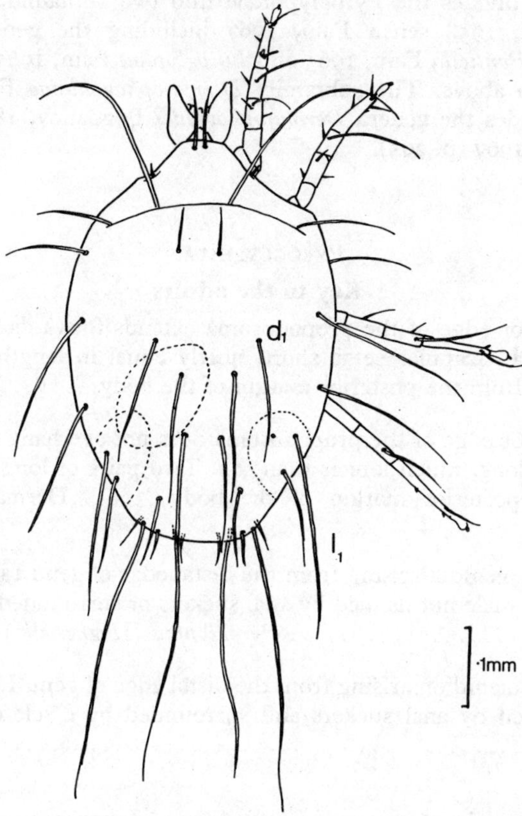

FIG. 222. *Carpoglyphus munroi* Hughes. Dorsal view of ♀. Setae of the idiosoma: d_1 and l_1.

PYROGLYPHIDAE Cunliffe, 1958 sensu Fain, 1965

Free-living mites sometimes found in the nests of rodents and birds.

The anterior edge of the propodosoma may be extended forward to overhang the gnathosoma.

A transverse groove separates the propodosoma from the hysterosoma.

A dorsal propodosomal shield is present and a dorsal hysterosomal shield may also be developed.

Vertical setae are absent.

All the legs terminate in a pretarsus. Legs III and IV of the female are usually of equal length and width; in the male legs IV may be slightly shorter than legs III.

The female genital opening is in the form of an inverted U; a sclerotized genital shield is present, as well as lateral genital plates (Fig. 229).

In the male, anal suckers may be present and encircled by a sclerotized ring. Tarsal suckers in the male are represented by short, cylindrical structures.

On legs I in both sexes, omega$_1$ as well as omega$_3$ and the famulus arise from the distal end of the tarsus.

Fain, 1967b, divides the Pyroglyphidae into two subfamilies: the Pyroglyphinae Cunliffe, 1958 sensu Fain, 1967 including the genera *Pyroglyphus* Cunliffe, 1958, *Bontiella* Fain, 1965 and *Euroglyphus* Fain, 1965 and having the characters given above. The subfamily *Dermatophagoidinae* Fain, 1963 sensu Fain 1965 includes the genera *Dermatophagoides* Bogdanov, 1864 and *Sturnophagoides* Fain, 1967 (p. 205).

PYROGLYPHIDAE

Key to the adults

1. The anterior edge of the propodosoma extends forwards to overhang the gnathosoma. Scapular setae short, nearly equal in length. No long setae projecting from the posterior margin of the body........Pyroglyphinae 2

—. The anterior edge of the propodosoma does not overhang the gnathosoma. Setae *sc e* long, much longer than *sc i*. Two pairs of long setae projecting from the posterior margin of the body........Dermatophagoidinae 4

2. With two solenidia arising from the distal edge of genu I (Fig. 226a). Anal opening of male not flanked by anal suckers or surrounded by a sclerotized ring........................ *Pyroglyphus* (*Hughesiella*) *africanus* p. 195

—. With one solenidion arising from the distal edge of genu I. Anal opening of male flanked by anal suckers and surrounded by a sclerotized ring (Fig. 236) 3

3. Posterior margin of hysterosoma distinctly bilobed in the male (Fig. 236). Setae *sR* present on trochanters I to III..........................
..........................*Euroglyphus* (*Gymnoglyphus*) *longior* p. 203

—. Posterior margin of hysterosoma slightly concave in the male. Setae *sR* absent from trochanters I to III.......*Euroglyphus* (*E.*) *maynéi* p. 200

4. In the female, the cuticle in the region of the d_2 and d_3 setae is longitudinally striated. In the male, legs I are no thicker than legs II and the hysterosomal shield reaches to a point midway between the d_1 and d_2 setae........
.................................*Dermatophagoides pteronyssinus*, p. 211

—. In the female, the cuticle between the d_2 and d_3 setae is striated transversely. In the male, legs I are thickened and the hysterosomal shield does not surround the d_2 setae..5

5. Process *S* on tarsus I of the female is small (Fig. 246). In the male, *S* is absent from tarsus II (Fig. 247)......*Dermatophagoides microceras*, p. 210

—. Process *S* on tarsus I of the female is large (Fig. 246). In the male, *S* is present as a finger-like process on tarsus II (Fig. 247)..................
....................................*Dermatophagoides farinae*, p. 206

PYROGLYPHUS Cunliffe, 1958

Anal suckers are absent in the male.

Genu I has two solenidia.

Fain (1965) divides this genus into two subgenera: *Pyroglyphus* (*P.*) type species: *P. morlani* Cunliffe, 1958 and *P.* (*Hughesiella*) type species: *Dermatophagoides africanus* Hughes, 1954. In *P.* (*Hughesiella*) the cuticle is less sclerotized and the folded thickenings more pronounced, particularly in the female. In the female also, the genital plate is less well-developed than in *Pyroglyphus* (*P.*), and its anterior edge is distinctly cleft.

PYROGLYPHUS (HUGHESIELLA) AFRICANUS (Hughes, 1954)

Dermatophagoides africanus Hughes, 1954.

MALE (Figs. 223, 224). Length of idiosoma: 250 to 300 μm.

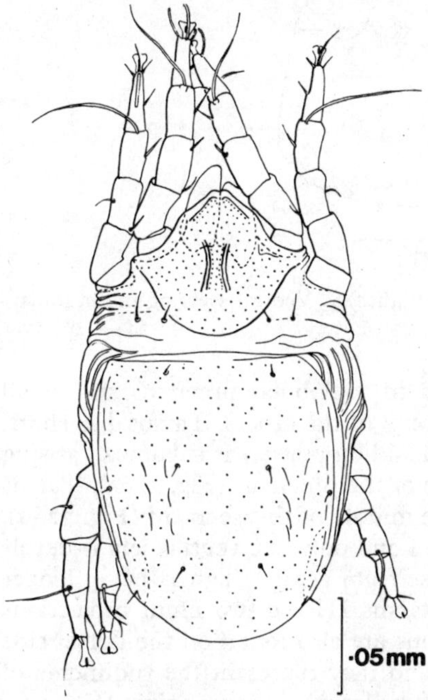

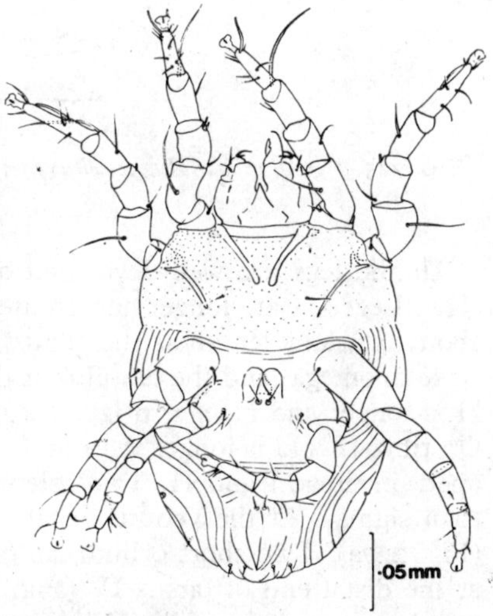

FIG. 223. *Pyroglyphus* (*Hughesiella*) *africanus* (Hughes). Dorsal view of ♂ (left).
FIG. 224. *Pyroglyphus* (*Hughesiella*) *africanus* (Hughes). Ventral view of ♂ (right).

The body is broadly oval, flattened and rounded posteriorly, the groove between propodosoma and hysterosoma accentuated by the folding of the cuticle. Except for the dorsal propodosomal and hysterosomal shields and the sternal region, the cuticle is thickened to form coarse wrinkles, the folds running longitudinally along the sides of the body and transversely in the propodosomal region. Both dorsal shields have a slightly punctate surface, the propodosomal extending laterally between the bases of legs I and II and having two longitudinal ridges running down its central region. Ventrally, the apodemes of legs I end freely in the midline. The penis is a small curved tube.

All the body setae are short and smooth. Owing to the absence of the vertical setae, the most anterior setae are the scapulars, *sc e* being slightly longer than *sc i*. Three pairs of dorsal setae arise on either side of the midline, and two pairs of laterals. Only one pair of humerals is inserted on a level with coxae III. The usual two pairs of coxal setae are present on coxae I and III. Two pairs of genital setae are posterior to the genital opening, one pair being immediately behind the genital opening, the other on a level with coxae IV. One pair of anal setae is situated at the anterior end of the anus, and two pairs at its posterior end.

The gnathosoma is partially concealed by the overhanging propodosoma whose anterior edge is slightly bifid. The chelicerae have well-developed teeth and the pedipalps (Fig. 225) are flattened.

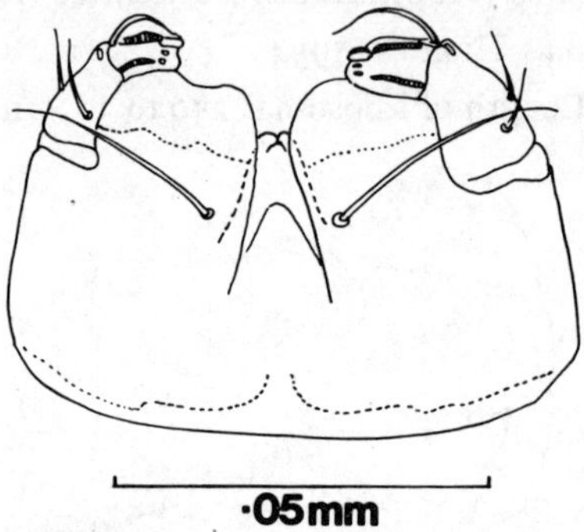

FIG. 225. *Pyroglyphus* (*Hughesiella*) *africanus* (Hughes) ♀. Ventral view of gnathosoma.

All the legs are well-developed and end in a globular pretarsus and small claw. Legs III are longer and stouter than any of the others. Tarsus I is short, about the same length as the genu. The solenidion omega$_1$ is subapical, arising close to omega$_3$ and the famulus at the base of the pretarsus (Fig. 226a). Tarsus II is longer and omega$_1$ originates near the middle of this segment (Fig. 226b). On tibia II, phi is longer than that on tibia I and only one ventral seta is developed on tibiae I and II. Two solenidia arise from genu I, sigma$_2$ being longer than sigma$_1$. At the ventral distal end of tarsus III are two blunt projections (Fig. 227a). Two short cylindrical projections are also found on the dorsal side at the distal end of tarsus IV (Fig. 227b) and may represent the rudiments of tarsal suckers. Phi on tibia III is about the same length as that on tibia IV.

FEMALE (Fig. 228). Length of idiosoma: 350 to 450 μm.

The dorsal shield is narrower in the female and only covers about half the width of the propodosoma. No hysterosomal shield is present, so that the pleat-like thickenings of the cuticle are more extensive in the female than in the male. The genital opening is shaped like an inverted U and covered by the genital plate hinged at its posterior end. The lateral walls of the opening are supported by genital plates on which can be seen traces of genital sense organs (Fig. 229). The anterior margin of the genital plate is indented. The opening of the bursa copulatrix is at the base of a small non-sclerotized pocket near the posterior end of the anus.

The arrangement and length of the body setae are the same as in the male, except that an additional pair of setae *sa e* is present.

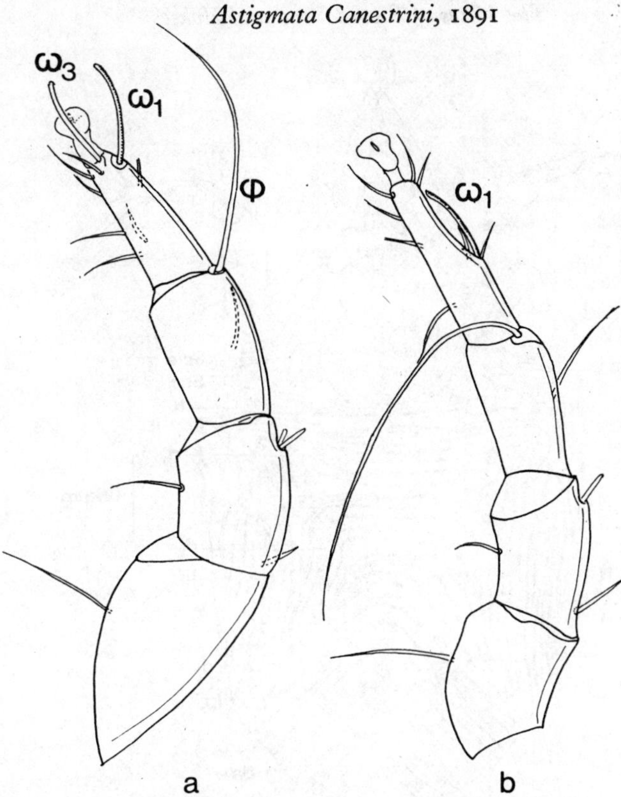

FIG. 226. *Pyroglyphus* (*Hughesiella*) *africanus* (Hughes) ♂. (*a*) Lateral view of left leg I; (*b*) lateral view of left leg II. Solenidia: ω_1, ω_2, Φ.

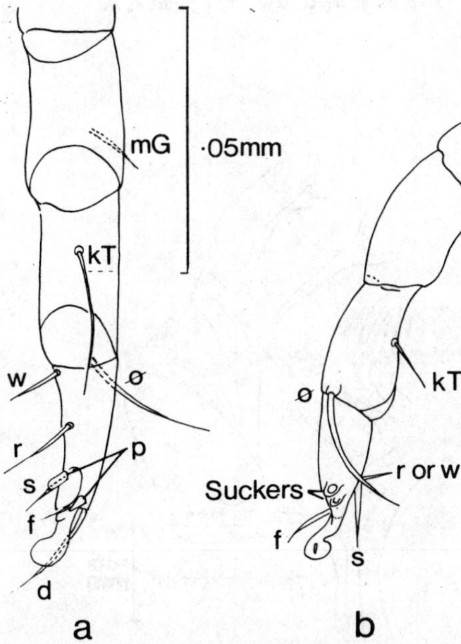

FIG. 227. *Pyroglyphus* (*Hughesiella*) *africanus* ♂. (*a*) Ventral view of left leg III; (*b*) dorsal view of left leg IV. Setae: *d*, *f*, *r*, *s*, *w*, *kT*, *mG*; solenidion: Φ. *p* horn-like projection.

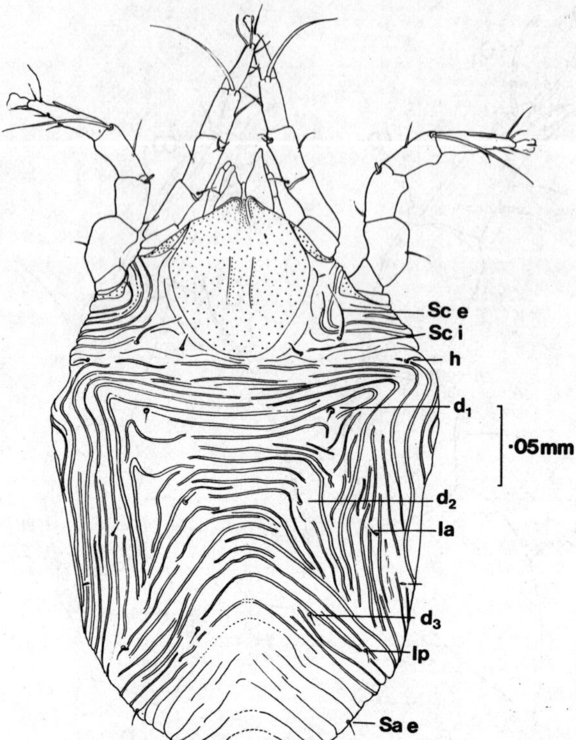

FIG. 228 *Pyroglyphus* (*Hughesiella*) *africanus* (Hughes). Dorsal view of ♀. Setae of idiosoma: *sc e*, *sc i*, d_1 to d_3, *l a*, *l p*, *sa e*, *h*.

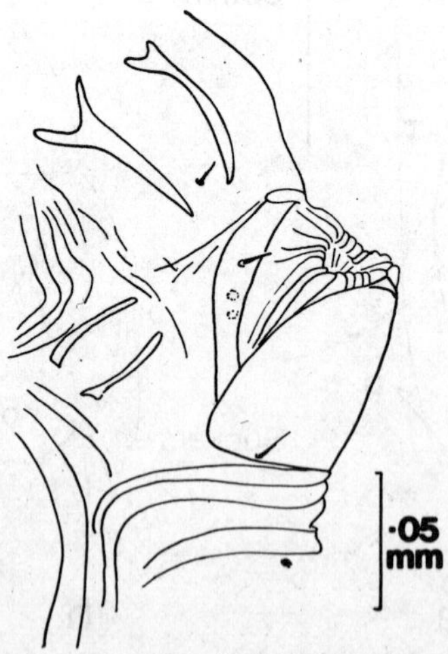

FIG. 229. *Pyroglyphus* (*Hughesiella*) *africanus* (Hughes) ♀. Lateral view of genital region.

Legs I and II closely resemble those of the male in length and chaetotaxy. Legs III and IV are considerably longer and more slender (Fig. 230a, b). The two processes and the vestigial suckers at the base of tarsi III and IV are not developed, but seta *e* is present. On tibia IV, phi is shorter in the female than in the male.

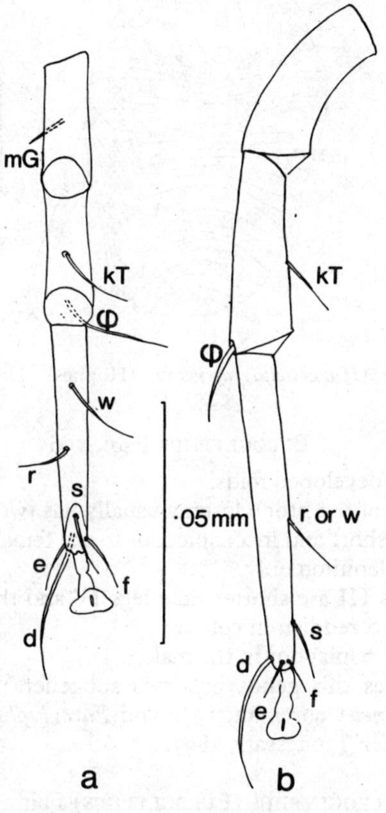

FIG. 230. *Pyroglyphus* (*Hughesiella*) *africanus* (Hughes) ♀. (*a*) Ventral view of right leg III; (*b*) dorsal view of left leg IV. Solenidion: *φ*; setae: *d, e, f, r, s, w, kT, mG*.

LARVA (Fig. 231).

This resembles the female. On tarsus I, the solenidion omega$_1$ is terminal, as it is in the nymphs. No coxal rods are present.

BIOLOGY AND DISTRIBUTION

Pyroglyphus (*H.*) *africanus* was found in large numbers on Angolan fishmeal stored in warehouses in Suffolk and Huntingdonshire, England, the fishmeal having been imported from West Africa. No cases of dermatitis were recorded in connection with the handling of the meal.

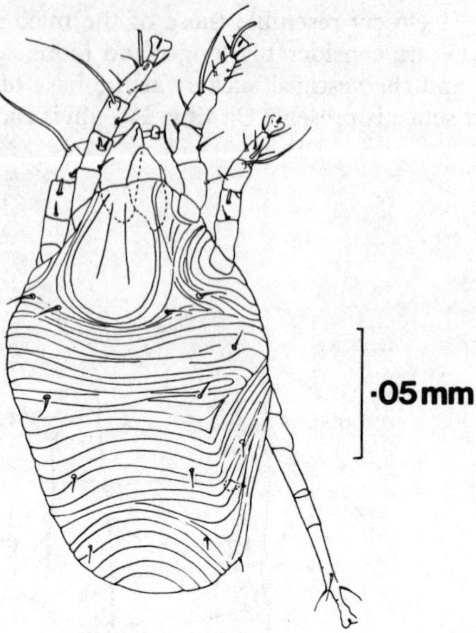

FIG. 231. *Pyroglyphus* (*Hughesiella*) *africanus* (Hughes). Dorso-lateral view of larva.

EUROGLYPHUS Fain, 1965

The cuticle has well-developed folds.

The anterior edge of the propodosoma usually has two projections.

Postanal setae are short and inconspicuous in the female (Fig. 234).

Genu I has one solenidion only.

In the female, legs III are shorter than legs IV and the receptaculum seminis is well-sclerotized and reddish in colour.

Anal suckers are conspicuous in the male.

Fain (1965) divides this genus into two subgenera: *Euroglyphus* (*E.*), type species: *Mealia maynéi* Cooreman, 1950 and *Euroglyphus* (*Gymnoglyphus*), type species: *Mealia longior* Trouessart, 1897.

EUROGLYPHUS (EUROGLYPHUS) Fain, 1965

The chaetotaxy is reduced in both sexes: setae are absent from trochanters I to III and tibia IV; tarsus III has only five setae, and tarsus IV three. One pair of anal and one or two pairs of genital setae are present in both sexes.

In the female, the posterior genital shield does not completely cover the genital opening.

Type species: *Mealia maynéi* Cooreman, 1950.

EUROGLYPHUS (EUROGLYPHUS) MAYNÉI (Cooreman, 1950)

Mealia maynéi Cooreman, 1950; *Dermatophagoides maynéi* (Cooreman, 1950) sensu Hughes, 1954.

MALE (Fig. 232). Length of idiosoma: about 200 μm.

The surfaces of the cuticle and dorsal shields are similar to those of the previous species. The dorsal propodosomal shield is smaller and pear-shaped. The longitudinal ridges extend to the anterior edge, sometimes giving this a bifid

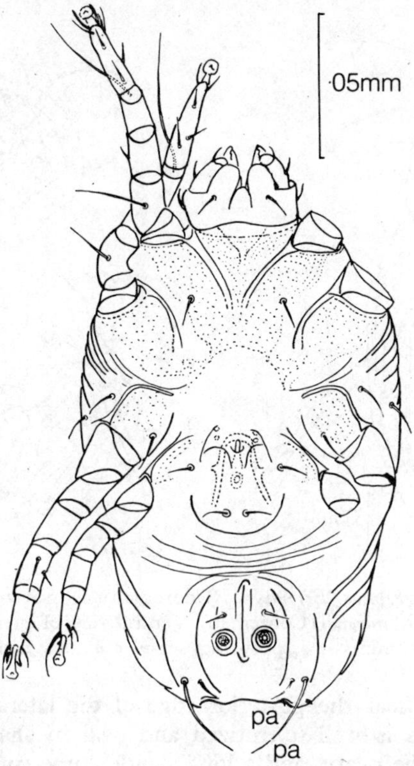

FIG. 232. *Euroglyphus* (*E.*) *maynéi* (Cooreman). Ventral view of ♂. Post-anal setae: *p a.*

appearance. The dorsal hysterosomal shield is less well-marked than in *P.* (*H.*) *africanus* and it only extends as far forward as the d_2 setae. The posterior margin of the body is distinctly excised in the mid-line (Fig. 232). The apodemes of legs I do not meet to form a sternum. The penis is a short straight tube and small genital sense organs are present. Conspicuous anal suckers are present, bounded by a sclerotized ring (Fig. 232).

All the body setae are smooth and short, except for the external pair of post-anal setae which is considerably longer than the internal.

All the legs terminate in a globular pretarsus, but no claw. Legs IV are rather shorter and narrower than legs III. Setae are missing from trochanters I to III and tibia IV. Tarsus III bears five setae, tarsus IV three setae, of which one is a peg-like structure at the distal end which may represent a vestigial sucker. A stout process is found at the end of tarsus III.

FEMALE (Figs. 233, 234). Length of idiosoma: about 280 μm.

The dorsal propodosomal shield is less well-marked than in the male and its anterior edge is smoothly rounded, not bifid as depicted by Fain (1965). No hysterosomal shield is delimited, but this region is devoid of cuticular ridges and its surface is punctate like that of the propodosomal shield and sternal regions.

The genital opening (Fig. 234) is only partly covered by the posterior genital plate whose anterior margin is pointed. None of my specimens has more than one pair of genital setae, although Fain (1965) states that a second pair may or

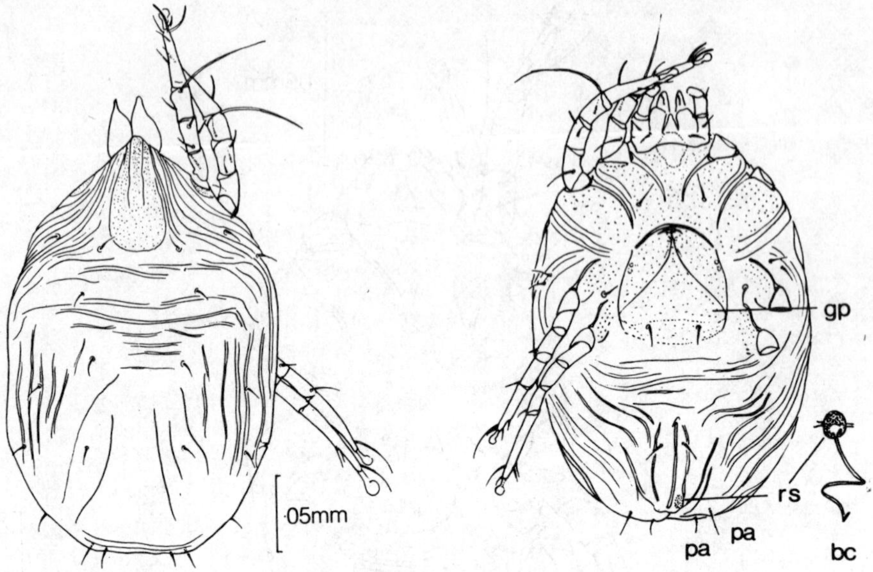

FIG. 233. *Euroglyphus (E.) maynéi* (Cooreman). Dorsal view of ♀ (left).
FIG. 234. *Euroglyphus (E.) maynéi* (Cooreman). Ventral view of ♀ (right). Bursa copulatrix: b c, receptaculum seminis: r s; post-anal setae: *p a*; posterior genital plate: g p.

may not be present near the posterior edge of the lateral genital plate. The receptaculum seminis is well-sclerotized and oval in shape. From it paired ducts connect with the ovary, and a long slender tube communicates with the opening of the bursa copulatrix which is near the posterior end of the anus.

The chaetotaxy of the idiosoma is similar to that of the male, except that the two pairs of postanal setae are of about the same length.

The legs are slender, legs IV being longer than legs III.

BIOLOGY

Cooreman (1950) found *E. (E.) maynéi* on decomposing cotton seed cake at Gembloux, Belgium. It has since been found in house dust (Fain, 1965, Spieksma and Spieksma-Boezeman, 1967) and mattress dust (Maunsell, Wraith and Cunnington, 1968). In any sample of mattress dust infested with this species, females greatly outnumber the males.

DISTRIBUTION

England, Holland, Belgium, Italy, Denmark, Japan.

EUROGLYPHUS (GYMNOGLYPHUS) Fain, 1965

Setae are present on trochanters I to III and tibia IV; tarsus III bears six setae and tarsus IV three.

Two pairs of anal setae are present in the female and one pair in the male, and three pairs of genital setae in both sexes.

In the female, the posterior genital shield completely covers the genital opening.

Type species: *Mealia longior* Trouessart, 1897.

EUROGLYPHUS (GYMNOGLYPHUS) LONGIOR (Trouessart, 1897)

Mealia longior Trouessart, 1897; *Dermatophagoides longior* (Trouessart) sensu Hughes, 1954; *Dermatophagoides delarnaensis* Sellnick, 1958.

MALE (Figs. 235, 236). Length of idiosoma: about 265 μm.

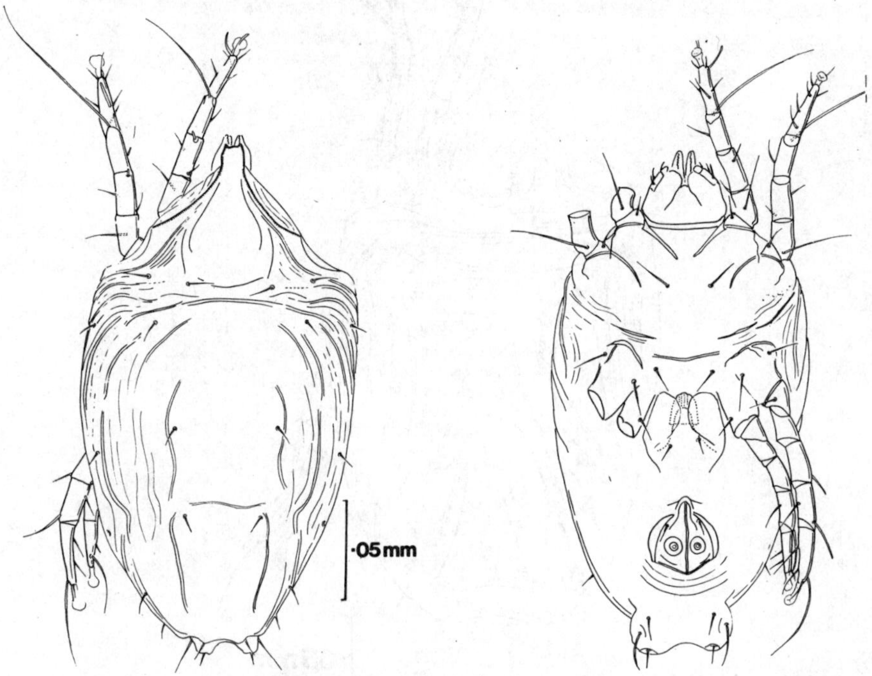

FIG. 235. *Euroglyphus* (*Gymnoglyphus*) *longior* (Trouessart). Dorsal view of ♂ (left).
FIG. 236. *Euroglyphus* (*Gymnoglyphus*) *longior* (Trouessart). Ventral view of ♂ (right).

The body is more elongated and spindle-shaped than in the previous species, the anterior margin of the propodosoma being raised into a median ridge which projects forward over the gnathosoma and ends in a toothed edge, sometimes asymmetrical. The posterior ventral end of the body is prolonged into a bifid shelf, from which arise the posterior setae. The dorsal shields are ill-defined, the propodosomal being narrow anteriorly and expanding towards the scapular setae, the hysterosomal covering most of the dorsal region. The remaining cuticle is finely wrinkled and also thrown into a few irregular coarse folds on the sides of the body. The sternal and ventral regions are almost smooth. The apodemes of the legs end freely, those of legs IV being ill-defined; apodemes III have an inwardly directed projection. The anal opening is remote from the posterior end of the body, it is flanked by anal suckers and encircled by a sclerotized ring (Fig. 236). The setae arising from the idiosoma are short and fine. Three pairs of genital setae encircle the genital opening.

All the legs are the same thickness and end in a pretarsus and very small claw. Legs III are slightly longer than legs IV. On legs I, omega$_1$ and omega$_3$ are terminal and one solenidion arises from genu I. The tibial solenidia are well-developed. Tarsus IV bears three setae and two subapical peg-like structures which may represent vestigial suckers.

FEMALE (Fig. 237). Length of idiosoma: about 280 to 320 μm.

FIG. 237. *Euroglyphus* (*Gymnoglyphus*) *longior* (Trouessart). Dorsal view of ♀.

The female closely resembles the male, except that the thickening of the cuticle is more pronounced and covers almost the whole of the dorsal surface. The posterior margin of the body is slightly concave. The genital opening is completely covered by a sclerotized triangular plate. Small genital sense organs are present and three pairs of genital setae (Fig. 238b). The opening of the bursa copulatrix is near the posterior end of the anus and this leads into an oval well-sclerotized receptaculum seminis.

BIOLOGY

E. longior was found originally in the dust collected from prepared mammalian skins which had been attacked by insects and mites (quoted from Fain, 1965). It was also found by G. E. Woodroffe at Slough, Berks, England, in the dust and debris of a granary. Sellnick (1958) and Johnston (Fain, 1965) also found it in granary dust.

DISTRIBUTION

England, Sweden, France, Ohio U.S.A.

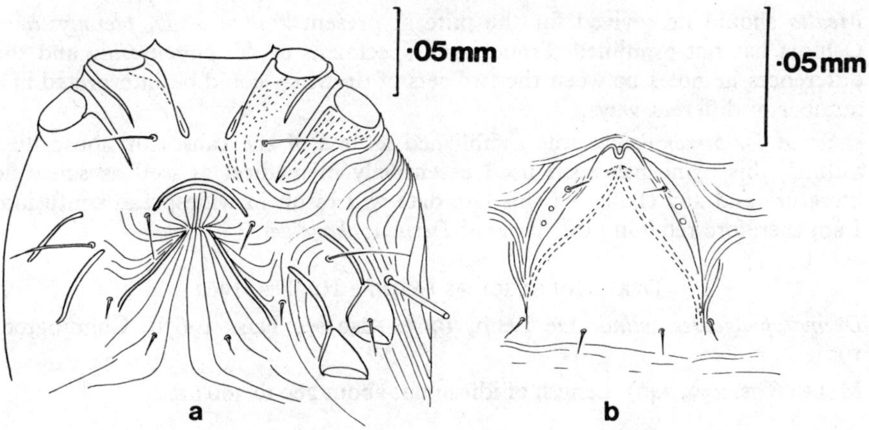

FIG. 238. Genital region of ♀ of (*a*) *Dermatophagoides farinae* Hughes; (*b*) *Euroglyphus* (*Gymnoglyphus*) *longior* (Trouessart).

DERMATOPHAGOIDINAE Fain, 1963 sensu Fain, 1967

The cuticle is less densely sclerotized than in the subfamily Pyroglyphinae.

The anterior border of the propodosoma does not overhang the gnathosoma.

Two pairs of long setae arise from the posterior margin of the body.

In the female, the genital shield is usually not sclerotized and legs IV are usually shorter and thinner than legs III.

Tarsus IV of the male has two disc-shaped tarsal suckers.

Fain (1967b) includes two genera in this subfamily: *Dermatophagoides* Bogdanov, 1864 and *Sturnophagoides* Fain, 1967. In *Dermatophagoides* (type species: *D. scheremetewski* Bogdanov, 1864) the cuticle is finely striated and in the female, the posterior genital shield is of average size, not sclerotized and lacking a bifid anterior margin. No hysterosomal shield is present. In *Sturnophagoides* (type species: *S. bakeri* Fain, 1967), the cuticular striations are finely punctate, the posterior genital shield in the female is well-developed and sclerotized and its anterior edge is deeply incised. A hysterosomal shield is present in the female.

The genus *Dermatophagoides* was established by Bogdanov, 1864 for *Dermatophagoides scheremetewski*, one specimen of which (a female) was found on human skin. A second specimen (male) was found on another patient and may belong to another species. The original specimens cannot be traced and no neotype has been established.

The genus *Mealia* was erected by Berlese (1897) under Trouessart's name. The specimens were found on prepared animal skins and *Mealia pteronyssina* was listed as the type species. Fain (1966) has compared specimens of house dust mite, now known as *Dermatophagoides pteronyssinus*, with those found by Trouessart. He believes them to be the same species and has published a description of a female lectotype.

In 1952 Baker and Wharton established the synonymy of *Mealia* and *Dermatophagoides*, and since then the name *Mealia* has lapsed.

Oshima (1968) compared Bogdanov's drawings of *D. scheremetewski* with Trouessart's drawings of *Mealia pteronyssina* and lists a number of characters in which they differ from one another. He concludes that these two species do not belong to the same genus and suggests that Trouessart's generic name of

Mealia should be revived for the mite at present known as *D. pteronyssinus*. Oshima has not examined Trouessart's specimens of *M. pteronyssina* and the differences he notes between the two sets of drawings could be interpreted in a number of different ways.

Since *D. pteronyssinus* was established as one of the causes of house dust asthma, this name has been used extensively in medical as well as scientific literature and any change at this late date will result in widespread confusion. I am therefore retaining the name of *Dermatophagoides*.

DERMATOPHAGOIDES FARINAE Hughes, 1961

Dermatophagoides culinae De Leon, 1963. Also see Fain, 1967b, Cunnington, 1971.

MALE (Figs. 239, 240). Length of idiosoma: about 260 to 360 µm.

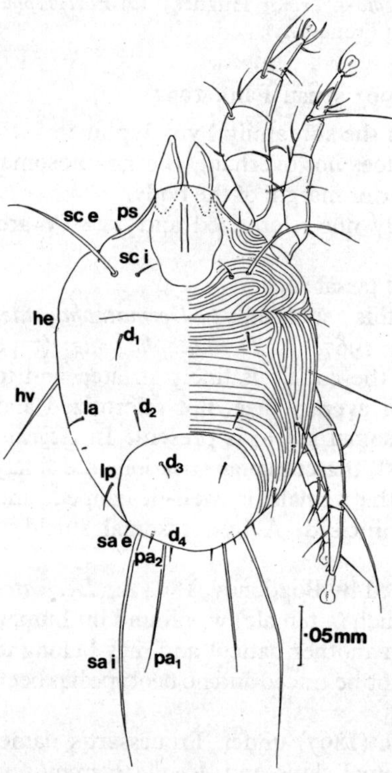

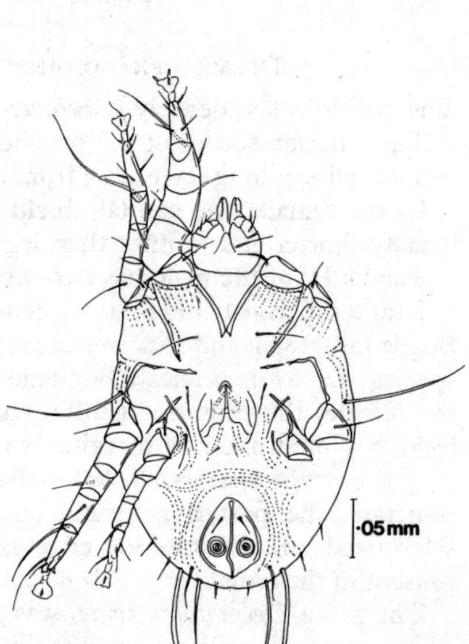

FIG. 239. *Dermatophagoides farinae* Hughes. Dorsal view of ♂ (left). Setae of the idiosoma: *sc c, sc i, h e, h v,* d_1 to d_4, *l a, l p, sa e, sa i,* $p\ a_1$, $p\ a_2$. *ps* supracoxal seta.

FIG. 240. *Dermatophagoides farinae* Hughes. Ventral view of ♂ (right).

The idiosoma is rounded, the constriction between propodosoma and hysterosoma ill-defined. The propodosomal shield is variable in shape and may have the posterior edge extended laterally to encircle the scapular setae. The hysterosomal shield does not extend as far forward as the d_2 setae. The coxal region is well sclerotized and finely punctate. Apodemes I may remain separate or fuse in the midline to form a short sternum. Apodemes III are long and bent sharply at a right angle. The genital opening is between coxae III and IV and the anus is encircled by an oval perianal ring which also encloses prominent anal suckers and a pair of preanal setae. The hind edge of this ring is convex posteriorly. The penis is long and tapering and arises from a triangular base (Fig. 241).

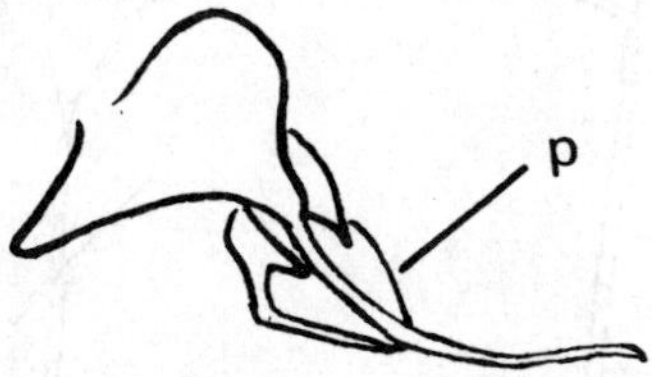

FIG. 241. *Dermatophagoides farinae* Hughes. Penis (*p*) of ♂.

All the idiosomal setae are smooth. Setae *sc e* are more than four times longer than *sc i* and a supracoxal seta is present. Two pairs of humeral setae are present, *h v* arising lateral to coxae III and being about the same length as *sc e*. Four pairs of short dorsal setae are arranged in two longitudinal rows which converge posteriorly, the setae are about the same length as *l a*, *l p*, *sa e* and $p\,a_2$; $p\,a_1$ and *sa i* are considerably longer, *sa i* exceeding half the length of the idiosoma. The usual coxal setae are present on coxae I and III and three pairs of setae surround the genital opening, the posterior pair is much shorter than the remaining pairs.

All the legs end in prominent pretarsi and small claws. Legs I (Fig. 242) are considerably thickened and compressed bilaterally, so that when alive, the animal resembles *Acarus siro*. On femur I a blunt process arises from the ventral surface.

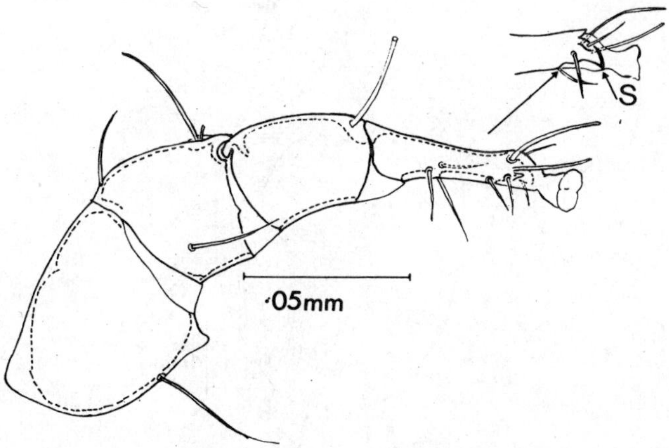

FIG. 242. *Dermatophagoides farinae* Hughes. ♂. Internal view of right leg I and lateral view of apex of tarsus. *S* blunt process.

On tarsus I, $omega_1$ arises at the base of the pretarsus on the same level as $omega_3$. The famulus is subapical. The lateral distal end of this tarsus is produced into a blunt process and on tarsus III, a similar process has a bifid extremity and a smaller process is present on the opposite side of the segment (Fig. 243a). $Omega_1$ arises from the base of tarsus II. One ventral seta is present on tibiae I and II. Two solenidia arise from genu I, one much longer than the other. Legs

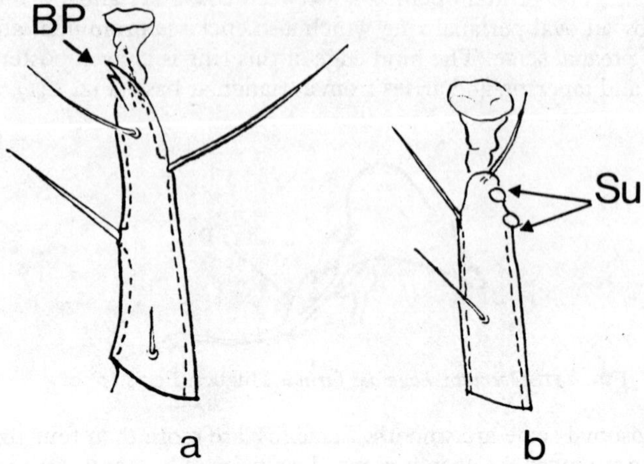

FIG. 243. *Dermatophagoides farinae* Hughes. ♂. Apex of (*a*) tarsus III, (*b*) tarsus IV. BP bifid process; Su suckers.

III are noticeably longer and thicker than legs IV. On legs IV two small stalked suckers are present at the end of the tarsus (Fig. 243b).

FEMALE (Fig. 244). Length of idiosoma: about 360 to 400 µm.

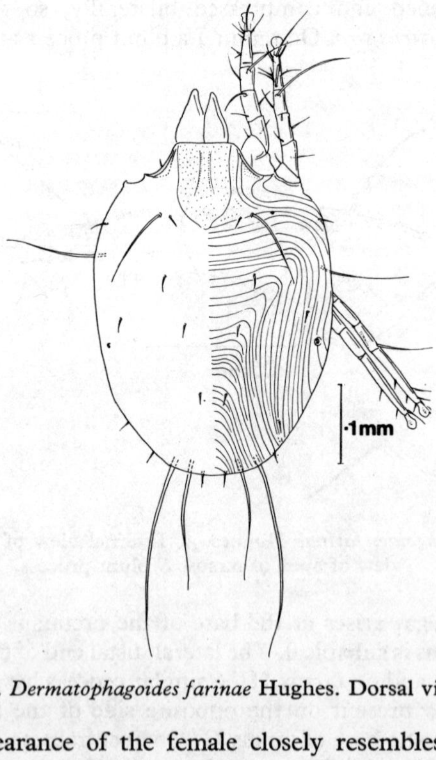

FIG. 244. *Dermatophagoides farinae* Hughes. Dorsal view of ♀.

The general appearance of the female closely resembles that of the male, except that a hysterosomal shield is not present. The dorsal region of the idiosoma is covered by cuticle which is transversely striated. The sternal region is not so heavily sclerotized. Apodemes I are widely separated from one another and apodemes III are not bent at right angles. The genital opening is ⅄-shaped and bounded anteriorly by a crescentic epigynium. The posterior genital plate (Fig. 238a) is only well-sclerotized along its lateral edges. The opening of the bursa copulatrix is on the dorsal surface above the anal region, it communicates with the flask-shaped sclerotized region of the receptaculum seminis by a fine duct (Fig. 245a).

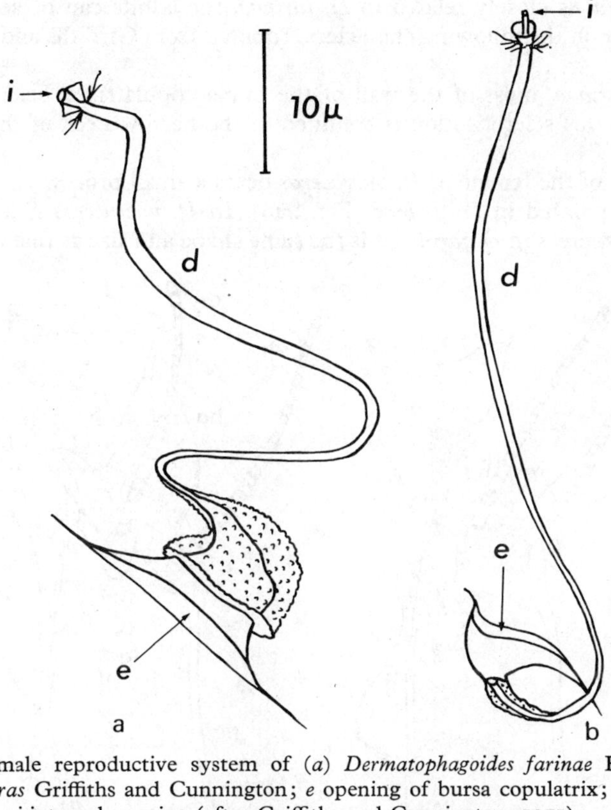

FIG. 245. Female reproductive system of (*a*) *Dermatophagoides farinae* Hughes and (*b*) *D. microceras* Griffiths and Cunnington; *e* opening of bursa copulatrix; *d* fine duct; *i* internal opening (after Griffiths and Cunnington, 1971).

Legs I and II are about the same length and width, legs IV are longer than legs III. The chaetotaxy is similar to that of the male, except that two short setae replace the vestigial suckers found on tarsus IV of the male.

BIOLOGY

Large numbers of these mites were isolated from poultry and pig-rearing meal near Bristol (G. C. Williams coll.). They have also been found in a culture of *Attagenus piceus* on dogmeal and yeast in Manchester, on wheat pollards from Argentina (quoted from Fain, 1967b) and on biscuit meal (De Leon, 1963). It has been recorded from house dust by Spieksma (1967) and others from many parts of the world. In the U.S.A. *D. farinae* appears to be more plentiful in house and mattress dust than *D. pteronyssinus* (Wharton, 1970). Larson, Mitchell and Wharton (1968, 1969) have cultured *D. farinae* on Gainesburger dogfood at room temperature and 75 per cent relative humidity. Development from egg to egg takes 30 days under these conditions with the egg stage lasting six to eight days. Females do not lay eggs unless fertilized, mating occurs more than once, and one to two eggs per day are laid for at least 30 days.

DISTRIBUTION

England, Holland, U.S.A., Japan, U.S.S.R., Sierra Leone.

DERMATOPHAGOIDES MICROCERAS Griffiths and Cunnington, 1971

D. microceras is closely related to *D. farinae*, the adults can be separated from one another on the following characters: (quoted from Griffiths and Cunnington, 1971).

In *D. farinae*, most of the wall of the bursa copulatrix is sclerotized, in *D. microceras*, this sclerotization is confined to the narrow neck of the bursa (Fig. 245b).

Tarsus I of the female of *D. microceras* bears a small process *S*. This is much larger and pointed in *D. farinae* (Fig. 246). In *D. microceras S* is absent from tarsus II, whereas in *D. farinae* it is the same shape and size as that on tarsus I.

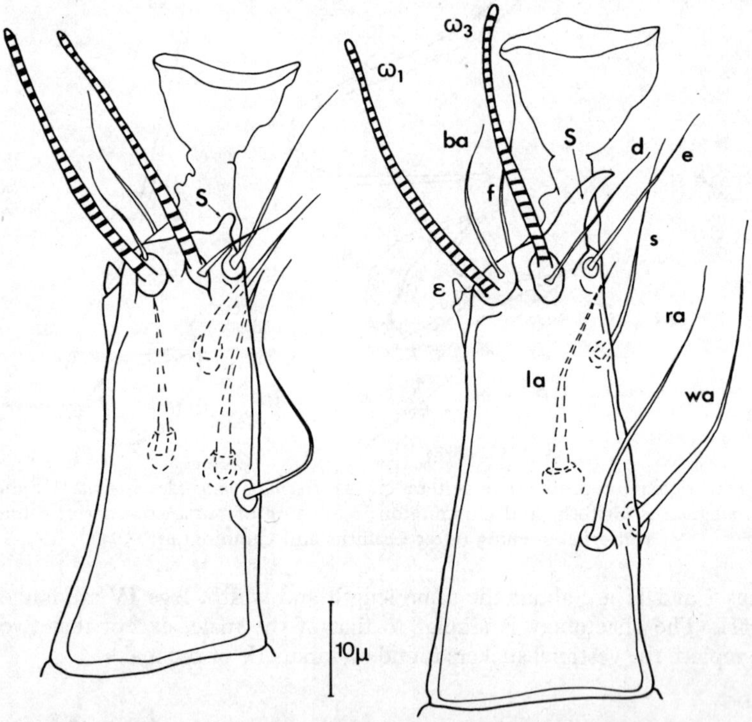

FIG. 246. Tarsus I of *Dermatophagoides microceras* Griffiths and Cunnington (left) ♀ and *D. farinae* Hughes (right) ♀. Solenidia: ω_1, ω_3; setae: *d, e, f, ba, la, ra, wa*; famulus: ε; chitinous process: S (after Griffiths and Cunnington, 1971).

Tarsus I of the male of both species terminates in a large claw-like process which is not homologous with process *S*. On tarsus I of the *D. microceras* male, process *S* is absent and on tarsus I in *D. farinae* it is present as a small blunt protuberance external to the large claw-like structure (Fig. 247).

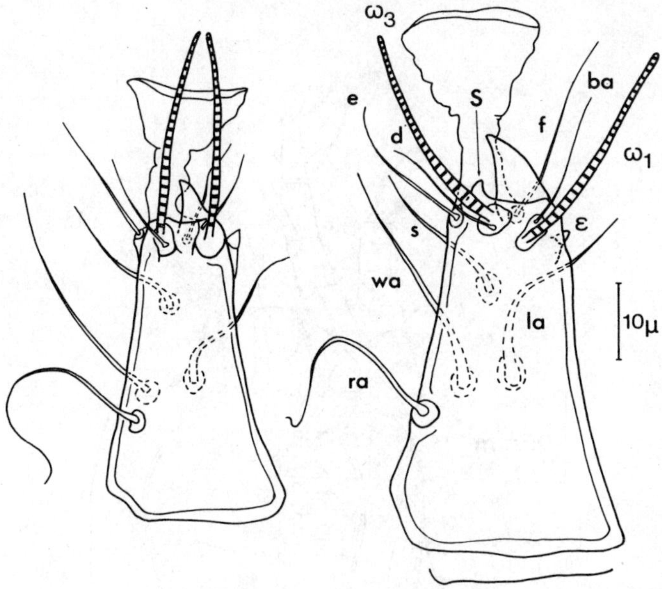

FIG. 247. Tarsus I of *Dermatophagoides microceras* Griffiths and Cunnington ♂ (left) and (*b*) *D. farinae* Hughes ♀ (right). Solenidia: ω_1, ω_3; setae: *d, f, s, ba, la, ra, wa*; famulus: ε; chitinous process: S (after Griffiths and Cunnington, 1971).

On tarsus II of the male *D. microceras*, process *S* is absent, whereas it is present as a distinct finger-like projection on tarsus II of *D. farinae*.

BIOLOGY

This species has been recorded from house and mattress dust, from inside a feather mattress and inside a feather cushion. It is reproductively isolated from *D. farinae*.

DISTRIBUTION

England, Spain, New Orleans U.S.A.

Dermatophagoides evansi Fain, Hughes and Johnston has also been found in feather pillows which were being made in Boston, England (see Fain, 1967b).

DERMATOPHAGOIDES PTERONYSSINUS (Trouessart, 1897)

Mealia toxopei Oudemans, 1928, *Visceroptes saitoi* Sasa, 1948 (Fain, 1966).

MALE (Fig. 248). Length of idiosoma: about 285 μm.

This differs from the previous species as follows: the hysterosomal shield is more extensive and extends forwards to a point mid-way between d_1 and d_2 (Fig. 248). Apodemes I are always widely separated from one another and do not meet to form a sternum. The coxal regions of legs I to IV are less extensively sclerotized and the posterior genital setae are vestigial.

The first and second pairs of legs are almost equal in length and width and the blunt process at the end of tarsus I is less conspicuous (Fig. 249a). Tarsi III and IV (Figs. 249b, c) are similar to those of *D. farinae*.

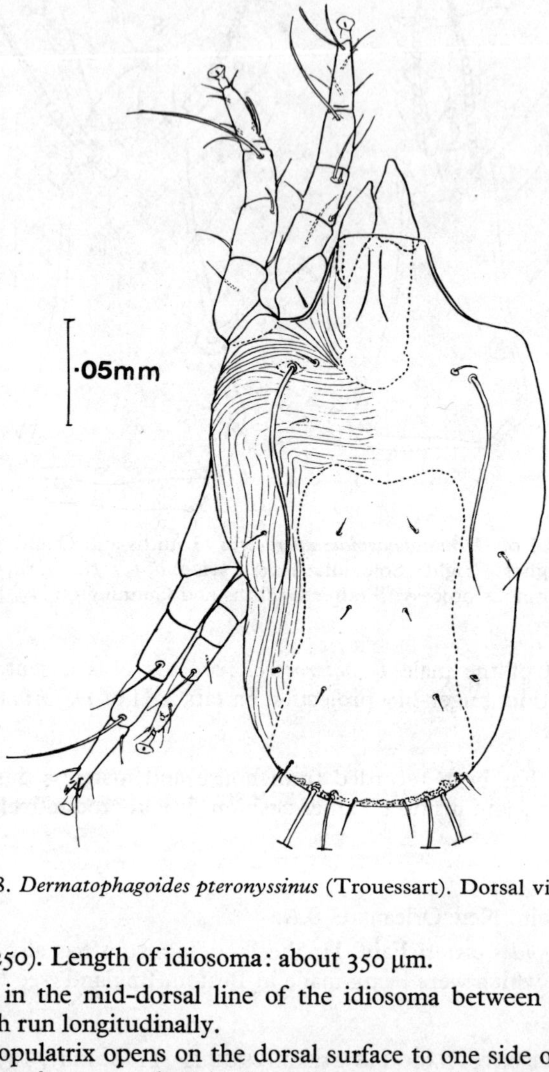

FIG. 248. *Dermatophagoides pteronyssinus* (Trouessart). Dorsal view of ♂.

FEMALE (Fig. 250). Length of idiosoma: about 350 μm.

The cuticle in the mid-dorsal line of the idiosoma between d_2 and d_3 has striations which run longitudinally.

The bursa copulatrix opens on the dorsal surface to one side of the posterior end of the anus. A narrow duct leads to the receptaculum seminis and opens at the base of a depression which has sclerotized walls (Fig. 251a). This depression has a characteristic appearance when seen in end-on view (Fig. 251b).

BIOLOGY

D. pteronyssinus was first discovered in the dust which was shaken from tanned mammal skins (Trouessart, 1901). Recently it was found that it is widely distributed in house dust (Voorhoorst, Spieksma, M.I.A. and Spieksma, F.Th.M., 1964) and the surface dust of mattresses (Maunsell, Wraith and Cunnington, 1967) where it feeds on desquamated skin scales. It has been shown by Voorhoorst *et al.*, (1964) to be the agent mainly responsible for house dust allergy.

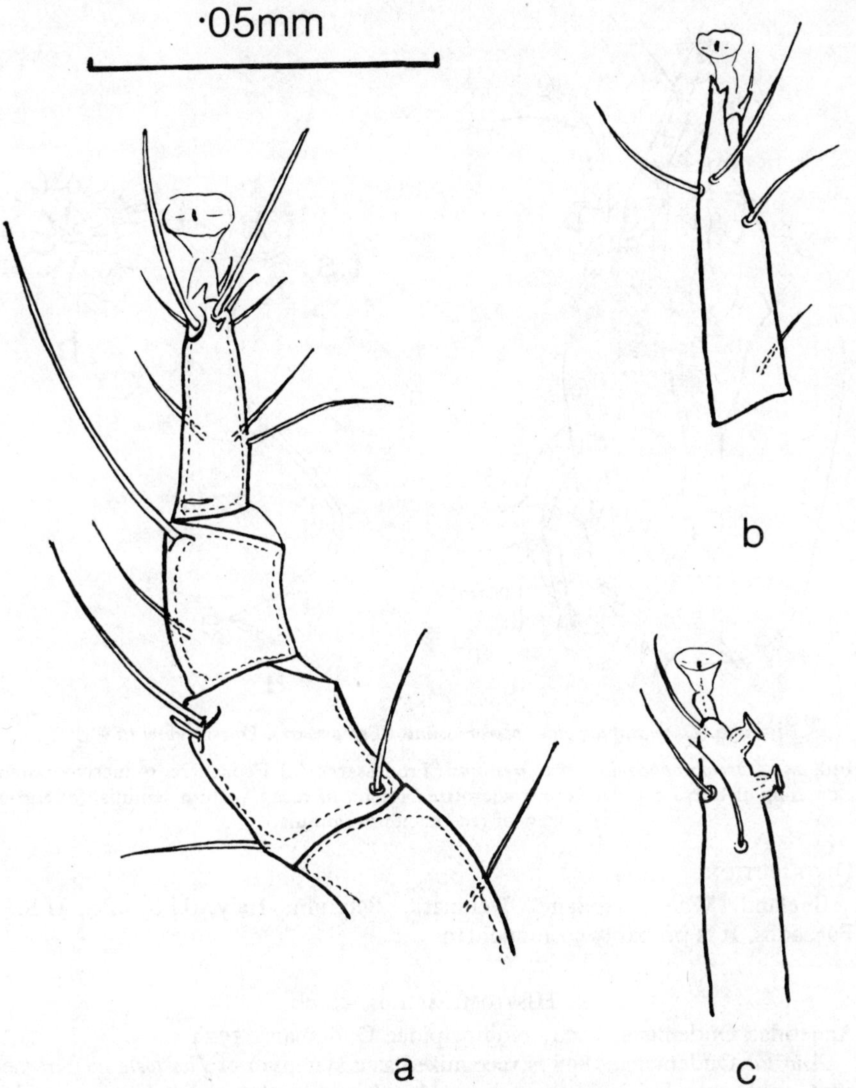

FIG. 249. *Dermatophagoides pteronyssinus* (Trouessart) ♂. (*a*) Distal segments of right leg I; (*b*) right tarsus III, (*c*) right tarsus IV.

The numbers of mites present in house dust is related to the dampness of the house, the damper houses harbouring the largest numbers of mites (Spieksma and Spieksma-Boezeman, 1967). These authors showed that there is a seasonal increase in the number of mites, the numbers starting to increase in early summer and reaching a maximum in early autumn. The same increase in numbers is shown by the mites living in mattress surface dust. The numbers remain relatively constant during the winter months, but in England start to increase in early May when a rise in relative humidity coincides with a rise of temperature (Plate 4, Fig. 2). (Hughes and Maunsell, 1973).

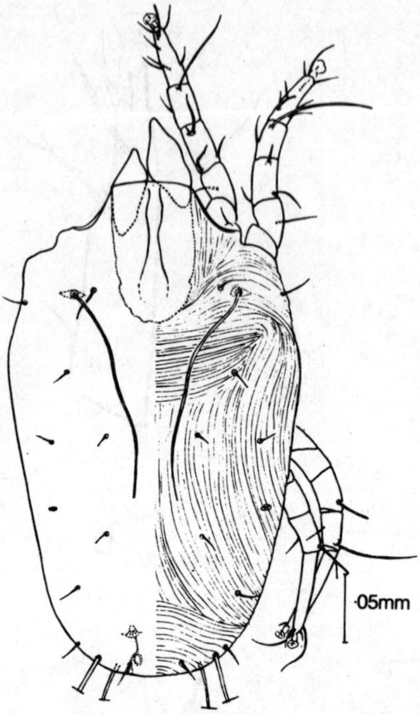

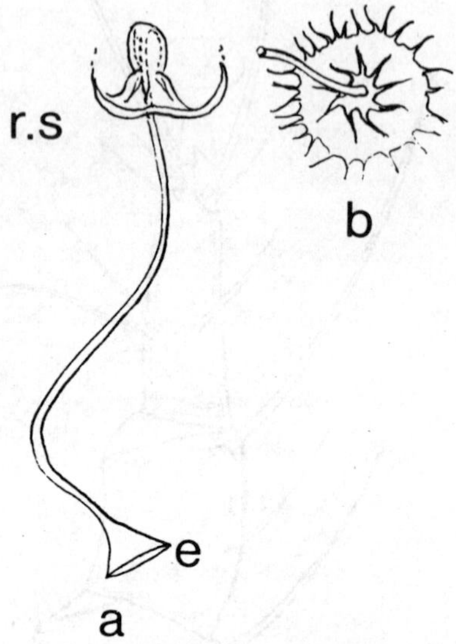

FIG. 250. *Dermatophagoides pteronyssinus* (Trouessart). Dorsal view of ♀ (left).

FIG. 251. *Dermatophagoides pteronyssinus* (Trouessart). (*a*) Female reproductive system. *e* opening of bursa copulatrix; *r s* sclerotized region of receptaculum seminis, (*b*) end-on view of receptaculum seminis.

DISTRIBUTION

England, Wales, Holland, Denmark, Belgium, Italy, U.S.S.R., U.S.A., Barbados. It is probably cosmopolitan.

HISTIOMIDAE NOV. comb.

Anoetidae Oudemans, 1904; Nodipalpidae Oudemans, 1924.

Anoetus Oudemans, 1898 is recognized as a synonym of *Histiostoma* Kramer, 1876 so that the family name of Anoetidae should be changed to Histiostomidae.

The histiostomid mites have a highly modified gnathosoma adapted for feeding on minute particles in suspension. The chelicerae are saw-shaped and the pedipalps have a freely movable and flattened distal segment. Two pairs of oval or circular sclerotized rings are found on the ventral side of the idiosoma. These mites are free-living and semi-aquatic in habit. A motile hypopus is often developed which clings to insects and has legs III and sometimes IV directed forwards. The anteriorly-directed third pair of legs is considered by Hall (1959) to be responsible for the jumping of the hypopi of *Histiostoma laboratorium*. Hall noticed that the coxal articulation of the third pair of legs has much greater freedom of movement and, when these legs are directed anteriorly, a quick downward push would throw the hypopus into the air and enable it to cling to a passing insect.

Many species are found on wet vegetable matter where decomposition is taking place. Some are found on decaying mushrooms (e.g., *Glyphanoetus fulmeki* Oudemans, 1929), on cow dung (*G. phyllotrichus* Berlese, 1881), others live a semi-aquatic existence in the resins and saps which exude from the damaged bark of trees (e.g., *Histiostoma pulchrum* Kramer, 1886). *H. laboratorium* R. Hughes, 1950 is a common pest of *Drosophila* cultures. In all these habitats, the food is of a fluid or semi-fluid nature and the mouth parts have become modified to gather up particles suspended in the fluid (T. Hughes, 1953). *Histiostoma feroniarum* (Dufour, 1839) is occasionally found on decaying food material such as grain or flour, too wet to support a population of *Rhizoglyphus* and *Caloglyphus*; *H. sapromyzarum* (Dufour, 1839) has also been found, on one occasion, on some damp timber which had been imported into this country.

R. S. Scheucher (1957) has revised this family and R. D. Hughes and C. G. Jackson (1958) have, in addition, extensively studied the biology of the group.

HISTIOSTOMA Kramer, 1876

Anoetus Oudemans, 1898; *Phyllostoma* Kramer, 1876; *Zschachia* Oudemans, 1929.

The female genital opening is a transverse slit lying between the anterior pair of chitinous rings. The posterior chitinous rings are on the same level as coxae IV, or between coxae III and IV.

The male has no genital sense organs and the penis is barely protrusible.

On tarsus I, all the setae except *d* are thickened as spines. Solenidion phi on tibiae I and II is short and inconspicuous.

The hypopi usually have a sucker plate with eight pairs of suckers, and suckers are often present on coxae I and II. Legs III and IV are usually directed anteriorly.

Type species: *Hypopus feroniarum* Dufour, 1830.

HISTIOSTOMA FERONIARUM (Dufour, 1839)

Hypopus feroniarum Dufour, 1839; *Hypopus dugèsi* Claparède, 1868; *Tyroglyphus rostro-serratum* Mégnin, 1873; *Histiostoma pectineum* Kramer, 1876; *Acarus mammilaris* Canestrini, 1878; *Histiostoma sapromyzarum* (Dufour, 1839) sensu Cooreman, 1944.

FEMALE (Figs. 252, 253). Length of idiosoma: 400 to 700 μm.

The idiosoma is whitish in colour and usually contains large quantities of guanin in the form of a chalky deposit. The gnathosoma (Fig. 254a) is rather small in proportion to the rest of the body. Each chelicera consists of the elongated movable digit which has a serrated edge, and can move freely backwards and forwards inside an extensive pre-oral trough. The lateral wall of this trough is formed by the base of the pedipalp, and the terminal segment of the palp itself is a flattened bilobed chitinous plate bearing a pair of spines, one directed laterally, the other laterally and posteriorly. The chitinous plate is freely movable, and its movement serves to sweep particles in the fluid medium towards the anterior end of the gnathosoma. The alternate backward and forward movement of the chelicerae then rakes the solid particles towards the mouth (T. Hughes, 1953).

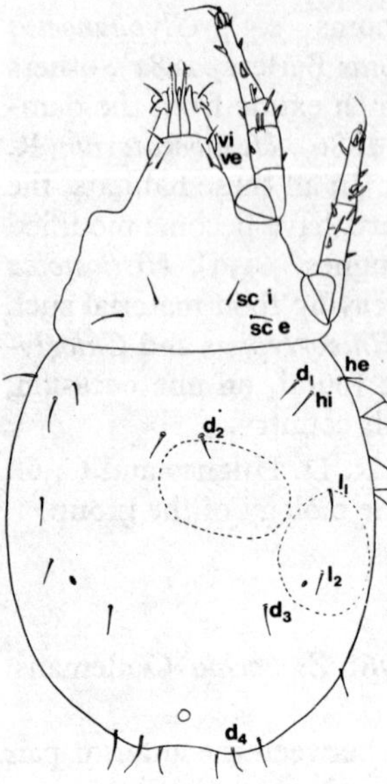

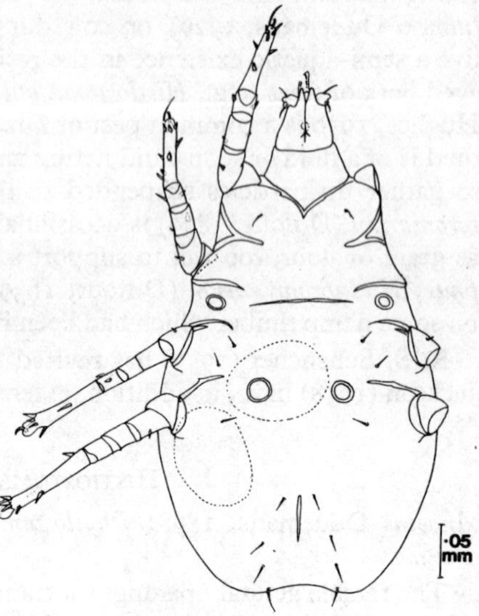

FIG. 252. *Histiostoma feroniarum* (Dufour). Dorsal view of ♀ (left). Setae of the idiosoma: *v e, v i, sc e, sc i, h e, h i*, d_1 to d_4, l_1, l_2.

FIG. 253. *Histiostoma feroniarum* (Dufour). Ventral view of ♀ (right).

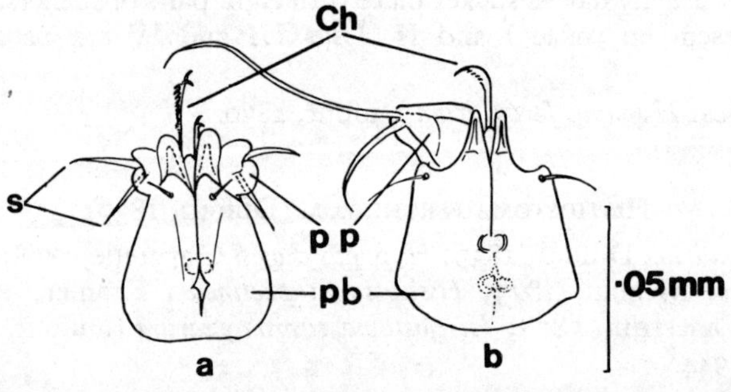

FIG. 254. Ventral view of gnathosoma of (*a*) *Histiostoma feroniarum* (Dufour) ♀ and (*b*) *H. sapromyzarum* (Dufour) ♀. Ch chelicera, pp terminal segment of pedipalp, pb base of pedipalp, S spines on pedipalp.

The cuticle of the idiosoma is covered with fine projections to which small particles of fungal hyphae and fine debris readily adhere. A transverse groove separates the propodosoma from the hysterosoma, and the posterior margin of the body is slightly indented. On the ventral surface, the two pairs of chitinous rings are circular or almost circular, the anterior pair lying between coxae II and III on either side of the transverse genital opening, the posterior being nearer together and lying at the same level as coxae IV. According to Perron (1954), these may function as osmo-regulating organs; alternatively, they could indicate the position of muscle insertions. The apodemes of legs I just meet in the midline, those of legs II to IV are short, their internal ends widely separated from one another. Possibly correlated with the nature of the food, the anus is unusually small and well-removed from the posterior margin of the body.

All the setae of the idiosoma are short, i.e., about the same length as tibia I. Setae *v i* are wide apart, about the same distance from each other as the more posterior *v e*. The scapular setae are widely separated from *v e* whereas *h e*, *h i* and d_1 arise close together in the anterior lateral corners of the hysterosoma and d_2 are nearer together than are d_3 and d_4. Setae d_4 are inserted near the posterior edge of the body and two pairs of lateral setae (l_1 and l_2) are in front of the latero-abdominal glands. Coxal setae are found on coxae I and III, and two pairs of genital setae arise in front of and behind the posterior chitinous rings. Four pairs of setae are arranged on either side of and behind the anus.

The legs are short and thick, closely resembling those of *Rhizoglyphus*. Each ends in a stout claw supported on paired struts, its base enveloped in a short pretarsus. As in *Rhizoglyphus* also, the setae are thickened to form spines, one (*ba*) lying immediately in front of omega$_1$ on tarsi I and II (Fig. 255a, b).

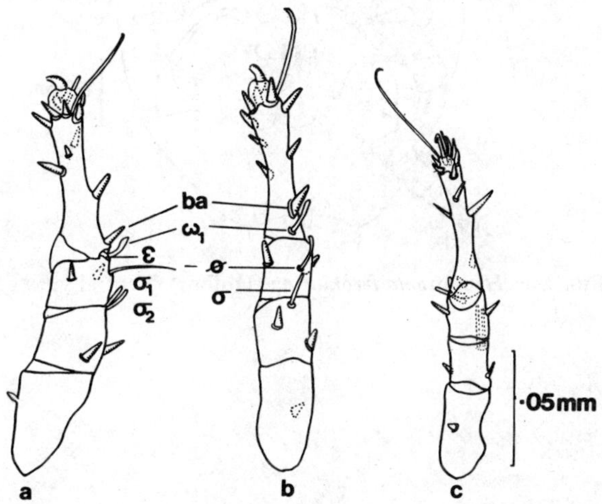

FIG. 255. *Histiostoma feroniarum* (Dufour) ♀. Lateral view of (*a*) right leg I and (*b*) right leg II. (*c*) *H. sapromyzarum* (Dufour) ♀. Ventral view of right leg I. Solenidia: ω_1, σ_1, σ_2; famulus: ε; *ba* tarsal seta.

Omega$_1$ arises from the extreme base of tarsus I and curves backward to overhang the anterior edge of tibia I, the famulus arising from the same rather deep pit. On tarsus II (Fig. 255b) omega$_1$ occupies its normal position and is only slightly curved. At the distal ends of all the tarsi, the ventral spines are well-developed. On tibia I and II, phi is shorter than tibia II, and on genu I, sigma$_1$ and sigma$_2$ are equal in length. Sigma is missing from genu III.

MALE (Fig. 256). Length of idiosoma: 250 to 500 μm.

The males vary in size and in thickness of the legs; in some, the legs and the spines are about the same size as in the female, in others they are considerably thickened and bear correspondingly well-developed spines which are particularly noticeable on tarsus II (Fig. 257). On the ventral side of the idiosoma, the apodemes are better developed than in the female, those of legs I meeting to form a well-developed sternum. The apodemes of legs II extend almost to the

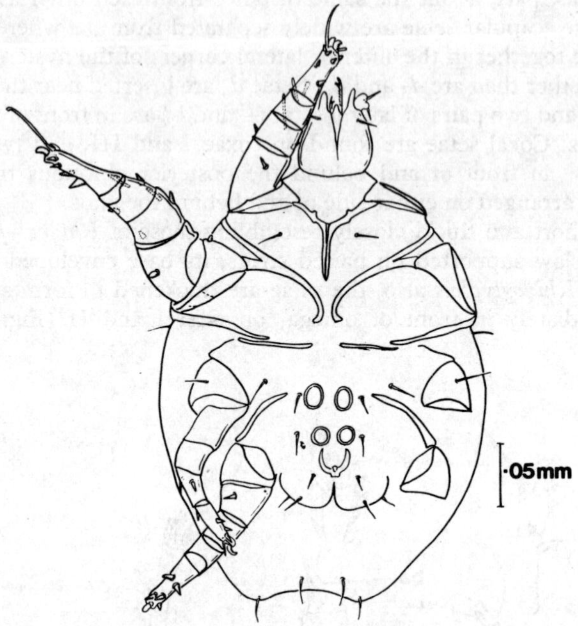

FIG. 256. *Histiostoma feroniarum* (Dufour). Ventral view of ♂.

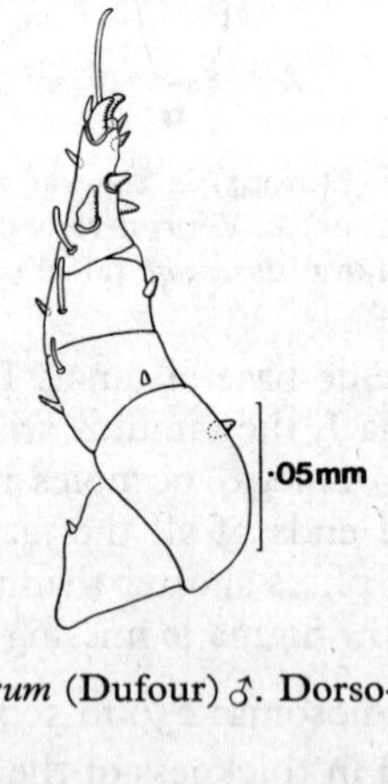

FIG. 257. *Histiostoma feroniarum* (Dufour) ♂. Dorso-lateral view of right leg II.

midline and are directed posteriorly at their internal ends. The four circular chitinous rings are close together, anterior to the genital opening. The genital sclerites are poorly developed and lie between coxae IV; immediately behind them are two lobes, possibly used as copulatory suckers. The chaetotaxy of the dorsal surface is similar to that of the female.

HYPOPUS (Figs. 258, 259). Length of idiosoma: 120 to 190 μm.

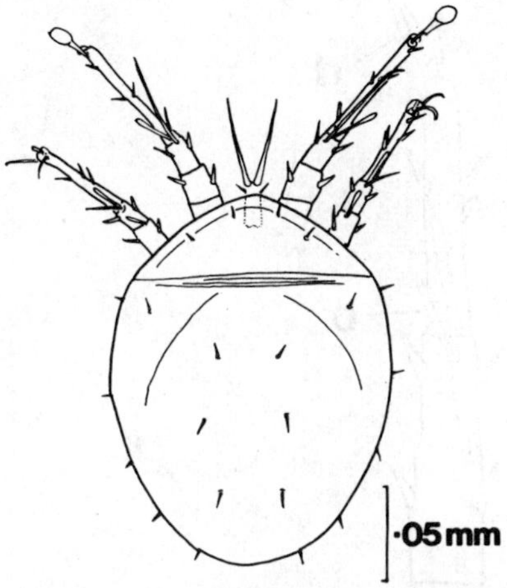

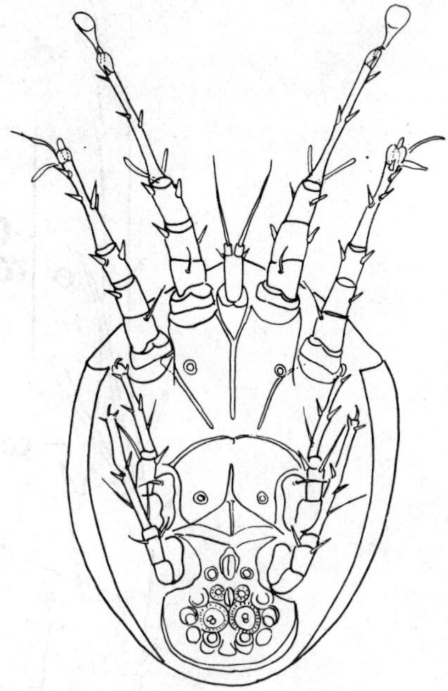

FIG. 258. *Histiostoma feroniarum* (Dufour). Dorsal view of hypopus (left).

FIG. 259. *Histiostoma feroniarum* (Dufour). Ventral view of hypopus (right).

The hypopus is rather flat, narrowing towards the posterior end of the body. The cuticle is well-sclerotized. The propodosoma is almost triangular in outline. Six pairs of very fine setae arise from the dorsal side of the idiosoma, corresponding in position to the scapulars and dorsals. Ventrally, apodemes III touch in the midline, so that the sternal and ventral shields are separated by an arched line. Coxal shields I and II are almost distinct from one another, as apodemes II nearly reach this line. Coxal shields III are also almost enclosed and, like coxal shields I, bear suckers.

The sucker plate is well-developed and through it project eight suckers arranged in the following order: 2 4 2. According to Perron (1954) who worked with *H. laboratorium*, contact with the host is first made with the two anterior suckers, the large central suckers then playing a major part in attaching the hypopus to its host.

All the legs are long and slender, the two hind pairs being directed forward and helping to anchor the mite in position. Claws are present on all the legs, and from the end of leg I projects an expanded seta, at the base of which arises a transparent leaf-shaped seta *d* (Fig. 260a). A leaf-shaped seta also arises at the end of leg II (Fig. 260b). On leg I, omega$_1$ is straight and expanded distally, only slightly shorter than phi of the same leg. Sigma on leg I is shorter than the spine-like setae arising from genu I. On leg II, omega$_1$ is only slightly longer than phi and sigma on the tibia and genu respectively.

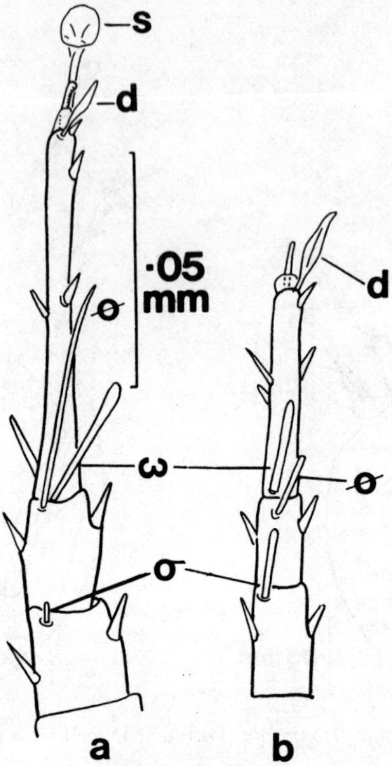

FIG. 260. *Histiostoma feroniarum* (Dufour) hypopus. Dorsal view of (*a*) right leg I, (*b*) right leg II. Solenidia: ω_1, Φ, σ; tarsal setae: *d*, *s*.

NYMPHS

The protonymphs and tritonymphs resemble the female, and can be distinguished from one another by the number of chitinous rings, the protonymph possessing one pair, the tritonymphs two pairs.

LARVA (Fig. 261).

The larva has one pair of chitinous rings between coxae I and II. From the dorsal side of the idiosoma project a series of lobes from which arise the dorsal setae.

BIOLOGY

According to R. Scheucher (1957), the life cycle is completed very rapidly and takes from three to three and a half days, the optimum temperature being between 25 to 30°C. This species, like *H. laboratorium* R. D. Hughes, 1950, is parthenogenetic and according to R. D. Hughes and C. G. Jackson (1958), two types of female are produced: one giving rise to males, the other to females. Scheucher, on the other hand, says that unfertilized eggs give rise to males. J. Cooreman (1944) also worked out the life-history and gives from two to four days at 20 to 25°C as the time required for development to be completed.

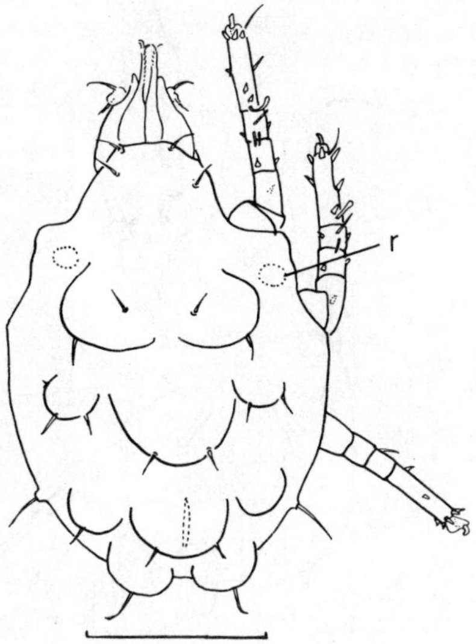

FIG. 261. *Histiostoma feroniarum* (Dufour). Dorsal view of larva. Chitinous ring: r.

H. feroniarum is by far the most common mite to be found in sewage bacteria beds where it feeds on the zoogloeal film covering the clinker surface, and thus, helps to prevent the beds from becoming clogged by too much film (Baker, 1964a). The hypopi will attach themselves to many different kinds of Arthropoda, and are particularly plentiful on filter beds where the zoogloeal film has been allowed to dry. Moulting was found to occur in two to three days when the hypopi came in contact with a film of water (Baker, 1964b).

H. feroniarum is also widely-distributed on all kinds of decaying vegetable matter and is occasionally found on wet grain and decaying mushrooms (M. E. John coll.).

DISTRIBUTION

England, Holland, France, Italy, Germany, U.S.A., Australia, New Zealand.

HISTIOSTOMA SAPROMYZARUM (Dufour, 1839)

Hypopus sapromyzarum Dufour, 1839; *Anoetus sapromyzarum* Oudemans, 1914; *Anoetus humididatus* Vitzthum, 1927 sensu Scheucher, 1957.

FEMALE (Fig. 262). Length of idiosoma: 300 to 650 μm.

In general appearance, this closely resembles the preceding species. The pedipalp differs in having its distal end flattened and entire (Fig. 254b), not bilobed, and in one of the setae arising from this lobe being more than twice as long as the other. On the ventral surface of the body, the two pairs of chitinous rings are oval and constricted in the middle, resembling the sole of a shoe.

The legs are more slender than in *H. feroniarum*. The chaetotaxy of leg I (Fig. 255c) is very similar in the two species.

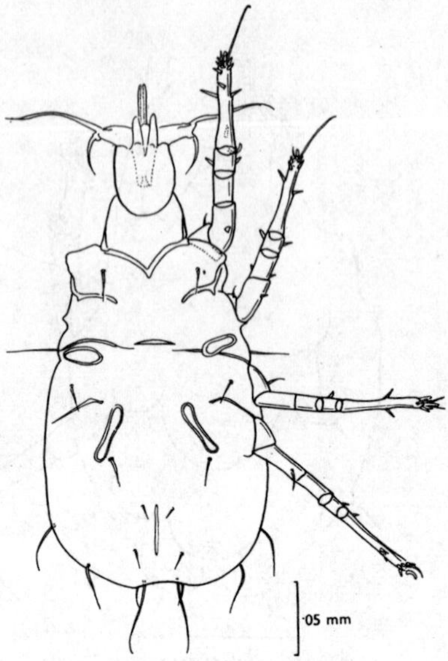

FIG. 262. *Histiostoma sapromyzarum* (Dufour). Ventral view of ♀.

MALE

According to R. Scheucher (1957) the male resembles the female. The chitinous rings are kidney-shaped, the convex side directed externally.

The *hypopus* closely resembles that of *H. feroniarum*.

BIOLOGY

The adults are frequently found on decaying fungi, e.g., species of *Lactarius*, *Russula*, *Tricholoma*, *Amanita* and *Scleroderma*. It has also been found on decaying hyacinth bulbs and on damp timber. The hypopus is carried by several kinds of beetles, flies and Myriapoda.

DISTRIBUTION

England, Germany, Holland, France, Italy, Brazil, Bolivia, Philippines, Australia.

Cryptostigmata Berlese, 1896

Except for such primitive forms as the Paleacaridae Grandjean, 1932 which have a colourless soft cuticle, the cryptostigmatid mites are darkly coloured and covered with a rigid exoskeleton: hence the popular name of 'beetle' mites.

The gnathosoma can usually be withdrawn into the body and is concealed by the overhanging rostral region of the propodosoma. The chelicerae are often prominent and chelate and lie above the infracapitulum or hypostome. Unlike the Astigmata, the palp typically has five free segments; in addition, the infracapitulum bears a pair of prominent toothed structures or rutella which assist in the breaking-up of food.

The idiosoma is usually clearly divided into propodosoma and hysterosoma, the two regions sometimes hinged on one another. The rigid exoskeleton is more elaborately constructed than in other mites and the surface cuticle may be elaborately sculptured and covered with a fine reticulate pattern or coarse irregular ridges. A number of projections usually arise from the body, of which the most conspicuous are the lamellae and pteromorphae. The lamellae are paired lateral folds of the propodosoma (Fig. 263), each terminating in a lamellar

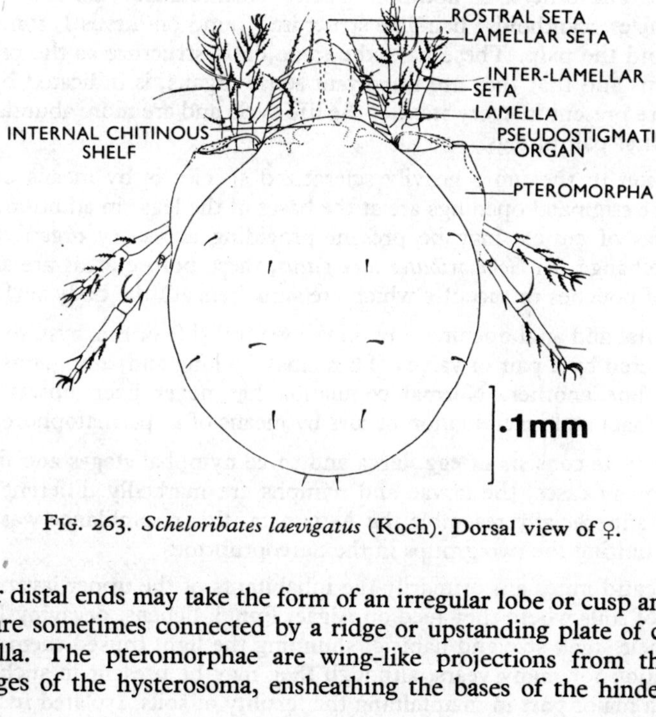

FIG. 263. *Scheloribates laevigatus* (Koch). Dorsal view of ♀.

seta; their distal ends may take the form of an irregular lobe or cusp and the two lamellae are sometimes connected by a ridge or upstanding plate of chitin: the translamella. The pteromorphae are wing-like projections from the antero-lateral edges of the hysterosoma, ensheathing the bases of the hinder pairs of legs.

Characteristically, the dorsal surface of the propodosoma is covered by one shield and another: the notogaster covers the dorsal and lateral sides of the hysterosoma. A plate also protects the ventral surface and is pierced by genital and anal openings which are covered by valve-like shields. The leg coxae have also become incorporated into the exoskeleton, so as to form a series of coxal plates.

In addition to the lamellar setae, rostral setae project forward from the overhanging propodosoma, these two pairs of setae corresponding in position to the internal and external verticals of the Astigmata. Posterior to these are the interlamellar setae and the pseudostigmatic organs which are homologous with the internal and external scapulars. According to Grandjean (1939b), the pseudostigmatic organs or trichobothria are composed of a hollow core of actinochitin containing the protoplasmic extension of a sensory cell. Each trichobothrion arises from a cup-shaped depression in the integument, and its outer surface may be pectinate and the distal end swollen and clavate. On the notogaster, the setae are variable in number and often inconspicuous. Ventrally, setae are also associated with the anal and genital shields. No anal suckers are developed in the males and little sexual dimorphism is apparent. Three pairs of genital sense organs are usually present in both sexes.

The majority of the Cryptostigmata have legs with five movable segments, the terminal one with a claw which may become subdivided into three in the adult. In many families, the basal segment or trochanter is sunk into a cup-shaped acetabulum whose free edge is produced into a pedotecta. The chaetotaxy is more elaborate than that of the Astigmata and much more reminiscent of that found in the Prostigmata. It has been extensively studied by F. Grandjean (1935b, etc.) and others. In addition to setae, solenidia and famulus, Grandjean also recognizes acanthoids: hair-like structures found on tarsus I, sometimes on tarsus II and the palp. These have the same basic structure as the pseudostigmatic organs and that they are important sense organs, is indicated by the fact that they are present in every stage of the life cycle and are more abundant on the palps and first pair of legs.

Respiration in the more heavily sclerotized species is by means of tracheal tubes whose stigmatal openings are at the bases of the legs; in addition, localized porose areas of cuticle may be present providing accessory organs for direct gaseous exchange. In *Scheloribates laevigatus*, these porose areas are situated in the walls of pouches or saccules which are sunk beneath the body surface.

The genital and anal openings are on the ventral side of the hysterosoma and each is covered by a pair of valves. These may be long and contiguous or separated from one another. Normal copulation has never been observed in the Cryptostigmata and insemination occurs by means of a spermatophore.

The life cycle consists of egg, larva and three nymphal stages and finally, the adult. In many cases, the larvae and nymphs are markedly different from the adults and superficially resemble the Astigmata; this resemblance was the main reason for uniting the two groups in the Sarcoptiformes.

The oribatid mites are primarily the inhabitants of the upper layers and surface litter of soils where they feed on Algae, Fungi, lichens, decaying leaves and shoots. Their small size and habit of shunning the light caused them to receive little attention for many years, although they may be present in such numbers as to play a major part in maintaining the fertility of soils. Isolated records exist of their behaviour as predators and some species have been shown to act as vectors of various tapeworms. (see Kates and Runkel, 1948).

Mites belonging to the Cryptostigmata are found occasionally in houses and stored food as stray visitors which have entered through windows, or been carried on clothes or straw. *Scheloribates laevigatus*, *Trichoribates novus* and *Phauloppia lucorum* have been found more often than others.

A. D. Michael wrote the first monograph on the Oribatei (1883, 1887) and others have been compiled by M. Sellnick, 1929 and C. Willmann (1931). F. Grandjean has studied the group extensively and his work is published in a large number of separate papers. The biology and ecology of certain species is discussed by J. Travé (1963).

SCHELORIBATES LAEVIGATUS (C. L. Koch, 1836)

Zetes laevigatus C. L. Koch, 1836; *Scheloribates pallidulus* (C. L. Koch, 1840) (Hughes, 1948).

Length of idiosoma of male and female: about 480 to 610 µm.

The idiosoma is brown and glossy in texture with a hard cuticle and with rather paler legs. The rounded shape, dark colour and ponderous movements of *S. laevigatus* distinguish it sharply from the astigmatid mites.

The propodosoma narrows anteriorly and completely overhangs the gnathosoma. To its lateral edges are attached narrow upstanding lamellae (Fig. 263), from whose truncate ends arise the pectinate lamellar setae. A transverse line across the propodosoma connects the anterior ends of the lamellae. The pseudostigmatic organs are recurved over the back of the animal and have slender stalks with spindle-shaped heads which are covered with fine 'hairs'. The division between propodosoma and hysterosoma is marked internally by a chitinous shelf, and the posterior internal processes of this (seen by transparency), are asymmetrical and directed backwards and inwards. There are no processes on the central region of the shelf.

The hysterosoma is arched, its antero-lateral margins produced into wing-like pteromorphae, triangular in shape and transparent. Inserted on the dorsal surface are a few very short setae, corresponding in position to d_1 to d_4, $l\ a$, $l\ p$ and one pair of posterior setae. The ventral plate is pierced anteriorly by the genital opening, posteriorly by the anal. Porose areas are present which are situated in the walls of pouches.

The four pairs of legs arise close together, the two front pairs being of equal length and having enlarged femora and, as in astigmatid mites, long solenidia arising from the anterior edges of the tibiae. The tarsi terminate in one large claw and two smaller ones.

BIOLOGY

The life cycle takes from 42 to 115 days at 25 °C and a relative humidity of 100 per cent, chicken faeces being supplied as food. At laboratory temperatures, it may take over a year. The mites will eat moss, but a diet of this is not favourable to egg production (Cleat, 1952). In the laboratory, a total of 20 eggs is laid, 3 to 8 at a time, and deposited very deeply in the tissue of decayed grass and anything soft. The immature stages live in and feed on the soft decaying debris and do not appear to eat fungal mycelium. The adults are surface dwellers and are generally distributed amongst moss, humus and rotten wood (Woodring and Cook, 1962).

In Holland, *S. laevigatus* is commonly found in wet and dry grassland, in alder and birch marshes (Van der Hammen, 1952) and it is one of the three most common species found in meadows in the St Paul, Minnesota area (Woodring and Cook, 1952). Frenzel (1936) (quoted from Van der Hammen), regards it also as a characteristic meadow form. It is also occasionally found in the damp debris of warehouse floors, feeding on mouldy grain. K. C. Kates and C. E. Runkel (1948) name *S. laevigatus* as the intermediate host of various tapeworms: *Moniezia expansa* in the U.S.A. and U.S.S.R., *M. benedeni* in the U.S.S.R., *Bertiella studeri*, *Cittotaenia ctenoides* and *C. denticulata* in Germany, and of *Anoplocephala perfoliata*, *A. magna* and *Thysaniezia giardi* in the U.S.S.R. *S. laevigatus* has also been recorded from the nests of *Talpa* and *Arvicola*.

DISTRIBUTION

England, Germany, France, Holland, U.S.A., U.S.S.R.

TRICHORIBATES NOVUS (Sellnick, 1929)

T. trimaculatus (Koch, 1836), (Hughes, 1961).

Length of idiosoma of male and female: about 620 μm.

In general appearance, this closely resembles *S. laevigatus*. It is a glossy brown in colour, a light greyish patch appearing on the dorsal side of the hysterosoma and possibly functioning as a light-sensitive organ.

The lamellae have distinct terminal cusps and are widely separated from one another (Fig. 264). They are connected by a distinct shelf-like translamella. The

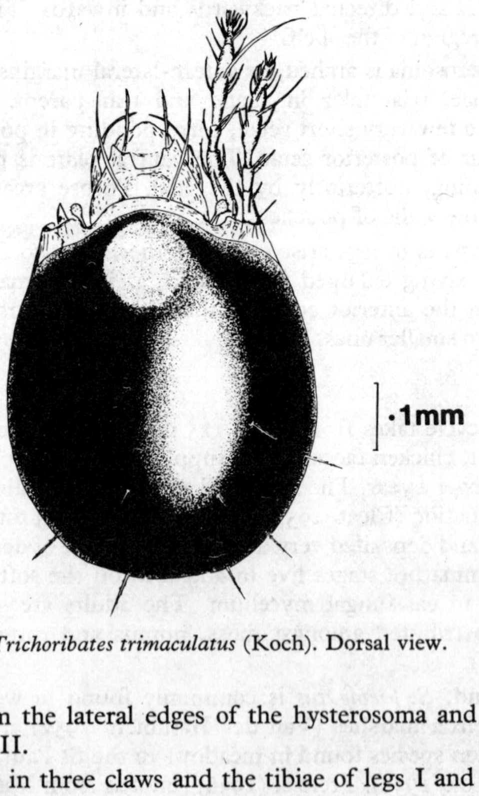

FIG. 264. *Trichoribates trimaculatus* (Koch). Dorsal view.

pteromorphae arise from the lateral edges of the hysterosoma and are curved around the bases of legs II.

All the legs terminate in three claws and the tibiae of legs I and II are constricted towards their bases.

BIOLOGY

The nymphs and adults may be found in considerable numbers in meadows in June by sweeping the long grass (A. D. Michael, 1883). It is also found beneath tree bark, in moss, under haystacks and in mouldy grain. On one occasion, it was present in a sparrow's nest at a depot (G. Hosie, coll.).

DISTRIBUTION

England, Germany, Holland, France, Switzerland, Italy, U.S.A.

PHAULOPPIA LUCORUM (C. L. Koch, 1840)

Notaspis burrowsi Michael, 1890.

Length of idiosoma: about 660 µm.

The body is heavily sclerotized, the cuticle rough and covered with a faint reticulate pattern.

The lamellae are slight ridges and are convergent, the translamella only visible as a faint line (Fig. 265). The pseudostigmatic organs are directed forwards and

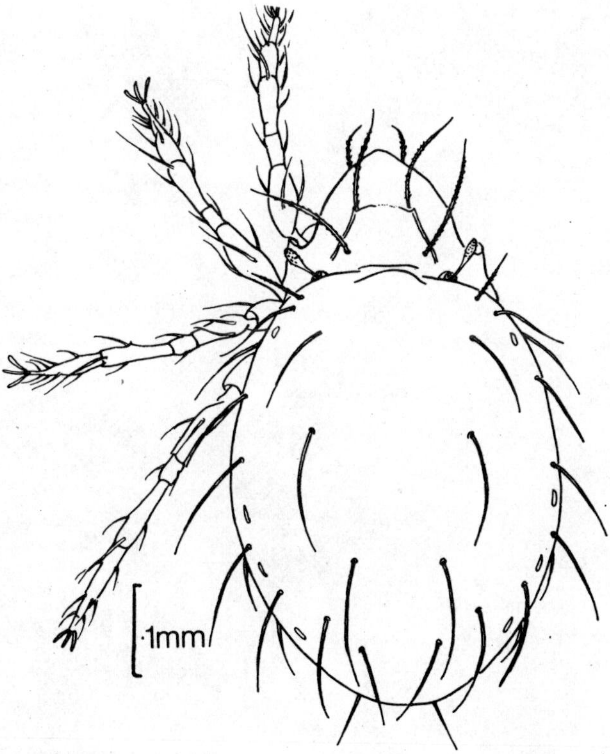

FIG. 265. *Phauloppia lucorum* (Koch). Dorsal view.

have spindle-shaped heads covered with small dark-coloured bristles. No pteromorphae are present.

The dorsal side of the hysterosoma (notogaster) bears 15 pairs of long setae, four pairs arising on either side of the midline, and the remaining from the lateral margins. Porose areas are visible along the sides of the body. The legs have three claws each.

Biology

P. lucorum is known as the 'window sill mite' owing to its habit of appearing in large numbers on window sills inside houses. It is also common in moss and is said to be associated with lichens (D. Macfarlane, pers. comm.). It has been found on home grown barley in Shropshire (W. J. Simpson coll.), from the surface of home-milled premium flour in Newark, Nottinghamshire (J. B. M. Tulloch coll.) and in a flour sample (K. G. Smith coll.).

Distribution

England, Germany, Bulgaria, U.S.S.R. It is widespread around the Mediterranean region (C. Willmann, 1931).

Prostigmata Kramer, 1877

The order Prostigmata or Trombidiformes Reuter, 1909 includes a large number of diverse forms varying in size from 100 to 1,600 µm. They are usually weakly sclerotized and characterized by having a tracheate respiratory system opening by a pair of stigmata lying at the base of the gnathosoma and often associated with a groove or peritreme. In some families, however, a tracheate respiratory system is absent.

Many species, such as those belonging to the Cheyletidae and Tetranychidae, live by sucking the juices of animals and plants, and in these, the mid-gut ends blindly and the hind-gut forms the single 'excretory canal' opening by a median anus or uropore. Sometimes, only the larva is parasitic in habit; for instance the water mites (Hydrachnellae) prey on freshwater mussels and aquatic insects, and the chiggers (Trombiculidae) and species of Erythraeidae suck the tissue fluid of vertebrates and arthropods respectively. Some species are endoparasites and the tubular bodies of the Demodicidae and Syringophilidae are admirably tailored to burrow into the hair follicles and sebaceous glands of mammals, and the feather quills of birds.

The prostigmatid mites vary greatly in structure and biology; indeed, taxonomists tend to include in this order, those mites which do not fit readily into the remaining ones, and it is probable that the group has a polyphyletic origin.

The order Prostigmata is divided into a large number of families, and members of the Cheyletidae, Bdellidae, Cunaxidae, Tydeidae, Tarsonemidae, Pyemotidae and Tetranychidae are found in stored food and in storage premises and houses.

PROSTIGMATA

Key to the families

1. Small mites (100 to 400 µm) with the hysterosoma showing traces of segmentation, particularly at the sides of the body 2

—. Body usually larger and without traces of segmentation 4

2. Body rounded. Anterior edge of the propodosoma dome-shaped and projecting over the gnathosoma . . . . *Acarophenax*, p. 274

—. Body oval. Anterior edge of propodosoma not dome-shaped 3

3. Tarsus IV of female terminating in claws and pretarsus. Legs IV of male curved inwards. Females often found with the opisthosoma distended with developing embryos . . . . *Pyemotes*, p. 271

—. Tarsus IV of female terminating in two long setae of unequal length. Legs IV of male bent inwards sharply at an acute angle. Females usually lay oval eggs . . . . *Tarsonemidae*, p. 266

4. Gnathosoma elongated to form a long snout. Chelicerae chelate or ending in a hook 5

—. Gnathosoma of normal dimensions or, if elongated, chelicerae are shaped like stylets 6

5. Chelicerae ending in small shears. The terminal segment of the palp bears two or three long setae, but no claw. With three pairs of genital suckers .... *Bdellidae*, p. 253

—. Chelicerae terminating in a hook. The terminal segment of the palp without long setae, but with a claw. With two pairs of genital suckers .... *Cunaxidae*, p. 257

6. Pedipalps enlarged, the penultimate segment (tibia) bearing a claw, the tarsus usually one or two comb-shaped setae (Fig. 271). Chelicerae shaped like stylets .... *Cheyletidae*, p. 230

—. Pedipalps of normal dimensions 7

7. Basal part of chelicerae fused to form a stylophore. Movable digit very long and styliform (Fig. 300). Peritreme present on anterior dorsal side of gnathosoma .... *Tetranychidae*, p. 263

—. Basal part of chelicerae fused. Movable digit short and sickle-shaped (Fig. 296). Peritreme absent .... *Tydeidae*, p. 258

CHEYLETIDAE, Leach, 1815

The body of a cheyletid mite is often colourless, sometimes red. The gnathosoma is clearly divided from the idiosoma and freely movable from side to side; its base is a broad structure formed by the fusion of the pedipalpal coxae with the tergites of the gnathosoma, and the anterior dorsal region is prolonged forward as a rostrum. The rostrum is a tube formed by the fusion of the cheliceral bases or stylophore (Snodgrass, 1948) with the endites of the pedipalp (T. E. Hughes, 1958) and at its distal end is the mouth (Fig. 266). From it

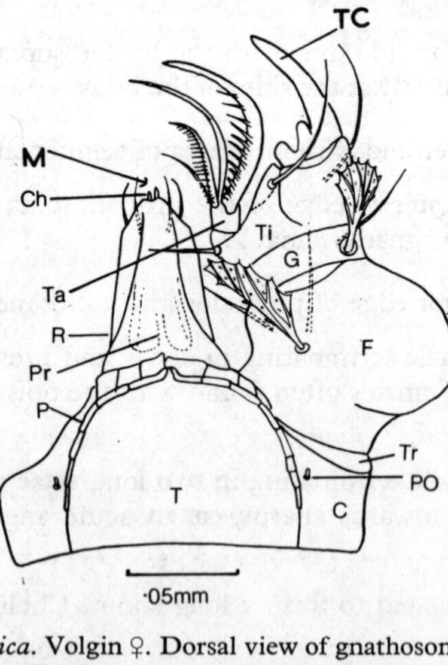

FIG. 266. *Eucheyletia taurica*. Volgin ♀. Dorsal view of gnathosoma. Segments of pedipalp (PP): C coxa, Tr trochanter, F femur, G genu, Ti tibia, Ta tarsus; Ch chelicera; M position of mouth; T tergite of gnathosoma or tegmen; Pr protegmen; P peritreme; PO palpostigmatic organ; R rostrum; TC tibial claw.

arise three pairs of setae: dorsal, ventral and also a very short pair in the region of the mouth. Attached to depressions on the ventral surface of the fused cheliceral bases and articulating with them by an arthrodial membrane, are the movable digits of the chelicerae: stylet-like structures which can move independently of one another and can be pushed in and out. Summers and Price (1970) have proposed the name 'tegmen' for that part of the cheliceral base or stylophore which is proximal to the peritreme (i.e., for the tergite of the gnathosoma). Protegmen is the name used for that part which lies distal to the peritreme.

The peritreme extends transversely across the rostrum and has backward projections on either side. It consists of a gutter running in the thickened exoskeleton, and is covered externally by a thin membrane and constricted at intervals by internal partitions. The stigmata open near the posterior end of each lateral peritrematal limb (T. E. Hughes, 1958) and two air sacs are continuous with the peritreme in the midline (Fig. 271a). Into these air sacs open numerous fine tracheae. External to the lateral limb of the peritreme is the palpostigmatic organ (Rohdendorf, 1940): a small rod or rounded swelling.

The palp of the pedipalp is five-segmented, the tibia enlarged into a claw-like structure whose base is sometimes produced internally into a variable number of teeth. In some specimens of cheyletids, the number of tibial teeth differs on the two sides of the body of the same individual. The base of the tibia bears three setae: two internal and one external. The pedipalpal tarsus is reduced to a small cushion arising from the tibia. From it usually arise two or three smooth falcate setae, one or two comb-shaped setae and a peg-shaped solenidion. The palpal trochanter is reduced and telescoped into the coxal region, whilst the femur may be enlarged, often to enormous dimensions, in heteromorphic males. Three or four setae arise from the femur and one or two from the genu. The trochanter has no setae. The two pedipalps are freely movable from side to side and can be used as a pair of pincers for catching the prey. In life, they are held so that they curve downwards. The tarsus is also movable, the muscles inserted on it originating from the external surface of the femur.

The dorsal surface of the idiosoma in the adult may be covered by one or two shields, distinguishable because they have a finely pitted surface, whilst the rest of the body is wrinkled. The propodosomal shield is trapezoidal in shape and usually bears three pairs of setae in each anterior corner and one or two pairs in the posterior (dorsolateral setae—Summers and Price, 1970). A pair of corneae may also be present. In males and some females additional setae (dorsomedian setae—Summers and Price, 1970) may be found in the centre of the shield. The dorsomedian setae are sometimes markedly different in shape in the females although they resemble the remaining dorsal setae in the deutonymphs. When present, the hysterosomal shield is narrow and oblong. Inserted on it are three or more pairs of lateral peripheral setae and one or two pairs arising from the anterior edge in males only. In nymphs the posterior dorsal plate may be replaced by two or more, or it may be absent. A long pair of scapular setae arises from the side of the body at the junction of the two dorsal plates and, like the remaining dorsal setae, may be simple, pectinate, or flattened and leaf-like in appearance. The ventral surface of the body has no plates, except for a small sternum extending as far back as coxae II in some males, and a ventral plate which may be present between coxae IV. The ventral region of the body anterior to the anus usually bears five pairs of ventral setae (Fig. 267).

The females possess a single elongated opening or uropore at the ventral

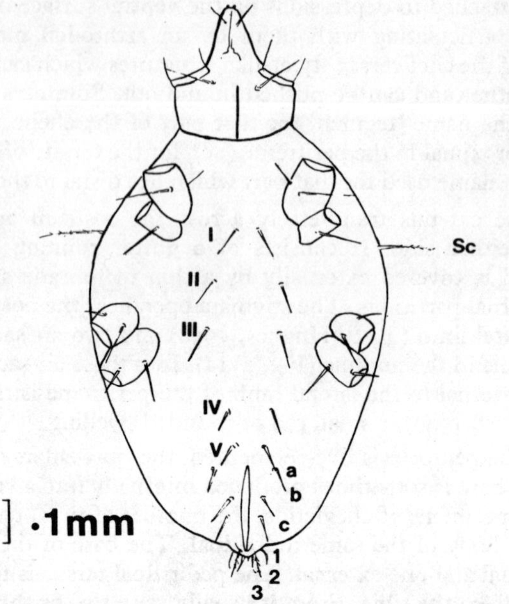

FIG. 267. *Cheyletus eruditus* (Schrank). Ventral view of ♀. Ventral setae: I to V; anal setae: *a* to *c*; postanal setae: 1 to 3; *Sc* scapular setae.

posterior end of the body, into which open genital and excretory ducts. On either side of this are inserted anal and postanal setae. Males have a small terminal opening through which a curved penis can be protruded. On the dorsal posterior end of the body, there are usually three pairs of minute papillae (Fig. 268), each ending in a small spine.

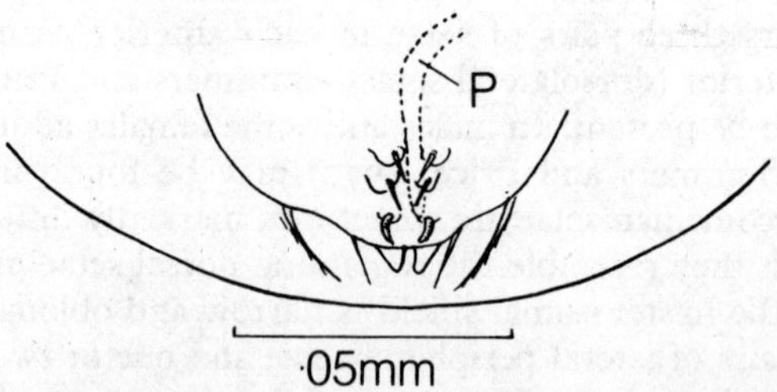

FIG. 268. *Cheyletus trouessarti* Oudemans ♂. Dorsal view of posterior end of body. P penis.

The legs have five free segments, the coxae having become incorporated into the ventral body wall. With the possible exception of legs I, the tarsus narrows to a pedicel at the distal end and terminates in a pretarsus bearing a pair of claws. Between these is a cushion-like pulvillus from which arise two comb-shaped setae whose barbs have blunt expanded ends. Encircling the ambulacrum is a number of setae, of which the most conspicuous are two long addorsal terminal setae and one ventral azygos (Summers and Price, 1970). Four shorter setae arise from the pedicel. Inserted on the external dorsal side of tarsi I and II is a solenidion omega which frequently arises from a hump (Fig. 269) and is associated with a supporting or guard seta. Two smaller solenidia, phi and sigma, are inserted in a similar position on the tibia and genu of leg I, as well as a number of setae.

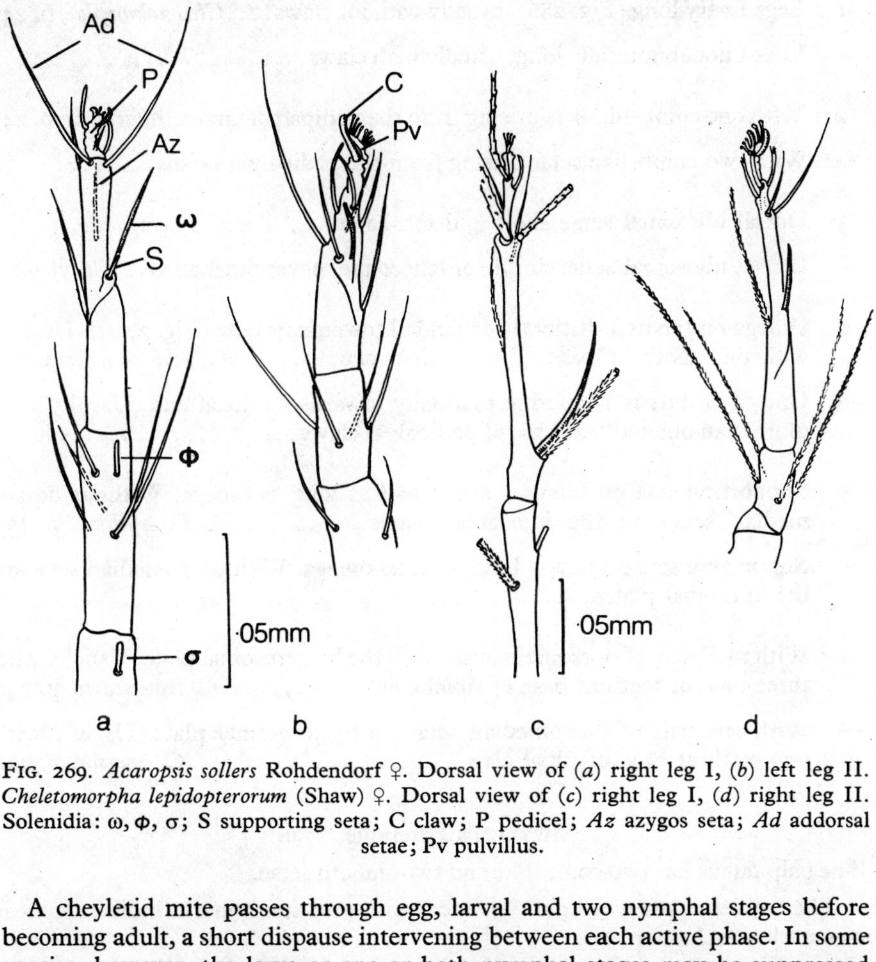

FIG. 269. *Acaropsis sollers* Rohdendorf ♀. Dorsal view of (*a*) right leg I, (*b*) left leg II. *Cheletomorpha lepidopterorum* (Shaw) ♀. Dorsal view of (*c*) right leg I, (*d*) right leg II. Solenidia: ω, Φ, σ; S supporting seta; C claw; P pedicel; *Az* azygos seta; *Ad* addorsal setae; Pv pulvillus.

A cheyletid mite passes through egg, larval and two nymphal stages before becoming adult, a short dispause intervening between each active phase. In some species, however, the larva or one or both nymphal stages may be suppressed (see Volgin, 1969). Parthenogenesis is of common occurrence and, in those few cases that have been studied, unfertilized eggs give rise to males (A. R. Edwards, pers. comm.). Some genera produce one or more heteromorphic males (*Cheyletus*, *Acaropsis*) in which the length of pedipalp and sculpturing on the rostrum are exaggerated.

Species of the Cheyletidae are mainly free-living predatory mites who catch their prey by means of the tibial claws, pierce it with the chelicerae and then suck out its contents. They are often found associated with acarid mites in granaries, warehouses and barns, as well as in leaf litter and topsoil inhabited by a variety of micro-arthropods. The bark and leaves of trees and nests of insects, birds and mammals also harbour these animals. In a few cases, the association is a more intimate one and *Cheyletiella parasitivorax* can pierce the skin of cats and rabbits and give rise to a form of mange.

CHEYLETIDAE

Key to the adult females

1. Legs I very long (Fig. 286), usually without claws....*Cheletomorpha*, p. 247

—. Legs I not abnormally long, usually with claws........................2

2. With one comb-like seta arising from the pedipalpal tarsus. *Acaropsis*, p. 245

—. With two comb-like setae arising from the pedipalpal tarsus............3

3. Dorsal idiosomal setae short and fanshaped..........*Eucheyletia*, p. 250

—. Dorsal idiosomal setae simple or lanceolate, never fanshaped...*Cheyletus* 4

4. Omega on tarsus I distinctly expanded towards its base (Fig. 272d). Usually with one tooth at base of pedipalpal claw........*C. malaccensis*, p. 242

—. Omega on tarsus I tapering gradually towards its distal end. Usually with more than one tooth at base of pedipalpal claw..........................5

5. Supporting seta on tarsus I about half as long as omega. Without dorsomedian setae on the idiosomal plates..............*C. eruditus*, p. 235

—. Supporting seta on tarsus I longer than omega. With dorsomedian setae on the idiosomal plates..6

6. With two pairs of dorsomedian setae on the hysterosomal plate. Usually with three or four teeth at base of tibial claw............*C. trouessarti*, p. 240

—. With one pair of dorsomedian setae on hysterosomal plate. Usually with two teeth at base of tibial claw......................*C. aversor*, p. 242

CHEYLETUS Latereille, 1796

The palp tarsus has two comb-like and two smooth setae.

The tegmen is plain or lightly striated, with prominent marks indicating areas of muscle attachment.

The peritreme is often M-shaped.

The propodosomal shield covers most of the propodosoma.

No corneae are present.

The hysterosomal shield only partially covers the hysterosoma.

The marginal setae of the dorsal shields are comparatively short and pectinate, never fan-shaped.

Tibiae I to IV have one dorsal seta much longer than the other setae of this segment.

Type species: *Acarus eruditus* Schrank, 1781.

CHEYLETUS ERUDITUS (Schrank, 1781)

Eutarsus cancriformis Hessling, 1852; *Cheyletus eburneus* Hardy, 1867; *Cheyletus ferox* Banks, 1906; *Cheyletus rabiosus* Rohdendorf, 1940; *Cheyletus butleri* Hughes, 1948.

FEMALE (Fig. 270). Length of idiosoma: 450 to 620 µm.

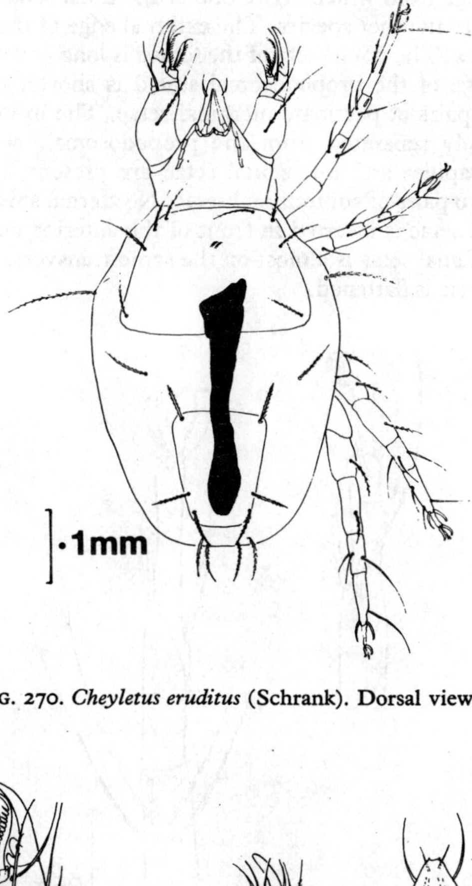

FIG. 270. *Cheyletus eruditus* (Schrank). Dorsal view of ♀.

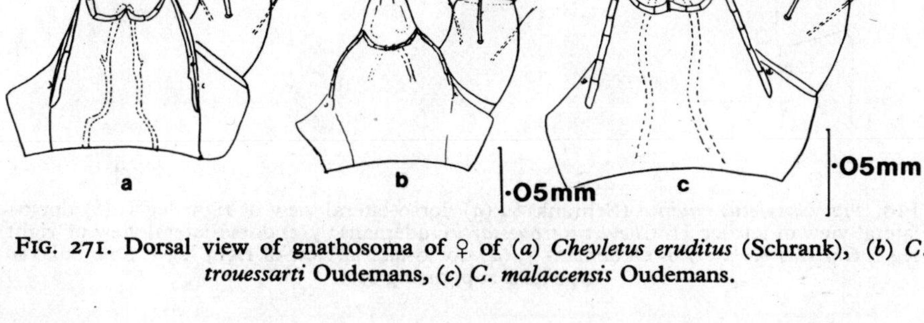

FIG. 271. Dorsal view of gnathosoma of ♀ of (*a*) *Cheyletus eruditus* (Schrank), (*b*) *C. trouessarti* Oudemans, (*c*) *C. malaccensis* Oudemans.

The idiosoma is colourless and diamond-shaped. The gnathosoma (Fig. 271a) is relatively long and narrow, the tegmen ornamented with striae radiating from its base. The peritreme is shaped like the letter M, with four to five internal partitions dividing it between the midline and lateral limb. On the pedipalpal tarsus, the external 'comb' has 13 to 15 strong teeth, the internal (which is characteristically curved) 16 or 17. The tibial claw usually has two prominent teeth at its base, although I have seen a specimen with two large teeth on one side of the body and three small ones on the other. The internal edge of the tibia has a thin flange from which arises one of the tibial setae, but this flange is not so pronounced as in other species. The external edge of the pedipalpal femur is moderately convex. The dorsal seta of the femur is long and sparsely barbed.

The anterior edge of the propodosomal shield is shorter than the posterior and there are four pairs of pectinate marginal setae. The hysterosomal shield is narrower and widely separated from the propodosomal, with three pairs of setae. A pair of scapular and interscutal setae are present, arising from small plates, as well as two pairs of subterminal setae. No sternal shield is present. The fifth pair of ventral setae is inserted in front of the anterior end of the uropore and the first pair of anal setae is almost on the same transverse line as its anterior end. None of the setae is flattened.

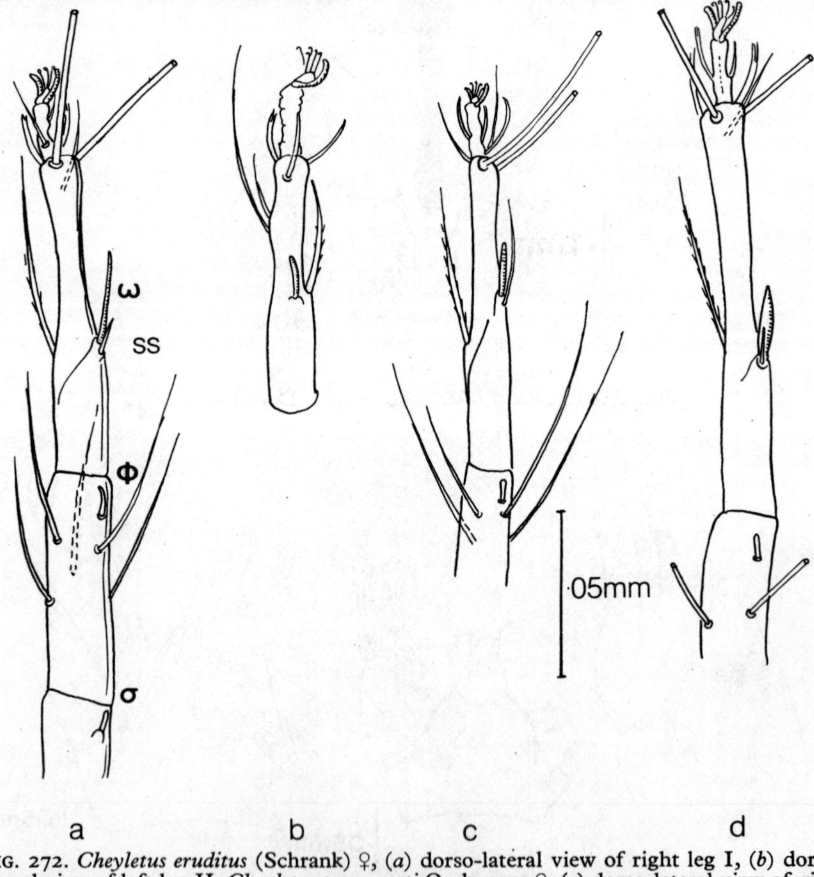

FIG. 272. *Cheyletus eruditus* (Schrank) ♀, (*a*) dorso-lateral view of right leg I, (*b*) dorso-lateral view of left leg II. *Cheyletus trouessarti* Oudemans ♀, (*c*) dorso-lateral view of right leg I. *Cheletus malaccensis* Oudemans ♀, (*d*) dorso-lateral view of right leg I. Solenidia: ω, Φ, σ; *SS* supporting seta.

The first pair of legs is of normal length (Fig. 272a). The solenidion omega arises from a mound on the external side of the leg and does not expand towards its base. The supporting seta is pressed tightly against it and is considerably shorter. Solenidia are also present on the tibia and genu. On tarsus II (Fig. 272b) omega arises ventrally midway along the segment. Femur IV bears two setae.

HOMOMORPHIC MALE (Fig. 273). Length of idiosoma: about 280 μm.

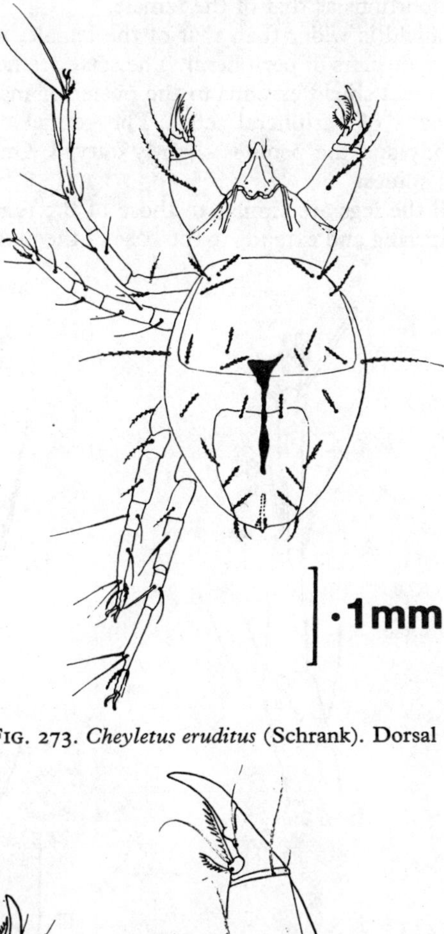

FIG. 273. *Cheyletus eruditus* (Schrank). Dorsal view of ♂.

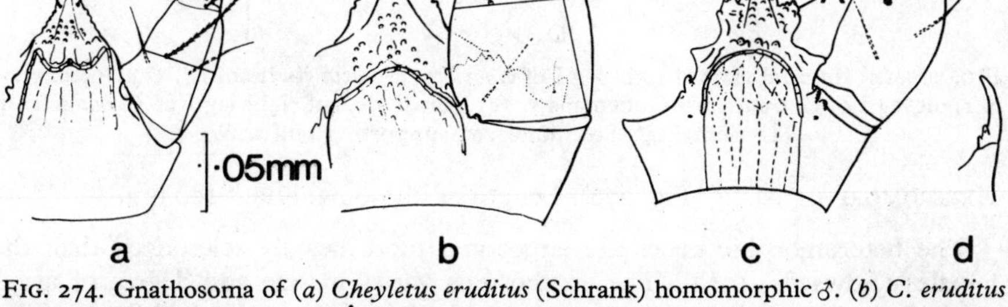

FIG. 274. Gnathosoma of (*a*) *Cheyletus eruditus* (Schrank) homomorphic ♂. (*b*) *C. eruditus* (Schrank) heteromorphic ♂, (*c*) *C. trouessarti* Oudemans ♂. (*d*) Abnormal tibial claw of *C. trouessarti* Oudemans. A ala.

The gnathosoma (Fig. 274a) is broader than in the female and more massively sclerotized. The protegmen is longitudinally ridged and it may bear a few tubercles. There are no prominent lateral protrusions or alae. The tegmen is also longitudinally striated. The external 'comb' of the pedipalpal tarsus has eight or nine teeth, the internal six to nine. The pedipalpal tibia usually has two teeth at its base, but the shape and the number of teeth are variable. The pedipalpal femur has the same proportions as that of the female.

The propodosomal shield is wider than that of the female, with two pairs of central setae, as well as four pairs of peripheral. The setae are needle-shaped and pectinate. The hysterosomal shield extends to the posterior margin of the body and bears five or six pairs of peripheral setae. The genital opening is at the posterior end of the body and the penis is slightly curved. On either side of it are three pairs of small spines.

The ambulacra of all the legs are similar to those of the female. On tarsus I (Fig. 275a), omega is tapering and extends to the base of the pedicel.

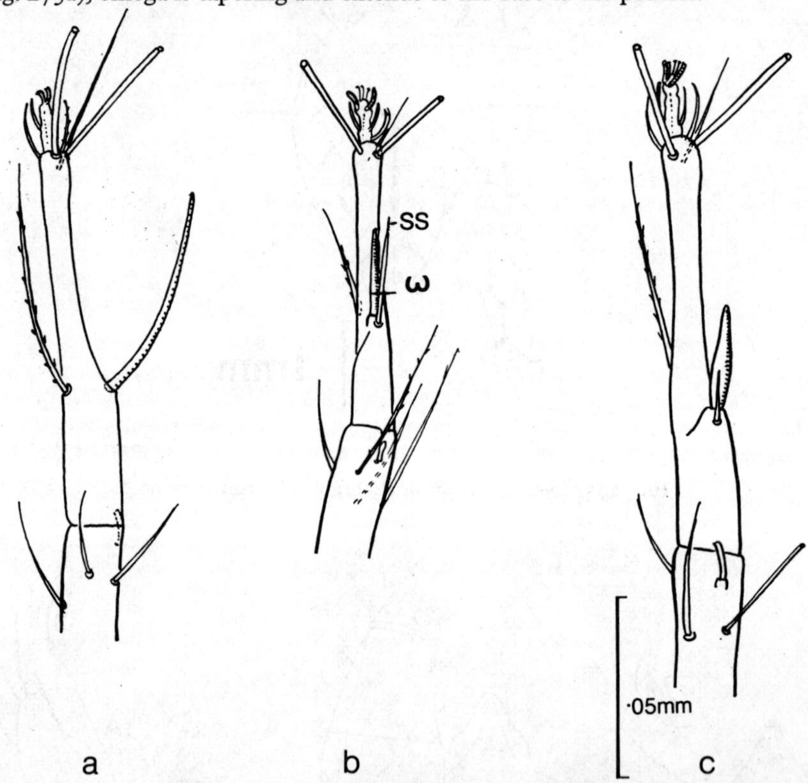

FIG. 275. (*a*) Internal view of right leg I of *Cheyletus eruditus* (Schrank) ♂. (*b*) Dorsal view of right leg I of *C. trouessarti* Oudemans ♂. (*c*) Dorsal view of right leg I of *C. malaccensis* Oudemans ♂. Solenidion: ω, supporting seta: *SS*.

HETEROMORPHIC MALE (Fig. 276). Length of idiosoma: about 400 μm.

The heteromorphic males are larger and more heavily sclerotized than the females (Edwards, 1952). The gnathosoma varies in size and degree of ornamentation of its dorsal surface and in the length of the pedipalpal femur. This may be two to four times longer than its width at the level of the dorsal femoral seta, so that the ventral femoral setae become widely separated from one another. Increase in length of the femur is associated with increased ornamentation of the dorsal surface of the gnathosoma, increased length of the rostrum and an anterior prolongation of the medial region of the peritreme (cf Figs. 274b and 277).

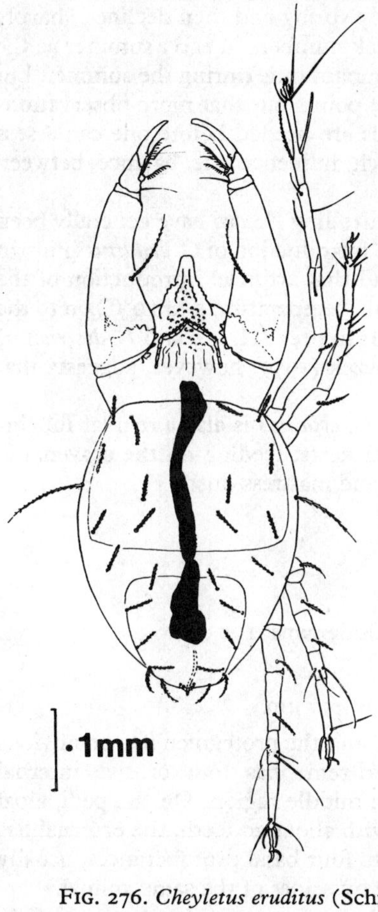

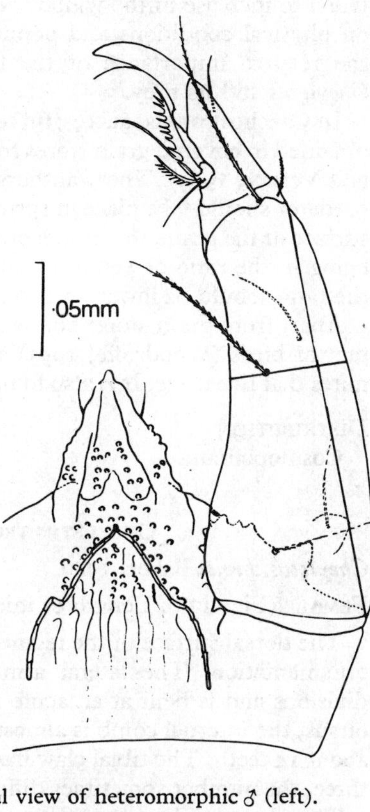

FIG. 276. *Cheyletus eruditus* (Schrank). Dorsal view of heteromorphic ♂ (left).

FIG. 277. *Cheyletus eruditus* (Schrank). Dorsal view of gnathosoma of heteromorphic ♂ (right).

BIOLOGY

The life cycle takes from 19 to 30 days at 26.7 to 29.4°C (Beer and Dailey, 1956), the eggs being laid in heaps and guarded by the female. In spite of this maternal solicitude, many of the newly emerged larvae are eaten by the female if they are kept in a confined space. The females are normally parthenogenetic, the unfertilized eggs giving rise to females. Males are rarely found, but when they do occur, several appear together. Pairing between heteromorphic males and females has been observed by Edwards (1952).

C. eruditus is commonly found in association with the acarid mites on which it feeds, although it is also capable of feeding on the very young stages of storage moths and beetles and on psocids (Coombs and Woodroffe, 1968). When short of food it behaves as a cannibal. Certain mites are preferred to others, and when given the choice between the two species, it will consume *A. siro* rather than

Lepidoglyphus destructor. Solomon (1946) found that under summer conditions, *C. eruditus* may be of considerable value in controlling the numbers of *A. siro*, but during the winter it tends to be ineffective, since its prey reproduces at a faster rate. Coombs and Woodroffe obtained the same results with *A. farris*. They found that *A. farris* remained few in number during the summer, increased during the winter to reach a maximum in early spring and then declined sharply during May and June. *C. eruditus* reached peak numbers in early summer as the *Acarus* declined in number, remained the dominant mite during the summer, but failed to increase in the winter. Norris (1958) points out that more observations on physical conditions and population trends are needed before one can assess the relative importance of the factors which influence the balance between *Cheyletus* and its prey.

In Czechoslovakia, successful results in controlling *A. siro* have generally been obtained in covered grain stores following the introduction of *C. eruditus* (Pulpán and Verner, 1965). These authors recommend that artificial introduction of the predator should take place in spring or autumn (temperature 5 to 10°C) on to the surface of the grain, the proportion of predator to prey (*A. siro* and *L. destructor*) being in the ratio of 1:100 to 1 to: 1000. Boczek (1959) however, suggests that the ratio should be lower : 1 to 20 or 25.

Apart from grain stores and farm detritus, *C. eruditus* is also a regular inhabitant of birds (Woodroffe, 1953) and mammal nests, feeding on the scavenging mites that live there. It is also found in house and mattress dust.

Distribution

Cosmopolitan.

Cheyletus trouessarti Oudemans, 1903

Cheyletus davisi Baker, 1949.

Female (Fig. 278). Length of idiosoma: about 450 µm.

The dorsal surface of the tegmen is striated and the protegmen has no obvious ornamentation. The lateral arm of the peritreme has four or five internal divisions and is bent at an acute angle to the middle region. On the pedipalpal tarsus, the internal comb is almost straight, with about 20 teeth, the external has about 14 teeth. The tibial claw has from two to four basal protuberances, usually three, the number sometimes differing on the two sides of the same animal.

The shape and arrangement of the dorsal plates are similar to those of *C. eruditus*, the peripheral setae arising from them are slightly flattened and pectinate (Fig. 278). In addition, thin-walled and inconspicuous dorsomedian setae are present, shaped like 'very small collapsed vesicles' (Summers and Price, 1970), one pair on the propodosomal plate, two pairs on the hysterosomal.

The ventral and anal setae are arranged as in *C. eruditus*, except that the fifth pair of ventral setae is more posterior in position, arising on either side of the anterior end of the uropore.

On leg I (Fig. 272c), the solenidion omega is internal to the supporting seta of almost twice its length.

Male (Fig. 279). Length of idiosoma: about 350 µm.

The sides of the rostrum are expanded into wing-like alae and the tegmen is covered with a reticulate pattern. The peritreme may be smoothly arched (Fig. 274c) or else slightly indented in the middle (Fig. 279). The internal comb of the

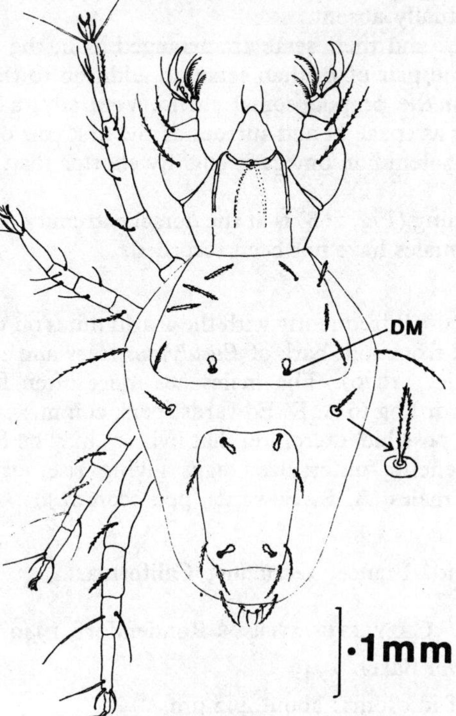

FIG. 278. *Cheyletus trouessarti* Oudemans. Dorsal view of ♀. DM dorsomedian seta.

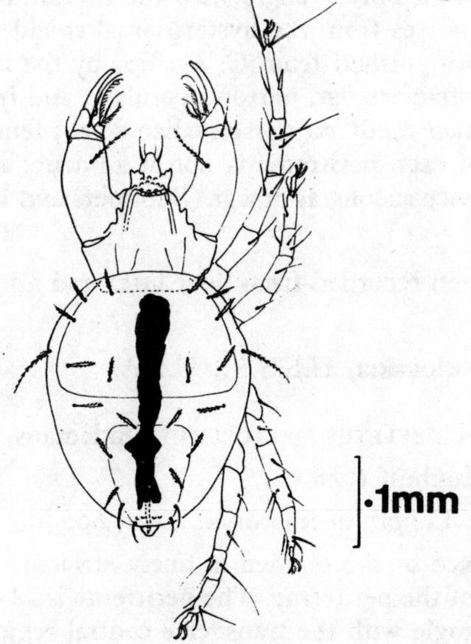

FIG. 279. *Cheyletus trouessarti* Oudemans. Dorsal view of ♂.

pedipalpal tarsus has about 12 fine teeth, the external about 14. The pedipalpal tibia has one or two protuberances at its base (Fig. 274c, d) or these may be so slight as to be virtually absent.

The dorsal plates and their setae are arranged as in the male of *C. eruditus*, except that only one pair of median setae, in addition to the four pairs of peripheral, arises from the propodosomal shield. Ventrally, a short sternal shield extends as far back as coxae II and surrounds the first pair of ventral setae.

On tarsus I the solenidion omega is slightly shorter than the supporting seta (Fig. 275b).

The genital opening (Fig. 268) is at the dorsal extremity of the body.

Heteromorphic males have not been recorded.

BIOLOGY

C. trouessarti occurs infrequently with the acarid mites on which it feeds. It has also been recorded from soil, bark of *Eucalyptus* trees and nests of *Neotoma sp.* (Summers and Price, 1970). The males are more often found than those of *C. eruditus* and, according to A. R. Edwards (pers. comm.) can pair with females of *C. eruditus*; it is possible, therefore, that hybrids may be found. Development can be parthenogenetic, unfertilized eggs giving rise, not to females as in *C. eruditus*, but to males (A. R. Edwards, pers. comm.).

DISTRIBUTION

England, Holland, France, Germany, California.

CHEYLETUS AVERSOR Rohdendorf, 1940

Cheyletus beauchampi Baker, 1949

FEMALE. Length of idiosoma: about 492 μm.

This can be distinguished from *C. trouessarti* by the following characters: the palp claw typically has two basal teeth and the inner 'comb' of the pedipalpal tarsus, has about 30 teeth. The lateral limb of the peritreme typically has six 'links' and it forms an obtuse angle with the middle region. Only one pair of dorsomedian setae arises from the hysterosomal shield.

C. aversor is distinguished from *C. eruditus* by the following characters: the dorsolateral body setae are flat, narrow spatulate, and fringed with barbs; there are two pairs of small modified dorsomedian setae; femur IV has one seta; the rearward flexure of each peritreme is not at an acute angle; the guard seta on tarsus I is nearly twice as long as omega (Summers and Price, 1970).

BIOLOGY

C. aversor has been recorded from floor litter and also from broiler houses.

DISTRIBUTION

England, Czechoslovakia, U.S.S.R., U.S.A.

CHEYLETUS MALACCENSIS Oudemans, 1903

Cheyletus munroi Hughes, 1948.

FEMALE (Fig. 280). Length of idiosoma: about 600 μm.

The dorsal surface of the tegmen is finely striated, the striations becoming indistinct in front of the peritreme. The peritreme is M-shaped, the lateral arms forming an acute angle with the transverse central region. Four or five internal divisions are present between the midline and the lateral limb of the peritreme.

FIG. 280. *Cheyletus malaccensis* Oudemans. Dorsal view of ♀.

The internal 'comb' of the pedipalpal tarsus is quite straight, with 24 to 30 fine teeth, the external 'comb' has about 18 teeth. The base of the tibial claw has a bilobed projection which is variable in shape, and a conspicuous thin flange extends from the base of this claw to the base of the tarsus (Fig. 271c).

The arrangement of setae and shape of the dorsal plates are similar to those of *C. eruditus*, the setae being flattened and pectinate. At the ventral extremity of the body (Fig. 281b), the fifth ventral seta is on a level with the anterior end of the uropore, as in *C. trouessarti*.

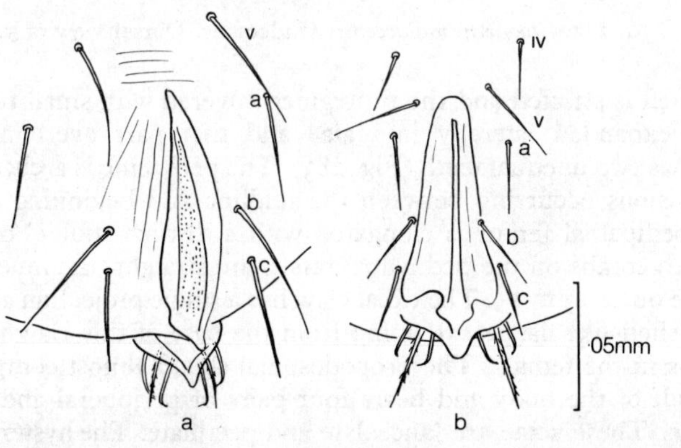

FIG. 281. Anal region of ♀ of (*a*) *Cheyletus eruditus* (Schrank) and (*b*) *C. malaccensis* Oudemans. Ventral setae: IV, V; anal setae: *a* to *c*.

On tarsus I (Fig. 272d), the solenidion omega is swollen at the base and the supporting seta is short and pressed against it.

MALE (Fig. 282). Length of idiosoma: about 320 μm.

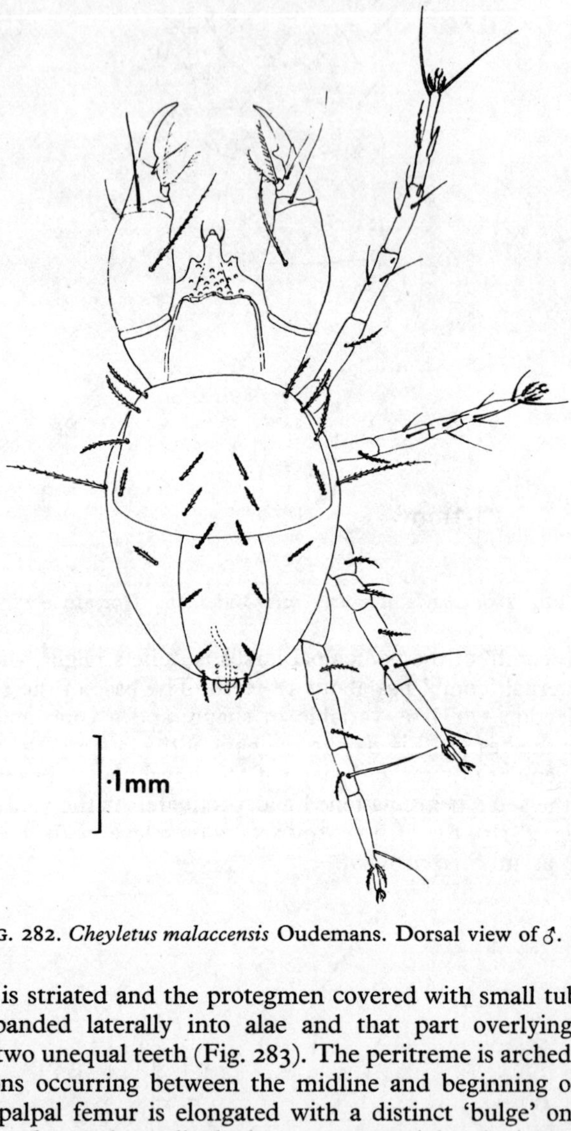

FIG. 282. *Cheyletus malaccensis* Oudemans. Dorsal view of ♂.

The tegmen is striated and the protegmen covered with small tubercles. The rostrum is expanded laterally into alae and that part overlying the palpal trochanter has two unequal teeth (Fig. 283). The peritreme is arched, four to five internal divisions occurring between the midline and beginning of the lateral arm. The pedipalpal femur is elongated with a distinct 'bulge' on its internal surface. Both combs on the pedipalpal tarsus are straight, the inner having 8 to 11 teeth, the outer 13 to 15. The tibial claw has a single projection at its base, but there is no shelf-like flange extending from the base of this claw to the base of the tarsus as in the female. The propodosomal shield almost completely covers the front half of the body and bears four pairs of peripheral and two pairs of central setae. These setae are lanceolate and pectinate. The hysterosomal shield narrows slightly posteriorly and bears four pairs of setae. A short sternal plate extends as far back as coxae II and surrounds the first pair of ventral setae.

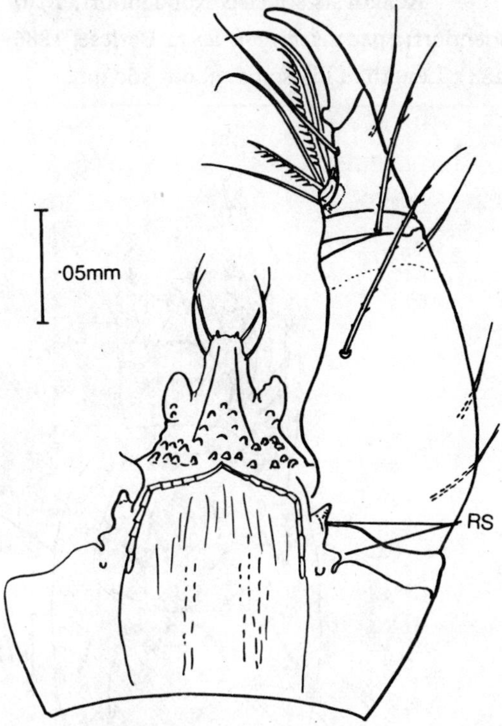

FIG. 283. *Cheyletus malaccensis* Oudemans ♂. Dorsal view of gnathosoma. RS rostal teeth.

On leg I (Fig. 275c) the solenidion omega is expanded towards the base as in the female, the supporting seta being short and inconspicuous.

BIOLOGY

C. malaccensis is found on many different kinds of dried foodstuffs in association with the acarid mites on which it feeds. Males and females are usually found in the same population. Oudemans (1906b) also records it from wheat screenings and on *Ephestia elutella*.

DISTRIBUTION

England, Portugal, Poland, Czechoslovakia, Germany, Iran, Turkey, U.S.A., U.S.S.R., Australia.

ACAROPSIS Moquin Tandon, 1863

The palp tarsus has one comb-like and three smooth setae.

The peritreme is smoothly rounded.

The idiosoma is incompletely covered by two dorsal plates.

Cornea are present.

The marginal setae of the dorsal surface are never fanshaped and no modified dorsomedian setae are present.

The scapular setae are noticeably longer than the remaining dorsal setae.

The supporting seta of tarsus I is minute or may be absent.

Tarsi I to IV bear claws and pulvilli.

Type species: *Tyroglyphus mericourti* Laboulbène, 1851.

Acaropsis sollers Rohdendorf, 1940

A. callida Rohdendorf, 1940; *Acaropsis docta* Berlese, 1886 sensu Hughes, 1961.

Female (Fig. 284). Length of idiosoma: about 560 μm.

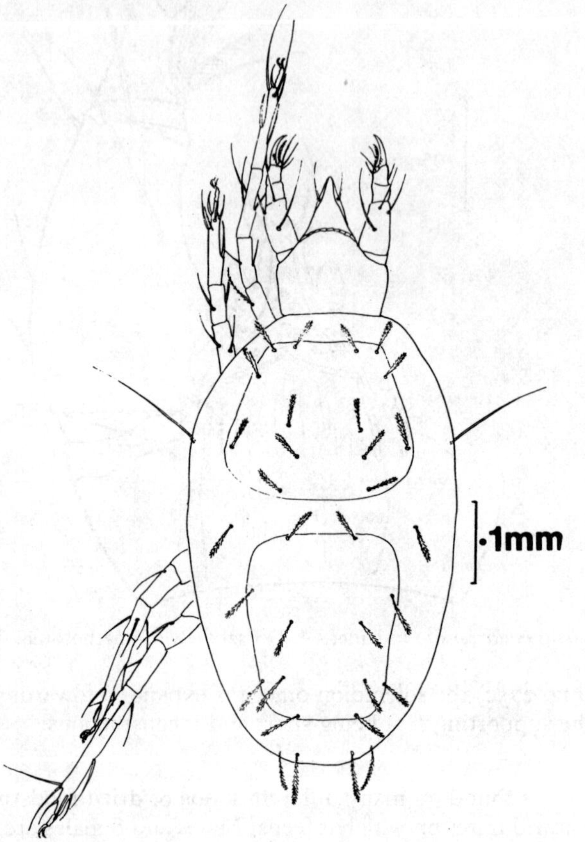

Fig. 284. *Acaropsis sollers* Rohdendorf. Dorsal view of ♀.

The idiosoma is narrow and elongated and orange in colour. The gnathosoma is narrow in relation to the rest of the body, the protegmen and rostrum cone-shaped and ornamented with faint longitudinal striae. The peritreme is smoothly arched and evenly segmented (Fig. 285a). The single comb-like seta has about 12 strong teeth. The tibial claw has three or four irregular teeth on its internal edge. The palp femur is about twice as long as it is wide and its dorsal seta is longer than the femur.

The two dorsal shields are lightly sclerotized. The corneae are inconspicuous. On the propodosomal plate, there are four pairs of peripheral setae and three pairs of median ones. The hysterosomal shield is widely separated from the propodosomal. It is longer than wide and bears six pairs of setae. In addition, a pair of setae arises midway between the two shields and two pairs project from the posterior end of the body. All these setae are lanceolate and densely pectinate. A single pair of long smooth scapular setae projects from the sides of the body between legs II and III.

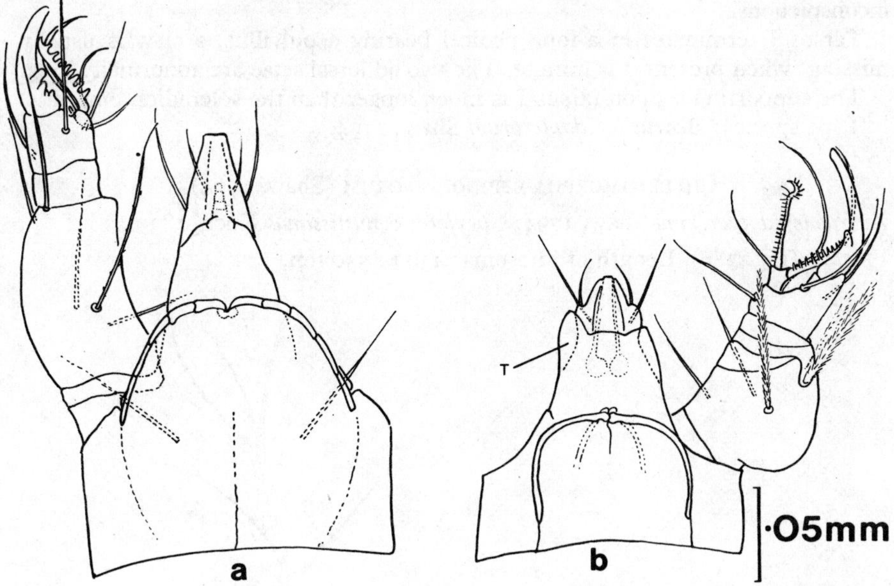

FIG. 285. Dorsal view of gnathosoma of (*a*) *Acarposis sollers* Rohdendorf and (*b*) *Cheletomorpha lepidopterorum* (Shaw). T transparent fold.

The claws and pretarsus of leg I are smaller than those of the remaining legs (Figs. 269a, b). On tarsus I, omega is long and cylindrical reaching to the bases of the addorsal setae. The supporting seta is short and pressed tightly against its base. On tarsus II, omega is shorter and lies against the tarsus. The setae of the legs are sparsely pectinate or smooth, except for one on the femur of each leg which is long and pectinate.

MALE

A heteromorphic male is described by Volgin, 1969.

BIOLOGY

A. sollers has been recorded from cattle pens and feed trash in swine barns (Summers and Price, 1970). It has also been found in the debris from barns and buffer depots where it feeds on acarid mites and their eggs.

DISTRIBUTION

England, Scotland, California U.S.A., Leningrad U.S.S.R.

CHELETOMORPHA Oudemans, 1904

The palp tarsus bears two comb-like and two smooth setae.

The protegmen is extended forward as a transparent fold covering about two thirds of the rostrum.

The greater part of the dorsal surface of the idiosoma is covered by the propodosomal and hysterosomal shields.

Two corneae are present.

The lateral setae of the dorsal shields and body are long and densely pectinate.

The median setae of the dorsal shields are aberrant in structure and inconspicuous.

Tarsus I terminates in a long pedicel bearing a pulvillus, a claw is usually missing; when present it is minute. The two addorsal setae are abnormally long.

The supporting seta on tarsus I is much longer than the solenidion omega.

Type species: *Acarus lepidopterorum* Shaw, 1794.

CHELETOMORPHA LEPIDOPTERORUM (Shaw, 1794)

Acarus lepidopterorum Shaw, 1794; *Cheyletus venustissimus* Koch, 1839.

FEMALE (Fig. 286). Length of idiosoma: 450 to 550 μm.

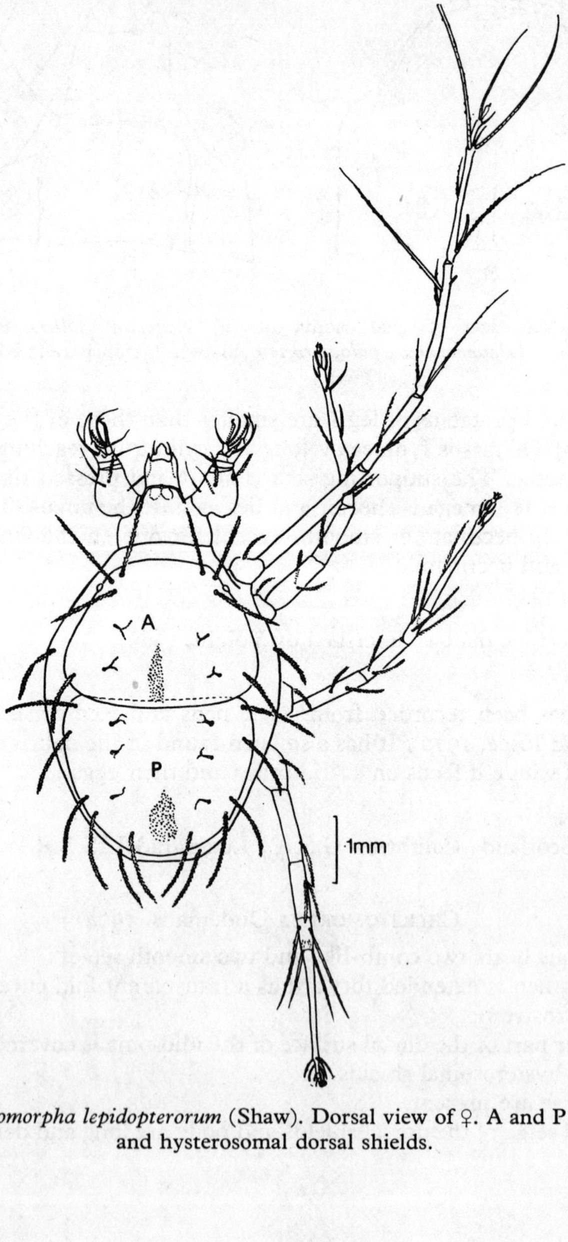

FIG. 286. *Cheletomorpha lepidopterorum* (Shaw). Dorsal view of ♀. A and P propodosomal and hysterosomal dorsal shields.

This mite is more easily recognized than many others on account of its long wispy front legs and orange colour. The orange pigment is in the haemocoel, the exoskeleton being colourless. The protegmen is extended anteriorly as a transparent fold, whose edge may be prolonged into lobes. These lobes may be asymmetrical in outline. The peritreme is shaped like a hoop, without conspicuous subdivisions. The internal comb-shaped seta of the tarsal pedipalp has a sickle-shaped end and numerous fine teeth. The outer comb is longer with about 20 teeth. The tibial claw has one basal projection, and a thin transparent flange extends from the base of this to the base of the tarsus. The pedipalpal femur has a strongly swollen external edge and an almost straight internal; its dorsal seta is pectinate, the two ventral are smooth.

The idiosoma is diamond-shaped, almost completely covered by the dorsal shields whose surfaces are finely pitted. On each of the anterior corners of the propodosomal shield is a transparent cornea adjacent to three long densely pectinate setae. A similar pair of setae arises from the posterior corners of the same shield. Five pairs of setae are also inserted around the edge of the hysterosomal shield. The aberrant dorsomedian setae consist of two pairs on the propodosomal and three pairs on the hysterosomal plates. Each has a short stalk terminating in a bunch of moniliform strands. Scapular and posterior setae are also present.

The first pair of legs is considerably elongated, largely owing to the great length of the femur. It ends in a reduced pretarsus and two very long segmented addorsal setae (Fig. 269c). The solenidion omega and its supporting seta are also unusually long, the supporting seta being barbed and about as long as the tarsus. Claws are usually absent, although Summers and Price (1970) state that some specimens may possess small claws. Tibia I bears a small solenidion, but none is present on genu I. The long anterior legs are not used in locomotion, but are held bent slightly at an angle in front of the gnathosoma. Their hesitant exploratory movements suggest that they function as antennae. The claws and pretarsus of legs II are well developed (Fig. 269d) and omega arises almost midway along the tarsus.

Male

This has been described by Michael (1878), Baker (1949) and Van Eyndhoven (1964). The idiosoma varies in length from 314 μm (Baker) to 472 μm (Michael). The rostrum is longer than that of the female and wider at its base. The external edge of the pedipalpal femur is less swollen. The tibial claw is thin, with from two to four projections at its base. The peritreme is smoothly arched.

The idiosomal setae are longer than those of the female, but arranged in the same way. The five pairs of dorsal median setae are of unequal length (Van Eyndhoven, 1964). The genital opening is on the dorsal surface with three short spines on either side. The first pair of legs is longer than those of the female. On tarsus I omega is half the length of the supporting seta.

Biology

The eggs of *C. lepidopterorum* are laid in isolated clusters and fixed to the ground by a few strands of silk (Michael, 1878, Hughes, 1957), the silk coming through the mouth. The female makes no attempt to protect the eggs. The young larvae (Fig. 287) are colourless, the orange colour beginning to appear at the end of the larval period and becoming more intense at each succeeding stage. The

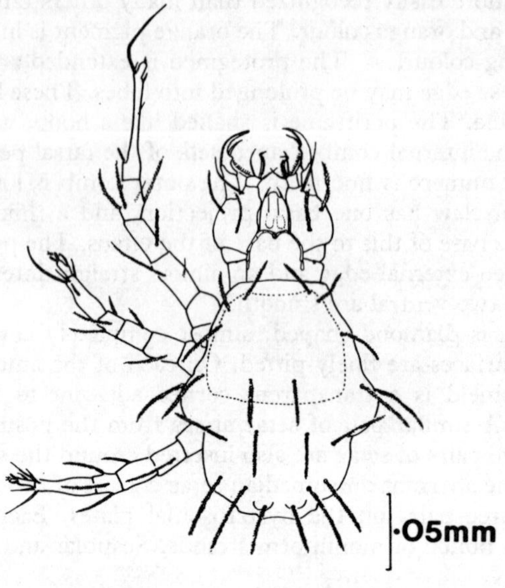

FIG. 287. *Cheletomorpha lepidopterorum* (Shaw). Dorsal view of larva.

first pair of legs is not abnormally long and each terminates in a small claw and is occasionally used in walking. The life cycle takes from two to three weeks at 23°C and a relative humidity of 87 per cent. All stages will feed on acarid mites and their eggs. Males are rarely formed but, according to Michael (1878), several may appear at the same time. Usually, the eggs develop parthenogenetically and then give rise to females.

C. lepidopterorum is often found in association with acarid mites. In one case, it was found in large numbers with *Cheyletus eruditus* on bulk oats. The *Cheyletus* was also orange in colour, since it moves more rapidly than the *Cheletomorpha* and was feeding on it. The orange pigment was unaffected by digestion and accumulated in the haemocoele of the *Cheyletus*. *C. lepidopterorum* was originally found by Shaw (1794) clinging to a moth's wing and it has also been found on the wings of *Hoffmannophila pseudospretella* (J. D. Ethelston coll.). Van Eyndhoven (1964) cites a number of cases where this mite has been found clinging to the wings of moths, but never to those of butterflies (Rhopalocera). He notes that it is always the adult that is found, the mite attaching itself to the scales of the moth wings possibly by a secretion from the anus. In one instance, 227 mites were found on the wings of one specimen of *Peridroma saucia*. D. W. Price (1970) found it on the bark of ponderosa pine, Yosemite Valley, U.S.A. and it has also been recovered from a sparrow's nest (G. Hosie coll.).

DISTRIBUTION

Cosmopolitan.

EUCHEYLETIA Baker, 1949

Cheyletia Haller, 1884 (part) Hughes, 1961.

The palp tarsus has two comb-like and two smooth setae.

The palp tibial claw has from two to four projections at its base.

The protegmen is extended into a cone-shaped process which merges with the rostrum.

The idiosoma is almost or completely covered by two dorsal plates.

No corneae are present.

The lateral setae of the dorsal shields and body are shaped like fans, the median region of the idiosomal shield has from six to 15 pairs of squamate, amoeboid or cloudlike setae (Baker, 1949). One pair of anal setae is also fan-like (except in *E. hardyi*).

Tarsi I to IV have claws and pulvilli.

Type species: *Eucheyletia bishoppi* Baker, 1949.

EUCHEYLETIA TAURICA Volgin, 1963

Cheyletia flabellifera (Michael, 1878) sensu Hughes, 1961.

FEMALE (Fig. 288). Length of idiosoma: 360 to 480 µm.

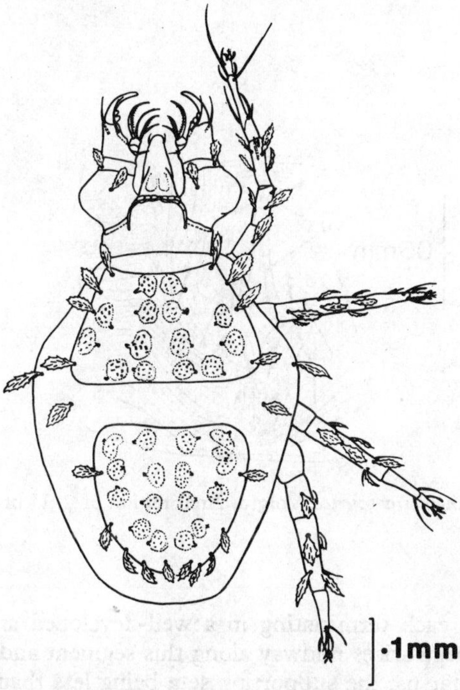

FIG. 288. *Eucheyletia taurica* Volgin. Dorsal view of ♀.

Eucheyletia taurica is more massively built than many other cheyletids. The body is pale yellow and diamond-shaped. The protegmen is well sclerotized and has a segmented flange extending transversely across it. The peritreme is a smoothly curved arc with distinct internal divisions (Fig. 266). The external comb-shaped seta of the pedipalpal tarsus has 13 or 14 distinct teeth, the internal 25 to 30. The tibial claw has two distinct protuberances at its base and the internal edge of this basal region is produced into a thin ridge, from which arises one of the three tibial setae. The dorsal seta of the genu is fan-shaped, the ventral is smooth. The external edge of the femur is strongly swollen and only the dorsal seta is fan-shaped. On each palpal coxa is a small cylindrical palpostigmatic organ.

The dorsal idiosomal plates have a pitted surface and are separate from one another whilst the surrounding cuticle is covered with pitted wrinkles, the pits being absent from the ventral surface. The anterior edge of the propodosomal shield is shorter than the posterior and the shield itself bears four pairs of peripheral setae. In addition the central region is largely covered by seven pairs of irregularly shaped 'cloudlike' setae. No corneae are present. The hysterosomal shield bears from three to five pairs of peripheral setae and eight pairs of 'cloudlike' setae. Scapular setae are present and also a pair of setae arising between the two shields. All the dorsal setae (with the exception of the 'cloudlike' setae) are fan-shaped, each consisting of a thin membrane supported by delicate thickenings like the veins of a leaf. No sternal shield is present and the uropore (Fig. 289) is removed from the posterior margin of the body and surrounded by six pairs of smooth setae and one pair of terminal fan-shaped setae.

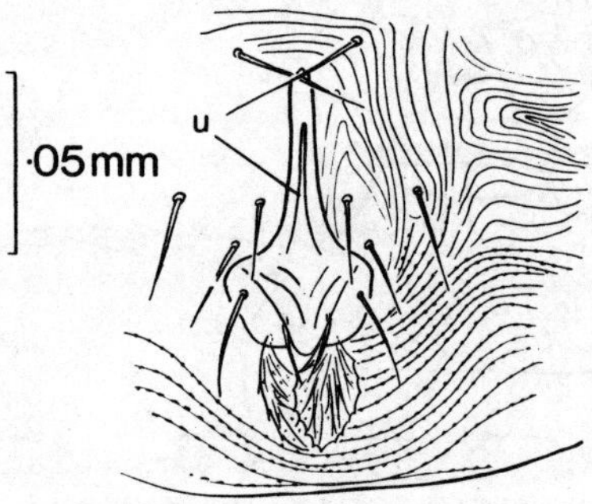

FIG. 289. *Eucheyletia taurica* Volgin. Anal region of ♀. U uropore.

The legs are short, each terminating in a well-developed ambulacrum. On tarsus I (Fig. 290) omega arises midway along this segment and extends almost to the base of the pretarsus, the supporting seta being less than half its length. Short cylindrical solenidia also arise from the tibia and genu. Most of the setae arising from the legs are fan-shaped, but slightly pectinate setae are found on the tarsi and ventral regions of the tibiae.

Biology

E. taurica is occasionally found in floor dust, in association with acarid mites and their eggs. Volgin (1969) records it from the nests of *Microtus arvalis* Pall. and the long-tailed field mouse (*Apodemus sylvaticus* L).

Distribution

England, the Crimean reservation, near Gursuf, U.S.S.R.

FIG. 290. *Eucheyletia taurica* Volgin ♀. Dorsal view of right leg I.

BDELLIDAE Dugès, 1834

This family is characterized by its members having an elongated gnathosoma, thus earning them the name of snout mites. They are often red or black in colour and vary in size from 0.5 to 3 mm in length. The colour is due partly to a pigmented cuticle and partly to the colour of the internal organs. The cuticle is soft, smooth or wrinkled, and almost without plating.

The gnathosoma or rostrum is bulbous at the base and tapers towards its distal end. The ventral region or hypostome is a gutter-shaped structure whose lateral edges, towards the base, meet in the midline to form a sub-cheliceral shelf or epistome. The distal end of the hypostome is truncated and terminates in two flattened lobes of varying structure. From the ventral region arise a number of setae which are of taxonomic importance. The chelicerae lie close together above the hypostome, separated in the midline by a thin low partition which may allow a certain amount of independent movement. Each terminates in a pair of small chelae devoid of teeth. From the dorsal side of the chelicerae arise setae whose number and position are of generic importance (Fig. 291).

In life, the long palps are held extended forwards and upwards. Each has six segments (Fig. 292). The coxa is fused with the base of the hypostome, the trochanter is small and devoid of setae, the femur is partially divided into a long proximal basifemur with a variable number of setae, and a short telofemur with one dorsal seta. The genu varies in length and may have from five to seven setae, and the tibiotarsus bears typically two long sensory setae and a number of shorter setae and tricobothria (Atyeo, 1960).

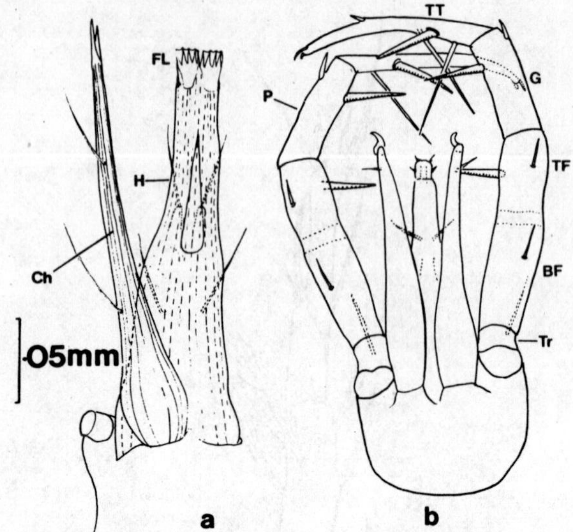

FIG. 291. Dorsal view of gnathosoma of (*a*) *Spinibdella* sp. and (*b*) *Cunaxa setirostris* (Hermann). Ch chelicera, FL flattened lobes at distal end of hypostome (H), P palp, Tr trochanter, BF basifemur, TF telofemur, G genu, TT tibiotarsus.

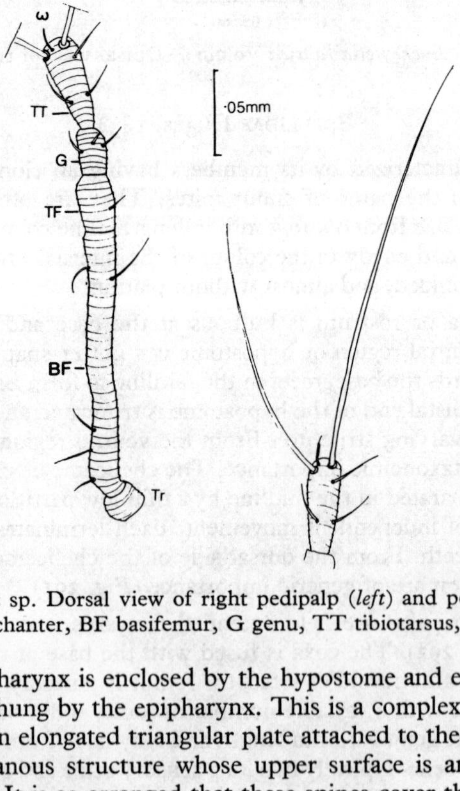

FIG. 292. *Spinibdella* sp. Dorsal view of right pedipalp (*left*) and pedipalpal tibiotarsus (*right*). Tr trochanter, BF basifemur, G genu, TT tibiotarsus, ω solenidion.

The muscular pharynx is enclosed by the hypostome and epistome, its opening or mouth overhung by the epipharynx. This is a complex structure consisting essentially of an elongated triangular plate attached to the base of which is a crescentic membranous structure whose upper surface is armed with a large number of spines. It is so arranged that these spines cover the mouth and stop the entrance of solid particles. The ventral region of the pharynx is extended anteriorly as the hypopharynx or lingua which starts as a flat tongue whose edges curl and fuse to form a membranous tube which can be extended beyond the end of the gnathosoma. At its distal end the hypopharynx is slightly enlarged and it is capable of movement in every direction. The bdellids prey on other mites and small insects whose cuticle is pierced by the slender blades of the chelicerae and the body contents are then sucked up the tubular hypopharynx by means of suction exerted by the muscular pharynx (Michael, 1896). The alimentary canal is modified to store a large bulk of fluid by the development of a diverticulum opening into the oesophagus.

The propodosoma is like an inverted triangle, its apex tapering to meet the base of the gnathosoma. From its dorsal side usually arise two pairs of setae—lateral and median propodosomals (Atyeo, 1960) and two pairs of long tapering sensillae or trichobothria (Grandjean, 1938b) which arise from cup-shaped pseudostigmatic organs (Fig. 293). Although the limits of a dorsal propodosomal shield are not apparent, the cuticle in the regions of the pseudostigmatic organs is thickened and internal apodemes are sometimes developed. The integument in this region is also distinctively ridged, the crests of the ridges broken at regular intervals to form a distinct pattern. Two pairs of eyes are usually present, external to the posterior sensillae, and a median eye is sometimes developed between the anterior sensillae, as in the genus *Cyta*.

The hysterosoma is elongated and square in outline and bears a number of smooth or plumose setae which are arranged in five transverse rows, the most anterior of these are the internal and external humerals. The genital opening is posterior to the bases of legs IV. In both sexes it is covered by genital folds, each of which bears a row of genital setae. External to these arise the paragenital setae. The female possesses an extrusible ovipositor at the base of which are three pairs of genital discs or sense organs. The male genitalia are more complicated, consisting essentially of a thin, muscular penis, surrounded by a sheath and supported by basal sclerites (Michael, 1896). The uropore or opening to the excretory canal is a longitudinal slit at the posterior end of the body. A number of anal setae is associated with it.

The legs have six free segments, the coxa being partially fused with the body. Sometimes a transverse constriction across the femur gives the appearance of an additional segment. The tarsus is usually long, terminating in a short pretarsus bearing a pair of claws and a median pad-like empodium or pulvillus. The claws may have small dorsal ridges bearing fine projections. From the legs arise many setae and solenidia whose arrangement and distribution are of taxonomic importance; these are particularly numerous on the two anterior pairs of legs. The true setae may be smooth or pectinate, the solenidia long and tapering (trichobothria) or short and rounded distally.

A pair of stigmata is situated in the mid-dorsal line between the bases of the chelicerae. These lead into spirally thickened tracheal tubes. In the subfamilies Spinibdellinae and Cytinae genital tracheae are also present and open beneath the genital folds.

According to Womersley (1933) the eggs of the Bdellidae are slightly elliptical and covered with clavate spines. They are brown in colour and are deposited on the ground or in decaying vegetable fibres. The life cycle consists of larva, three nymphal stages and adults.

The bdellid mites form a small, well-defined family containing about 10 genera.

All its members are free-living and found in grass, moss and on the sea shore, where they are conspicuous on account of their bright colour. They can stand intense cold and long immersion in water, and many specimens have been brought back from polar expeditions. The family is cosmopolitan and is the subject of monographs by Sig Thor (1931), Baker and Balock (1944) and Atyeo (1960). The internal and external anatomy have been studied in great detail by Michael (1896) and Grandjean (1938b) respectively.

SPINIBDELLA Sig Thor, 1930

The tibiotarsus of the pedipalp is short and truncate, the genu bears three or four setae and the basifemur more than two setae. The cheliceral setae may be minute or of medium length, with the distal seta not extending beyond the tips of the chelae.

Lateral propodosomal setae are present.

Type species: *Spinibdella reducta* Sig Thor, 1930.

SPINIBDELLA SP.

Bdella lignicola Canestrini, 1885 sensu Hughes, 1961.

FEMALE (Figs 293, 294). Length of gnathosoma and idiosoma: about 1,120 µm.

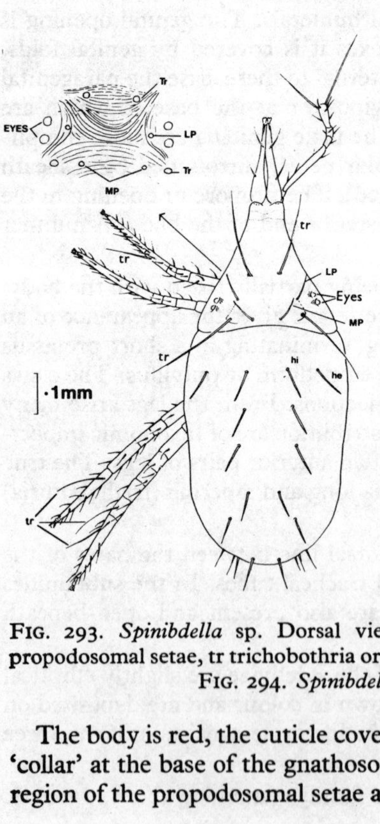

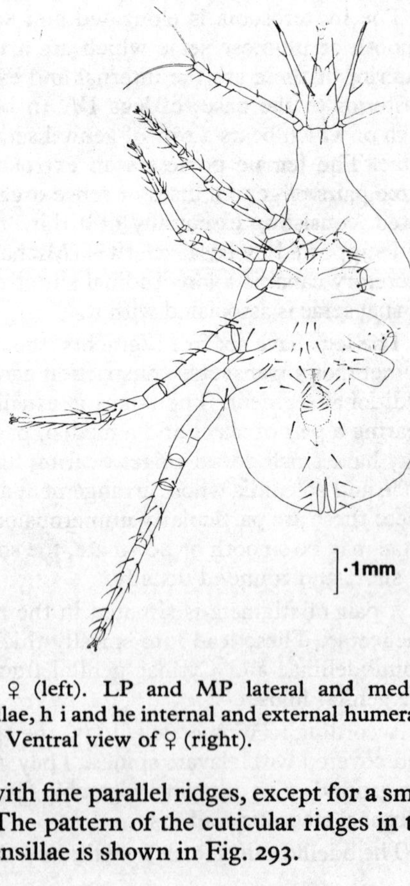

FIG. 293. *Spinibdella* sp. Dorsal view of ♀ (left). LP and MP lateral and median propodosomal setae, tr trichobothria or sensillae, h i and he internal and external humerals.

FIG. 294. *Spinibdella* sp. Ventral view of ♀ (right).

The body is red, the cuticle covered with fine parallel ridges, except for a small 'collar' at the base of the gnathosoma. The pattern of the cuticular ridges in the region of the propodosomal setae and sensillae is shown in Fig. 293.

The gnathosoma (Figs. 291a, 292) is long and slender, about one quarter the length of the idiosoma. The chelicerae have tapering shears and on their external surface, each bears two setae which are about equidistant from the base and apex of this appendage. The ventral region of the hypostome also bears two pairs of setae, at the same level as those on the chelicerae. The basifemur of the pedipalp exceeds the combined lengths of the two succeeding segments and bears four setae on its internal side, one on the external. The palpal telofemur has one seta, the genu three and the tibiotarsus is short and expanded distally. Two long sensory setae of unequal length, three short setae and a short blunt solenidion arise from the tibiotarsus.

The propodosoma bears two pairs of eyes, each consisting of a small mass of pigment covered by a cornea. The anterior and posterior sensillae are about equal in length, the lateral and median propodosomal setae considerably shorter. Seven pairs of short slightly pectinate setae arise from the dorsal side of the hysterosoma, arranged as in Fig. 293.

The genital opening lies behind coxae IV. It is covered by two genital plates and communicates with an internal chamber or vestibule, from the lateral walls of which arise three pairs of genital discs. Each genital plate bears a row of genital setae and is surrounded by a small number of paragenital setae. The uropore is terminal.

A number of setae and solenidia arise from the leg segments. Tarsus I bears two blunt solenidia and a short peg-shaped seta (Fig. 295) and long trichobothria arise from pits on the dorsal surface of tibia I, tarsi III and IV and tibia IV.

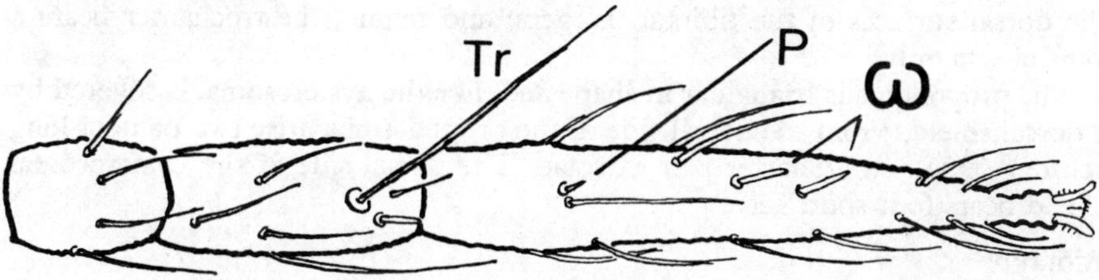

FIG. 295. *Spinibdella* sp. Dorsal view of right leg I. Blunt solenidion: ω, Peg-shaped seta: P, Trichobothrion: Tr.

MALE

Unknown.

BIOLOGY

A few specimens of *Spinibdella* sp. are sometimes found in debris surrounding flour and grain bags. They are probably feeding on other mites.

DISTRIBUTION

England, Northern Ireland.

CUNAXIDAE Sig Thor, 1902

The Cunaxidae and Bdellidae have many characters in common and are sometimes included in the superfamily Bdelloidea. They differ from one another mainly in the structure of the gnathosoma. In the Cunaxidae the chelicerae terminate in a hook and the distal ends of the palps end in a claw used for grasping. Apophyses and small spines may also be found on the internal surfaces of the palps.

Two pairs of genital sense organs are present in tritonymphs and adults.

Members of this family are usually brightly coloured, often found in humus and on the surface of leaves. They are predaceous on small Arthropoda and are cosmopolitan in distribution.

According to Sig Thor and Willmann, 1941 and Baker and Hoffmann (1948), the family Cunaxidae is divided into a small number of genera. The genus *Cunaxa* differs from the remaining genera in having a five-segmented pedipalp longer than the chelicera, and the second and third segments (basi- and telofemur) are partially fused with one another.

Cunaxa setirostris (Hermann, 1804)

Cunaxa taurus (Kramer, 1881) sensu Hughes, 1961.

FEMALE. Length of idiosoma: about 500 µm.

The idiosoma is red in colour and covered with a soft finely-wrinkled cuticle. The gnathosoma (Fig. 291b) is elongated to a cone-shaped structure, the chelicerae being exposed dorsally. Each chelicera has the movable digit modified to form a hook, the fixed digit being absent. The ventral region of the gnathosoma ends distally in a small membranous lobe, from which arise two pairs of small setae; another pair of longer setae is more proximal in position. The palp extends well beyond the end of the chelicera. The palpal tibiotarsus ends in a small claw overhung by a seta. A prominent spine or apophysis arises from its internal surface and a longer seta is proximal to this. Similar apophyses arise from the internal sides of the palpal genu and telofemur. Additional setae are inserted on the dorsal surfaces of the tibiotarsus, genu and femur; the trochanter bears a ventral seta only.

The propodoma is triangular in shape and, like the hysterosoma, is covered by a dorsal shield. From the dorsal side of the propodosoma arise two pairs of long sensory setae and a shorter pair of setae. The dorsal side of the hysterosomal shield bears four short setae.

Biology

C. setirostris is often found in soil, moss and on the leaves of plants. It is sometimes found in stored food such as barley meal, and also in wheat (R. G. Symes coll.).

Distribution

Europe, North Africa (Morocco), Perth (South Australia), United States, Philippines, Guam, Brazil, India.

Tydeidae Kramer, 1877

Small mites (less than 500 µm long) with oval or pear-shaped bodies, white, yellow, red, black or brown in colour. The cuticle is weakly or non-sclerotized and often characteristically ornamented.

The cheliceral bases are fused together. The fixed digit is small and the movable is styliform (Fig. 296). The pedipalp is four-segmented, the femur and genu united together.

The idiosoma is usually divided by a furrow into propodosoma and hysterosoma. The propodosoma bears three pairs of dorsal setae, and one pair of sensory setae arising from pits or pseudostigmata (Fig. 297). Two or three pigmented retinae may be present.

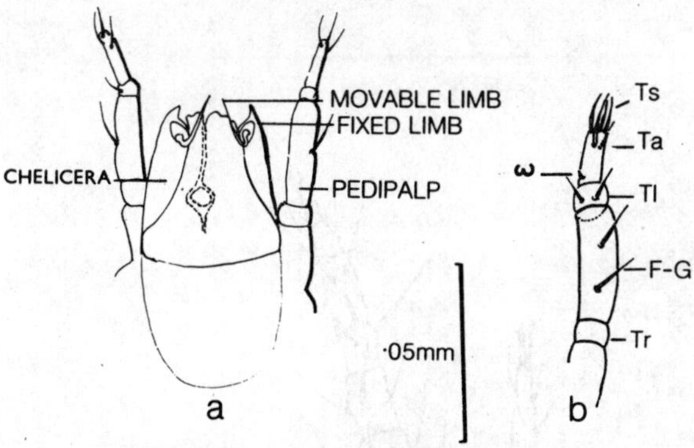

FIG. 296. *Tydeus interruptus* Thor ♀. (*a*) Ventral view of gnathosoma, (*b*) dorsal view of pedipalp. TS tarsal setae, Tr trochanter, F-G femur-genu, Ti tibia, Ta tarsus, ω solenidion.

From the hysterosoma arise five pairs of dorsal propodosomal setae and four or five pairs of laterals (Baker, 1965). Three pairs of ventrals are present and setae are also associated with the genital and anal openings. No genital sense organs are present.

The legs are five-segmented and each tarsus bears a pair of conspicuous claws and a pad-shaped pectinate empodium or pulvillus (Fig. 298).

The sexes only differ from one another in the chaetotaxy and structure of the genital opening. In the male, the genital opening is small and eugenital setae are present lying beneath the genital folds (Grandjean, 1938a). In the female, the opening is larger and no eugenital setae are present.

Respiration is by means of tracheal tubes opening by spiracles at the base of the gnathosoma. They are present at all stages of the life cycle (Grandjean 1938a).

The life cycle consists of a larva and three nymphal stages as well as the adults. The nymphal stages differ from one another in the chaetotaxy of the genital region (see Grandjean, 1938a).

The family Tydeidae includes a number of species often found running about the surface of leaves or moss. They feed on other mites and are also said to suck the juices of plants. According to Sig Thor (1933), *Tydeus molestus* Moniez, 1889 will also attack man and domestic animals. A number of species of Tydeidae may be found in grain residues and barn debris.

TYDEUS Koch, 1835

The idiosoma is constricted between the propodosoma and hysterosoma.

The cuticle is finely striated, the striae longitudinal on the propodosoma, transverse on the dorsal side of the hysterosoma.

The sensory setae of the propodosoma are simple.

Two pigmented retinae are present on the propodosoma.

The hysterosoma bears five pairs of dorsal setae and four pairs of laterals, L_2 (Fig. 297) being lateral in position and slightly posterior to D_2.

All the tarsi bear claws and empodia.

Type species: *Tydeus kochi* Oudemans, 1928.

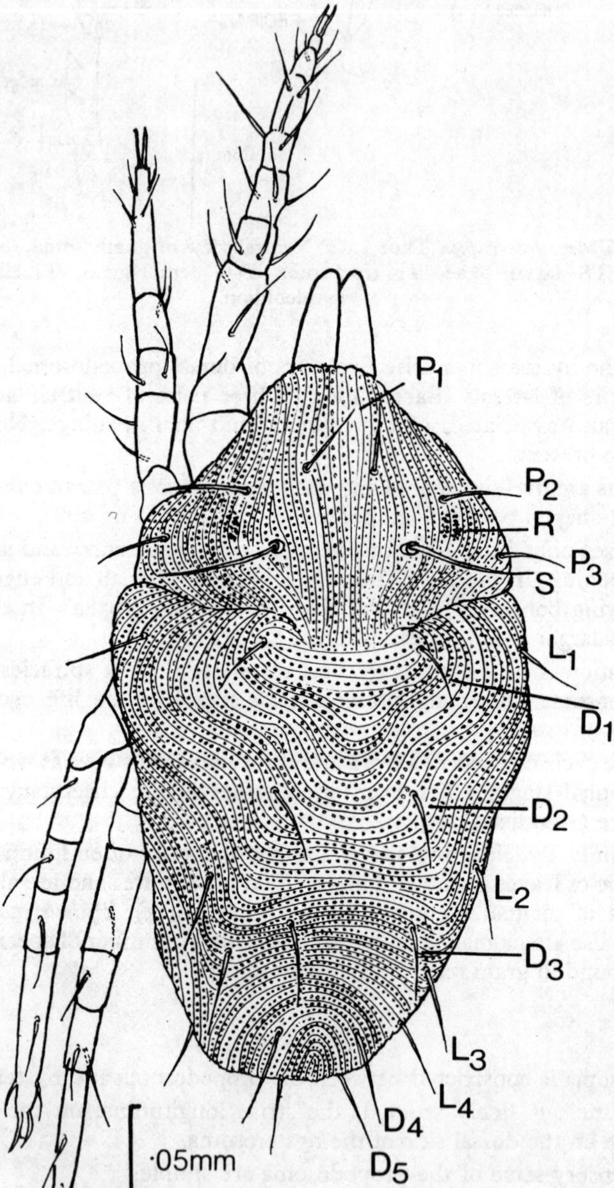

FIG. 297. *Tydeus interruptus* Thor. Dorsal view of ♀. Propodosomal setae: P_1 to P_3, sensory seta: S, dorsal setae: D_1 to D_5, lateral setae: L_1 to L_4. R pigmented retina.

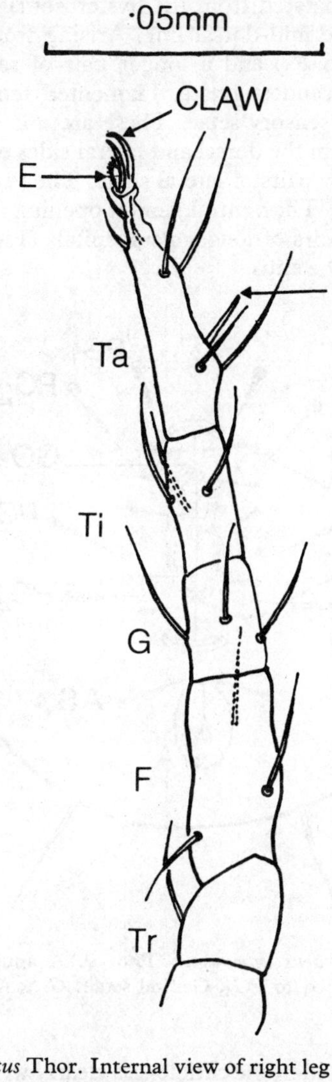

FIG. 298. *Tydeus interruptus* Thor. Internal view of right leg I. E empodium, ω solenidion, Tr trochanter, F femur, G genu, Ti tibia, Ta tarsus.

TYDEUS INTERRUPTUS Sig Thor, 1932

FEMALE (Fig. 297). Length of idiosoma: 250 to 280 μm.

The idiosoma is diamond-shaped and yellowish, the colour being due to the contents of the haemocoele. The cuticle is wrinkled, the grooves being accentuated by lines of evenly-spaced tubercles.

The gnathosoma is short and broad (Fig. 296), its ventral apex bilobed. The movable digit of the chelicera is a fine curved spike which fits into a shallow groove scooped out of the fixed arm. The pedipalpal tarsus bears a terminal cluster of five setae and a small basal solenidion; the tibia and femur-genu each have two setae.

The propodosoma is separated from the hysterosoma by a transverse groove which is incomplete in the mid-dorsal line. Arising from it are three pairs of propodosomal setae (P_1 to P_3) and a longer pair of sensory setae which are inserted into prominent pseudostigmata. Pigmented retinae lying beneath the cuticle are external to the sensory setae. These are not visible in preparations mounted in lactic acid. From the dorsal and lateral sides of the hysterosoma arise five pairs of dorsal and four pairs of lateral setae. The ventral side has the usual three pairs of ventral setae. The ventral genital opening is bordered by six pairs of genital setae and four pairs of longer paragenitals (Fig. 299). A pair of setae is inserted on the sides of the anus.

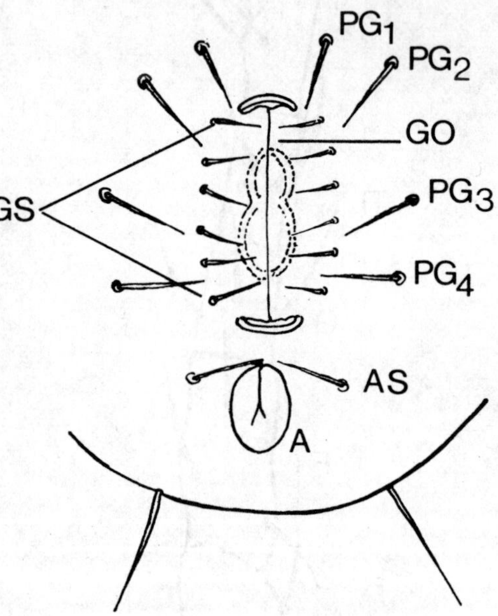

FIG. 299. Anal region of *Tydeus interruptus* Thor ♀. A anus, G O genital opening. Paragenital setae: PG_1 to PG_4, Genital setae: *G S*, Anal setae: *A S*.

The legs are long and slender, the cuticle less clearly wrinkled. Each terminates in a pair of claws and a pad-shaped empodium or pulvillus. The tarsus of leg I has a hump on its dorsal side from which arise a tapering solenidion and two setae of unequal length (Fig. 298). Two additional setae are more distal in position and three shorter setae encircle the tip of the tarsus. A whorl of three setae is inserted into the tibia, genu and femur respectively and a single seta arises from the trochanter.

BIOLOGY

T. interruptus was first found in moss. It is not uncommon in grain debris in association with acarid mites and sometimes occurs in large numbers.

DISTRIBUTION

Norway, England.

TETRANYCHIDAE Donnadieu, 1875

The Tetranychidae are soft-bodied mites often green, yellow, orange or red in colour. They possess a smooth or ridged cuticle and feed on the juices of plants, the mouth parts being modified for piercing and sucking. The bases of the chelicerae are fused to form a cheliceral plate or stylophore and attached to this are the curved proximal ends of the movable digits of the chelicerae which are shaped like flexible needles. The pedipalp is short and five-segmented, the tibia bearing a claw (Fig. 300). Setae and solenidia arise from the pedipalp and one of the distal tarsal setae may be modified as a spinneret.

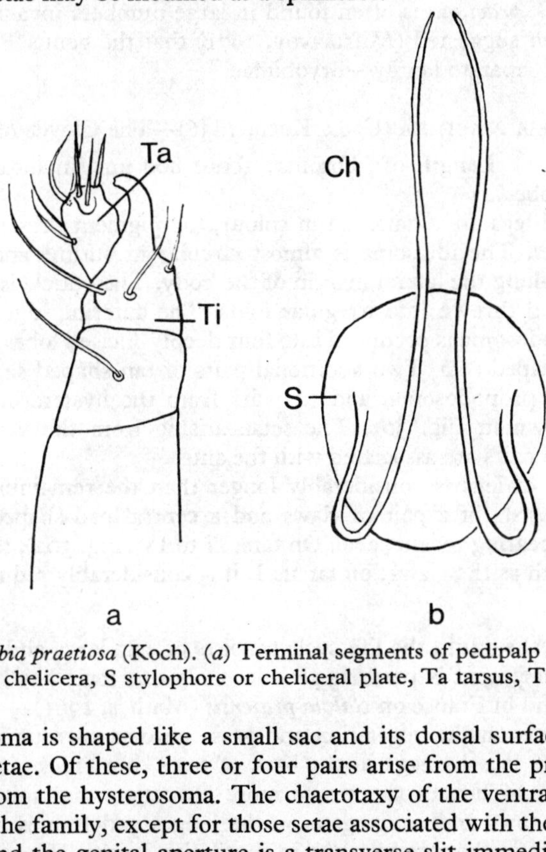

FIG. 300. *Bryobia praetiosa* (Koch). (*a*) Terminal segments of pedipalp and (*b*) chelicerae. Ch chelicera, S stylophore or cheliceral plate, Ta tarsus, Ti tibia.

The idiosoma is shaped like a small sac and its dorsal surface bears at most 16 pairs of setae. Of these, three or four pairs arise from the propodosoma, the remainder from the hysterosoma. The chaetotaxy of the ventral surface is constant within the family, except for those setae associated with the anus. The anus is terminal and the genital aperture is a transverse slit immediately anterior to it. Males possess a penis whose shape is of taxonomic importance. No genital discs or sense organs are present.

Respiration is tracheate, the tracheae opening into peritremes which arise at the anterior end of the body in the mid-dorsal line and extend laterally over the surface of the body in the region of the stylophore.

The legs are six-segmented and terminate in an ambulacrum consisting of a pair of claws and central empodium or pulvillus. Tenent setae, i.e., setae with knobs or hooks at their distal ends and originating from a number of 'roots', may be associated with the claws. Setae and solenidia arise from the legs, the solenidia mainly arising from tarsi I to IV and tibiae I and II, sometimes III and IV as well. Sometimes a long solenidion is associated with a short seta ('duplex seta') and located on the dorsal surface of the tarsus.

Many of the Tetranychidae are of economic importance. *Panonychus ulmi* (C. L. Koch) is the fruit tree red spider mite which feeds on apples, plums and pears in the orchards of this country. *Tetranychus urticae* C. L. Koch is a complex of closely related species which attack glasshouse crops of tomatoes and cucumbers, as well as outdoor hops and strawberries. The genus *Bryobia* includes a group of species not fully worked out in this country which differ in their biology as well as in their structure. Of these, *B. rubriocolus* (Scheuten) is restricted mainly to apples and pears, *B. kissophila* Van Eyndhoven to ivy and *B. praetiosa* sensu strictu to grasses and low-growing herbaceous plants. In the early summer, *B praetiosa* is often found in large numbers invading houses. It has recently been suggested (Mitrofavov, 1971) that the genus *Bryobia* should be included in a separate family—Bryobiidae.

Bryobia praetiosa (C. L. Koch, 1836)—The Clover Mite

Female (Fig 301). Length of idiosoma: about 800 µm (including the dorsal propodosomal lobes).

The body and legs are a dark red in colour, the pigment present in the fluid of the haemocoel. The idiosoma is almost circular in outline and compressed dorso-ventrally along the lateral margin of the body. The cuticle is covered with fine tubercles and thrown into irregular folds. The anterior edge of the dorsal side of the propodosoma is produced into four deeply-incised lobes, each terminating in a fan-shaped seta. Two additional pairs of fan-shaped setae arise from the sides of the propodosoma, and 12 pairs from the hysterosoma. They are arranged as shown in Fig. 301. The setae arising from the ventral side are simple and three pairs are associated with the anus.

The first pair of legs is considerably longer than the remaining ones. Each ambulacrum consists of a pair of claws and a central pad-shaped empodium, both structures bearing tenent setae. On tarsi II to IV (Fig. 302), the empodium is the same length as the claws; on tarsus I, it is considerably reduced.

Biology

In addition to the adult, the life cycle consists of egg, larva and two nymphal stages, and development is parthenogenetic. Only one male has been recorded and this was found in France on *Salvia pratensis* (Mathys, 1961).

B. praetiosa lives in the vegetation and grass borders surrounding trees and houses. The mite lays its eggs in the bark of trees, but cracks in the window frames or ledges of houses can prove more attractive, and mites may enter through the windows and congregate on the ledges and frames. The eggs are then laid in cracks in the woodwork or in ventilation bricks. Here the mite overwinters either as the adult or egg. In spring, when the temperature is suitable, the larvae emerge and migrate to any suitable vegetation on which they feed. When moulting is about to occur, the fully fed larvae return to the house and moult to form the protonymph which once more returns to its feeding ground. After two more moults, which involve the same in and out migration, the adult females continue feeding until ready to return to the building to lay their eggs. Egg laying may be repeated more than once, so that there is a continuous to and fro movement between the house and surrounding vegetation.

A high relative humidity and a temperature range of 7.2 to 18.3°C are the optimum conditions for *B. praetiosa*. The eggs fail to hatch below 7.2°C, and above 18.3°C they hatch more slowly, until at 29.4°C all remain dormant. In this country, if the early summer is wet and cool, as sometime occurs in May and

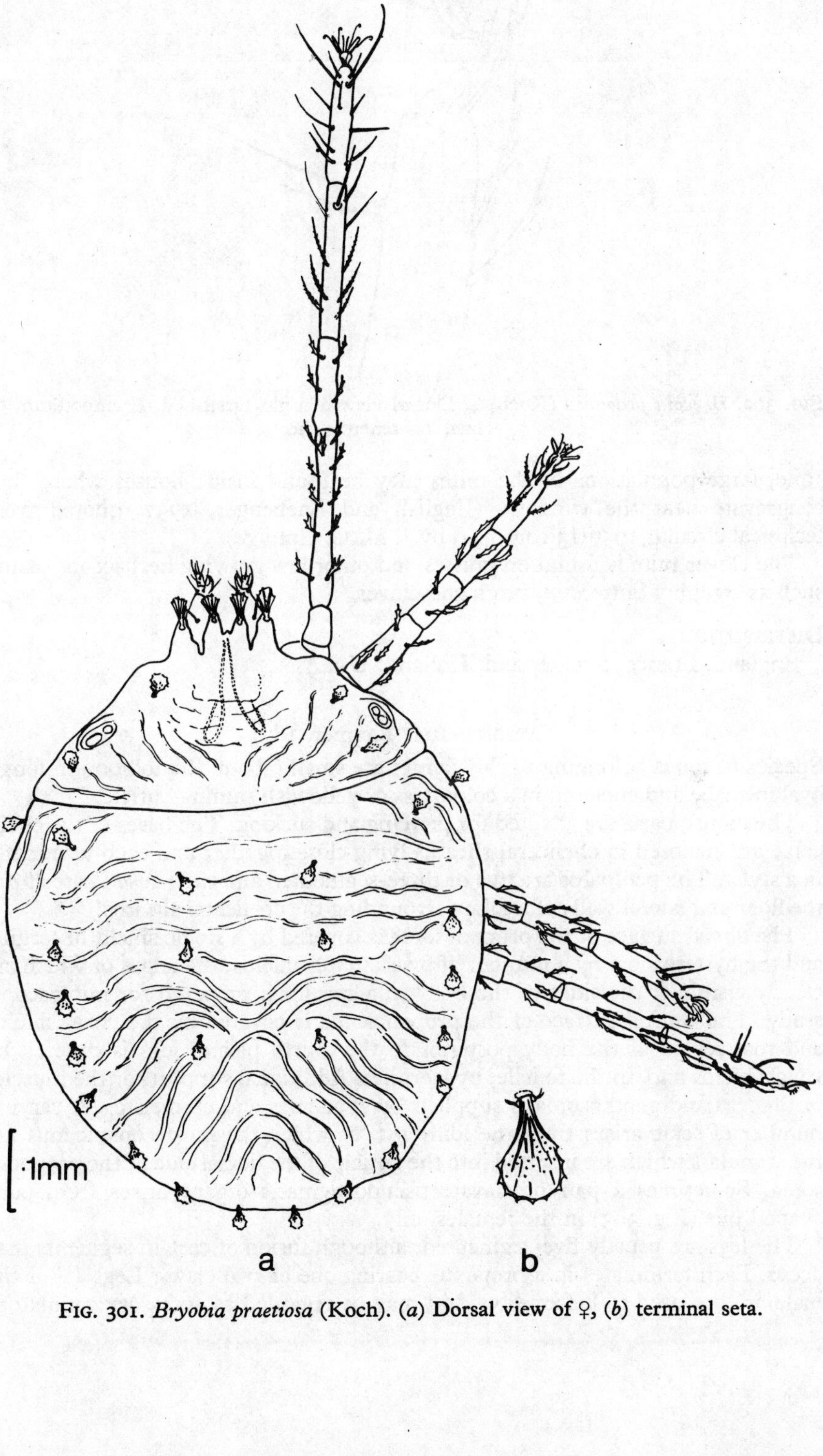

FIG. 301. *Bryobia praetiosa* (Koch). (*a*) Dorsal view of ♀, (*b*) terminal seta.

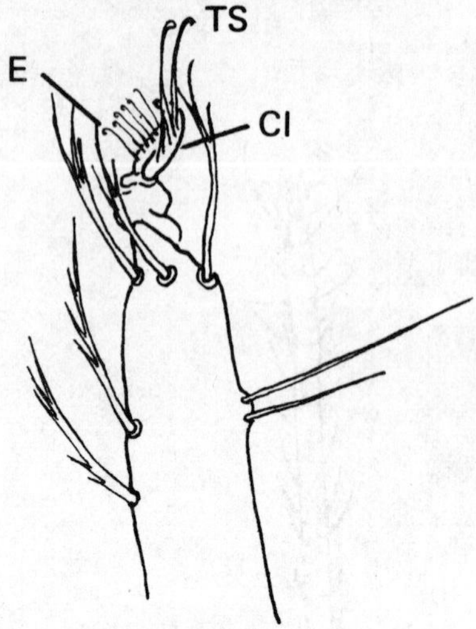

FIG. 302. *Bryobia praetiosa* (Koch) ♀. Dorsal view of right tarsus IV. E empodium, Cl claw, *TS* tenent setae.

June, large populations of the mites may be found inside houses where they congregate near the windows (English and Snetsinger, 1957, quoted from technical circular 1957/13 compiled by J. M. C. Gradige).

The clover mite is found on grasses and other low growing herbaceous plants such as creeping buttercup, vetch and clover.

DISTRIBUTION

England, France, Switzerland, Holland, U.S.A.

TARSONEMIDAE Kramer, 1877

Species of mites belonging to this family are small (about 100 to 400 μm long), oval in shape and enclosed in a colourless or yellowish shining cuticle.

The mouth parts are adapted for piercing and sucking. The bases of the chelicerae are enclosed in cheliceral sheaths lying close together and each terminates in a stylet. The pedipalps are two or three-segmented and their fused bases form the floor and lateral walls of a tube surrounding the cheliceral sheaths.

The dorsal surface of the propodosoma is covered by a single shield or tergite, and the hysterosoma by a number of overlapping shields, the edges of which are clearly visible at the sides of the body in a specimen prepared for microscopic study. The ventral surface of the propodosoma is covered by the coxal shields and that region of the body posterior to the fourth pair of legs is covered by coxal shields and, in the female, by sternites. Additional support for the muscles of the legs and gnathosoma is supplied by apodemes and epimerites. A varying number of setae arises from the idiosoma, of which the most conspicuous are the scapulars which are inserted into the cuticle of the dorsal side of the propodosoma. Sometimes a pair of clavate pseudostigmatic organs arises from cup-shaped pits (Fig. 303) in the females only.

The legs are usually five-segmented, although fusion of certain segments may occur. Each terminates in a pretarsus bearing one or two claws. Legs IV of the male are not used in locomotion, and may be regarded as accessory copulatory

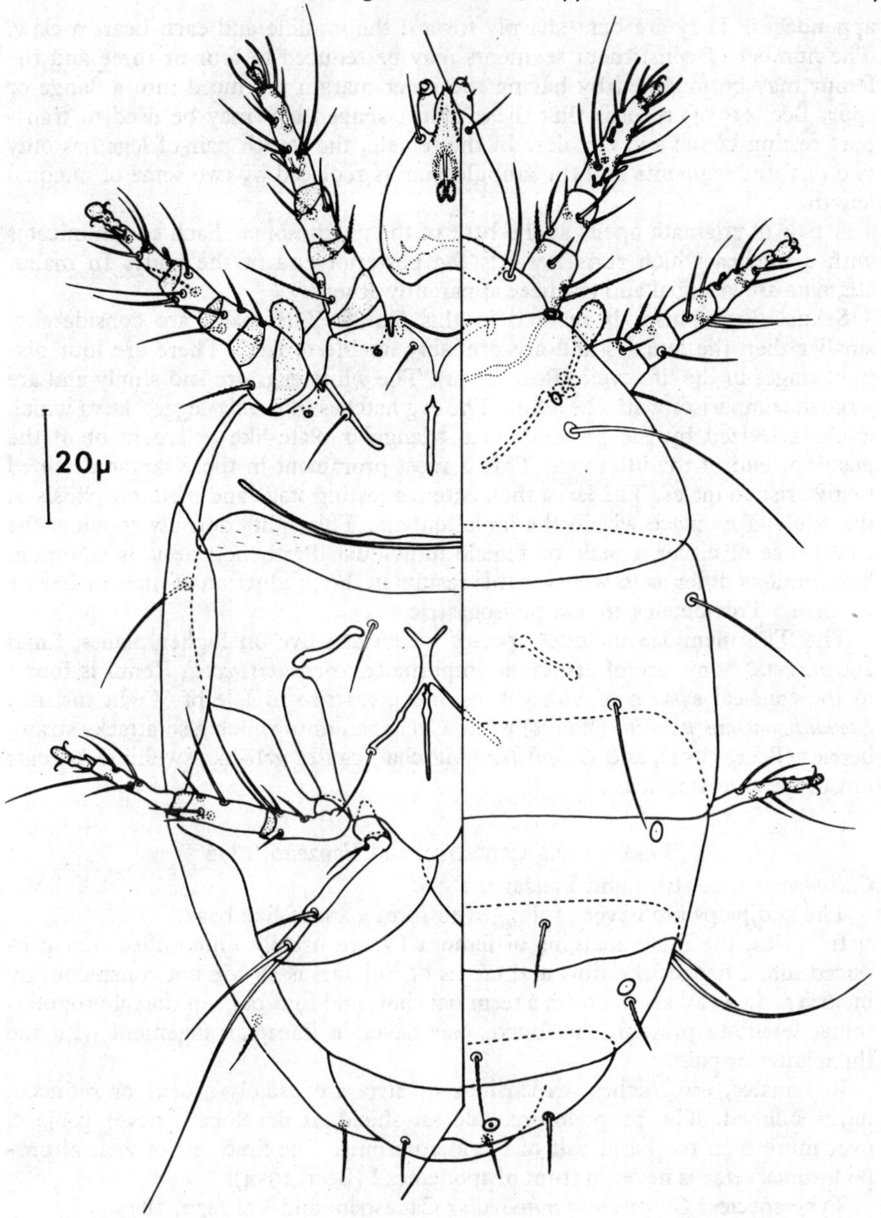

FIG. 303. *Tarsonemus granarius* Lindquist ♀. Ventral view (*left*), dorsal view (*right*); a-c details of bursa copulatrix: (*a*) lateral (sagittal) view, (*b*) ventral view, (*c*) dorsal view. d duct leading to ovary. s.r. seminal receptacle, t tergite IV (from Lindquist, 1972).

appendages. They are bent sharply toward the midline and each bears a claw. The number of constituent segments may be reduced to four or three and the femur may be modified by having the inner margin produced into a flange or spur. Beer (1954) records that these pincer-shaped legs may be used to transport resting larvae and females. In the female, the fourth pair of legs has only two or three segments and the ambulacrum is replaced by two setae of unequal length.

A pair of stigmata opens at the base of the gnathosoma. Each communicates with a trachea which runs towards the posterior end of the body. In males, stigmata are vestigial and tracheae apparently absent.

Sexual dimorphism is marked in this family. The males are considerably smaller than the females and are probably unable to feed. There are four distinct stages in the life cycle (Beer, 1954). The white eggs are laid singly and are large in comparison with the adult. The egg hatches into a six-legged larva which is characterized by the presence of a triangular plate-like enlargement of the posterior end of the idiosoma. This is most prominent in those larvae destined to give rise to males. The larva then enters a resting stage and metamorphosis to the adult takes place within the larval cuticle. This splits dorsally to allow the emergence of either a male or female individual. Parthenogenesis is common, but opinions differ as to whether this results in the production of male or female offspring. The females are not physogastric.

The Tarsonemidae includes species which can live on higher plants, fungi and insects. Some are of economic importance: *Acarapis woodi* Renni is found in the tracheal system of honey bees and gives rise to Isle of Wight disease; *Stenotarsonemus pallidus* (Banks) is the Cyclamen mite which also attacks strawberries (Beer, 1954), and *S. spirifex* (Marchal) causes gall-like swellings on oats in Europe and America.

TARSONEMUS Canestrini and Fanzago, 1876

Chironemus Canestrini and Fanzago, 1876.

The pedipalps are never prolonged to form a snout-like beak.

In males, the inner margins of femora IV are usually unmodified, not produced into a flange; the tibia and tarsus or tibiotarsus IV are not conspicuously incurved. Legs IV always have a terminal claw, and four pairs of dorsal propodosomal setae are present, the fourth pair never in linear arrangement with the three anterior pairs.

In females, the tracheal expansions or atria are usually absent or reduced, never bilobed. The propodosomal dorsal shield, if developed, never projects over more than the basal half of the gnathosoma. The first pair of ventral propodosomal setae is never in front of apodemes I (Beer, 1954).

Type species: *Chironemus minusculus* Canestrini and Fanzago, 1876.

TARSONEMUS GRANARIUS Lindquist, 1972

Tarsonemus sp. sensu Hughes, 1948 and 1961.

FEMALE (Fig. 303). Length of idiosoma: about 160 μm.

The idiosoma is yellow and covered with a shining cuticle. The gnathosoma is longer than wide with a dorso-median apodeme, but without a pair of ventral apodemes basally. The palps are cylindrical and nearly parallel, the cheliceral stylets short and inconspicuous. Dorsal and ventral setae arise near the base of the palps.

The dorsal propodosomal shield is almost triangular in shape, it partially covers the bases of the gnathosoma and pseudostigmatic organs. The idiosoma. is covered dorsally by four overlapping tergites, of which the first is the largest. The margins of the tergites overlap on to the ventral surface. The opisthogaster is also covered by coxal plates and a series of overlapping sternites, the posterior one forms a caudal lobe at the hind margin of the body and bears one pair of setae. The opening of the bursa copulatrix is at the end of the body. The apodemes of legs I meet to form a short sternum, those of legs II and III are free internally. An anterior median apodeme lies between legs II and another between legs III. The anterior extremity of this posterior apodeme is bifurcated and to it are joined the apodemes of legs IV. A transverse apodeme marks the posterior margins of the coxal plates and a strip of striated cuticle separates these from the anterior margin of the first ventral shield.

From the dorsal surface of the propodosoma arise one pair of vertical, one pair of scapular setae and a pair of clavate pseudostigmatic organs, the heads of these are covered with fine processes. The sigmata open at the base of legs I and a pair of pits is situated anterior to the scapular setae. Paired setae also arise from the hysterosomal tergites, arranged as in Fig. 303. Coxal setae arise from coxal shields I to IV.

The first two pairs of legs are directed forwards, the last two backwards. The ambulacra of legs I to III are well-developed, these of legs I with a membranous pulvillus and single claw. Ambulacra II and III have paired claws. Legs IV are close together and three-segmented, the subapical segment twice as long as the apical. The apical segment bears two setae, one about twice as long as the other. According to Lindquist (1972), the chaetotaxy in similar to that found in other species.

MALE (Fig. 304). Length of idiosoma: about 116 μm.

The male is not so frequently found as the female and is more difficult to see on account of its small size and slower gait. The dorsal propodosomal shield is weakly sclerotized. The hysterosoma is concealed dorsally by a large anterior tergite, and a smaller posterior one which overhangs the genitalia. The ventral surface is mainly covered by four coxal shields, the anterior ones separated from the posterior by an area of striated cuticle. The cuticle of coxal shields I and II has a striated area. Apodemes I unite to form a long sternum, apodemes II are free. Apodemes III and IV are elongated and unite with each other anteriorly and also with a median sclerite as in Fig. 304.

No stigmata or pseudostigmatic organs are present. The dorsal propodosomal shield bears two pairs of vertical setae and two pairs of scapulars. The anterior scapular setae are about twice as long as the posterior. The large anterior hysterosomal tergite has three pairs of setae and a pair of pores, the smaller posterior one a single pair of setae and pores. Coxal setae arise from coxal shields I to III, the third coxal shield has two pairs of setae.

The ambulacra of legs I to III resemble those of the female. The femora of legs IV are elongated, with almost straight internal, and slightly arcuate outer, margins. The terminal segments are short and bent sharply inwards and the tarsus terminates in a well-developed claw.

Both sexes of *T. granarius* are likely to be confused with *T. fusarii* Cooreman, 1941. They can be distinguished from the latter by the lack of the pair of ventral apodemes on the gnathosoma. Also, in the male, the anterior connections of apodemes III and IV with the postero-median apodeme are not interrupted in

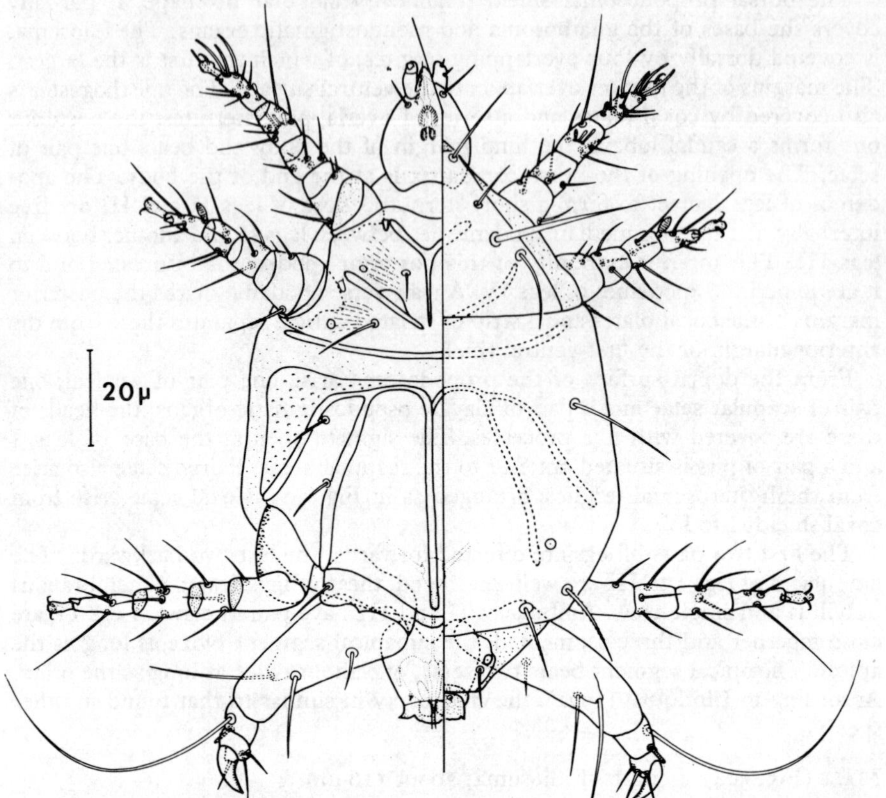

FIG. 304. *Tarsonemus granarius* Lindquist ♂. Ventral view (*left*), dorsal view (*right*) (from Lindquist, 1972).

the male of *granarius*, in contrast to that of *fusarii*, and the surfaces of coxae II have a striated area that is lacking in *fusarii*. There are also differences in leg chaetotaxy, setal length etc: (see Lindquist, 1972).

BIOLOGY

The biology of *T. granarius* and other tarsonemid mites has been extensively studied by Sinha and his associates. The relevant literature is reviewed by Lindquist (1972).

With other species of *Tarsonemus*, *T. granarius* is associated with and thrives on fungi, particularly species of *Penicillium*, *Aspergillus*, *Chaetomium* and *Hormodendrum*, which are likely to develop in stored grain (Sinha *et al.*, 1969a, b). It appears to be restricted to granaries, although the fungi on which it feeds will occupy many other habitats. In Canada, Sinha and Wallace (1966) found that tarsonemids, including *T. granarius*, were often present in bulk grain stored indoors and these showed a preference for wheat rather than oats, and were absent from barley. In Japan, however, the mites were more often found on granary floors than in the stored wheat or rice itself. Sinha (1968b) found that the age of the grain spill on the floors, the decaying organic debris and fungal flora played a larger part than temperature in the population fluctuation of these mites. Unlike *T. granarius*, as well as inhabiting granaries, *T. fusarii* is found in a great variety of niches supporting the fungi on which it feeds, such as hornet nests, humus of oaks and linden, withered grass etc. (quoted from Lindquist, 1972).

In this country, *T. granarius* and occasionally other species of Tarsonemidae, are found in cereal dust and spillage. They are easily distinguished from other mites by their small size, shining yellowish cuticle and smooth movement: as if they were running on casters. It is said to reproduce parthenogenetically (Sinha, 1968b, Sinha *et al.*, 1969a).

DISTRIBUTION

British Isles, Canada, Japan. *T fusarii* is cosmopolitan in distribution.

PYEMOTIDAE Oudemans, 1937

Pediculoididae Berlese, 1897.

In this family, the body is oval in shape and usually lacks the shelf-like prolongation of the anterior edge of the propodosoma which overhangs the gnathosoma and is characteristic of the Scutacaridae. It differs from the Tarsonemidae in that legs III and IV in the female have the same structure. Each is composed of four or five segments and terminates in a membranous pretarsus and two claws. The females are usually physogastric.

The family Pyemotidae contains a large number of species, mainly parasitic on insects but also able to suck the juices of grasses. They are also found in the nests of small mammals where they live on the other arthropods that also shelter there.

PYEMOTES Amerling, 1861

Heteropus Newport, 1850; *Pediculoides* Targioni-Tozzeti, 1878; also see Krezal, 1959.

The idiosoma is oval and the anterior edge of the dorsal propodosomal shield partially overhangs the gnathosoma.

In the female, legs I have five free segments and each terminates in a single claw. Legs II to IV terminate in two claws and a pretarsus. In the male, tarsus IV bears a single claw.

In the female, only the opisthosoma becomes distended with developing eggs.

Type species: *Pyemotes scolyti* Oudemans, 1936.

PYEMOTES HERFSI Oudemans, 1936

Pyemotes ventricosus (Newport, 1850) sensu Hughes, 1961.

FEMALE (Figs 305, 306). Length of non-gravid female: about 223 µm. The diameter of the gravid female may increase to 2 mm.

The idiosoma of the young female (Fig. 305) is flattened and oval, grey or yellowish in colour. The propodosoma has almost straight lateral margins and a pair of dorsal protuberances, posterior to which arise the pseudostigmatic organs. These have globular heads joined to fine stalks which are inserted into cup-shaped structures. The hysterosoma is more than twice as long as the propodosoma and is divided into five segments which decrease in size towards the posterior end of the body. The apodemes of legs I are joined to a long sternum, those of legs II are joined to epimerites I and almost reach the sternum. Apodemes III and IV and epimerites IV are inserted obliquely into the body and end freely at their internal extremities. The genital opening lies at the end of the body.

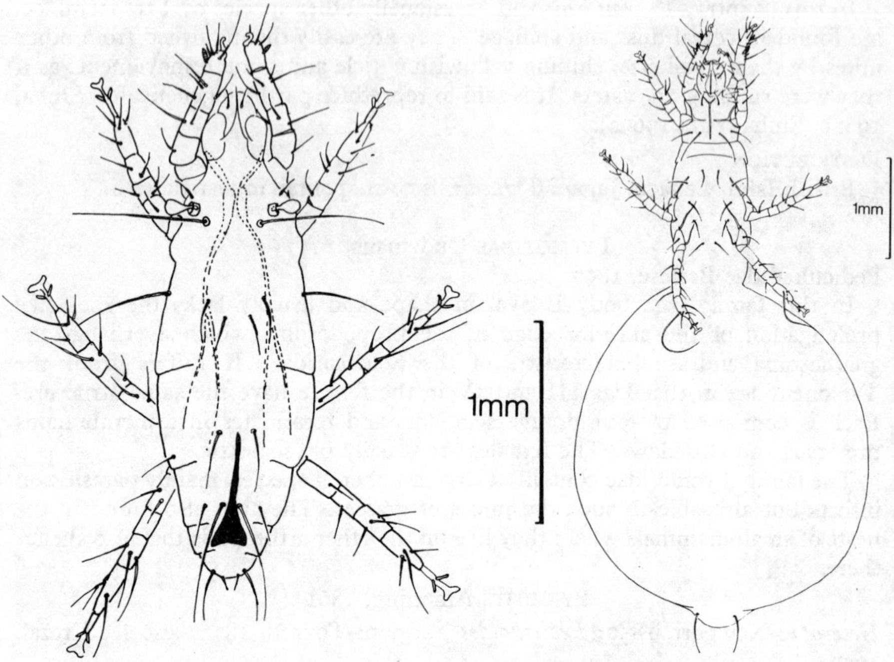

FIG. 305. *Pyemotes herfsi* Oudemans. Dorsal view of young ♀ (left).
FIG. 306. *Pyemotes herfsi* Oudemans. Ventral view of gravid ♀ (right).

The stigmata open laterally at the base of the gnathosoma. From them two tracheae converge and open into air reservoirs which supply the posterior region of the body.

The setae of the idiosoma are smooth and fine. On the dorsal side, the verticals project forward over the gnathosoma and the internal and external scapulars are inserted in front of and behind the pseudostigmatic organs, the external scapulars are longer than the internal. The first segment of the hysterosoma bears two pairs of setae, the second one pair, the third two pairs and two pairs of unequal length arise from the posterior margin of the body. A pair of long setae projects laterally in the humeral region. Ventrally, one pair of setae arises from coxae I, two pairs from coxae II and five pairs from the opisthosoma.

The gnathosoma is rounded. The chelicerae are stylet-shaped and the individual segments of the pedipalps are indistinguishable from one another.

The legs are five-segmented. The first pair of legs functions as an antenna, the remainder are used in walking. The tarsus of leg I (Fig. 307b) is blunt and bears a stout hooked claw. It bears several fine setae at its apex and one short, striated solenidion on its external margin; a similar, but narrower organ is found in the same position on tibia I. Tarsi II to IV have bifid claws arising from a slender pretarsus which is farther expanded into a bilobed pulvillus (Fig. 307a). The femur of leg IV is subdivided into a short basifemur and a longer telofemur (Fig. 307c). Those of legs II to IV are undivided.

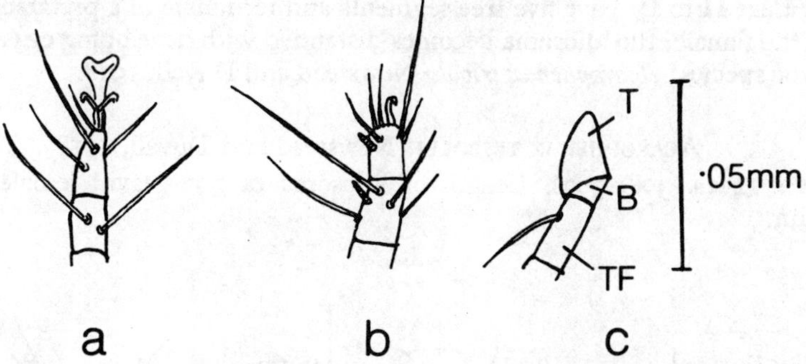

FIG. 307. *Pyemotes herfsi* Oudemans ♀. (*a*) Dorsal view of left leg II, (*b*) dorsal view of left leg I, (*c*) trochanter and femur of left leg IV. T trochanter; B basi-femur; TF telo-femur.

MALE

The males are usually found attached to the females and superficially resemble the males of *Tarsonemus* sp.

The hinder margin of the body has a telson-shaped appendage. On this is a sucker which is used in copulation.

The mouth parts and first three pairs of legs are similar to those of the female. The fourth pair has bent and elongated femora which are not sub-divided. Tarsus IV terminates in a massive claw.

BIOLOGY

Species of *Pyemotes* are ectoparasites of insects and the females feed by piercing the integument with their chelicerae.

The life-history of *P. ventricosus* (Newport) has been described in detail by Herfs (1926). The male spend its life wandering over the distended opisthosoma of the female, on which it is parasitic. The eggs inside the female continue their development until they become adults, when their sex can be determined by characters which can be seen through the transparent parent cuticle. When a young female is about to be born, it moves until it comes opposite the genital opening and begins to emerge. The male appears to be aware of what is happening and also moves towards the opening. With the aid of its powerfully built hind pair of legs, it seizes the young female and completes its birth by dragging it through the opening. Copulation then takes place. The male is said never to assist at the birth of young males, and the young females can emerge from the parent without his assistance. If all the males are removed from the opisthosoma of the gravid female, and the young females subsequently isolated, they give rise to young parthenogenetically, but these are all of the male sex.

P. ventricosus has been recorded by Nixon (1944), Mariani (1952) and Booth and Jones (1952) and others as giving rise to dermatitis in people handling infected material.

P. herfsi was found on *Corycyra* sp. (Talbot coll.).

DISTRIBUTION

England, Holland.

ACAROPHENAX Newstead and Duvall, 1918

The idiosoma is round and the anterior edge of the dorsal propodosomal shield completely overhangs the gnathosoma.

Legs I have four free segments, the tibia and tarsus being fused. Each bears a claw. Legs II to IV have five free segments and terminate in a pretarsus only.

In the female, the idiosoma becomes distended with developing eggs.

Type species: *Acarophenax tribolii* Newstead and Duvall, 1918.

ACAROPHENAX TRIBOLLII Newstead and Duvall, 1918

FEMALE (Figs. 308, 309). Length of idiosoma of non-gravid female: 160 to 190 μm.

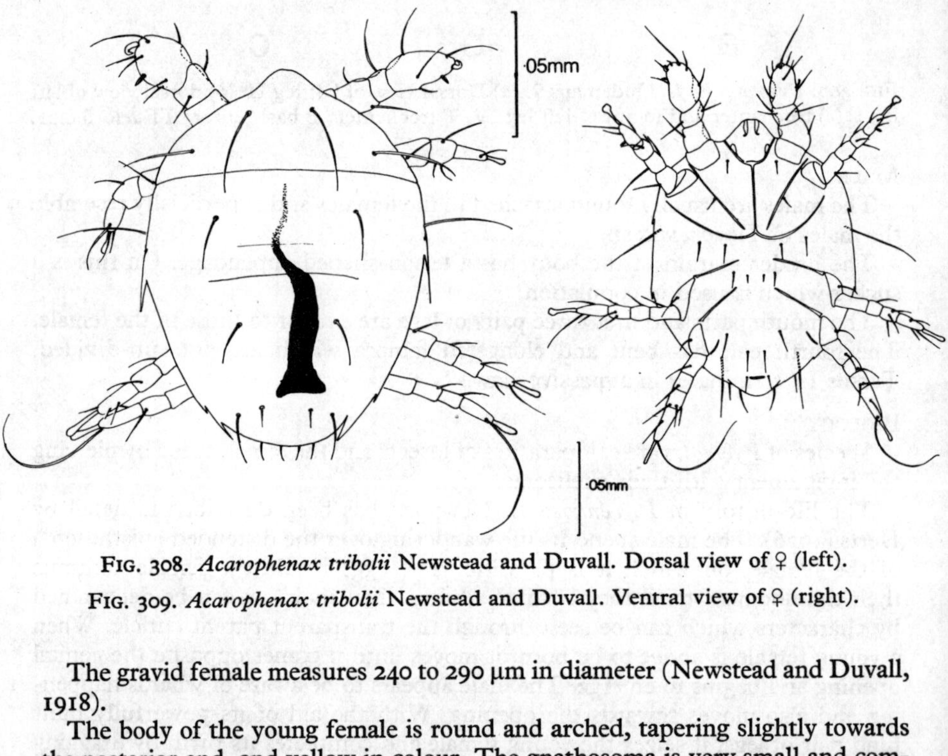

FIG. 308. *Acarophenax tribolii* Newstead and Duvall. Dorsal view of ♀ (left).

FIG. 309. *Acarophenax tribolii* Newstead and Duvall. Ventral view of ♀ (right).

The gravid female measures 240 to 290 μm in diameter (Newstead and Duvall, 1918).

The body of the young female is round and arched, tapering slightly towards the posterior end, and yellow in colour. The gnathosoma is very small and completely covered by the overhanging propodosoma. The mouth parts consist of two stylet-shaped chelicerae and rudimentary palps. The hysterosoma is covered dorsally with four tergites. The anterior pair of stigmata is situated on the propodosoma, the posterior pair opens directly behind; they communicate with each other and with tracheal trunks. The anterior ventral margin of the propodosoma is supported by a 'collar' of thickened cuticle encircling the base of the gnathosoma and joining the long sternum. Posteriorly, this sternum bifurcates and fuses with epimerites II. Apodemes II do not reach the sternum. Apodemes III and IV are directed obliquely forward.

Two pairs of scapular setae arise from the dorsal side of the propodosoma and a pair of dorsal setae arises from each tergite covering the hysterosoma; also a pair of long humeral setae projects from the lateral margin of the first hysterosomal segment. Two pairs of short setae project from the end of the body. Coxal setae arise from coxae I and II, two pairs are inserted between epimerites II and apodemes III, and another two pairs between apodemes III and IV. A genital seta is also present on either side of the genital opening.

The first two pairs of legs are widely separated from the posterior pairs. Leg I (Fig. 310a) is stout, with irregularly thickened walls and only four segments.

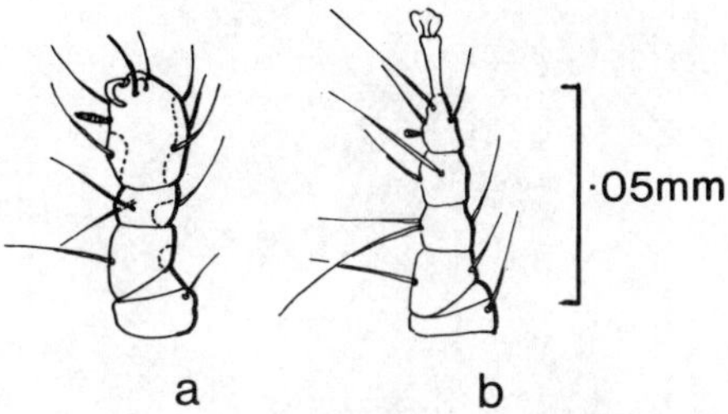

FIG. 310. *Acarophenax tribolii* Newstead and Duvall. Ventral view of (*a*) right leg I and (*b*) right leg II.

The tibio-tarsus is enlarged and bears a recurved claw. A blunt, striated solenidion arises from the external edge of the tibio-tarsus. The remaining pairs of legs have five free segments, the tarsi ending in long pretarsi. (Fig. 233b).

MALE. Length of idiosoma: about 170 μm.

According to Newstead and Duvall (1918), the idiosoma is rounder than that of the female and tapers from the third pair of legs to the posterior margin. It is flattened dorso-ventrally and the dorsal surface of the hysterosoma shows traces of three segments. The sternum is joined by the apodemes of the first and second pairs of legs, and bifurcates to encircle the base of the gnathosoma. Sinuous apodemes also articulate with legs III and IV. The genital opening lies near the posterior end of the body, it is a large, almost circular opening with a thickened rim. Near the centre of the orifice are two pairs of minute suckers and another pair is placed near the two long posterior marginal setae.

Two pairs of scapular setae arise from the dorsal side of the propodosoma and another pair is inserted at the junction of propodosoma and hysterosoma, in addition to a long pair of humerals. Two dorsal setae arise at the level of the fourth pair of legs and two long marginal setae project from the posterior margin of the body.

The first two pairs of legs are widely separated from the last two, and all have five free segments, often difficult to distinguish. The first leg is short, stumpy and terminates in a hooked claw, but the tarsus is not thickened as in the female. Legs II to IV are shorter and each ends in a large, membranous pretarsus.

BIOLOGY

A. tribolii is an ectoparasite of *Tribolium confusum* Fabr., *T. castaneum* Herbst. and *T. ferrugineum* Duv. It attaches itself by means of its chelicerae to parts of the body where the cuticle is soft, particularly beneath the second pair of wings, and feeds on the tissue fluid. According to Newstead and Duvall (1918), only females are ever found on the beetle and these leave the host before the idiosoma becomes distended with developing embryos. About 14 or 15 young mites develop at once, of which one is a male and this probably fertilizes the young females before they are born, as it is never found outside the body of the mother. When mature, the young females leave the parent through the enlarged genital opening and infect the beetles. The mother dies as soon as the young are born.

Distribution

England.

Mesostigmata Canestrini, 1891

In the Mesostigmata, the basal part of the gnathosoma or basis capituli is a sclerotized ring formed (according to T. Hughes, 1949) by the fusion of the pedipalpal coxae dorsally with the gnathosomal tergites and ventrally with the capitular groove or deutosternum (Fig. 311). Other authors consider that the gnathosomal tergites have become incorporated into the dorsal surface of the idiosoma. The chelicerae thus become covered by a dorsal plate whose anterior edge is produced forward as the tectum or epistome (Fig. 312). The capitular groove is a gutter-shaped structure, often lined with transverse rows of deutosternal denticles. To the lateral anterior walls of the basis capituli are attached the pedipalpal palps: five-segmented structures which bear a number of setae. One of the most conspicuous of these is a two to four tined clawlike structure which arises from the base of the pedipalpal tarsus and is considered by some authors to represent a sixth segment; it has been given the name apotele. The ventral anterior region of the gnathosoma or rostrum is formed by the paired endites of the pedipalps, each of which becomes subdivided into a corniculus maxillaris or external mala, and an internal mala, often elaborately shaped; these together form the floor of a pre-oral tube, the mouth opening being at the level of the base of the palp. The roof of this pre-oral tube is formed by a sub-cheliceral shelf lying immediately below the chelicerae, and a forward extension of the roof of the pharynx—the epipharynx or labrum. Paired lobes or paralabra may originate from the anterior end of the sub-cheliceral shelf and become associated with the epipharynx and enclose laterally the walls of the pre-oral canal (T. Hughes, 1949). From the ventral surface of the basis capituli and rostrum arise a number of setae: three pairs of hypostomals and one pair of capitular.

The chelicerae are three-segmented, each consisting of a short, basal segment attached to a much longer distal one, and ending in a fixed digit. This distal segment also bears ventro-laterally the third segment or movable chela. The fixed digit sometimes bears a seta—the pilus dentilis—and it may be scarred with two lyrifissures. Both digits are usually dentate, but they may be modified in accordance with the feeding habits of a particular species. Cuticular processes may be present encircling the arthrodial membrane at the base of the digits (Fig. 311). In males the movable digit may carry an accessory copulatory organ or spermadactyl, which may end freely or become fused with the tip of the digit. The chelicerae are enclosed within the basis capituli, inside a system of sheaths which are essentially a continuation of the cuticle of the body wall, and from which they can be protracted. Immediately below the chelicerae is a pair of fine styles which arise from the mesial walls of the pedipalpal coxae and through which run the salivary ducts. The salivary glands themselves are situated in the idiosoma.

The gnathosoma lies in a camerostome and is movably articulated with the idiosoma by an arthrodial membrane. The idiosoma itself is round or oval and the dorsal surface is sclerotized to form one or two plates which are smooth or elaborately sculptured; sometimes a marginal plate is present as well. From these plates and the surrounding interscutal membrane (when present) arise a number of setae whose number, position and shape may be of systematic importance. A number of 'pores' is also present on the dorsal shield.

The system of nomenclature of the setae of the dorsal surface of the idiosoma given here is that devised by Hirschmann (1957) and subsequently modified by Lindquist and Evans, 1965. They consider the normal setal complement of

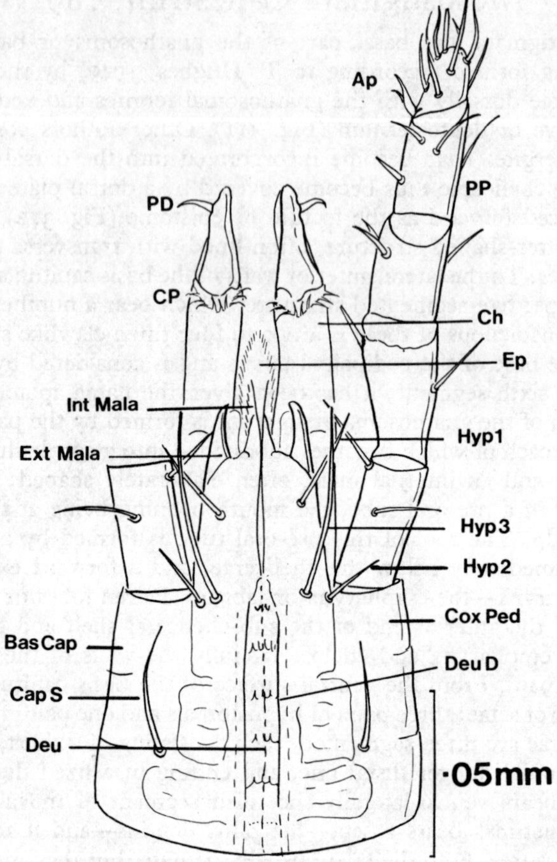

FIG. 311. *Androlaelaps casalis casalis* (Berlese) ♀. Ventral view of gnathosoma, Cox Ped pedipalpal coxa, Deu capitular groove or deutosternum, Deu D deutosternal denticles, PP pedipalp, Ap apotele, Ep epipharynx or labrum, Bas Cap basis capituli, Hyp. S 1 to 3 hypostomal setae 1 to 3, Cap S capitular seta, Ch chelicera, PD pilus dentilis, CP cuticular processes.

this region to consist of 44 pairs of setae arranged in four longitudinal rows of 11 setae each, a dorsal *jJ*, median *zZ*, lateral *sS* and marginal *rR* on each side of the median longitudinal axis. Setae of the anterior region (podonotum) are denoted by letters in lower case, those of the posterior region (opisthonotum) by capitals (Fig. 312). There may also be a fifth longitudinal row of six setae—the extra- or sub-marginal *UR*—which arise ventro-laterally from the opisthosoma. In the laelapines, an additional row of setae px_1 to px_3 may also be present between the *J* and *Z* series and unpaired setae are sometimes found between the *J* series.

A few of these setae, e.g., j_1, j_3 to j_6, z_2, z_4, z_5, s_4 associated with the anterior region, are present in all the developmental stages, the majority of them appear for the first time in the protonymph and deutonymph, the deutonymph having the same number of setae as the adult. Deviation from this normal setal complement frequently occurs and is due to the loss (hypotrichy) or addition of setae (hypertrichy).

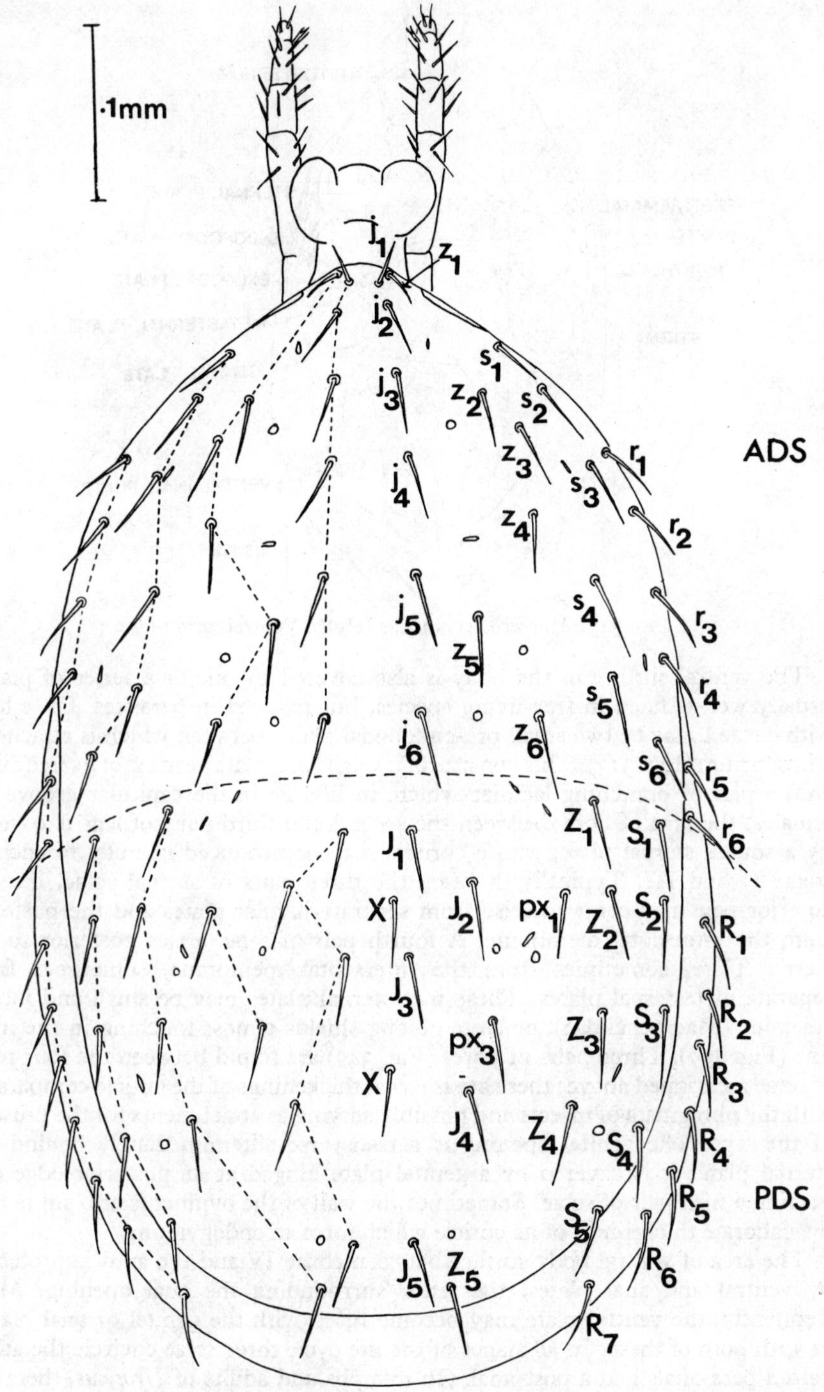

FIG. 312. *Androlaelaps casalis casalis* (Berlese). Dorsal view of ♀. For explanation of setal lettering, see text. ADS anterior region of dorsal shield, PDS posterior region of dorsal shield, X unpaired seta.

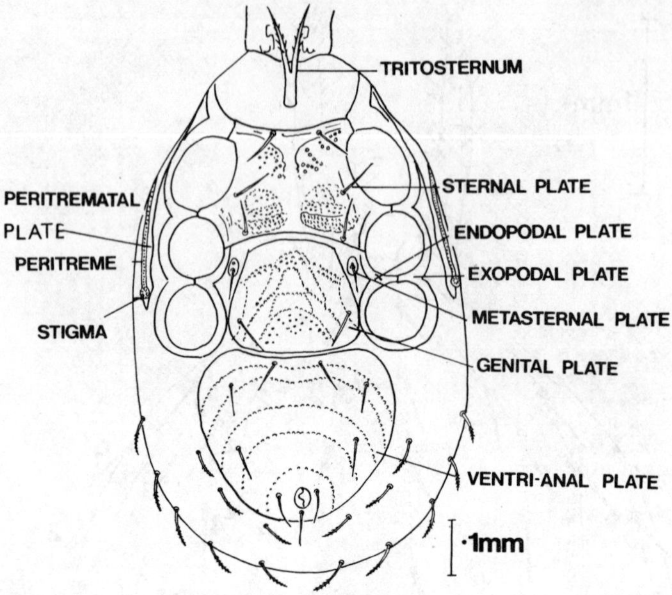

FIG. 313. *Macrocheles matrius* (Hull). Ventral view of ♀.

The ventral surface of the body is also covered by one or a series of plates, usually well-defined in free-living species, but reduced in parasites. On a level with coxae I may be two small pre-endopodal plates between which is attached a tritosternum (Fig. 313). This consists of a small basal plate bearing, at its free distal end, a pair of branching laciniae which, in life, lie in the capitular groove. In females, the area of body between the second and third pairs of legs is covered by a square sternal plate, whose corners may be produced laterally to encircle coxae II and III. Typically, it bears the three pairs of sternal setae, but the anterior pair may come to arise from separate jugular plates and the posterior from the interscutal membrane. A fourth pair of setae arises posterior to the sternal plate, sometimes from the interscutal membrane, sometimes from separate metasternal plates. These metasternal plates may be small and inconspicuous (Macrochelidae), or large oblong shields almost touching in the mid-line (Fig. 407). Three pairs of pores (Fig. 349) are found between the four pairs of setae mentioned above; these are internal thickenings of the cuticle comparable with the phragmata of insects and possibly serving as attachments for the muscles of the legs. The genital opening is a transverse slit immediately behind the sternal plate; it is covered by a genital plate hinged at its posterior edge and bearing a fifth pair of setae. Sometimes the wall of the oviduct is also supported by elaborate thickenings of its cuticle which form an endogynium.

The area of ventral body surface between coxae IV and the anus is protected by ventral and anal plates, the latter surrounding the anal opening. More frequently, the ventral plate may become fused with the genital or anal plates, or with both of these. In all stages of the life cycle three setae encircle the anus: paired para-anals and a post-anal. (In nymphs and adults of *Uroseius*, there are four setae). The internal edges of coxae II to IV may be encircled by a scalloped endopodal plate from which originate some of the leg muscles; the external surfaces of the coxae may also be supported by an exopodal plate.

In males, the genital opening is at the anterior edge of the sterno-genital shield, or in the middle of the same shield. The ventral plates tend to merge together, either to form one holoventral shield, or an anterior sterno-genital shield bearing typically five pairs of setae, and a posterior ventro-anal with three circum-anal setae.

Tracheal tubes lead to stigmata opening on the lateral or ventral surfaces of the body, between coxae II and III or III and IV. The stigmata are associated with peritremes which may be supported by a peritrematal plate. Tracheal tubes and stigmata first appear in the protonymph, the larvae obtaining oxygen by cutaneous respiration.

The legs are six-segmented, the basal segment or coxa being annular in shape and freely movable. Frequently the distal edge of the coxa is produced into cuticular processes, and fissures may be present on femora and tarsi dividing them into proximal and distal regions. The pretarsus, sometimes considered as an additional leg segment, is usually well-developed and bears an ambulacrum consisting of pulvillus—often deeply incised—and a pair of well-developed claws (Fig. 314a). In *Haemogamasus hirsutus* Berlese the pretarsus

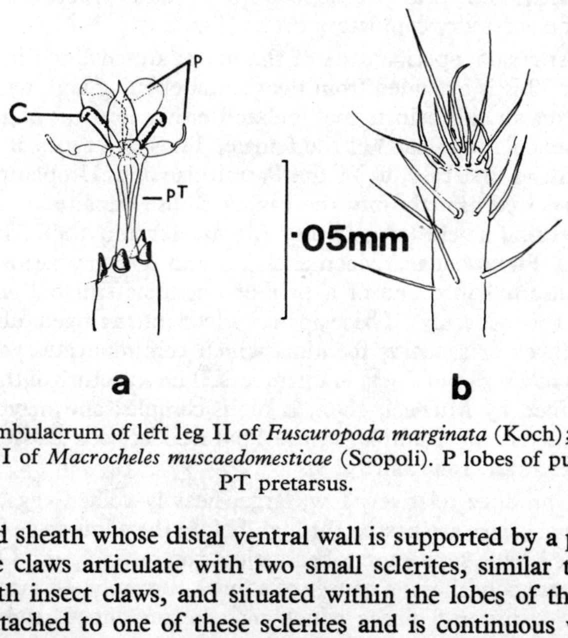

FIG. 314(a). Ambulacrum of left leg II of *Fuscuropoda marginata* (Koch); (*b*) dorsal view of right tarsus I of *Macrocheles muscaedomesticae* (Scopoli). P lobes of pulvillus, C claw, PT pretarsus.

is a semi-rigid sheath whose distal ventral wall is supported by a pair of slender sclerites. The claws articulate with two small sclerites, similar to the auxiliae associated with insect claws, and situated within the lobes of the pulvillus. A 'tendon' is attached to one of these sclerites and is continuous with a muscle originating within the tarsus. Contraction of this muscle results in withdrawal of the ambulacrum into the pretarsal sheath (Evans and Till, 1965).

With the exception of the pretarsi, all the leg segments have a well-defined chaetotaxy. The setae themselves are simple or pilose, sometimes thickened as spines; they are all optically isotropic and solid, although hollow structures comparable with solenidia are found on the distal region of tarsus I.

The arrangement and number of the leg setae are characters which are extensively used in the taxonomy of the Mesostigmata. Evans (1963) has devised a scheme for mapping the position of the setae on the leg segments. He assumes that the leg is at right angles to the longitudinal axis of the body and has four seta-bearing surfaces: dorsal, ventral, antero-lateral, postero-lateral. Where the setae on the dorsal and ventral surfaces form a discontinuous series, they are further subdivided into anterior and posterior groups.

Thus the position of a seta can be expressed as a formula:

$$\text{antero-lateral } al \Big/ \begin{matrix} \text{dorsal } d \\ \text{ventral } v \end{matrix} \Big/ \text{postero-lateral } pl$$

Where a larger number of setae is involved in the dorsal and ventral series, these can be further subdivided:

$$al \Big/ \begin{matrix} \text{antero-dorsal ad} \\ \text{antero-ventral } av \end{matrix} \Big/ \begin{matrix} \text{postero-dorsal } pd \\ \text{postero-ventral } pv \end{matrix} \Big/ pl$$

Fig. 315 shows the arrangement of the setae on genu IV of *Androlaelaps casalis* (Berlese) and the chaetotactic formula.

The anterior legs are not used in walking, even by the hexapod larva, but serve as sensory organs. This gives the larva a rather waddling gait. In some families, e.g. the Macrochelidae, pretarsus and ambulacrum are absent from legs I and replaced by setae (Fig. 314b). The more weakly tanned species often have the tarsus of leg I more deeply coloured, possibly making it sensitive to light. The second pair of legs in the male and more rarely, in the female may become thickened and bear thickened spines and processes, presumably functioning as an accessory copulatory organ (Fig. 405).

In the Mesostigmata, spermatozoa of the males are enclosed in a capsule or spermatophore. This is extruded from the genital opening and, by means of the modified chelicera or the palp is manipulated either into the vagina or into a spermatheca (vesicula seminalis) of the female. Thus two types of insemination are possible. In the first case, as in the Parasitidae and Uropodina, the spermatophore is inserted directly into the vagina whose unpaired opening is concealed by the genital shield (tocospermy Athias Henriot, 1968). In the second case, as in the Phytoseiidae, Macrochelidae and Dermanyssidae, the spermatophore is inserted into one of a pair of openings situated at the base of coxae III or IV (podospermy). These openings lead into a ringed tube continuous with a spermatheca or sacculus foeminis which communicates with the ovary often by means of a series of ducts or chambers. The structure of the spermathecae, first described by Michael, 1892, is often complex and may be used as a taxonomic character, for example, in the Phytoseiidae (see Dosse, 1958, Fain, 1963b, Athias Henriot, 1966 and 1968).

The females produce relatively few, large, heavily-yolked eggs, and a larval and two nymphal stages are passed through before the adult stage is reached. In any species that I have kept under observation, the resting period between each active stage is extemely short: a matter of a few hours only. In many species the deutonymph is a dispersal phase which may be prolonged, and this stage is frequently found clinging or attached to insects (the 'nympha coleoptrata' and 'nympha pedunculata' of Berlese). The transition from one stage to another usually results in an increase in the number of setae and the full complement is attained at the deutonymphal stages by most species (see Evans and Till, 1965).

The Mesostigmata have adapted themselves to a wide range of habitat. Many of them are predators and form an important part of the fauna of the surface litter of soils, where they feed on other more sluggish arthropods, and also on fungi. Some species are coprophagous, enlisting the service of dung-eating insects as a conveyance from one source of food to another. A number of species

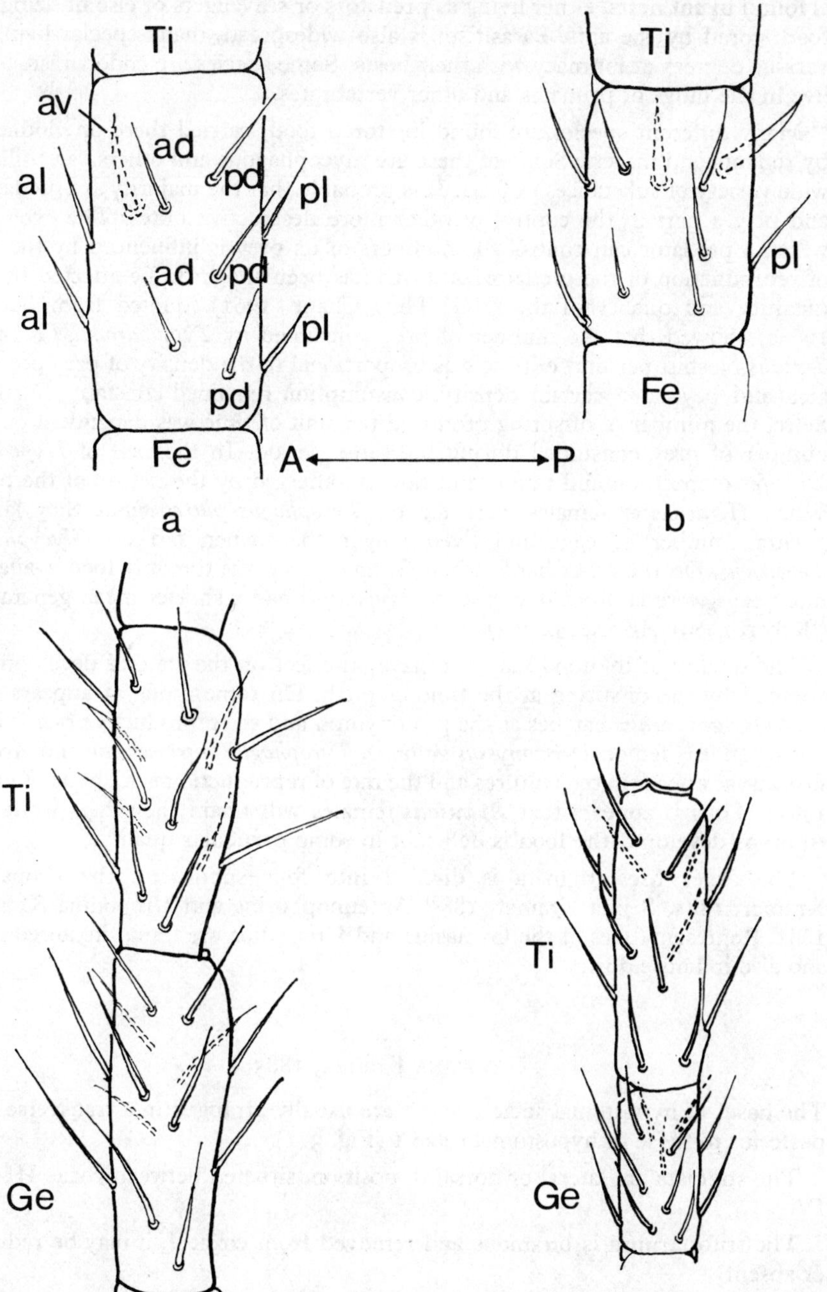

FIG. 315. Dorsal view of right genu IV of (*a*) *Androlaelaps casalis casalis* (Berlese). Chaetotactic formula: $2\ al \Big/ \frac{2\ ad\ 3pd}{1\ av\ 0pv} \Big/ 2\ pl$, (*b*) *Hypoaspis aculeifer* (Canestrini). Dorsal view of right genu and tibia I of (*c*) *Eugamasus butleri* Hughes and (*d*) *Androlaelaps casalis casalis* (Berlese). Fe femur, Ti tibia, Ge genu. The ventral setae are dotted.

is found in ant nests, either living as predators or scavengers or else utilizing the food stored by the ants. Parasitism is also widespread, many species living in various degrees of intimacy with their hosts. Some species are endoparasitic and live in the lungs of primates and other vertebrates.

Many different species are found in stored food, carried there on clothes or by rodents and insects. Some of these are mycophagous and others can utilize a wide variety of substances as food. It is probable that the majority are predatory and play a part in the control of other more destructive mites. The extent to which a predator can control the numbers of its prey is influenced by the rate of reproduction of the predator, and this has been shown to be affected by the quantity and quality of the food. Thus Chant (1961) (quoted from Barker, 1968a) showed that the number of prey consumed by *Typhlodromus* (*T.*) *occidentalis* Nesbitt per unit of time was proportional to the density of prey per unit area and beyond a certain density, consumption remained constant. Furthermore, the number of offspring produced per unit of time was dependent on the number of prey consumed during the same period. In the case of *Hypoaspis aculeifer* oviposition and rate of increase are affected by the nature of the prey. When *H. aculeifer* females were fed on *Tyrophagus putrescentiae* they laid a greater number of eggs and lived longer than when fed on *Glycyphagus domesticus*. On the other hand, when *G. domesticus* was the only food available, more eggs were laid per day, but the *Hypoaspis* had a shorter mean generation (Barker, 1969) (Also see p. 303).

The quality of the food may also have an effect on the state of development attained by the offspring at the time of birth. On some foods, it appears that *Androlaelaps casalis* hatches as the protonymph and the reproductive rate is low, but when it is fed on *Dermanyssus gallinae*, *Tyrophagus putrescentiae* and *Acarus siro*, larvae appear in the cultures and the rate of reproduction is high (McKinley, 1963). Thus it appears that *A. casalis* females will retain their eggs while the embryos develop if the food is deficient in some particular quality.

The order Mesostigmata is divided into four suborders: the Gamasina Kramer, 1885, Sejina Kramer, 1885, Antennophorina and Uropodina Kramer, 1881. Representatives of the Gamasina and Uropodina are found in stored food and also in house dust.

GAMASINA Kramer, 1885

The bases of hypostomal setae 2 and 3 are usually arranged in a transverse row posterior to those of hypostomal setae 1 (Fig. 311).

The stigmata are lateral or dorsal in position, situated between coxae III and IV.

The tritosternum is biramous and removed from coxae I, it may be reduced or absent.

The palpgenu usually bears six setae, rarely five or less.

The male chelicera may or may not bear a spermadactyl.

No marginal shield is ever present.

No leg grooves or scabellum are developed.

Tibia I usually has three ventral setae, less commonly two or four.

Mesostigmata

Key to suborders. Adults

1. Hypostomal setae 2 and 3 forming a transverse row posterior to hypostomal setae 1 (Fig. 311). The tritosternum is usually removed from coxae 1 (Fig. 311). No leg grooves or scabellum..............Gamasina, p. 285

—. Hypostomal setae 1, 2 and 3 in linear series (Fig. 416). Tritosternum usually partially or completely covered by the enlarged, flattened coxae 1 (Fig. 426). Leg grooves and scabellum usually developed (Fig. 426)..Uropodina, p. 359

Gamasina

Key to the families. Adults. (adapted from Evans, 1957).

1. Metasternal shields in the female large, flanking the anterior portion of the genital shield (Fig. 407). Spermadactyl of movable digit of male chelicera fused with the digit distally (Fig. 404a)..............Parasitidae, p. 350

—. Without large metasternal shields in the female. Spermadactyl of male chelicera free distally..2

2. Leg I in both sexes without an ambulacrum. Peritreme looped in the region of the stigma (Fig. 409)..........................Macrochelidae, p. 354

Femur of leg II of female without strong spurs. Ventro-anal shield of female with three pairs of pre-anal setae...............*Macrocheles*

(a) Setae j_1 close together, so that their bases are touching (Fig. 410a). Setae j_1, j_6, J_3, z_3, z_6 (Fig. 408) smooth..*M. muscaedomesticae*, p. 355

(b) Setae j_1 further apart (Fig. 410b). All the dorsal setae are plumose. ....*M. matrius*, p. 357

—. Leg I in both sexes with an ambulacrum. Peritreme usually not looped in the region of the stigma..3

3. Adults with two dorsal shields of approximately equal size. Females with the fourth pair of sternal setae arising from the sternal shield (Fig. 396). Male with the ventro-anal shield fused with the posterior dorsal shield and with legs II bearing strong spines or processes on its ventral surface (Fig. 399)..................................Digamasellidae, p. 346

—. Adults with an entire or incised dorsal shield. Females with the fourth pair of sternal setae arising from separate metasternal plates or from the inter-scutal membrane. Ventro-anal shield of male not fused with the dorsal shield, and with legs II usually unarmed..........................4

4. Dorsal shield in both sexes with 20 or less pairs of setae. Females with a truncate genital shield and a ventro-anal shield. Males with a sterno-genital shield separate from a ventro-anal......................Phytoseiidae

(a) Setae s_4, Z_3 and Z_5 (Fig. 386) long and whip-like, s_4 and Z_3 approximately equal in length..........................*Amblyseius sp.*, p. 341

(b) All the dorsal setae are relatively short, z_1, s_1, s_3 slightly longer than half the distance between their bases (Fig. 391). Male with four pairs of pre-anal setae..........................*Neoseiulus barkeri*, p. 343

—. Dorsal shield in both sexes with more than 23 pairs of setae. Females with ventro-anal or anal shields...5

5. Genital shield of female truncate posteriorly or very slightly convex, usually with one pair of setae and lying close to a ventro-anal shield or remote from an anal. Male with sterno-genital and a separate ventro-anal or anal shield 6

—. With genital or genito-ventral shield in female. Genital shield flask-shaped (Fig. 323) or if truncate, almost touching the anal shield. Male with holoventral shield (Fig. 325) or with sterno-genito-ventral shield separate from anal .... Dermanyssidae, p. 286

6. Adults with only two pairs of setae in the *J* series arising from the posterior region of the dorsal shield (Fig. 379). Corniculi forked distally (Fig. 384a) .... Ameroseiidae

(a) Dorsal setae leaf-shaped with a thickened mid-rib and with slightly serrated margins (Fig. 380a). Sternal shield of female usually with a central region devoid of reticulation (Fig. 381b)................ .... *Kleemannia plumosus*, p. 338

(b) Dorsal setae narrower without a thickened mid-rib and with pectinate margins (Fig. 380b). Sternal shield of female evenly reticulate (Fig. 381a) .... *Kleemannia plumigera*, p. 336

—. Adults with five pairs of setae in the *J* series arising from the posterior region of the dorsal shield (Fig. 352). Corniculi not forked distally...... .... Ascidae, p. 312

Dermanyssidae Kolenati, 1859 (sensu Evans and Till, 1966)

The lateral margin of the dorsal shield is entire or incised.

In the female, the genital opening is covered by a genito-ventral shield, or a flask-shaped genital shield. The metasternal setae are on discrete shields or on the striated cuticle. Paired spermathecae are present, opening between coxae III and IV.

In the male, the ventral surface is usually covered by one holoventral shield and the second pair of legs is usually not armed. The genital opening is usually presternal.

The chelicerae are variable in structure, usually dentate in free-living forms and modified in parasites.

The deutosternal denticles are arranged in single file or in transverse rows.

Tibia I has two or three ventral (Fig. 342b) setae.

The family Dermanyssidae sensu Evans and Till contains a vast assemblage of species adapted to free-living and parasitic ways of life. The free-living ones are mainly predatory, but some may be mycophagous or able to live on a variety of substances. Many species are nest-dwellers and it is probable that the parasites have evolved from these. The wide range of feeding habits shown by this family is accompanied by extensive morphological specialization, so that a comprehensive definition of the family as a whole tends to be vague. Evans and Till (1966) divide the family into a number of subfamilies which are easier to define. Representatives of the Dermanyssinae, Laelapinae and Haemogamasinae are often found in stored food or houses.

DERMANYSSIDAE

Key to the adults. (partly from Evans and Till, 1966)

1. Chelicerae in nymphs and females with the second segment enormously elongated, like stylets; digits minute (Fig. 344a). Males with the fixed digit reduced and the elongated spermadactyl entirely fused with the movable digit (Fig. 344b); chelicerae never chelate. Tarsi III and IV in male with seta pv_2 modified into a short, tooth-like projection (Fig. 347).......... ....Dermanyssinae......*Dermanyssus gallinae*, p. 309

—. Second segment of chelicera in nymphs and females never conspicuously elongated; digits prominent. Seta pv_2 on the tarsi of males never modified into a tooth-like projection..2

2. Dorsal shield in deutonymphs and adults markedly hypertrichous (Fig. 332). Genital shield and sometimes sternal and anal shields with unpaired accessory setae. Deutosternum with 10 or more transverse rows of denticles (Fig. 340). Tectum with margin conspicuouly fimbriated (Fig. 335).... ....Haemogamasinae

 (a) Female with large, subtriangular metapodal shields flanking a broad genito-ventral shield (Fig. 338); palptrochanter with sensory organ (Fig. 340a); anterior rows of deutosternal denticles with five to seven denticles per row in both sexes (Fig. 340a). Genito-ventral shield in female with lateral incisions posterior to the genital setae (Fig. 338) ....*Eulaelaps stabularis*, p. 304

 (b) Female with small, oval metapodal shields; genito-ventral shield relatively narrow; palptrochanter lacking a sensory organ; anterior rows of deutosternal denticles with two or three denticles per row in both sexes. Posterior edge of sternal shield markedly concave (Fig. 333) ....*Haemogamasus pontiger*, p. 301

—. Hypertrichy of dorsal shield, when present, restricted to the region of the *J* series. Deutosternum with five to seven, usually six, transverse rows of denticles (Fig. 311); rows usually compound. Tectum with the anterior margin denticulate, or membranous with a smooth anterior margin (Fig. 321)..Laelapinae....3

3. Sternal shield of female broader than long. Digits of male chelicera edentate. Pilus dentilis long and slender (Fig. 317b). Genu IV with two postero-lateral setae, pl_2 present (Fig. 315a)..*Androlaelaps casalis casalis*, p. 288

—. Sternal shield of female usually longer than broad. Digits of male chelicera dentate. Pilus dentilis short and setiform (Fig. 317a).....*Hypoaspis*....4

4. Dorsal setae leaf-like, spatulate (Fig. 330). Corniculi long and slender, extending almost to the anterior margin of the palpfemur............ ....*Hypoaspis miles*, p. 299

—. Dorsal setae simple. Corniculi shorter...................................5

5. Femur and genu of leg II each with a conspicuous blunt seta (av_1) in both sexes (Fig. 327)..........................*Hypoaspis sardoa*, p. 297

—. Femur and genu of leg II without conspicuous blunt setae.............6

6. Genital shield of female small. Pre-endopodal plates indistinct. Tectum with a denticulate margin (Fig. 321b).........*Hypoaspis aculeifer*, p. 292

—. Genital shield of female large. Pre-endopodal plates well-defined and connected by a narrow strip (Fig. 323). Tectum with a smooth anterior margin (Fig. 321a)............................*Hypoaspis lubrica*, p. 295

LAELAPINAE Berlese, 1892

Laelaptinae Berlese, 1892 sensu Hughes, 1961.

The dorsal shield is entire in both sexes; when hypertrichy is present, it is usually restricted to the *J* series.

The sternal shield of the female has three, rarely four, pairs of setae and two or three pairs of pores. The pre-sternal area is reticulated or has pre-endopodal plates.

The male has a holoventral shield, rarely discrete sterno-genito-ventral and anal shields.

The stigmata have well-developed peritremes, usually extending up to and beyond coxae I. The peritrematal shields are usually free posteriorly, sometimes fused with the exopodals posterior to coxae IV.

The chelicerae usually have prominent dentate digits. A pilus dentilis is present. The chelicera of the male has a long grooved spermadactyl which is usually free distally.

The margin of the tectum is denticulate, or membranous and with a smooth edge.

The deutosternum has from five to seven transverse rows of denticles, the transverse rows sometimes reduced to a single denticle.

Genu and tibia I normally have three ventral setae. Genu II has two ventral setae. Coxae II are without a prominent anterior spine. Seta pv_2 on the tarsi of males is never modified into a tooth-like projection.

This sub-family contains free-living species, as well as ectoparasites of birds and mammals.

ANDROLAELAPS Berlese, 1903

Fairly-well sclerotized mites, ranging in length from 400 to 1,600 μm.

The sternal shield in the female is usually broader than long. The peritrematal shields are free posteriorly and metapodal shields are present in the female.

The tectum may be rounded or membranous, but is never denticulate.

The movable digit of the female chelicera is usually bidentate. Both digits in the male chelicera are edentate; the movable digit is longer than the fixed and partially fused with the spermadactyl. The pilus dentilis is relatively long and may be swollen towards the base.

The apotele of the pedipalp is bipronged.

Two postero-lateral setae are usually present on genu IV and one seta (av_1) on femur I may be modified to form a stout spur in the male, or in both sexes.

The ambulaecra of all the legs have paired claws.

Type species: *Laelaps* (*Iphis*) *hermaphrodita* Berlese, 1887.

ANDROLAELAPS CASALIS CASALIS (Berlese, 1887)

Iphis casalis Berlese, 1887; *Haemolaelaps megaventralis* Strandtmann, 1947; *Hypoaspis freemani* Hughes, 1948; *Haemolaelaps casalis* (Berlese, 1887) sensu Hughes, 1961.

FEMALE (Figs. 312, 316). Length of idiosoma: 640 to 740 μm.

A. casalis casalis is a brown mite which moves rapidly when alive. The dorsal surface is almost completely covered by the dorsal shield which has a reticulate pattern. It bears 39 pairs of simple setae and two or three unpaired setae arranged as in Fig. 312. About nine pairs of *rR* setae arise from the

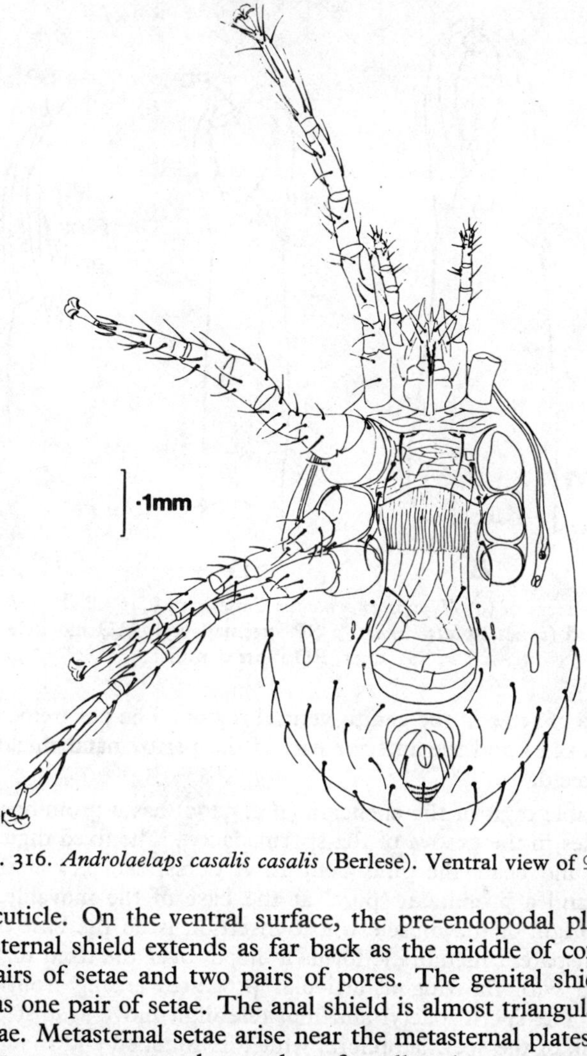

FIG. 316. *Androlaelaps casalis casalis* (Berlese). Ventral view of ♀.

surrounding cuticle. On the ventral surface, the pre-endopodal plates are ill-defined; the sternal shield extends as far back as the middle of coxae III and bears three pairs of setae and two pairs of pores. The genital shield is flask-shaped and has one pair of setae. The anal shield is almost triangular in shape, with three setae. Metasternal setae arise near the metasternal plates and pores. Metapodal plates are present and a number of smaller plates are situated behind coxae IV. The peritrematal plate is distinct from the exopodal and its anterior tip is not fused with the dorsal shield. Seven pairs of setae arise from the interscutal membrane on either side of the genital and anal plates (Fig. 316).

The tectum (Fig. 312) is shaped like a hood. The corniculi are well-sclerotized and the internal malae have fringed external edges. The fixed and movable digits of the chelicerae each have two small teeth and the pilus dentilis is expanded towards its base (Fig. 317). The deutosternum has six transverse rows of denticles.

The chaetotaxy of the legs differs from the normal pattern in having two postero-lateral setae on genu IV (Fig. 315a).

MALE (Fig. 318). Length of idiosoma: 460 to 570 µm.

The ventral surface is almost completely covered by a holoventral shield. Four pairs of setae and three pairs of pores are present in the sternal region

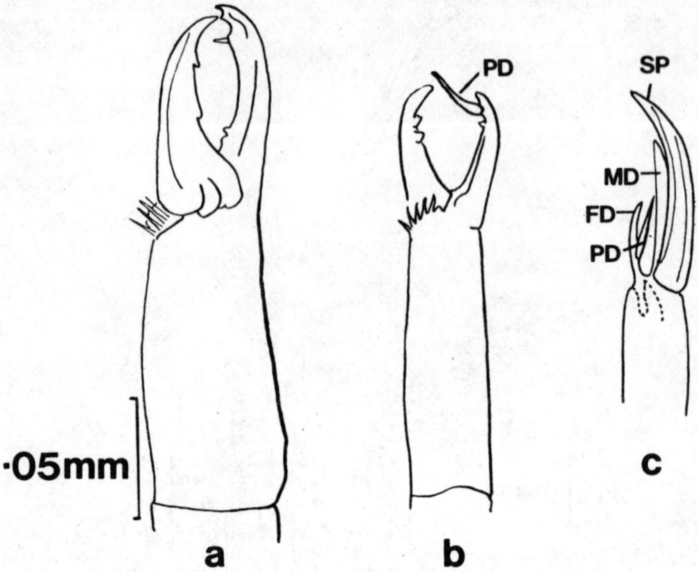

FIG. 317. Chelicera of (*a*) *Hypoaspis aculeifer* Canestrini ♀, (*b*) *Androlaelaps casalis casalis* (Berlese) ♀ and (*c*) *A. casalis casalis* ♂. SP spermadactyl, MD movable digit, FD fixed digit, PD pilus dentilis.

and six pairs of setae in the genito-ventral region. The peritreme extends to the middle of coxae I and the anterior part of the peritrematal shield is fused with the dorsal shield.

The movable digit of the chelicera (Fig. 317c) has a prominent basal articulation and lies in the groove of the spermadactyl. The fixed digit is curved and transparent and bears the pilus dentilis. A dorsal seta lies at the base of the fixed digit and a prominent 'pore' at the base of the movable. Possibly this marks the origin of the muscle whose insertion is on the base of the movable digit. A transparent membrane forms a 'hood' over the fixed digit and another is associated with the ring of articular processes arising from the arthrodial membrane. The spermadactyl and movable digit move against one another to grip the stalk of the spermatophore, whilst its rounded end is supported by the fixed digit and pilus dentilis. McKinley (1963) suggests that if the spermatophore is held like this and if the movable digit and spermadactyl are closed against the fixed digit, the spermatophore would lie first in the deep groove and then slide down the groove towards the tip. On emerging from the tip, the spermatophore could be introduced into the spermathecal aperture of the female.

BIOLOGY

A. casalis casalis is associated with a wide range of habitats. It has been collected from the bodies and also the nests of 14 species of mammals and 12 species of birds (Strandtmann and Wharton, 1958) and it has also been recovered from the bodies of noctuid moths (Treat, 1967) which it was probably using as a conveyance. Berlese, 1887 and Cooreman, 1962 found *A. casalis casalis* in houses and Rack, 1970 and Woodroffe, 1956 obtained specimens from a block of offices and some bat roosts respectively. It is common in the peat used for horticultural purposes and also in sievings from grain and hay. In the peat

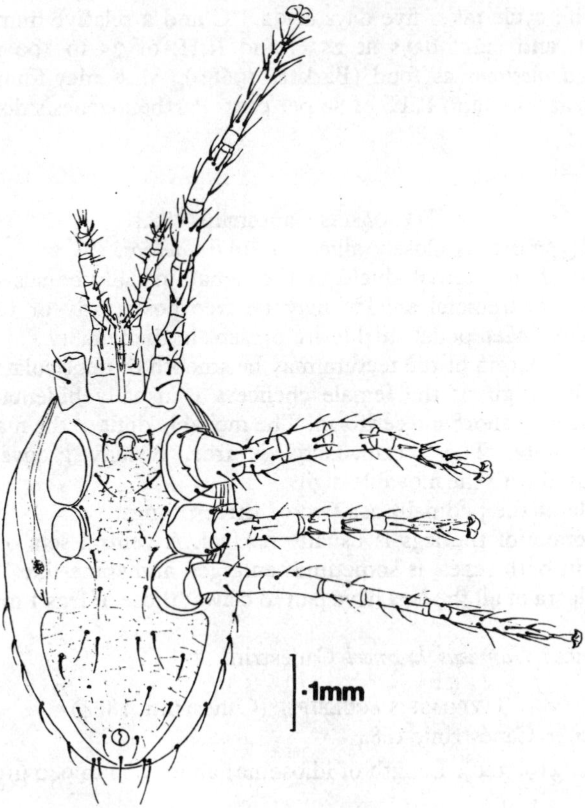

FIG. 318. *Androlaelaps casalis casalis* (Berlese). Ventral view of ♂.

moss and wood shavings used in broiler houses, it can increase to very large numbers, although it does not appear to harm the chickens (McKinley, 1963, Brady, 1970).

A. casalis casalis is a general feeder, thriving on a mixed diet of animal and plant origin. It is a predator of acarid mites and also of the young larvae of *Tribolium confusum* (Barker, 1968b). Flourishing cultures are easily maintained on a culture of *Acarus siro* feeding on wheat germ and kept at room temperature and a relative humidity of 87 per cent (personal observation). It has also been cultured on *Tyrophagus putrescentiae*, *Glycyphagus domesticus*, *Aëroglyphus robustus*, *Blattisocius keegani* (Barker, 1968b) and the young stages of *Dermanyssus gallinae* (McKinley, 1963).

A. casalis casalis will develop on wheat germ alone, but mortality is high (Barker, 1968b); on one occasion it was found in large numbers on Manitoba wheat free from Acaridae (H. C. Crozier coll.). Sinha (1964a) was unsuccessful in attempts to rear this mite on any of 21 species of micro-organisms associated with stored food, but it will feed on egg yolk (H. Dealey, personal communication), fresh meat and blood, although unable to penetrate mammalian and avian skin (McKinley, 1963). Barker and McKinley found that water must be provided to keep the mites alive, although my culture feeding on *A. siro* thrived without access to free water. The eggs are very large and hatching occurs at the larval or, if conditions are unfavourable, at the protonymphal stage (McKinley,

1963). The life cycle takes five days at 32.8°C and a relative humidity of 95 to 100 per cent, and eight days at 25°C and R.H. of 75 to 100 per cent with *Glycyphagus domesticus* as food (Barker, 1968b). McKinley found that it took 18 to 20 days at 19°C and R.H. of 86 per cent. Parthenogenesis does not occur.

DISTRIBUTION

Cosmopolitan.

HYPOASPIS Canestrini, 1884

The genus *Hypoaspis* is closely allied to *Androlaelaps*.

The length of the sternal shield of the female usually equals or exceeds its breadth. The peritrematal shields may be free posteriorly or fused with the exopodal shields. Metapodal shields are present in the female.

The anterior margin of the tectum may be smooth or denticulate.

The movable digit of the female chelicera is usually bidentate. The pilus dentilis is normally short and setiform. The movable digit of the male chelicera is usually unidentate. The spermadactyl is free, distally grooved and never completely fused with the movable digit.

The apotele of the pedipalp has two or three prongs.

The chaetotaxy of the legs is usually normal. A ventral seta on femur II in the male, or in both sexes, is sometimes enlarged and spine-like.

The ambulacra of all the legs have paired claws; those of legs I may be reduced or absent.

Type species: *Gamasus krameri* Canestrini, 1881.

HYPOASPIS ACULEIFER (Canestrini, 1884)

Laelaps aculeifer Canestrini, 1884.

FEMALE (Figs. 319, 320). Length of idiosoma: about 600 to 940 μm.

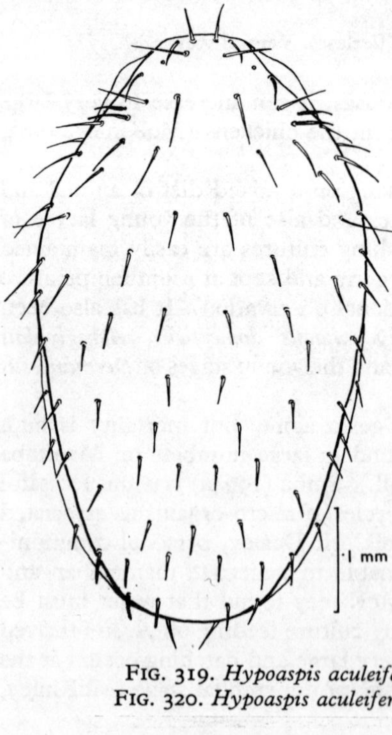

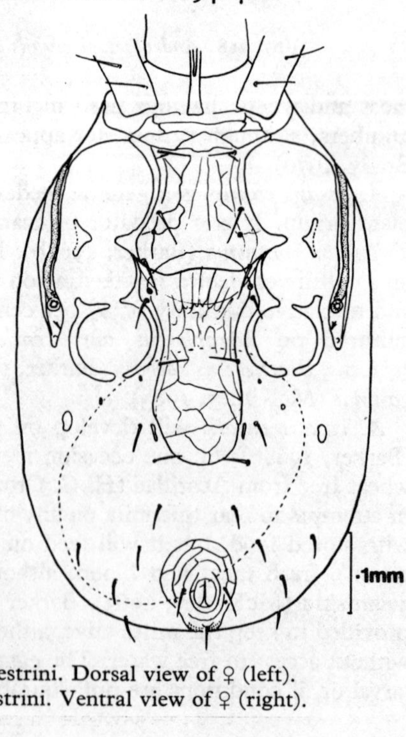

FIG. 319. *Hypoaspis aculeifer* Canestrini. Dorsal view of ♀ (left).
FIG. 320. *Hypoaspis aculeifer* Canestrini. Ventral view of ♀ (right).

The dorsal shield is well-sclerotized and covered with a faint reticulate pattern. It bears 39 pairs of fine, pointed setae arranged as in Fig. 319; in some specimens an unpaired seta is also present. The more anterior setae are longer than the posterior. The sternal shield with its three pairs of setae and two pairs of pores, extends as far back as the anterior margins of coxae IV. The genital shield is flask-shaped and has one pair of setae. The metasternal setae and associated pores are situated in the interscutal membrane. The anal plate has three circum-anal setae. Seven pairs of setae arise from the cuticle on either side of genital and anal plates. According to Kevan and Sharma, 1964, variations occur in the shape of the anal plate and the setal arrangement of the ventral surface. The peritrematal plate is distinct from the exopodal, its posterior end is free and the anterior is fused with the dorsal plate.

The tectum has a rounded denticulate edge (Fig. 321b). The chelicerae are well-tanned; the movable digit has two teeth, the fixed a number of smaller ones and a small pilus dentilis (Fig. 317a).

The legs are long, the second, third and fourth pairs (Fig. 322) bearing a number of thickened spines. The second pair of legs is thicker than the fourth and has shorter spines. All legs bear claws and pulvilli.

The male and immature stages are described and figured by Oudemans (in Strandtmann, 1963) and Evans and Till, 1966.

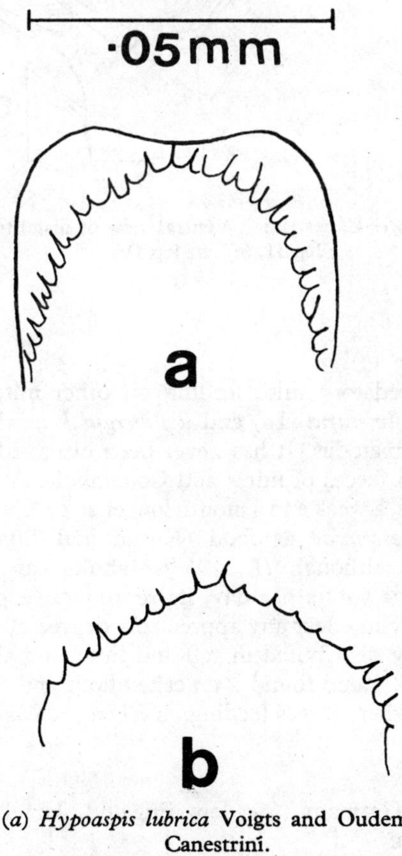

FIG. 321. Tectum of (*a*) *Hypoaspis lubrica* Voigts and Oudemans and (*b*) *H. aculeifer* Canestrinī.

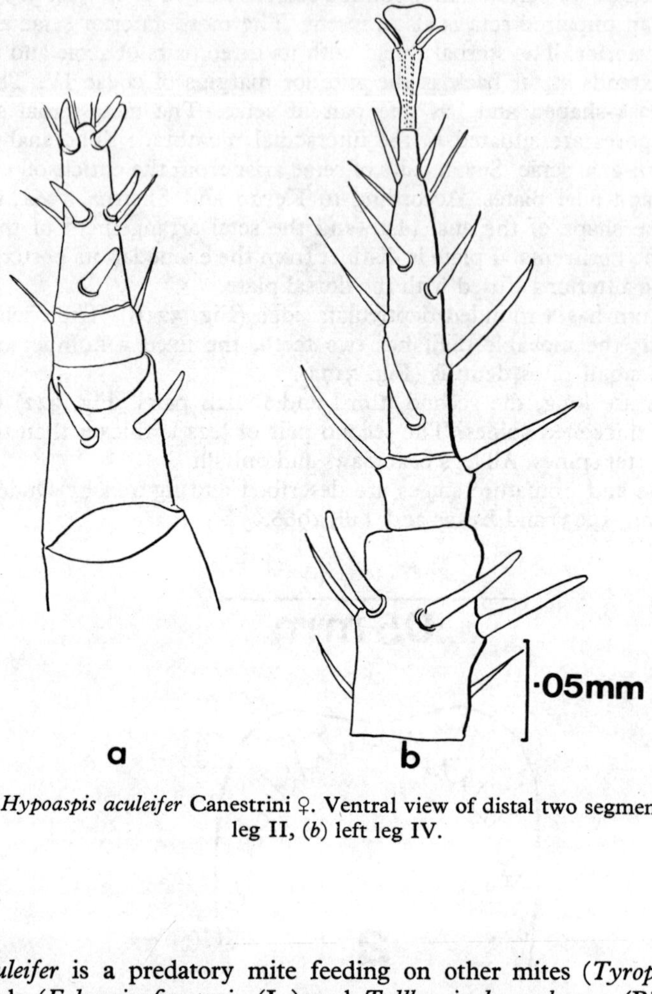

FIG. 322. *Hypoaspis aculeifer* Canestrini ♀. Ventral view of distal two segments of (*a*) left leg II, (*b*) left leg IV.

BIOLOGY

H. aculeifer is a predatory mite feeding on other mites (*Tyrophagus* sp.), Collembola (*Folsomia fimetaria* (L.) and *Tullbergia kraussbauer*, (Börner) and, to some extent, on Nematodes. It has never been observed to feed on detritus or the dead bodies and faeces of mites and Collembola. At 26°C, the life cycle takes from 10 to 13 days, a week and a month longer at 17°C and 11°C respectively with *Tyrophagus putrescentiae* as food (Kevan and Sharma, 1964). These authors consider that, although *H. aculeifer* adults can consume up to 10 *Tyrophagus putrescentiae* adults per day, its reproductive potential is so much lower, that it is unlikely to effect any appreciable degree of control.

H. aculeifer is widely distributed in soil and litter and also in the nests of a variety of rodents. J. U. Lund found it on caked flour and G. A. Brett collected it from daffodil bulbs where it was feeding on *Rhizoglyphus* sp.

DISTRIBUTION

Italy, Netherlands, Germany, Sweden, England, Israel, U.S.S.R., Eastern Canada, North America.

HYPOASPIS LUBRICA Voigts and Oudemans, 1904

Hypoaspis smithii Hughes, 1948; *H. murinus* Strandtmann and Menzies, 1948. Also see Evans and Till, 1966.

FEMALE (Fig. 323). Length of idiosoma: 600 to 700 μm.

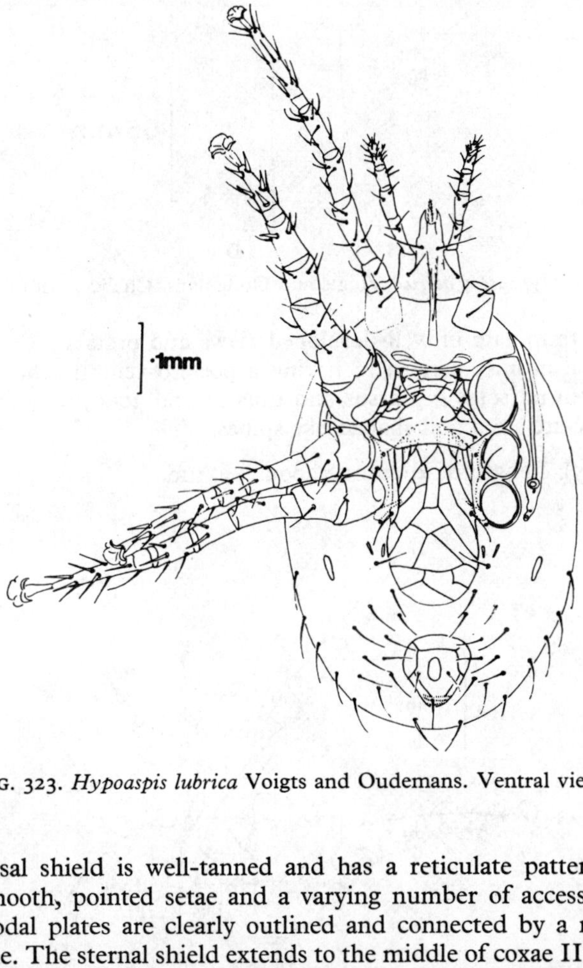

FIG. 323. *Hypoaspis lubrica* Voigts and Oudemans. Ventral view of ♀.

The dorsal shield is well-tanned and has a reticulate pattern. It bears 39 pairs of smooth, pointed setae and a varying number of accessory setae. The pre-endopodal plates are clearly outlined and connected by a narrow strip of thick cuticle. The sternal shield extends to the middle of coxae III, it bears three pairs of setae and two pairs of pores. Its anterior half has a reticulate pattern. The genital plate also has a reticulate pattern, and its posterior edge is rounded; one pair of genital setae is present. The anal plate is almost triangular, with three circum-anal setae. Metapodal plates and two pairs of narrow plates are present on either side of the genital shield. The peritrematal plate is distinct from the exopodal and is fused anteriorly with the dorsal shield, the posterior region is free. The chaetotaxy of the opisthogaster is shown in Fig. 323.

The tectum (Fig. 321a) is a smoothly rounded lobe with an irregular denticulate pattern. The deutosternum has six transverse rows of denticles. The movable digit of the chelicera (Fig. 324a) has two teeth, the fixed three teeth and a short, slender pilus dentilis. The apotele of the pedipalp has two prongs.

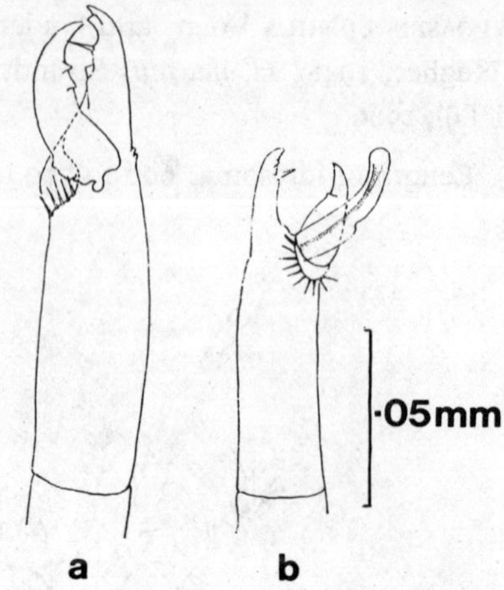

FIG. 324. *Hypoaspis lubrica* Voigts and Oudemans. Chelicera of (*a*) ♀, (*b*) ♂.

All the legs terminate in well-developed claws and pretarsi. The chaetotaxy differs from the normal pattern in having a postero-ventral seta on genu IV. Some of the ventral setae on tarsus and tibia II and genu, tibia and tarsus of legs III and IV are stout and shaped like spines.

MALE (Fig. 325). Length of idiosoma: about 440 μm.

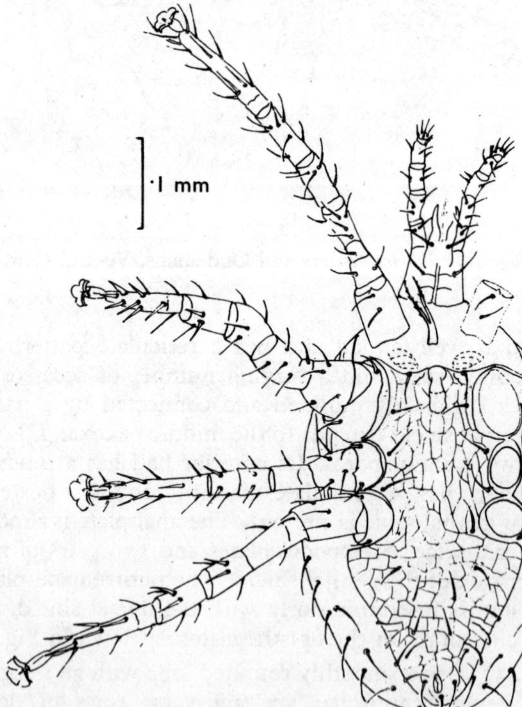

FIG. 325. *Hypoaspis lubrica* Voigts and Oudemans. Ventral view of ♂.

The male resembles the female. The pre-endopodal plates are less well-defined than those of the female and separate from one another. The ventral surface is covered by one reticulate holoventral shield.

The tectum is similar to that of the female. The chelicera (Fig. 324b) has a small pilus dentilis and a single tooth on each digit. The spermadactyl has a spoon-shaped tip and is grooved.

BIOLOGY

H. lubrica is found in association with acarid mites on grain debris, rotting oats, etc. Bregetova (1956) records it from the nests of small mammals and Evans and Till (1966) collected it from swallows' nests. Strandtmann and Menzies (1948) recovered it from various species of rats and mice and it has also been found in the deep litter of broiler houses (J. D. Ethelston coll., Brady, 1970).

DISTRIBUTION

England, Northern Ireland, Czechoslovakia, U.S.S.R., U.S.A.

HYPOASPIS SARDOA (Berlese, 1911)

Hypoaspis (*Androlaelaps*) *sardous* Berlese, 1911 sensu Hughes, 1961.

FEMALE (Fig. 326). Length of idiosoma: about 800 μm.

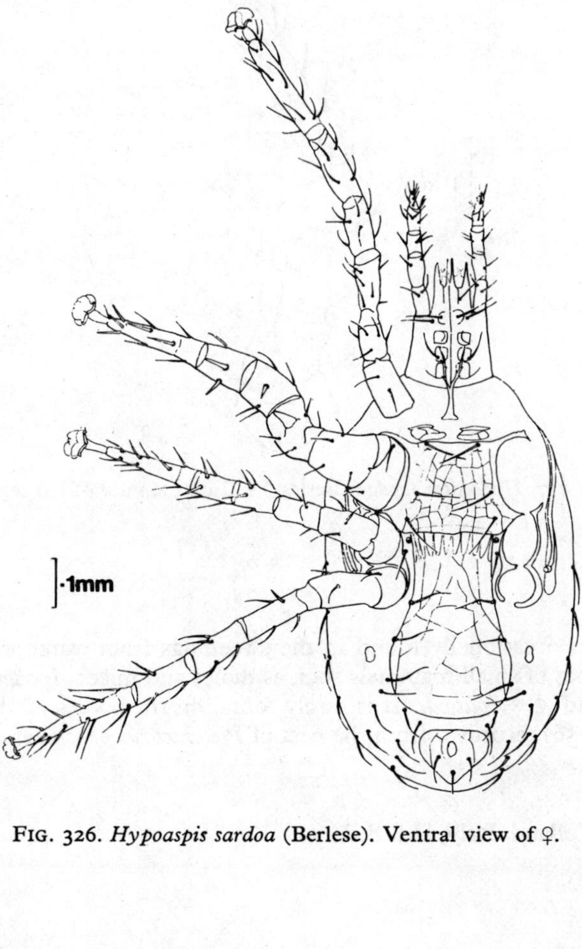

FIG. 326. *Hypoaspis sardoa* (Berlese). Ventral view of ♀.

The dorsal shield is dark brown in colour, smooth except for some pitted markings mainly concentrated in the central region. It bears 39 pairs of setae. Ventrally, the pre-endopodal plates have distinct outlines and are separate from one another. The sternal shield has three pairs of setae and two pairs of pores. The genital shield is flask-shaped and has two pairs of setae. The metasternal setae arise from the interscutal membrane. The peritrematal plate is separate from the exopodal, its anterior part is fused with the dorsal shield. The peritreme extends to the anterior margin of coxa II. The anal shield has three circum-anal setae. Small metapodal plates are present. The ventral plates are covered with a reticulate pattern.

The tectum is rounded, slightly pointed in the middle and covered with radiating lines. The digits of the chelicerae are dentate; a small pilus dentilis arises from the fixed limb.

The chaetotaxy of the legs differs from the normal pattern in that tibia III and genu IV each have a second posterolateral seta (Evans and Till, 1966). Ventrally, the femur of leg II (Fig. 327) bears a stout blunt seta (av_1), the genu: a smaller one (av_1), and tibia II has two thickened spines (av_1 and pv_1).

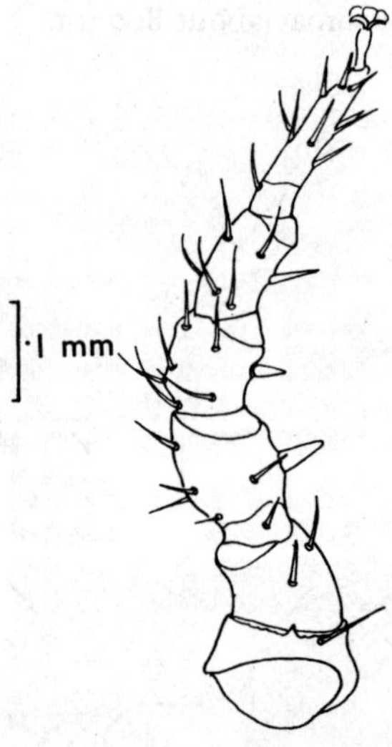

FIG. 327. *Hypoaspis sardoa* (Berlese) ♀. Internal view of left leg II.

Biology

H. sardoa is occasionally found in the sweepings from warehouse floors and also in the nests of small mammals such as moles and mice: *Apodemus flavicollis* (Melchior) and *A. sylvaticus*. It is rarely found on the bodies of these animals. Bregetova (1956) records it from the nest of *Pheonicurus phoenicurus* (L.).

Distribution

England, Holland, Italy, U.S.S.R.

HYPOASPIS MILES (Berlese, 1892)

Laelaps (*Iphis*) *milès* Berlese, 1892; *Cosmolaelaps gurabensis* Fox, 1946; *Stratiolaelaps gurabensis* (Fox, 1946) sensu Hughes, 1961.

FEMALE (Figs. 328, 329). Length of idiosoma: about 650 μm.

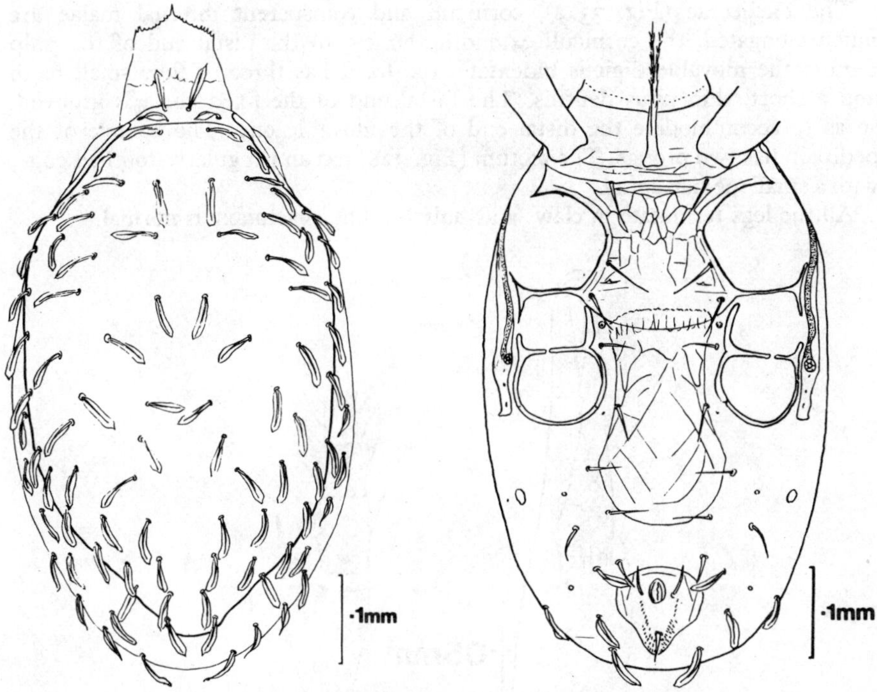

FIG. 328. *Hypoaspis miles* (Berlese). Dorsal view of ♀ (left).
FIG. 329. *Hypoaspis miles* (Berlese). Ventral view of ♀ (right).

The idiosoma and legs are brown in colour, paler than the chelicerae and corniculi. The dorsal surface is almost completely covered by the dorsal shield which tapers rather abruptly towards its posterior extremity. Its surface is finely reticulate and from it arise 37 pairs of setae; these setae (Fig. 330) are narrowly leaf-shaped and the distal end of each is pointed. Seven pairs of *R* setae arise from the surrounding cuticle. Ventrally, the pre-endopodal plates are ill-defined. The sternal shield extends to the middle of coxa III, it bears three

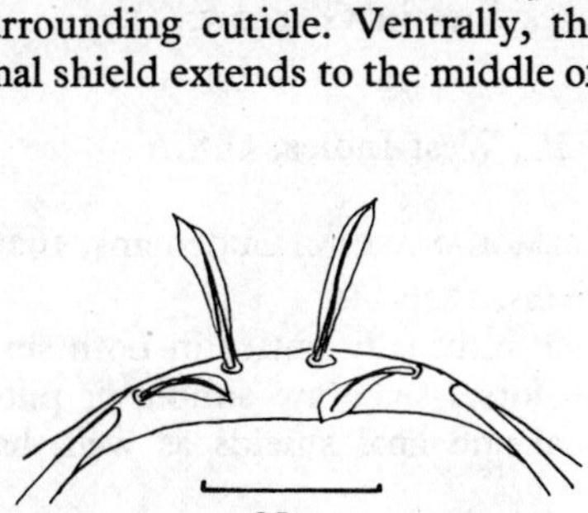

FIG. 330. *Hypoaspis miles* (Berlese) ♀. Vertex of idiosoma.

pairs of setae and two pairs of pores. The metasternal setae are free and posterior to the metasternal pores. The genital shield is flask-shaped, with one pair of setae. The anal shield bears three circum-anal setae. The peritreme is sinuous in outline and the peritrematal plate is free anteriorly and posteriorly. Metapodal plates are small. All the ventral setae are setiform, except for three pairs on either side of the anal plate.

The chelicerae (Fig. 331a), corniculi and transparent internal malae are much elongated, the corniculi extending almost to the distal end of the palp femur; the movable digit is bidentate, the fixed has three or four small teeth and a short, thin pilus dentilis. The distal end of the fixed chela is grooved, so as to accommodate the distal end of the movable one. The apotele of the pedipalp has two prongs. The tectum (Fig. 328) has an irregularly-toothed edge, with a small median spine.

All the legs terminate in claws and pulvilli. The chaetotaxy is normal.

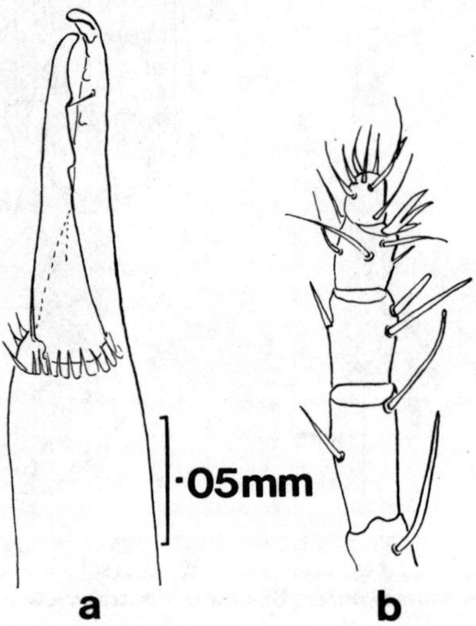

FIG. 331. Chelicera (*a*) and right pedipalp (*b*) of *Hypoaspis miles* (Berlese).

BIOLOGY

H. miles has been found associated with a variety of rodents and their nests. J. H. Cuthbert found it on decaying oatmeal spillage in Leith and G. Hosie collected it from wheat residues in Glasgow.

DISTRIBUTION

Scotland, Italy, U.S.S.R., West Indies, U.S.A.

HAEMOGAMASINAE Oudemans, 1926

Haemogamasidae Oudemans, 1926

The dorsal shield which is usually entire in both sexes, is markedly hypertrichous (Fig. 332). The dorsal setae are simple or pilose. The genital shield, and sometimes the sternal and anal shields as well, have unpaired accessory setae.

The peritremes may be well-developed or reduced. The peritrematal shields are free posteriorly or fused with the exopodals posterior to coxae IV.

The chelicerae are chelate and usually dentate. A pilus dentilis is present. The spermadactyl of the male is grooved, relatively short and free distally, but closely associated with the movable digit. The corniculi are horn-like and distinct.

The deutosternum has 10 or more transverse rows of denticles.

The margin of the tectum is conspicuously fimbriated (Fig. 335b).

Genu IV has one ventral seta (*av*) instead of two. The anterior spine on coxae II is small or absent.

The subfamily Haemogamasinae is represented by two genera in this country: *Eulaelaps* and *Haemogamasus*; species belonging to these two genera are sometimes found in stored food, although they occur more frequently in the nests of birds and mammals.

HAEMOGAMASUS Berlese, 1889

Euhaemogamasus Ewing, 1933.

In the female, the genito-ventral shield is relatively narrow and the metapodal shields are small and subovate.

The palptrochanter is without a sensory organ.

The anterior row of deutosternal denticles has two to three denticles per row.

Type species: *Haemogamasus hirsutus* Berlese, 1889.

HAEMOGAMASUS PONTIGER Berlese, 1904

Haemogamasus oudemansi Hirst, 1914; *Groschaftella pontiger* (Berlese, 1904) sensu Samšiňák, 1957; *Haemogamasus* (*Euhaemogamasus*) *pontiger* Berlese, 1904 sensu Hughes, 1961.

FEMALE (Figs. 332, 333). Length of idiosoma: 950 to 1,100 µm.

The idiosoma is almost pear-shaped in outline, the dorsal surface only partially covered by a brown shield with a reticulate pattern. From the shield and surrounding interscutal membrane arise numerous setae; setae j_1 are larger than the others and sparsely pectinate; some of the anterior and antero-lateral setae are also barbed. The sternal shield is deeply concave posteriorly and bears three pairs of setae and three pairs of pores. The metasternal setae are free. The genito-ventral shield is enlarged posteriorly and bears one pair of genital setae and about 16 additional setae on the posterior region of the shield. The anal shield has only three circum-anal setae. Small subovate metapodal shields are present and numerous setae arise from the surrounding interscutal membrane. The peritrematal shield is free anteriorly and fused posteriorly with the exopodal plate of coxa IV (Fig. 334). The peritreme is very short, only extending as far forward as the posterior edge of coxa II.

The tectum (Fig. 335b) is pointed, with an uneven fimbriate edge. The fixed limb of the chelicera (Fig. 336a) is pointed, with a tooth on each side of the pilus dentilis; the movable limb has a sharp end and two backwardly directed teeth. The deutosternum has 14 transverse rows of denticles.

The chaetotaxy of the legs differs from the normal pattern in that genu IV has two postero-lateral setae and tibia IV two antero- and two postero-dorsal setae (Evans and Till, 1966). Most of the leg setae are barbed.

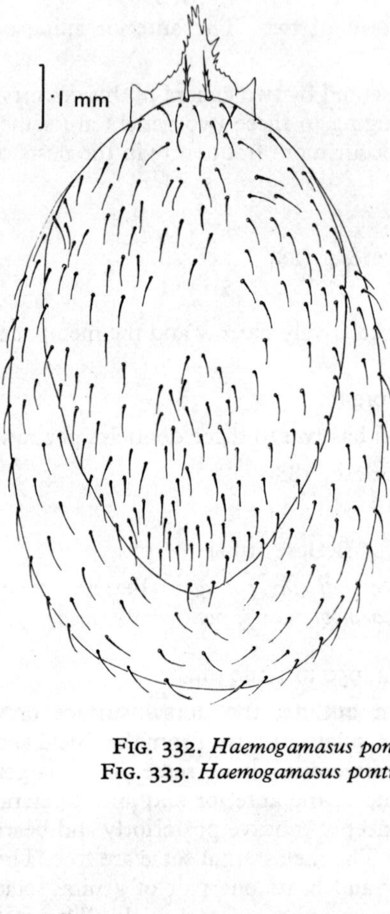

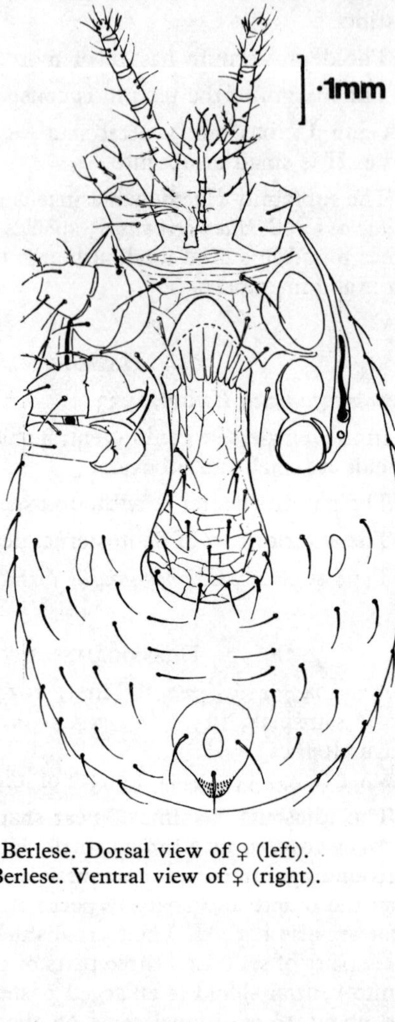

FIG. 332. *Haemogamasus pontiger* Berlese. Dorsal view of ♀ (left).
FIG. 333. *Haemogamasus pontiger* Berlese. Ventral view of ♀ (right).

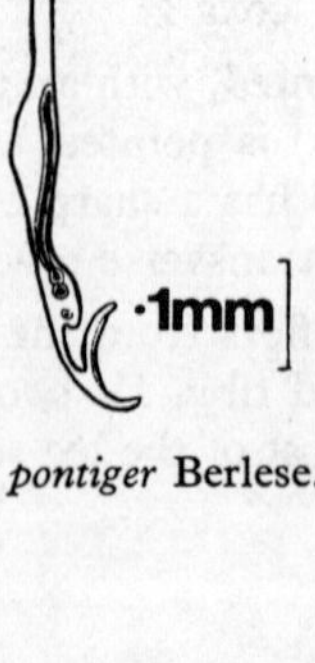

FIG. 334. *Haemogamasus pontiger* Berlese. Peritrematal shield of ♀.

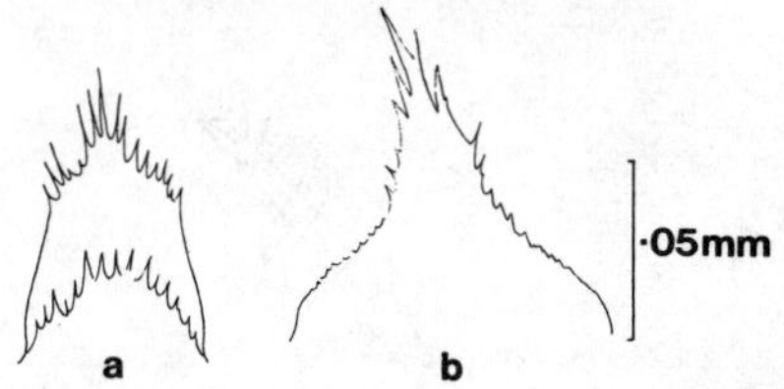

FIG. 335. Tectum of (*a*) *Eulaelaps stabularis* (Koch) ♀ and (*b*) *Haemogamasus pontiger* ♀

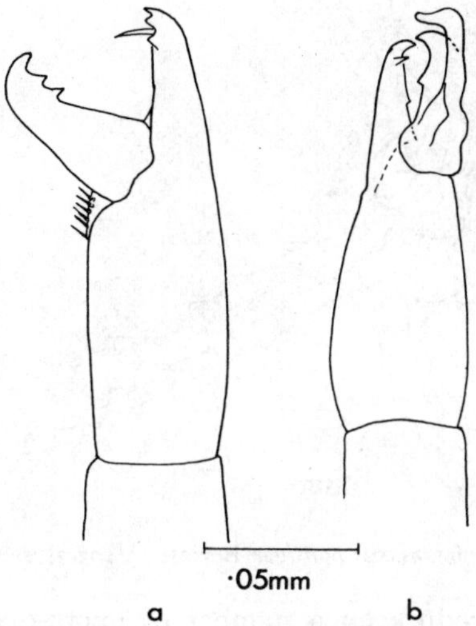

FIG. 336. *Haemogamasus pontiger* Berlese. Chelicera of (*a*) ♀, (*b*) ♂.

MALE (Fig. 337). Length of idiosoma: about 730 μm.

The ventral surface is covered by a single reticulate shield with four pairs of setae and three pairs of pores in the sternal region. Numerous setae arise from the genito-ventral region.

Both limbs of the chelicera (Fig. 336b) are pointed and unidentate. The pilus dentilis is short and setiform and the distal end of the spermadactyl is bent at a right angle.

BIOLOGY

H. pontiger is frequently found in the detritus of warehouse floors, as well as in oats, wheat, barley and a variety of other substances. It is a predator of astigmatid mites and of weakly sclerotized insect larvae and it will also feed on heparinized mouse and human blood (Barker, 1968c). Barker showed that at 24° and 27°C, the eggs of *H. pontiger* hatched in an average of 3.2 and 2.6 days respectively when the relative humidity was 95 to 100 per cent and *Glycyphagus domesticus* was supplied as food. Under these conditions, the life cycle of the female was completed in 7.9 and 7.6 days and that of the male in 7.0 and 6.3 days; i.e., males developed faster at 27°C than at 24°. Fewer eggs were laid when *G. domesticus* was the only prey available, but when the diet was supplemented with wounded *Tribolium confusum* larvae, the rate of oviposition was usually increased. *H. pontiger* can complete its development when wheat germ only is supplied as food (personal observation) and Sinha (1964a) found that it was

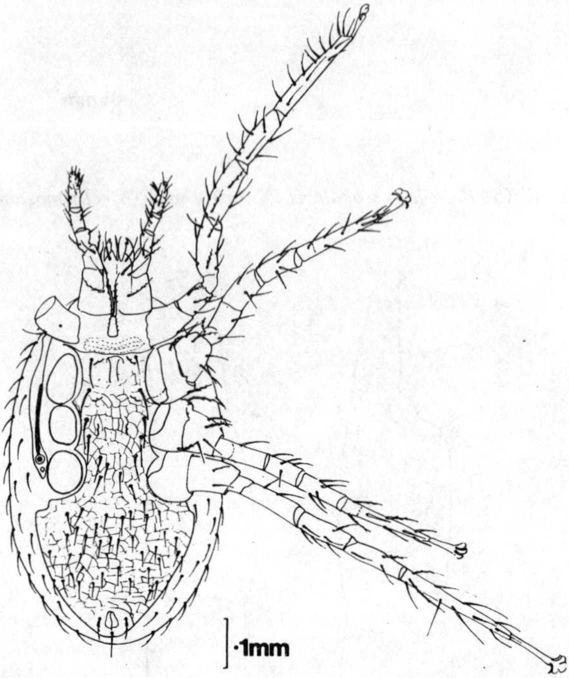

FIG. 337. *Haemogamasus pontiger* Berlese. Ventral view of ♂.

able to live but not reproduce on a number of micro-organisms commonly found in the granaries of Canada.

H. pontiger has also been found in association with rats, moles and a bat (Evans and Till, 1966) and also in the deep litter of broiler houses (Brady, 1970). I have never found specimens containing a large amount of blood; indeed, the mouth parts are not adapted for piercing skin, but it would probably feed on blood or tissue fluid if the host's skin was abraided.

DISTRIBUTION

Cosmopolitan.

EULAELAPS, Berlese, 1903

In the female, the genito-ventral shield is relatively broad and the metapodal shields are large and almost triangular in shape.

The palptrochanter has a unique sensory organ on its ventral surface.

The anterior row of deutosternal denticles has from five to seven denticles per row.

Type species: *Gamasus stabularis* Koch, 1836.

EULAELAPS STABULARIS (Koch, 1836)

FEMALE (Fig. 338). Length of idiosoma: about 1,120 μm.

The idiosoma is almost completely covered on its dorsal surface by a brown shield which shows marked hypertrichy. Ventrally, the pre-endopodal plates are ill-defined. The sternal shield has three pairs of setae and two pairs of pores, the metasternal setae are free and have associated pores. The genito-ventral

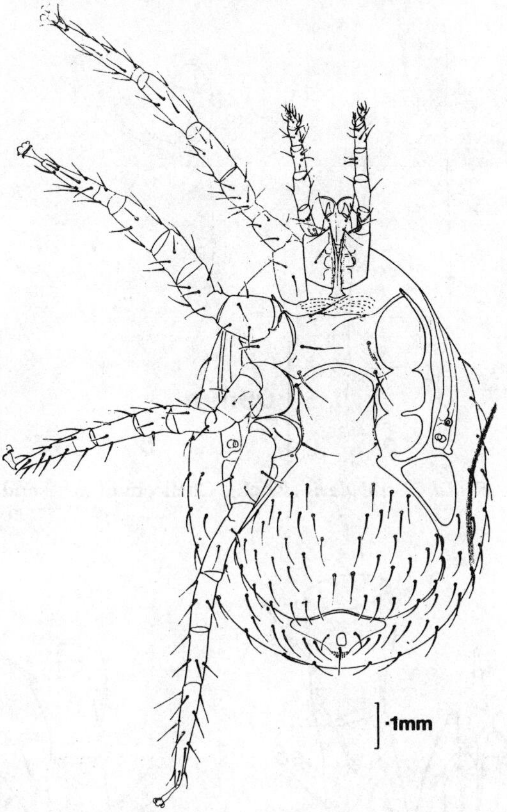

FIG. 338. *Eulaelaps stabularis* (Koch). Ventral view of ♀.

shield has lateral incisions posterior to the genital setae and is widely expanded behind coxae IV. It bears more than 20 pairs of opisthogastric setae. The anal shield is almost triangular in shape, its slightly convex anterior margin fitting into a concavity of the genito-ventral shield; it bears three circum-anal setae. The anterior tip of the peritrematal shield is fused with the dorsal shield, the posterior part is distinct from the exopodal. The peritreme extends on to the dorsal surface as far as the base of the gnathosoma.

The gnathosoma is rather small and narrow for the size of the body. The tectum (Fig. 335a) has an unevenly serrated margin. The movable digit of the chelicera (Fig. 339a) is bidentate, the fixed digit has three inconspicuous teeth and a small pilus dentilis. The deutosternum (Fig. 340) has 10 or more transverse rows of denticles, with four to seven denticles per row. The palp trochanter has a unique sensory organ on its ventral surface.

All the legs bear claws and pulvilli; the claws on legs I are sometimes reduced.

MALE (Fig. 341). Length of idiosoma: about 880 μm.

The idiosoma is more slender than that of the female. The ventral surface is completely covered by a single reticulate shield marked by a transverse line at the level of coxae IV. The sternal region bears four pairs of setae and three pairs of pores, the genito-ventral a pair of genital setae and more than 20 pairs of additional setae.

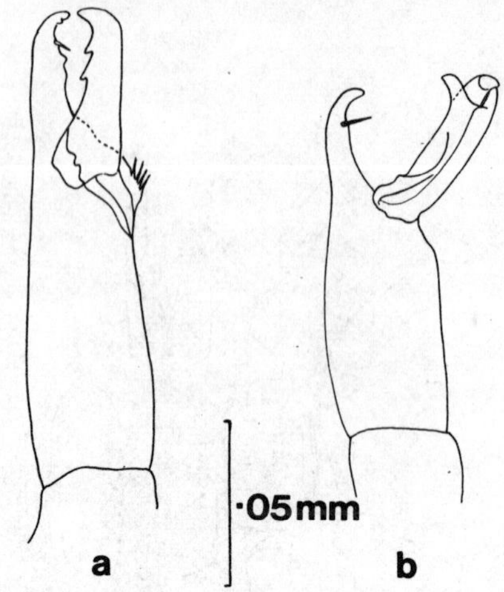

FIG. 339. *Eulaelaps stabularis* (Koch). Chelicera of (*a*) ♀ and (*b*) ♂.

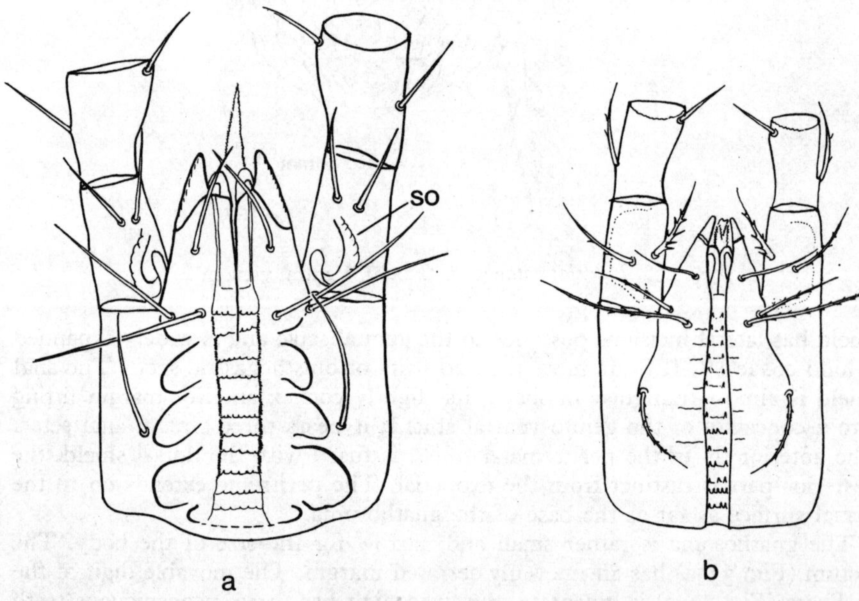

FIG. 340. Ventral view of gnathosoma of (*a*) *Eulaelaps stabularis* (Koch) and (*b*) *Haemogamasus pontiger* Berlese. SO sensory organ.

The tectum is more regularly serrated than that of the female. The movable digit of the chelicera (Fig. 339b) has a large, subterminal tooth-like projection. The spermadactyl is almost the same length as the movable digit. The fixed digit is edentate with a small pilus dentilis.

BIOLOGY

E. stabularis is widely distributed in the fur and nests of small mammals, also in the nests of birds and the deep litter of broiler houses (Brady, 1970). It is sometimes found distended with blood, but it is probable that it can also prey on soft-bodied mites and insects. It is possible that it can act as a vector of *Brucella tularensis*, since this was isolated from *E. stabularis* collected from infected water rats (Grzhebina, 1939, quoted from Baker *et al.*, 1956).

E. stabularis is sometimes found in the detritus of warehouse floors and is probably distributed by rats and mice. On one occasion, it was recovered from the bedding of a cat's basket (N. Hunter coll.).

DISTRIBUTION

Cosmopolitan.

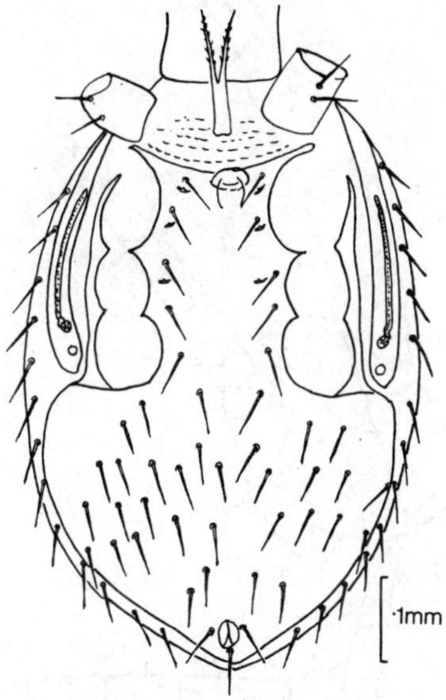

FIG. 341. *Eulaelaps stabularis* (Koch). Ventral view of ♂.

DERMANYSSINAE Kolenati, 1859

The dorsal shield of the female is poorly sclerotized and shows hypotrichy in some regions, the paravertical setae z_1 being absent and *J* series reduced.

The sternal shield of the female has two, rarely three pairs of setae and two pairs of pores.

The male has a holoventral shield or two separate shields.

The peritremes are variable in length and the peritrematal shields are fused with the exopodals behind coxae IV.

The chelicerae of the female are strongly modified and resemble stylets, the second segment enormously elongated. The digits are minute and weakly dentate. A pilus dentilis, dorsal seta and arthrodial processes are absent. In the male, the second segment is of normal length, the fixed digit of the chelicera is usually shorter than the long spermadactyl which is entirely fused with the movable digit. The corniculi are membranous.

The margin of the tectum is in the form of a triangular lobe.

The deutosternum has nine or more denticles arranged in a single file.

The chaetotaxy of the leg segments is diverse, often showing intraspecific variation. Tarsi III and IV of the male have seta pv_2 modified into a short, tooth-like projection. Coxae II usually have no anterior spine.

Members of this subfamily are ectoparasites of birds and *Dermanyssus gallinae* which is commonly found in sparrow and pigeon nests, is sometimes found in houses.

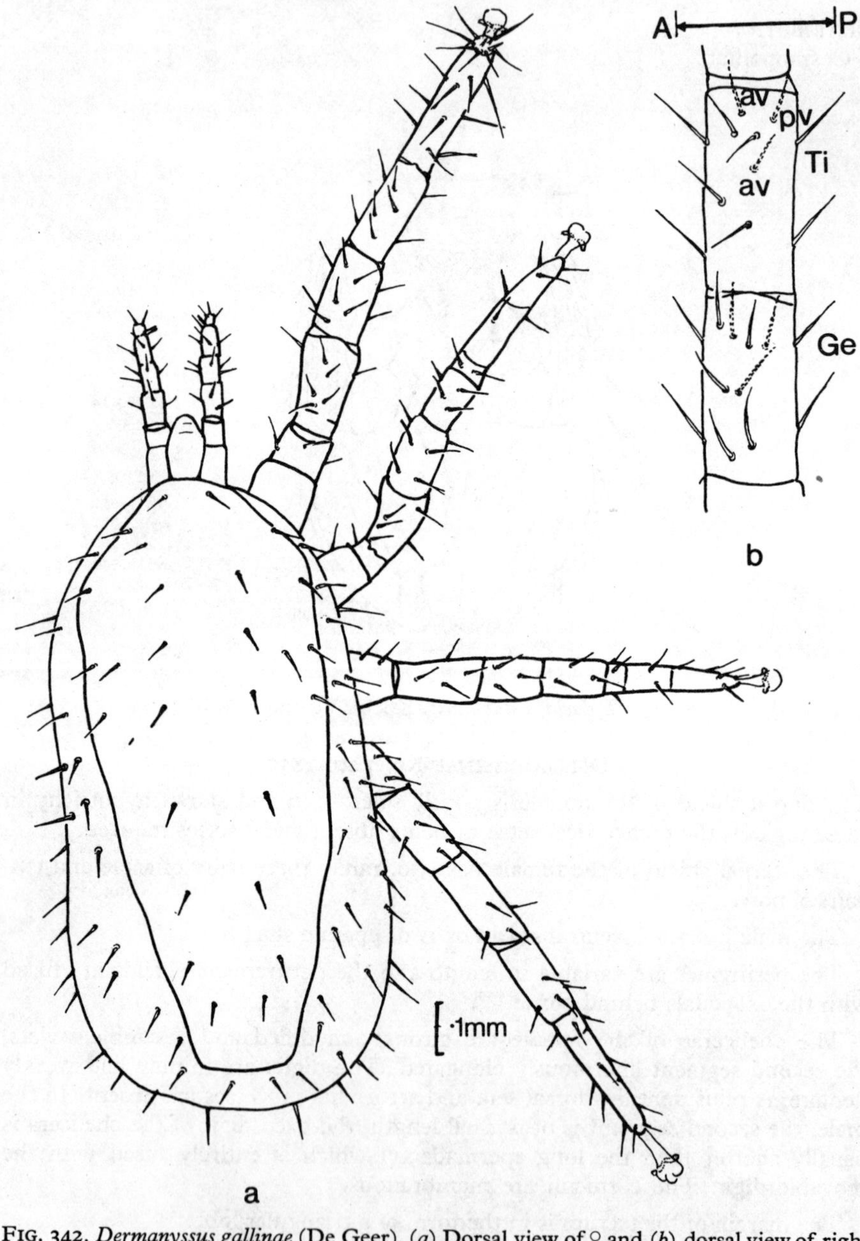

FIG. 342. *Dermanyssus gallinae* (De Geer). (*a*) Dorsal view of ♀ and (*b*) dorsal view of right tibia and genu I. Ventral setae: *av*, *pv*. Ti tibia, Ge genu.

DERMANYSSUS GALLINAE (De Geer, 1778)

Acarus gallinae De Geer, 1778.

FEMALE (Figs. 342, 343). Length of idiosoma: about 1,000 μm.

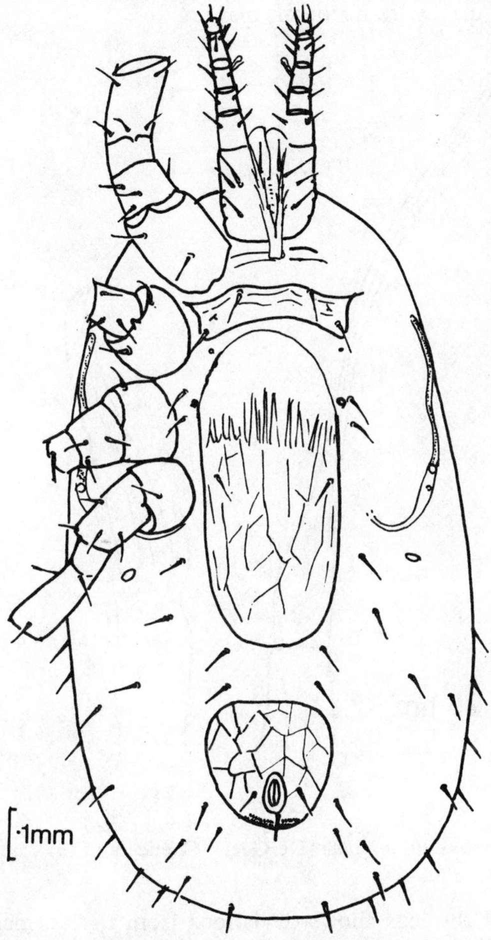

FIG. 343. *Dermanyssus gallinae* (De Geer). Ventral view of ♀.

The idiosoma is colourless or bright red when engorged with fresh blood. Specimens containing partially digested blood are smaller and black or grey in colour. The dorsal shield is feebly sclerotized and only partially covers the dorsal surface. It has a truncate posterior margin. Its surface is reticulate and from it arise 15 pairs of smooth setae, the posterior ones being rather shorter than the more anterior. Setae also arise from the interscutal membrane. The sternal shield is shaped as in Fig. 343. It bears two pairs of setae and one pair of pores, the third pair of sternal setae, the metasternal setae and two pairs of pores are situated on the unsclerotized cuticle. The genital shield has a rounded posterior margin and one pair of setae and the anal has the usual three circum-anal setae. The ventral shields have a reticulate surface. The peritreme extends to the anterior half of coxa II and the peritrematal shield is continuous posteriorly with the podal plate of coxa IV.

The chelicerae are about half as long as the idiosoma, segment II being about six times as long as segment I (Fig. 344a). The digits are small and weakly dentate. The deutosternum has 11 denticles arranged in single file (Fig. 345). The tectum is triangular, with a smooth margin.

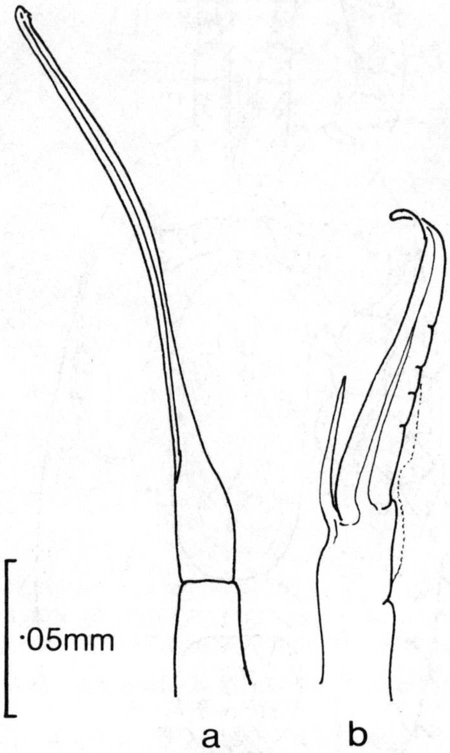

FIG. 344. *Dermanyssus gallinae* (De Geer). Chelicera of (*a*) ♀ and (*b*) ♂.

The chaetotaxy of the legs shows deviations from the normal. Coxa II has no anterior spine. All the legs have ambulacra with claws and pulvilli.

MALE (Fig. 346). Length of idiosoma: about 800 μm.

The dorsal shield is wider than that of the female and bears 19 or 20 pairs of setae. The holoventral shield has five pairs of setae in the sterno-genital region, two pairs in the ventral and three circum-anal setae. Additional setae also arise from the interscutal membrane.

Segment II of the chelicera (Fig. 344b) is less than twice as long as segment I and the fixed digit is reduced. The spermadactyl is long and grooved, completely fused with the movable.

The chaetotaxy of the legs resembles that of the female. Seta pv_2 on tarsi III and IV is modified to form a backwardly directed tooth-like projections (Fig. 347). Coxa II has no anterior spine.

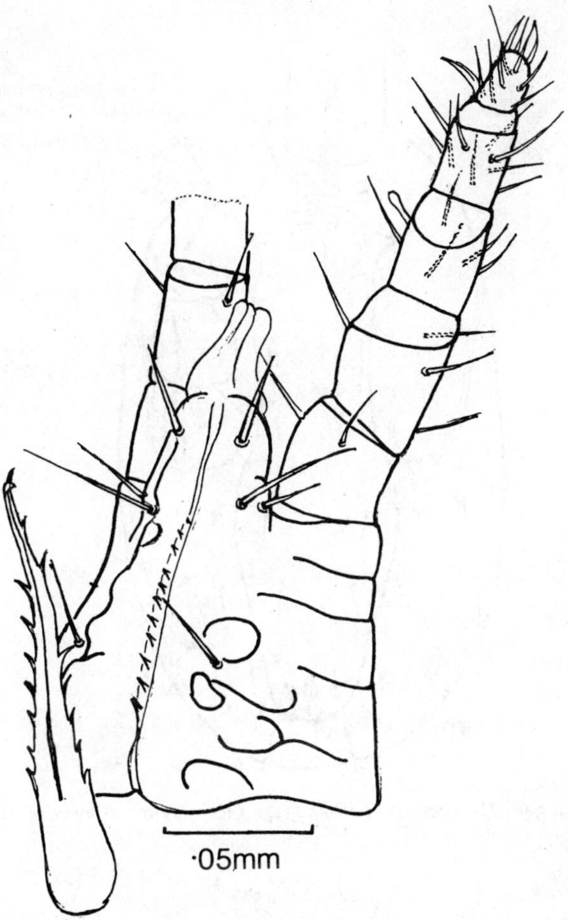

FIG. 345. *Dermanyssus gallinae* (De Geer) ♀. Ventral view of gnathosoma.

BIOLOGY

D. gallinae is an ectoparasite of chickens and also of many species of wild and domestic birds. As a parasite of chickens, it may give rise to severe illness and egg-laying is frequently affected. The adults and nymphs feed mainly at night and, during the day, are found in crevices in walls and ceilings of poultry houses.

The red chicken mite is also commonly found in sparrow, starling and pigeon nests which are frequently built in attics or chimneys or under the eaves of houses. When the birds leave the nests, the mites frequently invade the houses and will then bite man, giving rise to considerable irritation.

According to Baker *et al.*, (1956), oviposition starts at 12 to 24 hours after the female has fed, as many as seven eggs being deposited at one time. Successive batches of eggs are produced, each batch being laid after a blood meal has been taken. In a colony that was under observation in Texas, the entire life cycle was completed in seven days. The adults can survive without feeding for four or five months.

DISTRIBUTION

Cosmopolitan.

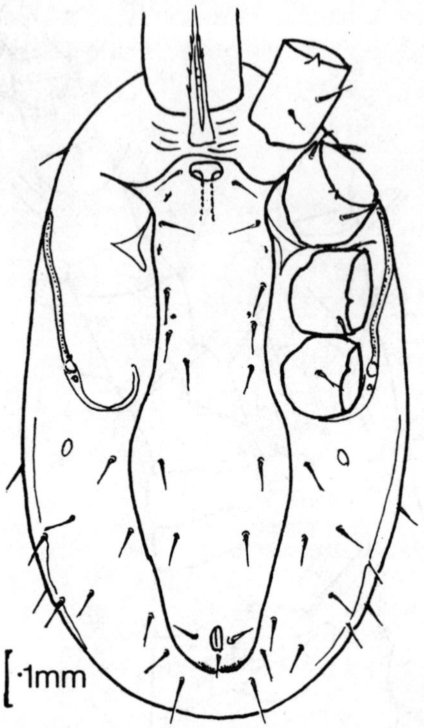

FIG. 346. *Dermanyssus gallinae* (De Geer). Ventral view of ♂.

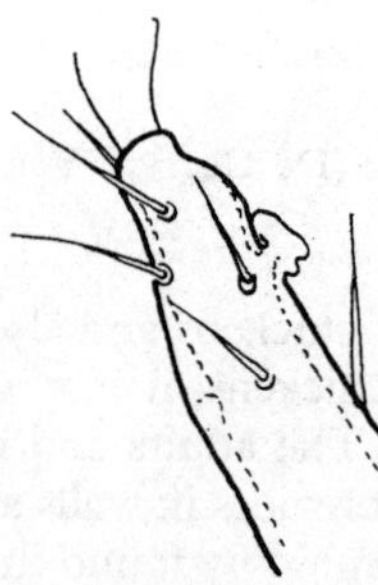

FIG. 347. *Dermanyssus gallinae* (De Geer). ♂ Distal half of tarsus IV.

ASCIDAE Voigts and Oudemans, 1905

Aceosejidae Baker and Wharton, 1952 sensu Evans, 1957. Also see Lindquist and Evans, 1965.

The adults have an entire or subdivided dorsal shield with more than 23 pairs of setae.

At least three pairs of marginal setae arise from the interscutal membrane surrounding the dorsal shield.

In the female, the metasternal setae arise from discrete shields or from the cuticle. The genital shield is longer than wide, with a truncate or rounded posterior margin and one pair of setae, which may arise from the interscutal membrane. Podal spermathecae open between coxae III and IV.

In the male, sterno-genital and ventro-anal shields are usually present.

The chelicerae are usually dentate, with either a pilus dentilis or a hyaline lobe on the masticatory surface. A spermadactyl is attached to the chelicera in the male. The apotele of the palp tarsus is usually two-tined.

Seven transverse rows of deutosternal denticles are usually present.

The tectum varies in structure, it is never produced into a single unbranched denticulate process.

The first pair of legs sometimes lacks pretarsus and claws.

Tibia I has three ventral setae. Femur I has a total of 11 or 12 setae.

The family Ascidae contains about 350 species of world-wide distribution. They are found in a wide variety of habitat, ranging from the nares of humming birds to the surface litter of forests and grasslands. Their biology is largely unknown, but the shape of the chelicerae indicates that they are predatory or mycophagous in habit.

ASCIDAE

Key to the adults (adapted from Evans, 1958)

1. Dorsal shield in both sexes with lateral incisions (Fig. 349b); with 14 pairs of setae arising from posterior region of dorsal shield. Tectum three-pronged. Anal shield present in female. Pre-endopodal shields fused with the sternum (Fig. 349a) .... *Arctoseius butleri*, p. 314

—. Dorsal shield in both sexes with entire lateral margins 2

2. Marginal (*R*) series of setae arising from posterior region of dorsal shield (Fig. 352); fixed digit of chelicera with a membranous lobe in place of a pilus dentilis (Fig. 355a) .... *Proctolaelaps*, p. 317

—. Marginal (*R*) series of setae arising from lateral interscutal membrane (Fig. 363) 3

3. Peritrematal and exopodal shields distinct from one another. Female with an anal shield bearing three anal setae only, genital shield rounded posteriorly .... *Melichares agilis*, p. 322

—. Peritrematal and exopodal shields touching in the region of coxae IV. Female with a ventro-anal shield bearing two to six setae in addition to the anals 4

4. Corniculi slender and convergent (Fig. 378b). Fixed digit of chelicera well-developed or reduced, with a few or no teeth. Female with humeral seta r_3 usually arising from interscutal membrane....... *Blattisocius*, p. 324

—. Corniculi stout and parallel with one another (Fig. 378a). Fixed digit of chelicera well-developed and multidentate. Female with humeral seta r_3 arising from dorsal shield .... *Lasioseius penicilliger*, p. 334

ARCTOSEIUS Thor, 1930

Tristomus Hughes, 1948

The dorsal shield is entire in both sexes or with lateral incisions midway along the lateral margin (Fig. 349b). From it in the female, arise from 31 to 34 pairs of simple setae, 13 or 14 pairs of which arise from the posterior region of the shield. Nine or 10 pairs of marginals (*r*, *R* series) arise from the interscutal membrane.

In the female, the sternal shield has three pairs of setae, the third pair of sternal pores is situated on its postero-lateral corners. The genital shield is narrow and the genital setae arise from the interscutal membrane. Normally, the anal shield bears only three circum-anal setae.

In the male, a sterno-genital and ventro-anal plate is present.

In the female, the pilus dentilis is setiform. The spermadactyl of the male chelicera has a characteristic shape (see Lindquist, 1961).

The tectum is bi- or triramous.

The deutosternum has seven rows of denticles confined by well-formed lines.

All the legs have pretarsi and claws. Tibia I has five dorsal seta. Tibiae III and IV each a maximum of seven setae.

Type species: *Arctoseius laterincisus* Thor, 1930.

ARCTOSEIUS BUTLERI (Hughes, 1948)

Tristomus butleri Hughes, 1948.

FEMALE (Fig. 348). Length of idiosoma: about 400 μm.

The idiosoma is uniformly brown and narrow, with parallel sides. The dorsal shield is covered with faint, irregular reticulations and is notched midway along its lateral margin (Fig. 349b). The 31 pairs of setae arising from it are smooth and fine. The sternal shield is only faintly reticulate and the pre-endopodal plates are continuous with its anterior margin (Fig. 349a). The metasternal setae and pores, as well as the genital setae, arise from the interscutal membrane. The endopodal plates between coxae III and IV are small and triangular in shape. The genital shield is narrow and elongate. The anal plate is almost circular and bears only pre- and post-anal setae. Posteriorly, the peritrematal plate does not extend beyond the posterior edge of coxae IV.

The tectum (Fig. 349c) has three pointed prongs. The chelicera (Fig. 350a) has a few small teeth on both digits and a small pilus dentilis. The ventral side of the gnathosoma (Fig. 350b) is feebly sclerotized, the deutosternal denticles arranged in distinct lines. The palp tarsus has no macroseta.

All the legs terminate in a pretarsus and claws and no macrosetae arise from tarsi II to IV.

MALE (Fig. 351). Length of idiosoma: about 300 μm.

The ventral surface is almost completely covered by the sterno-genital and ventro-anal shields, the latter bearing six pairs of setae in addition to the anals. The lateral margins of the ventro-anal shield are distinctly indented.

The fixed digit of the chelicera (Fig. 350c) is slender and pointed, with three or four small sharp teeth. The movable digit is broad at its base and has a pointed tip and single tooth near its distal end. The spermadactyl is slightly twisted and extends beyond the end of the fixed digit.

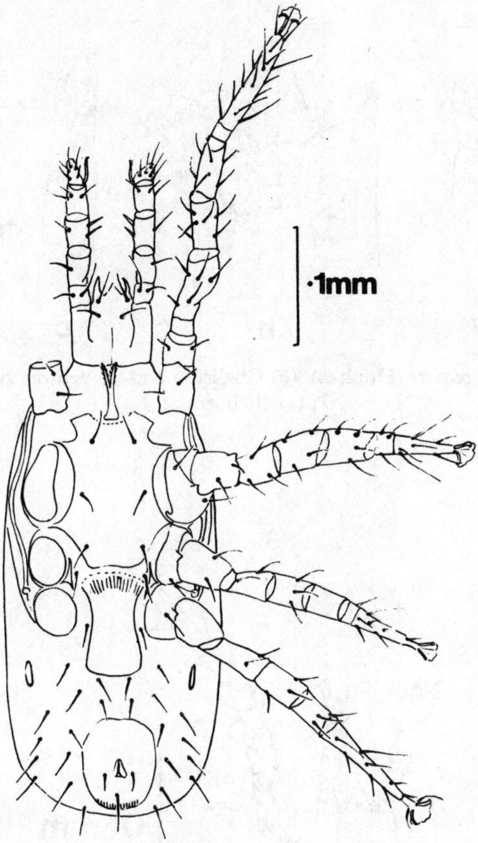

FIG. 348. *Arctoseius butleri* (Hughes). Ventral view of ♀.

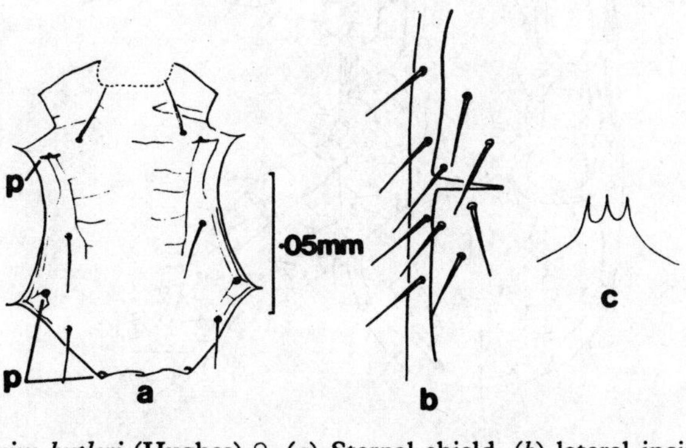

FIG. 349. *Arctoseius butleri* (Hughes) ♀. (*a*) Sternal shield, (*b*) lateral incision on dorsal shield, (*c*) tectum. P pores.

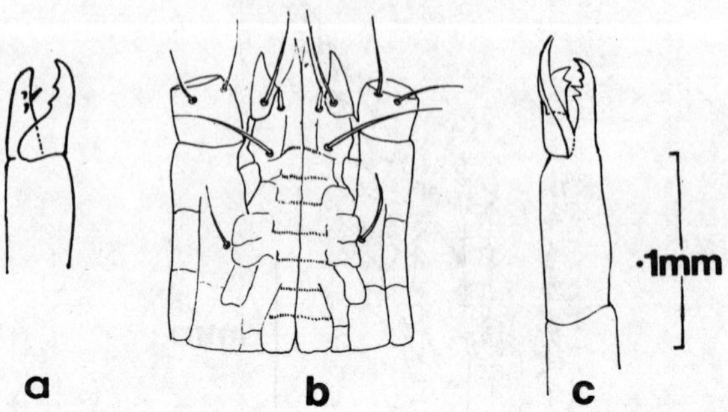

FIG. 350. *Arctoseius butleri* (Hughes). (*a*) Chelicera and (*b*) ventral view of gnathosoma of ♀; (*c*) chelicera of ♂.

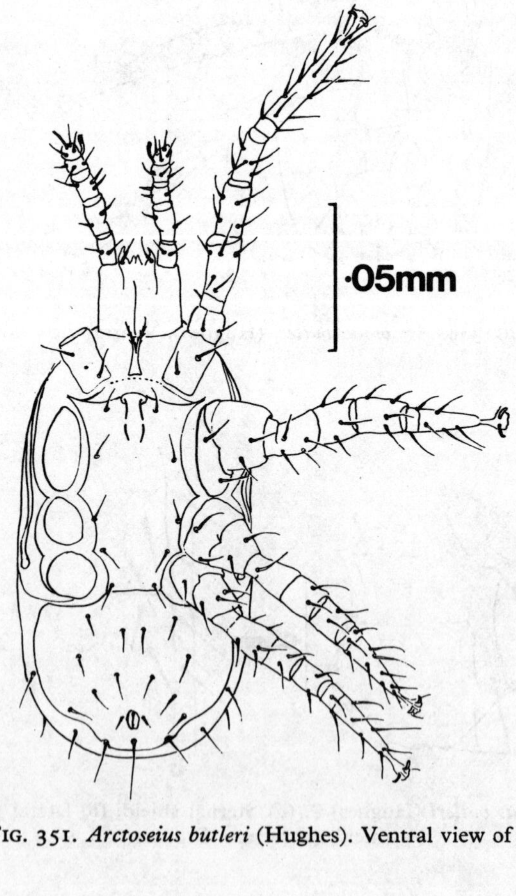

FIG. 351. *Arctoseius butleri* (Hughes). Ventral view of ♂.

BIOLOGY

This is unknown. O'Farrell and Butler (1948) found *A. butleri* preying on *Drosophila* larvae in rotten dried peaches. It has also been found with *Acarus siro* in sawdust near a window on the top of a silo, and J. G. Jeffrey collected it from the base of a wet stack of straw bales.

DISTRIBUTION

Belfast, Northern Ireland; Scotland.

PROCTOLAELAPS Berlese, 1923

The dorsal shield is entire in both sexes. In the female, it bears from 42 to 52 pairs of setae, of which 18 to 20 pairs arise from the posterior region. Three to seven pairs of marginals (*R series*) arise from the posterior region of the shield (Fig. 352) and nought to five submarginals are inserted on the interscutal membrane.

The genital shield usually has a truncate posterior margin. The anal shield usually bears three anal setae and is rarely extended anteriorly. The peritrematal shield in adults is free posteriorly or narrowly attached to the exopodal plate beside coxae IV.

In the male, the dorsal setation is the same as that of the female, except that the posterior marginals and submarginals are missing from the ventro-lateral membrane.

The fixed digit of the chelicera has a small membranous lobe in place of the pilus dentilis. The two digits are about the same length in the female.

All the legs end in pretarsi and claws. Tibia I has six dorsal setae, Tibia III and IV a minimum of eight and ten respectively.

Type species: *Proctolaelaps productus* Berlese, 1923.

PROCTOLAELAPS (adapted from Evans, 1958)
Key to the females

1. Fixed digit of chelicera with a row of closely-set teeth along its entire margin 2

—. Fixed digit of chelicera with, at most, three or four well-spaced teeth. Tectum with anterior margin denticulate. Without platelets between the genital and anal shields. Dorsal setae relatively long....*P. pomorum*, p. 321

2. First pair of hypostomal setae stout, spine-like (Fig. 354a). Corniculi sinuous..........*P. pygmaeus*, p. 317

—. First pair of hypostomal setae normal (Fig. 354b). Corniculi straight. Tectum with anterior margin minutely denticulate......*P. scolyti*, p. 320

PROCTOLAELAPS PYGMAEUS (Müller, 1859)

Gamasus pygmaeus Müller, 1859; *Hypoaspis hypudaei* Oudemans, 1902; *Typhlodromus bulbicolus* Oudemans, 1929; *Garmania bulbicola* (Oudemans, 1929) sensu Nesbitt, 1951; *Proctolaelaps* (*Proctolaelaps*) *hypudaei* (Oudemans, 1902) sensu Hughes, 1961.

In 1958, Evans divided the genus *Proctolaelaps* into two subgenera: *Proctolaelaps* (*P.*) and *P.* (*Neojordensia*). Chant, 1963 considers that *Proctolaelaps* and *Neojordensia* should be considered as separate genera.

FEMALE (Figs. 352, 353). Length of idiosoma: about 430 μm.

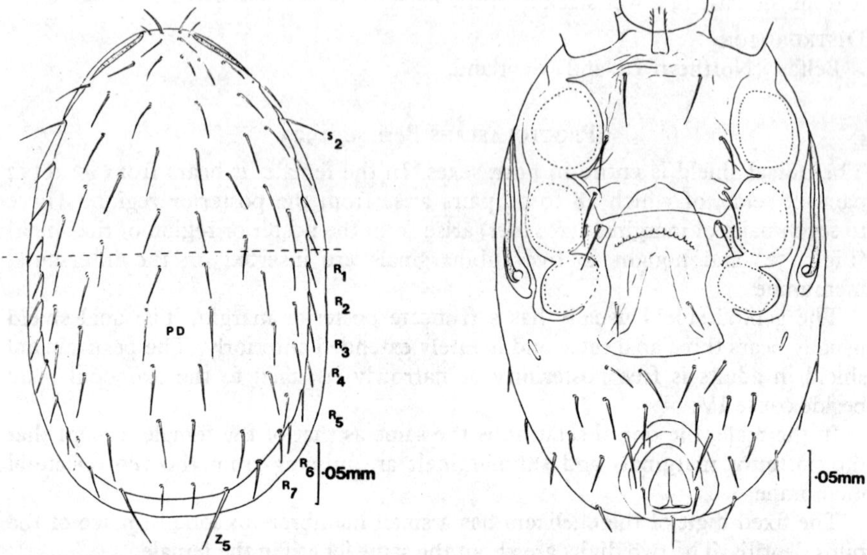

FIG. 352. *Proctolaelaps pygmaeus* (Müller) ♀. Dorsal view of idiosoma (left). PD posterior region of dorsal shield (opisthonotum); R_1 to R_7 marginal series of setae, s_2 lateral seta, Z_5 median seta.

FIG. 353. *Proctolaelaps pygmaeus* (Müller) ♀. Ventral view of idiosoma (right).

The dorsal surface is almost completely covered by a shield with a faint reticulate pattern. From it arise 43 pairs of fairly long setae, those of the *J* series being as long as or slightly longer than the distance between their bases. One pair of setae arising in the humeral region (s_2) and a posterior pair (Z_5) are longer than the others. The marginal (R) setae of the posterior half of the shield arise from the shield and not the interscutal membrane. The anterior margin of the sternal shield is ill-defined and the shield itself bears three pairs of setae and two pairs of pores. The metasternal seta arises from a small plate in association with the third sternal pore. The genital plate is shaped as in Fig. 353 and bears one pair of setae. The anal shield is almost round and the anus unusually large. It bears the three anal setae and nine pairs also arise from the cuticle on either side. The peritrematal and exopodal shields are separate from one another.

The gnathosoma is characterized by having sinuous corniculi and the anterior hypostomal setae are enlarged (Fig. 354a). The fixed digit of the chelicera (Fig. 355a) has an uneven row of fine teeth, at its base is a membranous lobe with a free serrated edge. The tectum is variable in outline.

BIOLOGY

P. pygmaeus is versatile in its feeding habits. It has been reared on fungal cultures (Muma, 1955) and it will also feed on *Tetranychus urticae*, devouring its prey entirely, rather than sucking it dry. (Mathys and Tencalla, 1959). It has been recorded many times from soil, leaf mould, mouldy wheat, barley and

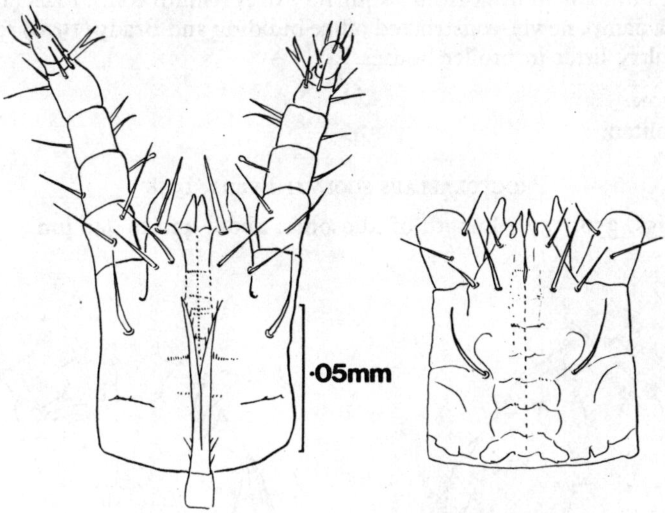

FIG. 354. Ventral view of gnathosoma of (*a*) *Proctolaelaps pygmaeus* (Müller) and (*b*) *P. pomorum* (Oudemans).

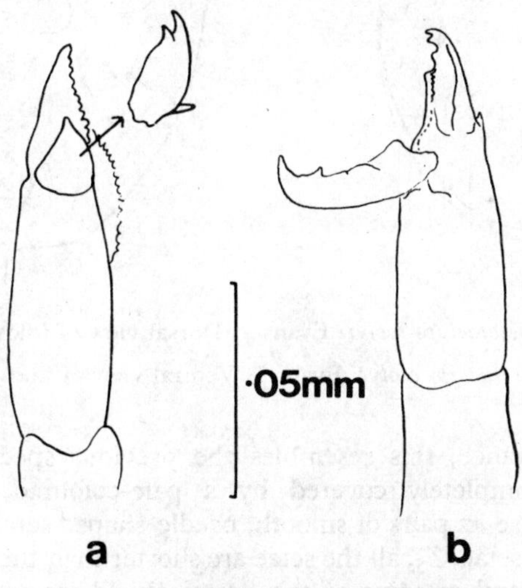

FIG. 355. Chelicera of (*a*) *Proctolaelaps pygmaeus* (Müller) ♀ and (*b*) *P. pomorum* (Oudemans) ♀.

timber, and also from rotting bulbs. Mathys and Tencalla showed that *P. pygmaeus* is common on the leaves of orchard trees and it has also been found on *Citrus* (Ehara, 1964) (where it was associated with the citrus rust mite *Aculus pelekassi* Keifer) and *Protea* flowers (Ryke, 1964). The female lays as many as 25 eggs which hatch in from one to three days, the larval stage lasts two to four days, the nymphal three to eight days (Mathys and Tencalla, 1959).

Evans (1958) records *P. pygmaeus* from the nests of small mammals and I found large numbers in litter from a squirrel's drey (Smith coll.). Rack (1970) has found it in a damp, newly-constructed office building and Brady (1970) recovered it from poultry litter in broiler houses.

DISTRIBUTION
Cosmopolitan.

PROCTOLAELAPS SCOLYTI Evans, 1958

FEMALE (Figs. 356, 357). Length of idiosoma: about 420 to 440 µm.

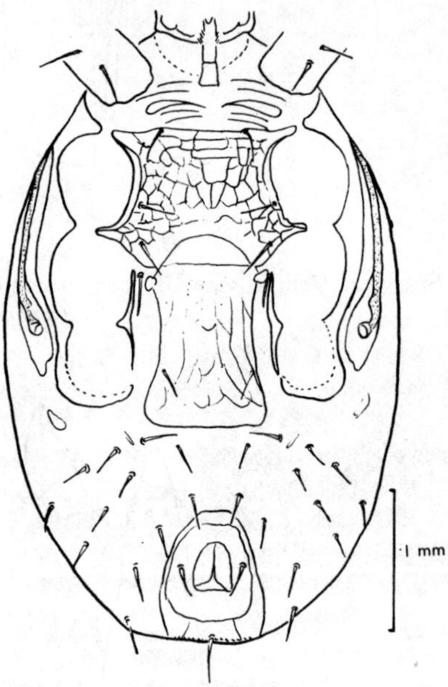

FIG. 356. *Proctolaelaps scolyti* Evans ♀. Dorsal view of idiosoma (left).

FIG. 357. *Proctolaelaps scolyti* Evans ♀. Ventral view of idiosoma (right).

In general appearance, this resembles the previous species in having the idiosoma almost completely covered by a pale-coloured, reticulate dorsal shield. From this arise 42 pairs of smooth, needle-shaped setae. With the exception of the posterior setae, Z_5, all the setae are shorter than the distances between their bases. The lateral margins of the dorsal shield are not so conspicuously serrated as in *P. pygmaeus*. The sternal and genital shields are distinctly reticulate. The metasternal setae do not arise from small plates, but from the interscutal membrane.

On the gnathosoma the corniculi are straight and convergent, the first pair of hypostomal setae not unusually enlarged. The fixed digit of the chelicera has an evenly serrated edge as in *P. pygmaeus* and the movable is bidentate. The pilus dentilis is replaced by a transparent membrane with a serrated margin which arises from the base of the digits.

BIOLOGY

P. scolyti was found by Evans (1958) on *Scolytus* sp. and the chateid *Entodon* sp. It has also been found on unsound onions imported from Egypt (M. D. Stevens coll.), on home-grown hay (H. P. Smith coll.) and in the deep litter of a broiler house in North Yorkshire where it was associated with *Caloglyphus berlesei* (M. Connell coll.).

DISTRIBUTION

England.

PROCTOLAELAPS POMORUM (Oudemans, 1929)

Typhlodromus pomorum Oudemans, 1929; *Garmania pomorum* (Oudemans, 1929) sensu Nesbitt, 1951; *Proctolaelaps* (*P.*) *pomorum* (Oudemans, 1929) sensu Hughes, 1961.

FEMALE (Figs. 358, 359). Length of idiosoma: about 420 µm.

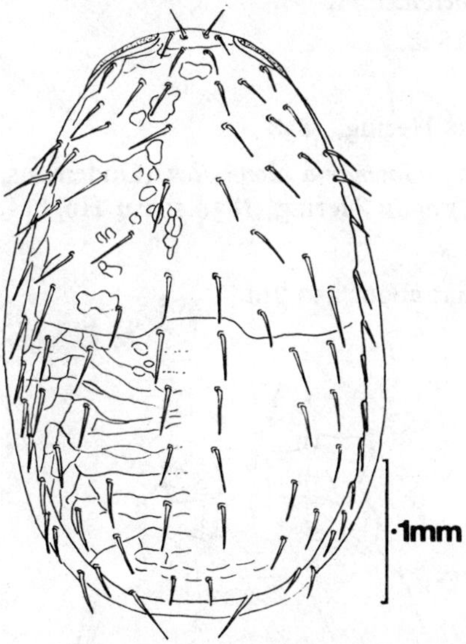

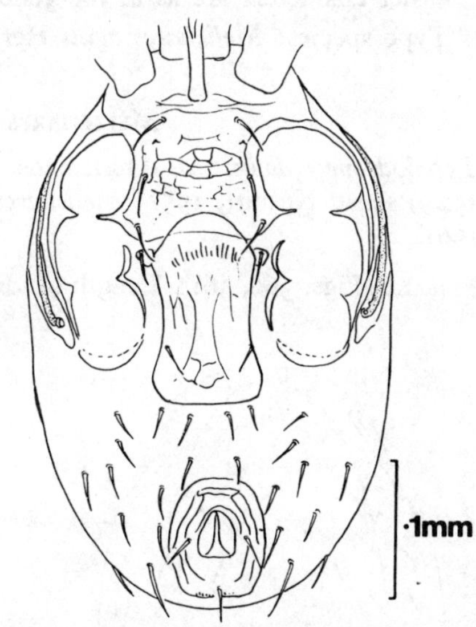

FIG. 358. *Proctolaelaps pomorum* (Oudemans) ♀. Dorsal view of idiosoma (left).

FIG. 359. *Proctolaelaps pomorum* (Oudemans) ♀. Ventral view of idiosoma (right).

The dorsal shield is divided by a transverse line between coxae II and III. Anterior to this line, the dorsal shield has a number of rugose patches; posterior to it, the shield has a reticulate pattern. From the dorsal shield arise 41 pairs of smooth setae.

The gnathosoma (Fig. 354b) differs from that of *P. pygmaeus* in having straight corniculi and the first pair of hypostomal setae is not enlarged. The chelicera (Fig. 355b) has from two to three teeth on the fixed digit and one distinct tooth on the movable. The tectum is rounded, with a denticulate edge.

MALE

This is described and figured by Evans, 1958.

BIOLOGY

Oudemans (1929) found this species on the fruit and bark of an apple tree. Evans (1958) records it from a house in Lancashire where it was associated with

Acarus siro and *P. pygmaeus*. It has also been found on home-grown oats with *A. siro* (J. Thompson coll.) and on bales of mouldy Algerian esparto grass heavily infested with *A. siro* and *Tyrophagus* sp. (K.G. Smith coll.).

DISTRIBUTION

England, Holland, Australia.

MELICHARES Hering, 1838

Garmania Nesbitt, 1951.

In the female, the dorsal shield bears from 33 to 49 pairs of setae. The posterior region of the shield bears from 10 to 15 pairs of setae, marginal setae of the *R* series arising from the interscutal membrane (Fig. 360). In the male, the dorsal setation is the same as that of the female, but additional marginal setae may arise from the edge of the dorsal shield.

Other characters are as in the genus *Proctolaelaps*.

Type species: *Melichares agilis* Hering, 1838.

MELICHARES AGILIS Hering, 1838

Typhlodromus domesticus Oudemans, 1929; *Garmania domestica* (Oudemans, 1929) sensu Nesbitt, 1951; *Melichares* (*M.*) *agilis* Hering, 1838 sensu Hughes, 1961.

FEMALE (Figs. 360, 361). Length of idiosoma: about 450 µm.

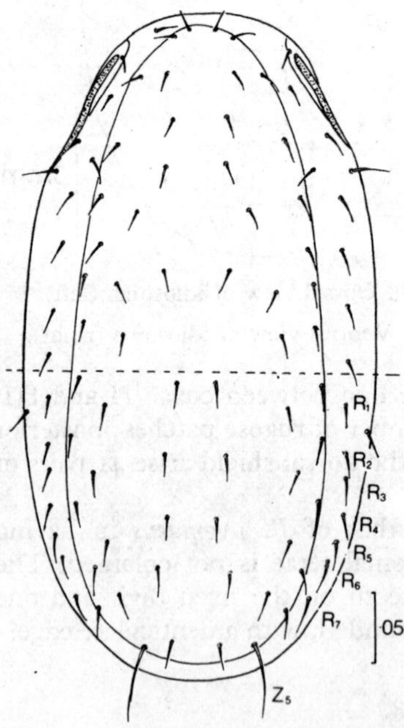

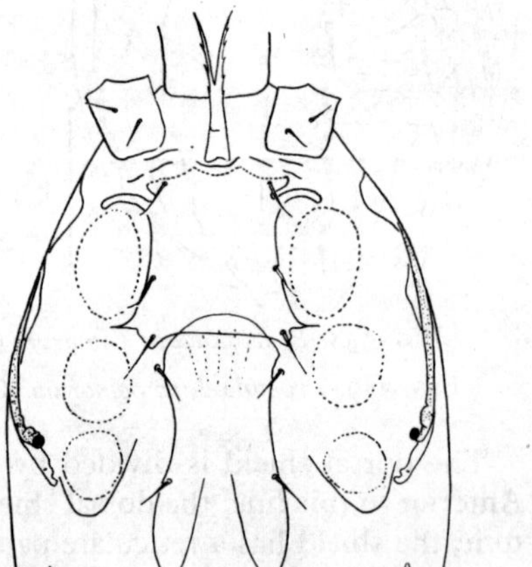

FIG. 360. *Melichares agilis* Hering. ♀. Dorsal view of idiosoma (left). R_1 to R_7 marginal setae. Z_5 lateral seta.

FIG. 361. *Melichares agilis* Hering ♀. Ventral view of idiosoma (right).

The dorsal shield is weakly sclerotized, pale yellow in colour. It has a reticulate pattern and is narrow, so that a certain amount of interscutal membrane is visible from the dorsal aspect. From the shield arise 29 pairs of needle-shaped setae, 15 pairs being inserted on the posterior part of the shield. The terminal pair (Z_5) is longer than the others. The marginal (R_1 to R_7) setae arise from the interscutal membrane on either side of the shield. The sternal shield has an ill-defined anterior margin, three pairs of setae and two pairs of pores; metasternal setae arise from the interscutal membrane in association with the third pair of pores. The genital shield is flask-shaped, with one pair of setae. The anal shield is well-removed from the genital and bears the three anal setae only. About 10 pairs of setae arise from the cuticle on either side of genital and anal shields. The peritreme is looped near its posterior end, and anteriorly it extends on to the dorsal surface as far as the posterior margin of coxae I. The peritrematal shield does not fuse with the exopodal in the region of coxa IV.

The chelicerae are dentate, the fixed digit with four or five small teeth, the movable with one. The corniculi are slender and convergent, the first pair of hypostomal setae of normal size. The tectum is smooth and almost triangular in shape.

All the legs have pretarsi and claws. No macrosetae arise from tarsus IV.

MALE (Fig. 362). Length of idiosoma: about 415 µm.

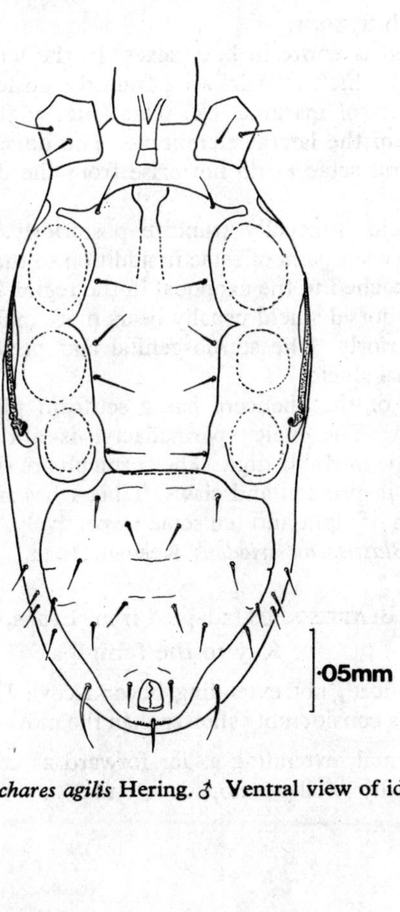

FIG. 362. *Melichares agilis* Hering. ♂ Ventral view of idiosoma.

The chaetotaxy of the dorsal shield is the same as that of the female.

The sterno-genital shield extends as far back as the posterior margin of coxae IV; it bears five pairs of setae. The ventro-anal shield is shaped as in Fig. 362, with six pairs of setae in addition to the three anals. The peritreme is more folded at its posterior end than in the female.

The chelicerae have toothed digits and the spermadactyl extends beyond their distal ends.

BIOLOGY

The life cycle takes from seven to nine days at 23°C and a relative humidity of 87 per cent, with *Acarus siro* living on wheat germ, as food. The eggs are unusually large, and only one is developed in the body of the female at the same time. The larvae feed on the larvae and eggs of *A. siro*, the larvae being sucked dry of their contents. The nymphs and adults will feed on the larvae and nymphs of *A. siro*, but prefer the eggs. These are not sucked dry, but the *Melichares* pierces the egg shell, takes a little of its contents and then passes to the next. I have never seen the adult stages of *A. siro* attacked by the *Melichares*.

M. agilis is commonly found associated with dried fruit infested with *Carpoglyphus lactis*. Oudemans, 1929 (quoted from Nesbitt, 1951) found it in house dust, on smoked beef and on hogs' bladders.

DISTRIBUTION

England, Germany, Holland.

BLATTISOCIUS Keegan, 1944

Paragarmania Nesbitt, 1951.

The dorsal shield is entire in both sexes. In the female, it bears from 32 to 36 pairs of setae, of which 15 pairs arise from the posterior region of the shield. Nine to twelve pairs of marginal (*R*) setae (Fig. 363) and five to eleven submarginals arise from the lateral membrane. The dorsal setae of the adults are simple, and humeral setae r_3 do not arise from the dorsal shield and are not differentiated.

The genital shield is usually truncate posteriorly. The ventro-anal shield bears from two to seven pairs of setae in addition to the anals. The peritrematal shield is usually attached to the exopodal in the region of coxa IV.

In the male, the dorsal shield usually bears more marginal (*r*) setae anteriorly, and *R* setae posteriorly. The sterno-genital and peritrematal shields are free from the ventro-anal shield.

The fixed digit of the chelicera has a setiform pilus dentilis and may be reduced in length. The male spermadactyl is abruptly angled downward beyond the tip of the movable digit. The corniculi are slender and convergent.

All the legs end in pretarsi and claws. Tibia I has six dorsal setae, Tibia III and IV a minimum of eight and ten setae respectively.

Type species: *Blattisocius triodons* Keegan, 1944.

BLATTISOCIUS (adapted from Evans, 1958)

Key to the females

1. Peritreme reduced, not extending beyond coxa II (Fig. 364). Fixed digit of the chelicera considerably shorter than the movable (Fig. 366a)........2

—. Peritreme normal, extending as far forward as coxa I. Both digits of the chelicera equally well-developed..................................3

2. Peritreme reaching to about the posterior margin of coxa II (Fig. 364). Fixed digit of the chelicera very short, without teeth. Pilus dentilis relatively long...*B. tarsalis*, p. 325

—. Peritreme short, barely reaching the middle of coxa III (Fig. 368). Fixed digit of chelicera with minute teeth and about two-thirds the length of the movable digit.....................................*B. keegani*, p. 328

3. Tarsus IV with a long whip-like macroseta. Sternal setae III and IV (metasternal) on platelets connected to the sternal shield by the endopodal shield (Fig. 369).................................*B. dentriticus*, p. 329

—. Tarsus IV without macroseta. Sternal shield with three pairs of setae. Metasternal setae free (Fig. 373)........................*B. mali*, p. 331

BLATTISOCIUS TARSALIS (Berlese, 1918)

Typhlodromus tineivorus Oudemans, 1929; *T. tineivorans* Oudemans, 1929 sensu Hughes, 1948; *Blattisocius triodons* Keegan, 1944; *Melichares* (*B.*) *tarsalis* (Berlese, 1918) sensu Hughes, 1961.

FEMALE (Figs. 363, 364). Length of idiosoma: 520 to 600 μm.

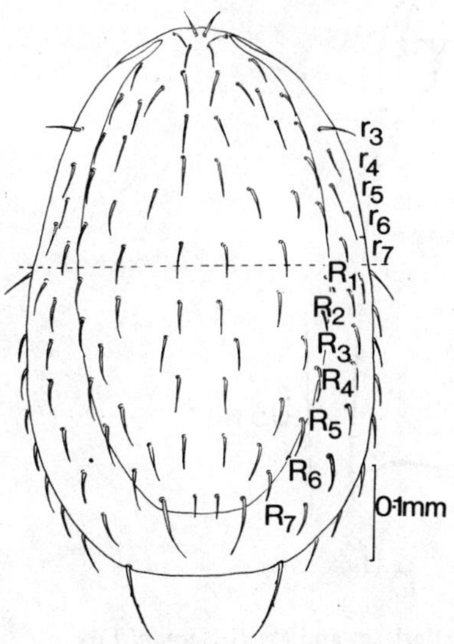

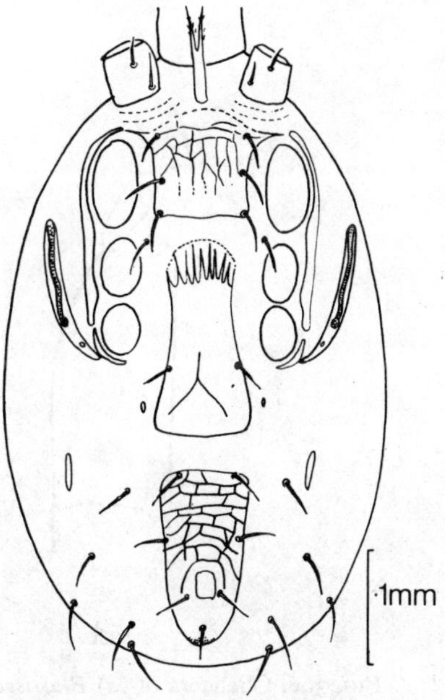

FIG. 363. *Blattisocius tarsalis* (Berlese) ♀. Dorsal view of idiosoma (left). r_3 to r_7, R_1 to R_7 marginal setae of idiosoma.

FIG. 364. *Blattisocius tarsalis* (Berlese). Ventral view of ♀ (right).

The body is pale yellow, the tarsi of the legs much darker in colour. The dorsal shield is faintly reticulate and partially covers the dorsal surface, it bears 33 pairs of slightly pectinate setae; of these, 15 pairs arise from the posterior part of the shield. Five pairs of marginal setae (r_3 to r_7) are inserted on the membrane on either side of the anterior part of the dorsal shield, and seven pairs (R_1 to R_7) occupy a similar position in the posterior region. The sternal shield is faintly reticulate and bears three pairs of setae. Metasternal setae arise from the interscutal membrane. The genital shield has one pair of setae. The anal shield is oblong in shape, with three pairs of setae in addition to the anals. The peritreme is short, extending as far forward as the posterior margin of coxa II. Posteriorly, the peritrematal plate fuses with the exopodal. The spermatheca is globular with a sinuous duct opening between coxae III and IV.

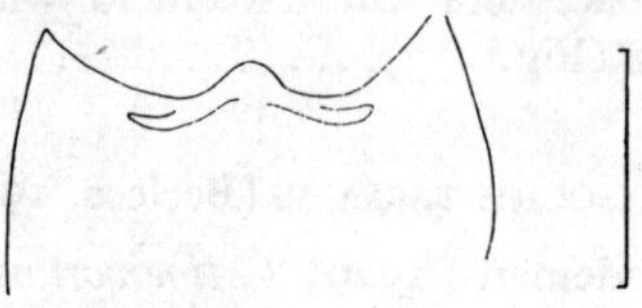

FIG. 365. *Blattisocius tarsalis* (Berlese). Tectum of ♀.

The tectum is smoothly curved, shaped as in Fig. 365. The corniculi are slender and convergent. The fixed digit of the chelicera (Fig. 366a) is without

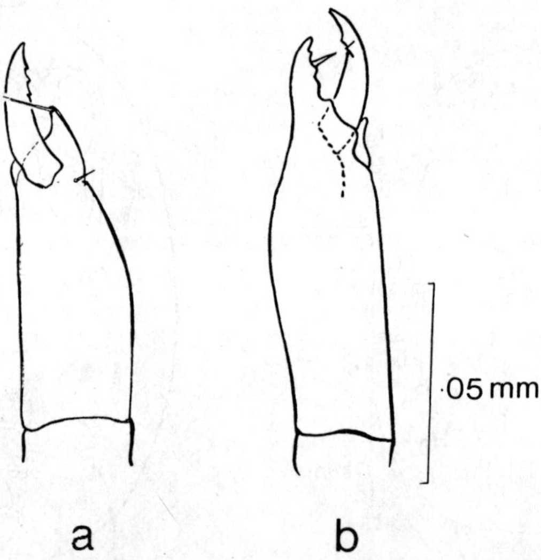

FIG. 366. Chelicera of (*a*) *Blattisocius tarsalis* (Berlese) and (*b*) *B. keegani* Fox.

teeth, much shorter than the movable digit, it has a terminal pilus dentilis. The movable digit has three small teeth.

MALE (Fig. 367). Length of idiosoma: about 450 µm.

This is similar to the female. The ventral surface is covered by two shields. The ventro-anal covers the greater part of the hysterosoma and bears six pairs of pre-anal setae. An additional 'peritreme' is present only in the male external to the normal peritreme (see Fig. 367). The fixed digit of the chelicera is longer

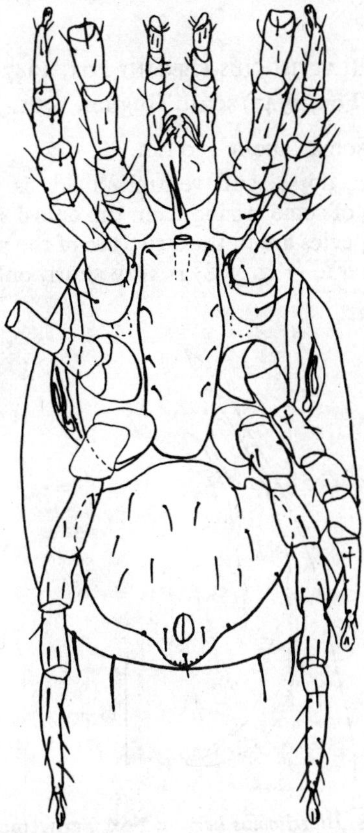

FIG. 367. *Blattisocius tarsalis* (Berlese). Ventral view of ♂. (After Oudemans).

than that of the female, but still shorter than the movable. The spermadactyl is bent at its distal end.

BIOLOGY

B. tarsalis is frequently found in stored grain infested with corn moths belonging to the genera *Anagasta*, *Plodia* and *Sitotroga*. The mites lay their eggs on the silken webbing and larvae and nymphs appear to feed on the moth eggs and caterpillars. The females are carried by the moths hanging on to the body, particularly at the base of the wings. It has also been found on cultures of *Tribolium confusum*, *T. castaneum*, *Attagenus piceus* and *Tineola bisselliella* (Nesbitt, 1951) and on Spanish muscatel raisins infested with *Carpoglyphus lactis* (A. M. Shenker coll.). Flanders and Badgeley (1963) (quoted from Barker 1967b) showed that *B. tarsalis* can only control *Anagasta kuehniella* under very special conditions, e.g., the depth of grain in which the moths lived must not exceed 8 mm; deviation from these conditions often eliminated the predators.

Treat (1969b) has frequently found *B. tarsalis* on the noctuid moth *Epizeuxis aemula* (Hübner) which feeds on the dead leaves of the forest floor. This does not commonly infest stored grain, but is on occasion, injurious to corn fodder in Mississippi (Crumb, 1934). The moths are probably used only as a vehicle.

DISTRIBUTION

England, Italy, Switzerland, Holland, North Africa, U.S.A., Hawaii, Australia, Israel.

BLATTISOCIUS KEEGANI Fox, 1947

Melichares (*B.*) *keegani* (Fox, 1947) sensu Hughes, 1961.

FEMALE. Length of idiosoma: about 470 μm.

The chaetotaxy of the dorsal and ventral shields is very similar to that of *B. tarsalis*, with 33 pairs of setae arising from the dorsal shield. The main differences between the two species are in the structure of the peritreme and chelicera. In *B. keegani*, the peritreme (Fig. 368) is very short, only extending as far for-

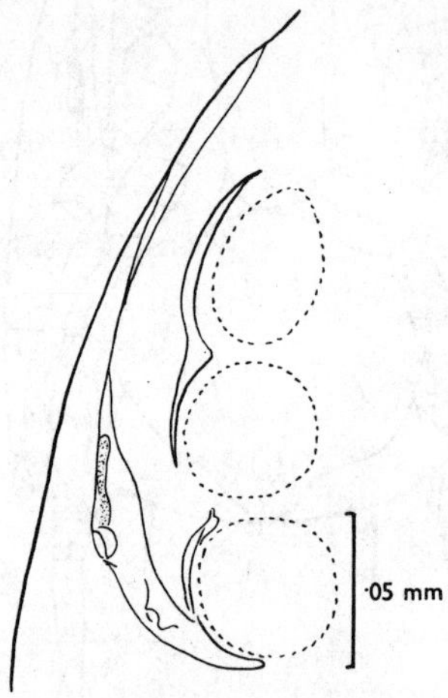

FIG. 368. *Blattisocius keegani* Fox. Peritreme of ♀.

ward as the middle of coxa III. The digits of the chelicera vary in length, but the fixed digit is proportionally longer than in *tarsalis* and the pilus dentilis is shorter (Fig. 366b). The movable digit has one or two teeth.

BIOLOGY

Adults, protonymphs and deutonymphs of *B. keegani* will feed readily on the eggs of a number of different grain beetles, e.g., *Cryptolestes turcicus* (Grouvelle), *C. ferrugineus* (Stephens), *Tribolium confusum* (J. duVal), *T. castaneum* Hbst. *Trogoderma parabile* Beal and *Oryzaephilus surinamensis* L., also on eggs of *Glycyphagus domesticus* and *Aëroglyphus robustus*. All stages of *Tyrophagus putrescentiae* were rejected. With an unlimited supply of food, a temperature of 26.7°C and relative humidity of 70 to 75 per cent, the life cycle can be completed in an average of 4.1 days (Barker, 1967b). He also found that three *Cryptolestes turcicus* eggs were all that was required to rear one *B. keegani* to the adult stage. When pairing occurs, the female carries the male clinging beneath its body (personal observations).

B. keegani is more versatile in habitat than *B. tarsalis*. Fox (1947) first recorded it from *Rattus norvegicus* in Puerto Rico and M. M. Day collected it from baby rats found on a ship coming from S. America. It has also been found in a sparrow's nest in Stirling, Scotland (G. Hosie coll.), on machinery in a flour mill, Dakar, Senegal, with *Suidasia nesbitti* (J. Freeman coll.), on nutmegs loaded at Penang (A. M. Shenker coll.), on the outside of a jar containing a culture of *Lasioderma serricorne* (K. M. Easton coll.) and from Plate maize and wheat infested with insects. Muma (1955) records it from *Citrus* trees in Florida and Chmielewski (1969) sieved it from sugar beet seed in Poland.

DISTRIBUTION

England, Poland, Mexico, S. America, U.S.A., Hawaii, Malaya, W. Africa, Israel.

BLATTISOCIUS DENTRITICUS (Berlese, 1918)

Lasioseius (*L.*) *dentriticus* Berlese, 1918; *Seiulus amboinensis* Oudemans, 1925; *Garmania* (*Paragarmania*) *amboinensis* (Oudemans, 1925) sensu Nesbitt, 1951; *Melichares* (*Blattisocius*) *dentriticus* (Berlese, 1918) sensu Hughes, 1961.

FEMALE (Figs. 369, 370). Length of idiosoma: about 470 μm.

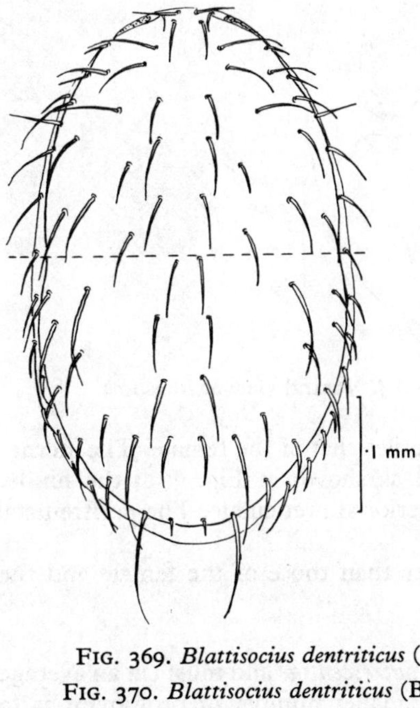

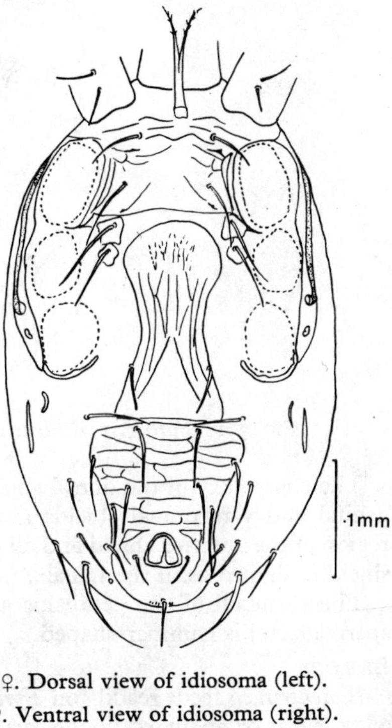

FIG. 369. *Blattisocius dentriticus* (Berlese) ♀. Dorsal view of idiosoma (left).
FIG. 370. *Blattisocius dentriticus* (Berlese) ♀. Ventral view of idiosoma (right).

The idiosoma is pale yellow, the dorsal surface almost completely covered with a faintly reticulate shield. From this arise 36 pairs of setae, 15 pairs arising from the posterior region. Except for the long pair arising from the posterior end of the dorsal shield, these setae are smooth and curved. Seven pairs of marginal setae are inserted into the interscutal membrane external to the posterior half of the shield. The anterior half of the sternal shield is faintly reticulate and bears two pairs of setae, the third pair, as well as the metasternal setae, arising from separate shields. These shields are joined to the sternal shield and to the endopodal which partially encircles the hind margins of coxae II. The genital shield, as well as the ventro-anal, is faintly reticulate, the ventro-anal shaped as in Fig. 370 and bearing four pairs of pre-anal setae. Three pairs of setae and a longer pair arise from the interscutal membrane external to the ventro-anal shield. The peritrematal plate is extended posteriorly to touch the exopodal encircling the posterior edge of coxa IV. Two narrow metapodal plates are present.

The tectum has a smoothly rounded edge and the corniculi are slender and convergent. Both limbs of the chelicerae are dentate and they are about the same length. A long macroseta arises from the base of tarsus IV.

MALE (Fig. 371). Length of idiosoma: about 370 μm.

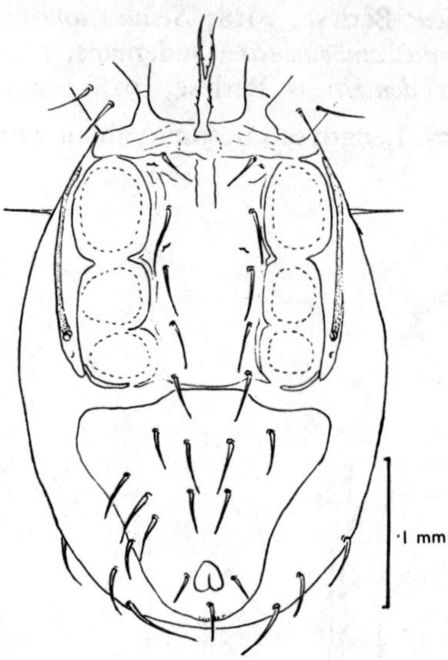

FIG. 371. *Blattisocius dentriticus* (Berlese) ♂. Ventral view of idiosoma.

The chaetotaxy of the dorsal shield resembles that of the female. The sterno-genital and ventro-anal shields are shaped as shown in Fig. 371; the hinder region of the anterior shield and all the posterior are reticulate. The peritrematal shield is shaped as in the female.

The corniculi of the pedipalps are longer than those of the female and the spermadactyl is hammer-shaped.

BIOLOGY

B. dentriticus feeds readily on *Tyrophagus putrescentiae* and must eat an average of 70 eggs, or about 50 young larvae or a smaller number of protonymphs to complete its development. At a temperature of 22.2°C and a relative humidity of 80 per cent, development of the male and of the female was completed in 9 and 10 days respectively, when *T. putrescentiae* larvae were supplied as food. Only the mated females laid eggs. It cannot develop to the adult stage when mould is the only source of food (Rivard, 1960).

B. dentriticus is frequently found on imported food material in the U.S.A. (Chant, 1963) and in this country, it is not uncommon on the surface of wheat and flour in association with acarid mites. It has also been recorded from the thorax of *Caradrina morpheus* (Evans, 1958), on mouldy coconuts imported from Ceylon (E. Powell coll.) and from sprouting potatoes (Halbert, 1923).

DISTRIBUTION

England, Italy, Eire, Holland, Amboina, U.S.A., Hawaii, Sri Lanka, Japan, Israel.

BLATTISOCIUS MALI (Oudemans, 1929)

Typhlodromus mali Oudemans, 1929; *Garmania* (*Paragarmania*) *mali* (Oudemans, 1929) sensu Nesbitt, 1951; *Melichares* (*B.*) *mali* (Oudemans, 1929) sensu Hughes, 1961.

FEMALE (Figs. 372, 373). Length of idiosoma: about 520 µm.

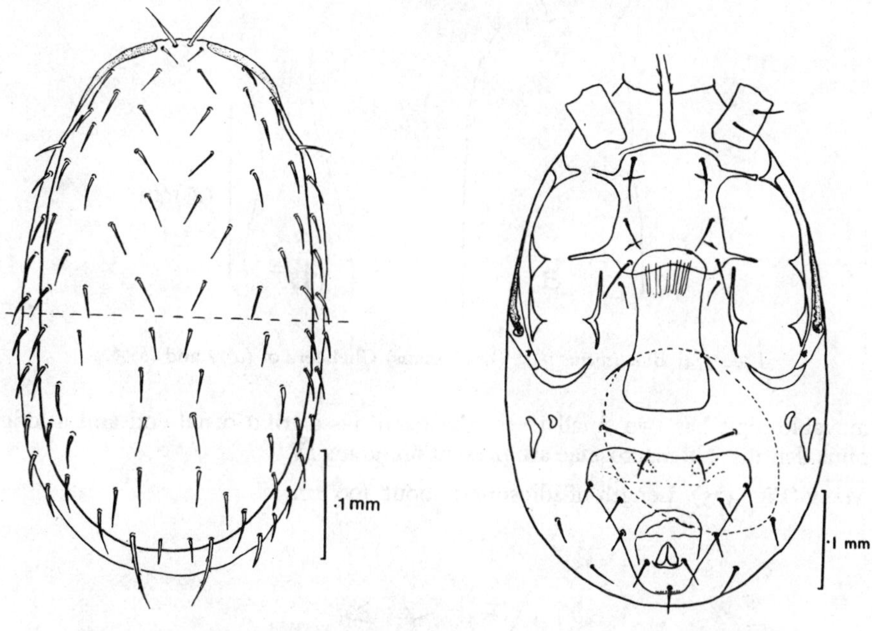

FIG. 372. *Blattisocius mali* (Oudemans) ♀. Dorsal view of idiosoma (left).
FIG. 373. *Blattisocius mali* (Oudemans) ♀. Ventral view of idiosoma. (right)

The dorsal shield is yellow-brown in colour, more heavily tanned in this species than in others; it nearly covers the idiosoma and, except for the central region, is faintly and evenly reticulate. From the dorsal shield arise 34 pairs of smooth setae, the apical j_1 and terminal Z_5 being longer than the remaining setae. The ventral shields are almost smooth, the sternal shield bearing three pairs of setae and conspicuous pores, the genital, one pair of setae. The metasternal setae arise from the interscutal membrane and no plates are present. The ventro-anal shield is shaped as in Fig. 373, with four pairs of setae in addition to the three anals. In some specimens, the anterior margin of the ventro-anal shield is excised, as shown by dotted lines. Specimens with ventro-anal shields having a continuous or excised anterior margin were found in the same population. Two pairs of metapodal plates are present. The peritreme is well-developed and almost completely covers the stigma. The peritrematal shield joins the exopodal in the region of coxae IV.

The tectum has a rounded margin and the corniculi are long and convergent (Fig. 378b). Each chelicera (Fig. 374a) is large and unusually well-tanned; the

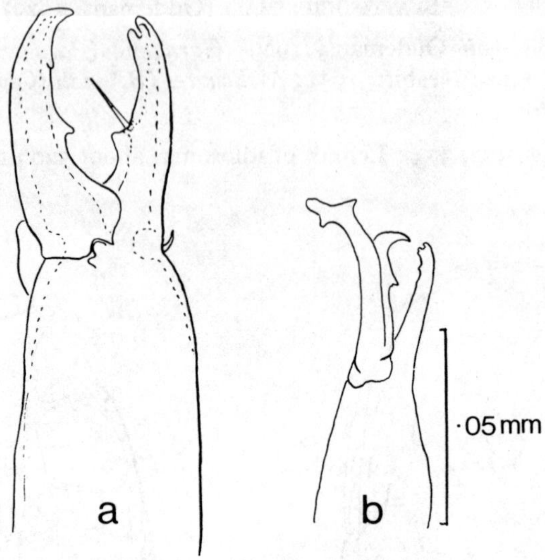

FIG. 374. *Blattisocius mali* (Oudemans) Chelicera of (*a*) ♀ and (*b*) ♂.

movable digit has two small teeth, the fixed has a trifid distal end and a long pilus dentilis. No macrosetae are present on tarsus IV.

MALE (Fig. 375). Length of idiosoma: about 400 μm.

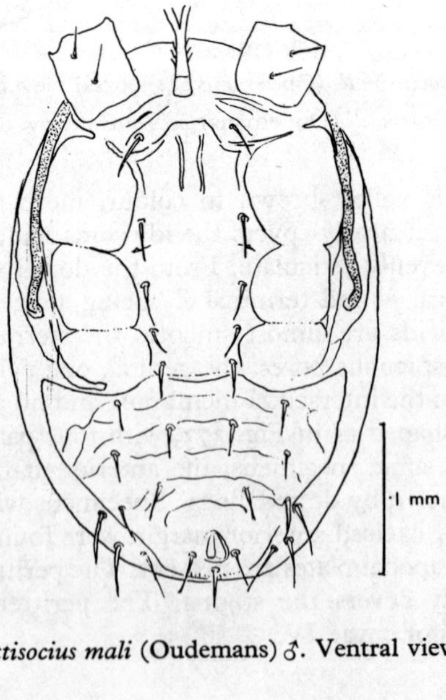

FIG. 375. *Blattisocius mali* (Oudemans) ♂. Ventral view of idiosoma.

The chaetotaxy of the dorsal surface resembles that of the female. Almost the whole of the ventral surface is covered by the sterno-genital and ventro-anal shields, the latter extended laterally to cover the ventral surface posterior to coxae IV. The ventro-anal shield is covered with a faint reticulate pattern and bears five pairs of setae in addition to the anals.

The corniculi are stouter than those of the female and set close together. The chelicera (Fig. 374b) has a well-developed spermadactyl with a recurved distal extremity.

Biology

B. mali is probably a predator of acarid mites and their eggs. It has been found on dried apricots with *Carpoglyphus lactis* (Mrs. Aitkin coll.), on South African stored raisins with *C. lactis* (H. C. Crozier coll.), on Burmese rice with *Tyrophagus putrescentiae* (J. Jocelyn coll.), on mouldy grain with *Acarus siro* (J. D. Norris coll.) and in a flour mill with *Tyrophagus* sp. (J. Thompson coll.). Türk and Türk (1952) found it on dried currants with Acaridae. Oudemans' specimens were found on the bark of *Pyrus malus* and in holes in an apple. Under the name of *Melichares* (*B.*) *daci*, Narayanan and Ghai (1961) record it from a culture of fruit flies *Dacus cucurbitae* and *D. dorsalis* bred in New Delhi, India.

Distribution

England, Holland, India.

Lasioseius Berlese, 1916

The dorsal shield is entire in both sexes. In the female it bears from 22 to 38 pairs of setae, of which 10 to 15 pairs arise from the posterior region of the shield. Three to nine pairs of marginals arise from the lateral membrane and nought to six pairs arise from the ventro-lateral membrane of the posterior region. Some of the dorsal setae may be tricarinate (Fig. 376). The humeral seta r_3 always arises from the dorsal shield.

The genital shield is truncate posteriorly. The ventro-anal shield bears from two to six pairs of setae in addition to the anals. The peritrematal shield is attached to the exopodal in the region of coxae IV.

In the male, the dorsal shield usually bears all the marginal setae of the anterior region on the edge of the dorsal shield; one or two of the marginals (*R*) in the posterior region may also arise from the shield. The total number of posterior (*R*) marginals is fewer than in the female. The sterno-genital and peritrematal shields are nearly always free from the ventro-anal shield.

The fixed digit of the chelicera may have a few teeth, more often it is multidentate. The movable digit is usually tridentate. In the male, the spermadactyl lacks a subapical pointed projection on its ventral surface. The corniculi are well separated from one another and parallel (Fig. 378a).

All the legs end in pretarsi and claws. Tibia I bears six dorsal setae. Tibia III and IV have a minimum of eight and ten setae respectively.

Type species: *Seius muricatus* Berlese, 1887.

LASIOSEIUS PENICILLIGER Berlese, 1916

FEMALE (Figs. 376, 377). Length of idiosoma: about 550 µm.

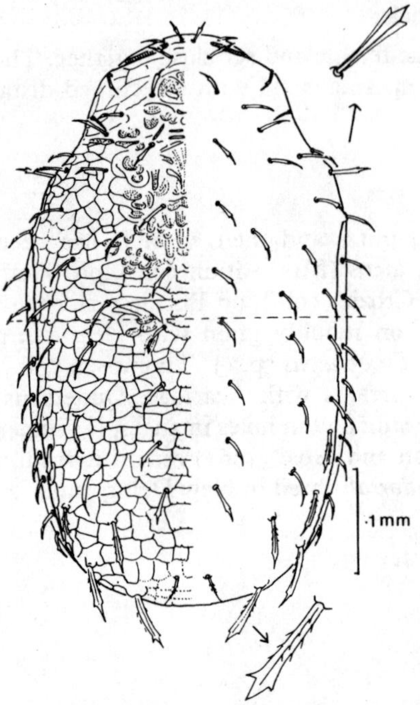

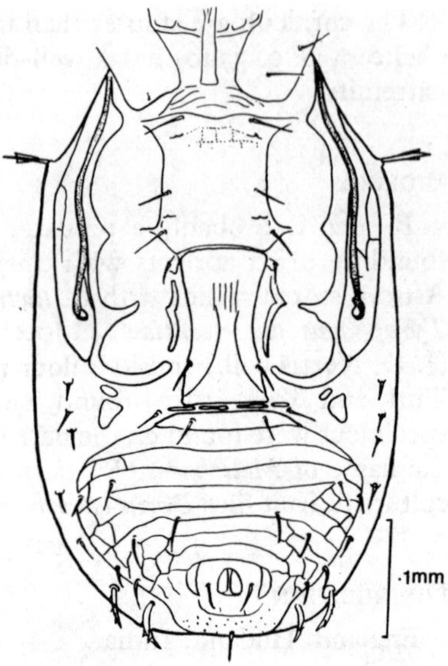

FIG. 376. *Lasioseius penicilliger* Berlese ♀. Dorsal view of idiosoma (left).
FIG. 377. *Lasioseius penicilliger* Berlese ♀. Ventral view of idiosoma (right).

The idiosoma is almost completely covered by a well-tanned dorsal shield. The surface of this shield has a reticulate pattern and from it arise 36 pairs of setae, 15 pairs from the posterior region of the shield. Setae belonging to the *R* series are inserted into the lateral interscutal membrane. The setae are tricarinate (Fig. 376) and the more posterior ones have pectinate edges. The humeral seta r_3 arises from the margin of the dorsal shield.

The sternal and genital shields are paler in colour and almost smooth. The sternal shield bears three pairs of setae, the genital, one pair. The metasternal setae arise from small plates and are associated with the third sternal pore. The ventro-anal shield is very large and well-sclerotized. It is covered with a reticulate pattern and has six pairs of pre-anal setae. Four narrow platelets lie in the gap between the straight margins of the genital and ventro-anal shields. Two pairs of conspicuous metapodal plates are present. Seven pairs of setae are inserted into small projections of the interscutal membrane on either side of the ventro-anal shield, the posterior two pairs are pectinate. The peritrematal shield is large and well-developed, fused with the exopodal in the region of coxa IV.

The gnathosoma (Fig. 378a) is well-sclerotized, the corniculi large and parallel with one another. The deutosternum has seven row of denticles. The tectum has a denticulate margin. As in other species of *Lasioseius*, the fixed digit of the chelicera is multidentate, the movable has three teeth.

The male is unknown.

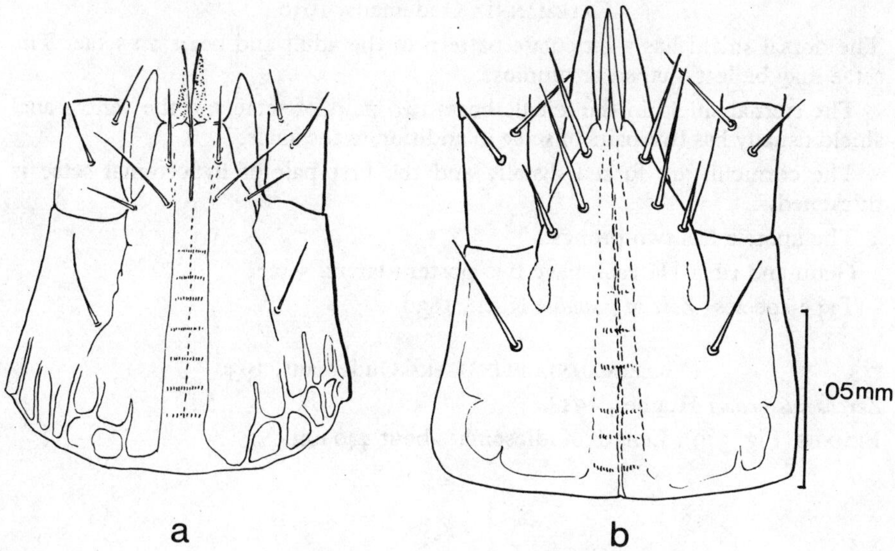

FIG. 378. Ventral view of gnathosoma of (*a*) *Lasioseius penicilliger* Berlese ♀ and (*b*) *Battisoclius mali* (Oudemans) ♀.

BIOLOGY

Sheals (1956) found this species in old grassland in Glasgow and Rack (1962) recovered it from a mole's nest. It has also been collected from sample bags in a Liverpool office (L. A. Walton coll.) and from damp barley in Dumfriesshire (J. M. Baynes coll.). Several hundreds were also found clustered together on the underside of the inspection lids of some machinery in a mill (D. Burgess coll.); no other mites and few insects were present and it is possible that *L. penicilliger* was feeding on moulds.

DISTRIBUTION

England, Scotland, Germany, U.S.A.

AMEROSEIIDAE (Evans, 1963)

The dorsal shield is entire in both sexes and usually bears 29 pairs of setae. The marginal (*R*) setae of the posterior region and also J_1 and J_5 are missing.

The sternal shield of the female bears two or three pairs of setae. Metasternal setae always arise from the interscutal membrane. The genital shield is usually wedge-shaped and always bears genital setae. An anal or ventro-anal shield may be present. Podal spermathecae are present.

The male possesses a sterno-genital and ventro-anal shield. The genital opening is presternal.

The corniculi are conspicuous and usually forked distally.

The pedipalpal apotele is usually two, rarely three-pronged.

The family Ameroseiidae contains five genera (Evans, 1963) and species belonging to the genus *Kleemannia* are often found in the detritus of warehouse floors and also beneath haystacks.

KLEEMANNIA Oudemans, 1930

The dorsal shield has a reticulate pattern in the adult and bears 29 setae. The setae may be leaf-shaped or plumose.

The sternal shield in the female bears two pairs of setae and the ventro-anal shield usually has two pairs of setae in addition to the anals.

The corniculi are forked distally and the first pair of hypostomal setae is thickened.

The apotele has two prongs.

Genu and tibia III each have two postero-lateral setae.

Type species: *Zercon pavidus* Koch, 1839.

KLEEMANNIA PLUMIGERA Oudemans, 1930

Zercoseius gracei Hughes, 1948.

FEMALE (Fig. 379). Length of idiosoma: about 440 μm.

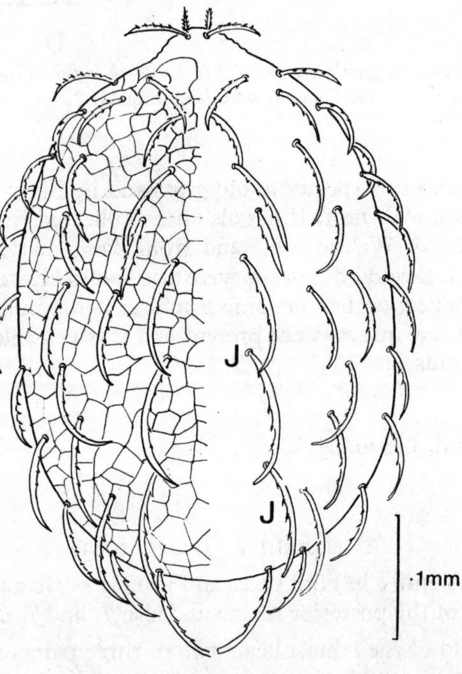

FIG. 379. *Kleemannia plumigera* Oudemans ♀. Dorsal view of idiosoma. Setae of the idiosoma: *J*.

The dorsal surface is completely covered by a single flattened shield which overhangs the gnathosoma. This shield is pinkish-brown in colour and ornamented with a reticulate pattern; from it arise 29 pairs of setae, only two pairs of the *J* series arising from the posterior region of the dorsal shield (Fig. 379). Each seta is sickle-shaped, with two rows of short pectinations (Fig. 380a). The sternal shield (Fig. 381a) is also covered with a reticulate pattern which is less well-marked on the posterior region of the shield. From it arise two pairs of setae, the third pair arising from independent platelets situated anterior to the

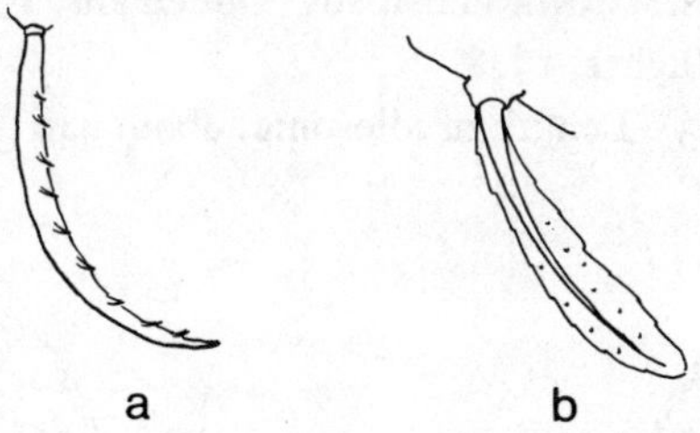

FIG. 380. Terminal seta of (*a*) *Kleemannia plumigera* Oudemans and (*b*) *K. plumosus* (Oudemans).

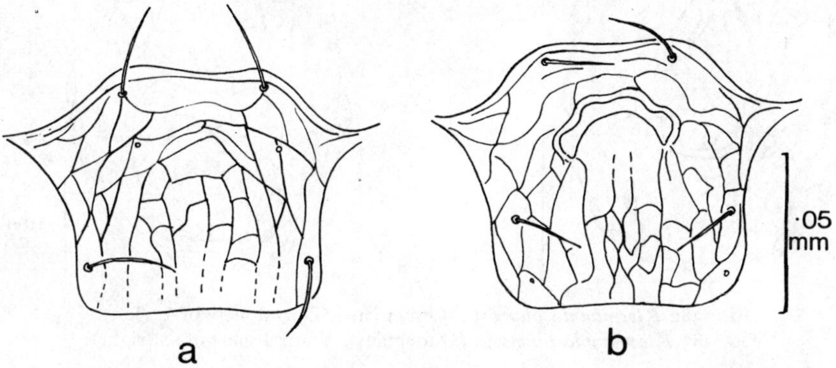

FIG. 381. Sternal shield of ♀ of (*a*) *Kleemannia plumigera* Oudemans and (*b*) *K. plumosus* (Oudemans).

metasternal setae. The posterior margin of the genital shield is wider than the anterior; it bears one pair of setae and lies close to the large ventro-anal shield. The latter bears two pairs of setae in addition to the anals. Four small smooth setae arise from platelets between genital and ventro-anal shields.

The corniculi of the gnathosoma are bifid and the first pair of hypostomal setae is thickened.

All the legs end in pretarsi and claws.

MALE. Length of idiosoma: about 360 μm.

This resembles the female. The ventral surface is almost completely covered by the sterno-genital and ventro-anal shields.

BIOLOGY

K. plumigera has been found in cracks in a damp stone wall near Belfast, N. Ireland, also in siftings from oats and the surface of baled hay (P. O. Hill coll.). Rack (1963, 1964, 1968, 1970, 1971) lists many cases of *K. plumigera* occurring in large numbers in new houses and office buildings. She considers it to be a mycophagous species and it is often found in very large numbers living on moulds which grow on vegetable material used as an insulating material. The infestation disappears as the new building gradually becomes drier. Brady (1970) records it from the deep litter of broiler houses.

DISTRIBUTION

England, Ireland, Germany, Holland.

KLEEMANNIA PLUMOSUS (Oudemans, 1902)

Zercoseius macauleyi Hughes, 1948.

FEMALE (Figs. 382, 383). Length of idiosoma: about 440 µm.

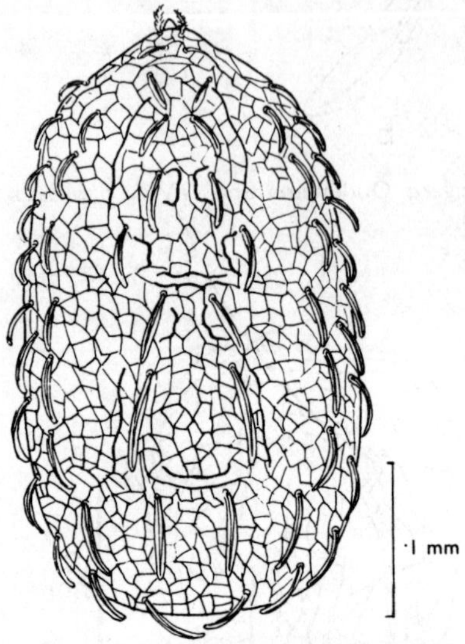

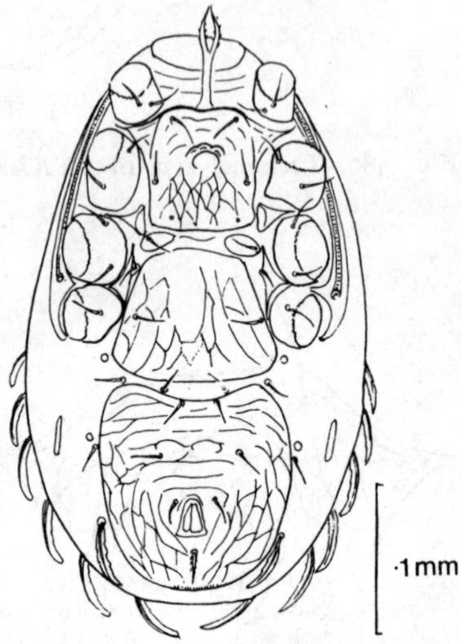

FIG. 382. *Kleemannia plumosus* (Oudemans). Dorsal view of ♀ (left).
FIG. 383. *Kleemannia plumosus* (Oudemans). Ventral view of ♀ (right).

This closely resembles the previous species. The pattern on the dorsal shield is deeply etched, particularly in the central region. Twenty-nine pairs of setae arise from the dorsal shield. Except for the most anterior setae which have pectinate edges, all the dorsal setae are leaf-shaped, with a thickened 'mid-rib' (Fig. 380b). The ventral shields are also covered with a reticulate pattern, the sternal shield (Fig. 381b) sometimes having a central smooth region bordered anteriorly by a scalloped edge. Some specimens do not have this marking and the shield is evenly reticulate.

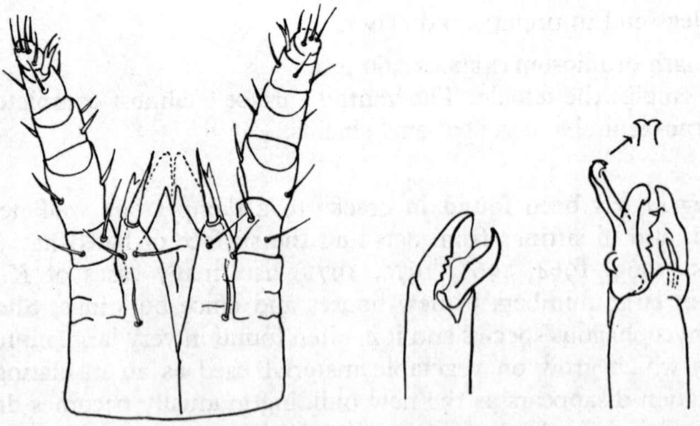

FIG. 384. *Kleemannia plumosus* (Oudemans). (*a*) Ventral view of gnathosoma of ♀, (*b*) chelicera of ♀ and (*c*) chelicera of ♂.

The gnathosoma (Fig. 384a) has bifid corniculi and enlarged hypostomal setae as in the previous species. The chelicera (Fig. 384b) has an irregularly-shaped fixed limb with three or four teeth; the movable digit has a row of fine teeth on its basal half.

MALE. (Fig. 385). Length of idiosoma: about 360 µm.

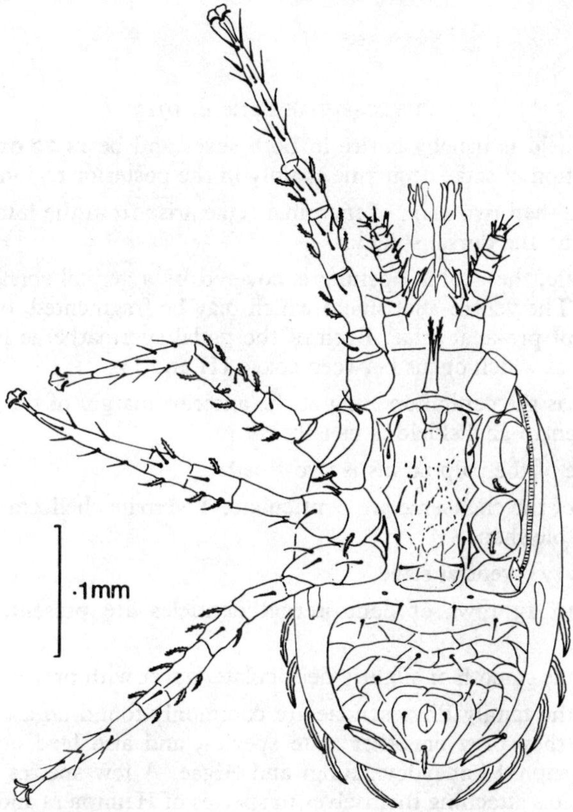

FIG. 385. *Kleemannia plumosus* (Oudemans). Ventral view of ♂.

The ventro-anal shield bears four pairs of smooth setae in addition to the anals. Two platelets lie between the sterno-genital and ventro-anal shields.

The movable digit of the chelicera (Fig. 384c) is twisted at its base and terminates in a sharp point, the fixed is narrower and has three or four teeth on its internal edge. The spermadactyl is spirally twisted.

BIOLOGY

K. plumosus is associated with substrates supporting a growth of mould and Sinha (1964a) found that it would reproduce and form large colonies on *Nigrospora sphaerica*, *Rhizopus nigricans* and *Scopulariopsis brevicaulis*. J. Conway collected large numbers from the damp surface layers of heating barley and W. J. Simpson found it on a mouldy parcel of Portuguese cork.

In Germany, vegetable material is used as an insulating material in new houses and office blocks. This absorbs moisture given off by drying brickwork or concrete and readily becomes mouldy, the moulds supporting a large population of mites, particularly *K. plumosus*. (Rack, 1971).

K. plumosus is widely distributed beneath haystacks, in the detritus of warehouse floors, in the nests of small mammals and bumble bees and in the deep litter of broiler houses (Brady, 1970).

DISTRIBUTION

England, Northern Ireland, Holland, Germany, Iceland, Israel, Australia, Canada.

PHYTOSEIIDAE Berlese, 1913

The dorsal shield is usually entire in both sexes and bears 20 or less pairs of setae, a reduction of setae occurring mainly in the posterior region of the shield.

Never more than two pairs of marginal setae arise from the interscutal membrane encircling the dorsal shield.

In the female, the genital opening is covered by a genital shield bearing one pair of setae. The ventro-anal shield which may be fragmented, bears from one to four pairs of pre-anal setae. Each of the podal spermathecae has sclerotized walls and a duct which opens between coxae III and IV.

The male has the genital opening at the anterior margin of the sterno-genital shield. The ventro-anal shield is usually entire.

The apotele of the palp tarsus is two-tined.

The digits of the chelicerae are denticulate. The male chelicera has a spermadactyl of variable shape.

Tibia I has two ventral setae.

From six to 13 rows of deutosternal denticles are present, often poorly sclerotized.

The tectum is smooth or slightly denticulate, never with processes.

Species of the family Phytoseiidae are commonly found on a wide variety of plants where they prey on other mite species, and also feed on plant juices, pollen grains, aphid honeydew, fungi and Algae. A few species are, however, external parasites attaching themselves to species of Hemiptera and Lepidoptera. During the past 30 years, phytophagous Acari (e.g. Tetranychidae, Eriophyidae) have increased in orchards (Chant, 1959) and also in glasshouses, and, as a result of this, the Phytoseiidae have been extensively studied in the hope of using them as agents for biological control (see Gould, Hussey and Parr, 1969).

The family is divided into two subfamilies, the free-living forms being grouped into the Phytoseiinae. Two species belonging to the genera *Amblyseius* and *Neoseiulus* are sometimes found in hay bales and on the surface of grain.

AMBLYSEIUS Berlese, 1915

The dorsal shield has from 13 to 15 pairs of setae of variable length, four pairs arising in the antero-lateral region (Fig. 386). The setae are smooth or finely pectinate.

The legs often have macrosetae (Fig. 390).

Type species: *Zercon obtusus* Koch.

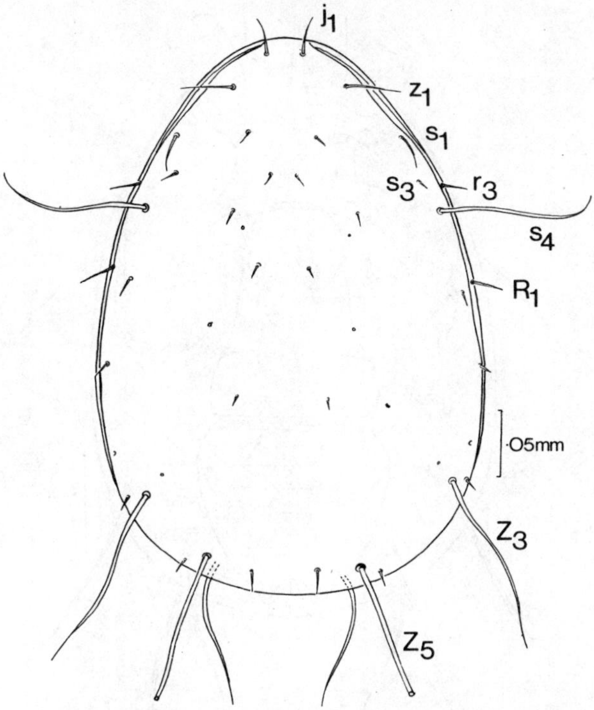

FIG. 386, *Amblyseius* sp. ♀. Dorsal view of idiosoma. Setae arising from dorsal shield: j_1, z_1, s_1, s_3, s_4, Z_3, Z_5; setae arising from interscutal membrane: r_3, R_1.

AMBLYSEIUS sp.

Typhlodromus (*Amblyseius*) *obtusus* (C. L. Koch, 1839) sensu Hughes, 1961.

This species differs from *A. obtusus* Koch sensu Chant, 1959 in possessing four narrow plates between the genital and ventro-anal shield and also in having two, not one, pairs of metapodal plates.

FEMALE (Figs. 386, 387). Length of idiosoma: about 450 µm.

The idiosoma is pear-shaped and reddish-brown, the reddish tinge being due to the colour of the body contents. Seventeen pairs of setae arise from the single dorsal shield, setae s_4, Z_3 and Z_5 being long, stout and whip-like; s_4 and Z_3 are approximately equal in length. Setae s_1 are longer than s_3 and, except for z_1 and j_1, the remaining setae are short, approximately equal to s_3 in length. Four pairs of antero-lateral setae (z_1, s_1, s_3, s_4) are present and two pairs of setae r_3 and R_1 arise from the interscutal membrane. The ventral plates are almost smooth. The sternal shield is wider than long and bears three pairs of setae, a fourth pair arises from the metasternal plates. The genital plate has a truncate posterior margin and one pair of setae and the ventro-anal has a straight anterior edge and is shaped as in Fig. 387. It bears three pairs of setae in addition to those associated with the anus. The peritrematal shield encircles the posterior edge of coxa IV and is continuous with the exopodal. The peritreme extends on to the dorsal surface. Four narrow plates lie between the genital and ventro-anal shield. Four pairs of setae arise from the interscutal membrane of the ventral surface, of which the posterior pair is the longest. Two pairs of unequal meta-

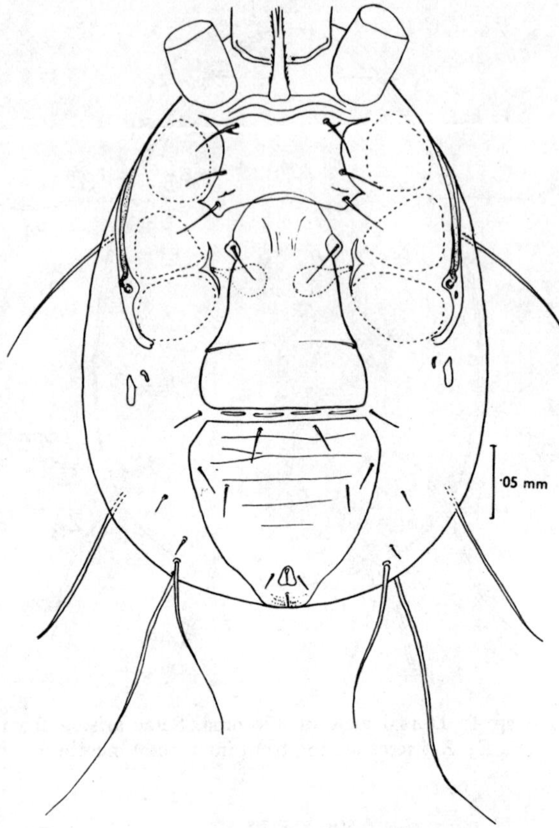

FIG. 387. *Amblyseius* sp. ♀. Ventral view of idiosoma.

podal plates are present. The spermathecae (Fig. 388) are conspicuous and open between coxae III and IV.

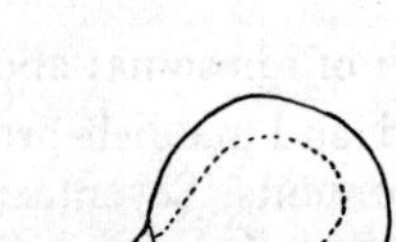

FIG. 388. *Amblyseius* sp. Spermatheca of ♀.

The movable digit of the chelicera (Fig. 389) has one very small tooth, the fixed digit has two teeth, as well as a small pilus dentilis.

On legs IV, extra long setae (macrosetae) arise from the base of the tarsus and from the tibia and genu (Fig. 390).

BIOLOGY

This mite was found in hay bales (F. A. Hunter coll.), also on the surface of bin-stored barley (A. S. C. Smith coll.).

DISTRIBUTION

England.

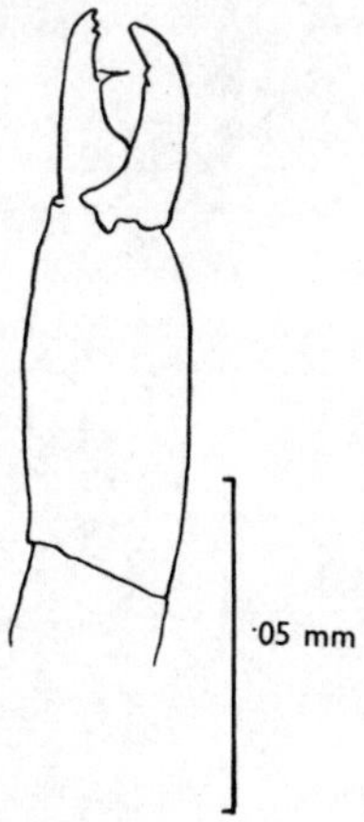

FIG. 389. *Amblyseius* sp. Chelicera of ♀.

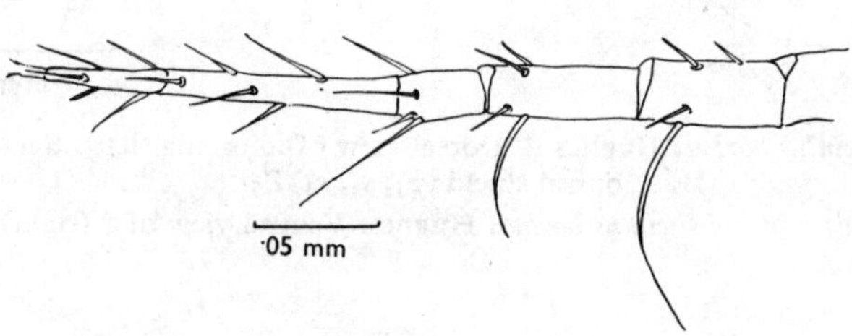

FIG. 390. *Amblyseius* sp. Leg IV of ♀. The ambulacrum is not shown.

NEOSEIULUS BARKERI Hughes, 1948

Typhlodromus (*Amblyseius*) *barkeri* (Hughes, 1948) sensu Hughes, 1961, *Amblyseius barkeri* (Hughes, 1948) sensu Athias Henriot, 1966.

According to Muma and Denmark, 1971, the genus *Neoseiulus* Hughes, 1948 (type species: *N. barkeri* Hughes, 1948) differs from *Amblyseius* mainly in setae Z_5 being shorter than the distance between their bases.

FEMALE (Figs. 391, 392). Length of idiosoma: about 380 μm.

This mite is pale brown in colour and moves rapidly when alive. The idiosoma is oval in shape. The dorsal shield bears 17 pairs of setae arranged as in Fig. 391. All the setae are about the same length, except for Z_5 which are slightly longer and pectinate. Setae z_1, s_1 and s_3 are slightly longer than half the distances between their bases. The anal plate is shaped as in Fig. 392, its lateral edges are not indented.

The tectum is feebly sclerotized, with a smoothly curved edge; the corniculi (Fig. 393a) are long and slender, converging towards the midline. The movable limb of the chelicera (Fig. 393b) has a single small tooth, the fixed has three small sharp teeth distal to the colourless pilus dentilis. Both digits are of approximately the same length.

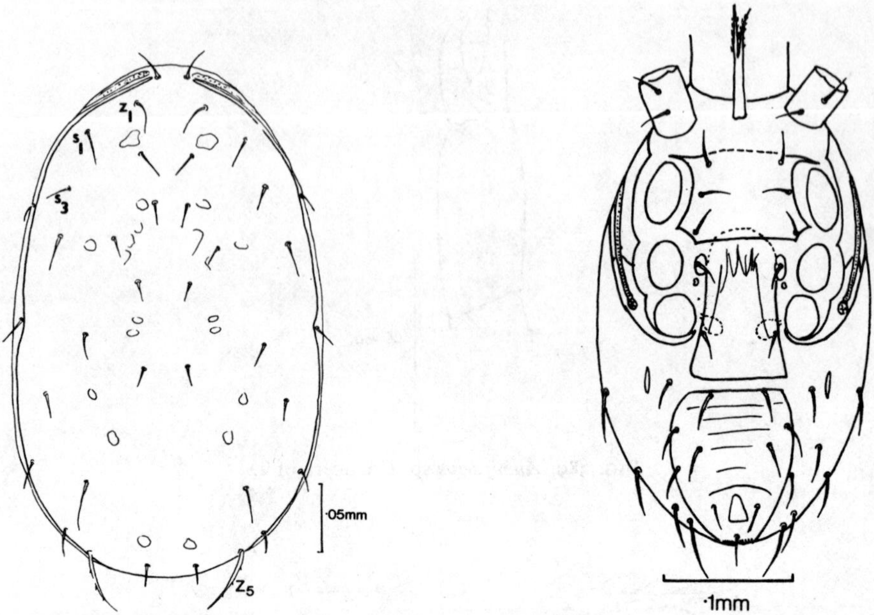

FIG. 391. *Neoseiulus barkeri* Hughes ♀. Dorsal view of idiosoma (left). Setae arising from dorsal shield: z_1, s_1, s_3, Z_5.

FIG. 392. *Neoseiulus barkeri* Hughes. Ventral view of ♀ (right).

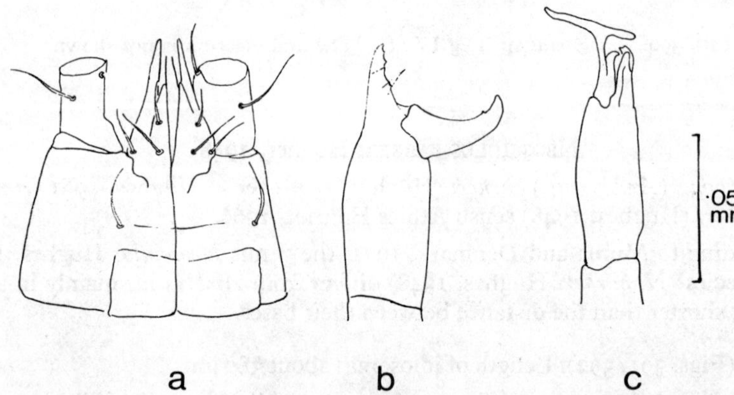

FIG. 393. *Neoseiulus barkeri* Hughes. (*a*) Ventral view of gnathosoma of ♀, (*b*) chelicera of ♀ and (*c*) chelicera of ♂.

MALE (Fig. 394). Length of idiosoma: about 330 μm.

The chaetotaxy of the dorsal surface is similar to that of the female, except that the extra-marginal setae r_3 and R_1 also arise from the dorsal shield. The ventro-anal shield is shaped as in Fig. 394, with four pairs of setae in addition to the anals.

The chelicera (Fig. 393c) is smaller than that of the female and the spermadactyl is T-shaped.

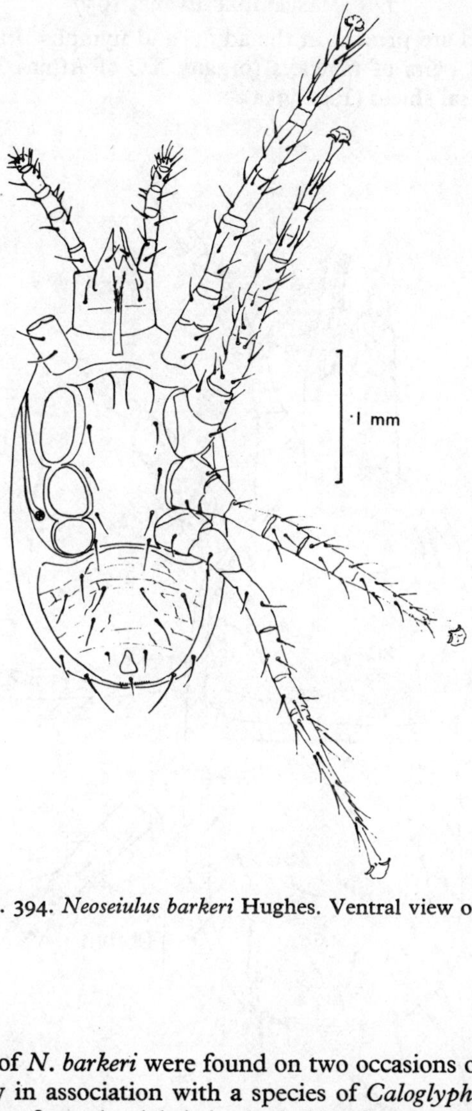

FIG. 394. *Neoseiulus barkeri* Hughes. Ventral view of ♂.

BIOLOGY

Large numbers of *N. barkeri* were found on two occasions on the plumules of germinating barley in association with a species of *Caloglyphus*. Nesbitt (1951) obtained this species from the debris in the holds of wheat-transporting vessels, together with *Acarus siro* and a species of *Tyrophagus*. In two samples taken by different collectors from a damp, newly-constructed office block in Berlin, Rack (1970, 1971) found large numbers of *N. barkeri* in association with six other species of mould-eating mites (*Tyrophagus putrescentiae*, *Glycyphagus domesticus*, *Kleemannia pavida*, *K. plumosus*, *K. plumigera* and *Proctolaelaps pygmaeus*. It is possible that *N. barkeri*, like other phytoseiids, is a predator and also supplements its diet with moulds.

DISTRIBUTION

London Docks, England; Germany. Athias-Henriot (1966) records it from Europe, North Africa, Israel and California.

DIGAMASELLIDAE Evans, 1957

Two dorsal shields are present in the adults and nymphs. In deutonymphs and adults one or two pairs of platelets (organs XC of Athias Henriot, 1961) are present on the dorsal shield (Fig. 395a).

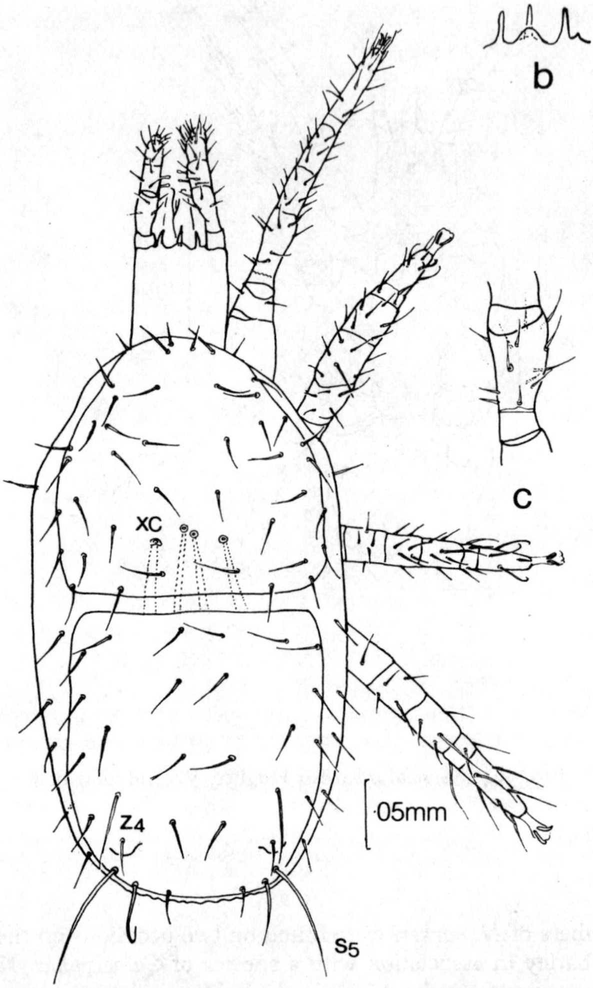

FIG. 395. *Digamasellus presepum* Berlese. (*a*) Dorsal view of ♀. XC one of four platelets, Z_4 seta arising from posterior dorsal shield, (*b*) tectum and (*c*) right femur IV.

In females, the fourth pair of sternal setae arises from the sternal shield and the transverse distance between the third sternal setae is usually considerably less than that between the second and fourth.

In males, the genital setae are on separate plates (Fig. 398). The antero-median region of the ventro-anal shield is usually separated from the main body of the shield by a pair of oblique incisions. The ventro-anal shield is fused laterally and posteriorly with the posterior dorsal shield, so that some of the marginal (*R*) setae are on its lateral extremities.

On the gnathosoma, the deutosternum has five transverse rows of denticles, the proximal row usually extended laterally (Fig. 397). The median branch of the triramous tectum has a more ventral origin than have the lateral branches (Fig. 395b).

Femur I has 13 setae, including v_4 (Fig. 395c).

In the male, the proximal part of the spermadactyl is S-shaped and leg II is usually thickened with some strong opposable spine-like setae on its ventral surface (Fig. 399).

The deutonymphs are usually well-sclerotized and adapted for dispersal by insects.

The family Digamasellidae, according to Lindquist and Evans, 1965, contains two genera: *Digamasellus* Berlese, 1905, type species: *Gamasellus* (*Digamasellus*) *perpusillus* Berlese, 1905 and *Longoseius* Chant, 1961. The genus *Longoseius* is associated with the galleries of bark-boring beetles.

DIGAMASELLUS PRESEPUM Berlese, 1918

FEMALE (Figs. 395, 396). Length of idiosoma: about 400 μm.

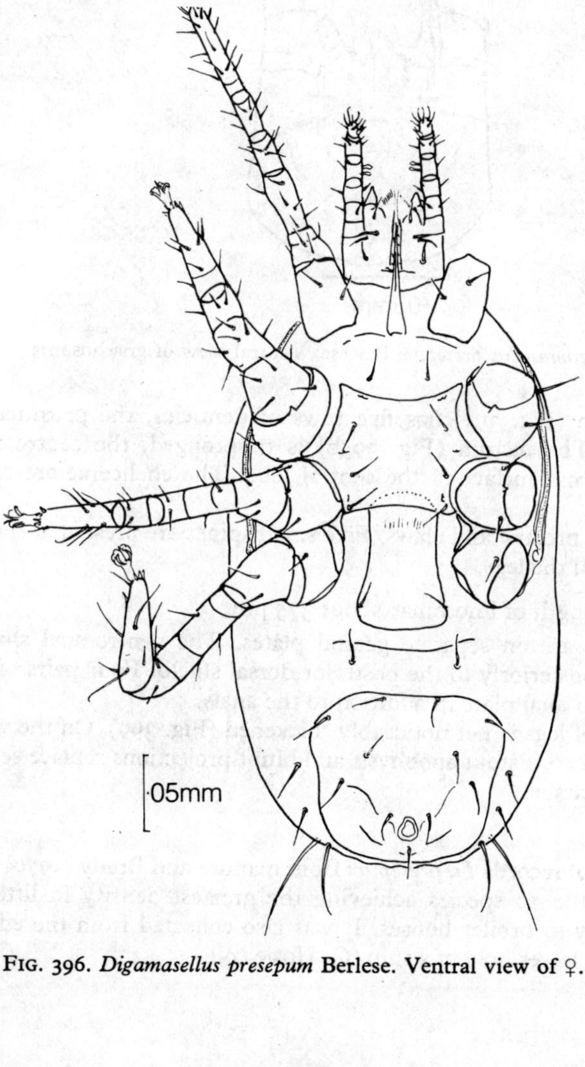

FIG. 396. *Digamasellus presepum* Berlese. Ventral view of ♀.

The dorsal surface is almost completely covered by two shields. Twenty-one pairs of setae arise from the anterior shield, 15 from the posterior. They are arranged as shown in Fig. 395. The setae are smooth. Seta Z_4 on the posterior shield arises from a knob and the terminal setae S_5 are longer than the other dorsal setae. Four small platelets (organs XC of Athias Henriot, 1961) are embedded in the anterior shield. From each platelet a pair of muscles extends posteriorly; these are inserted on the posterior dorsal shield and their contraction would result in an arching of the back.

The first pair of sternal setae is inserted into the cuticle in front of the sternal shield, the fourth pair originates from the sternal shield itself. The distance between sternal setae 2 and 4 only slightly exceeds that between setae 3. The genital shield is narrow and elongated and bears a single pair of setae. The ventro-anal shield is almost square in outline, with four pairs of setae in addition to the three anals. A pair of setae arises between the genital and ventro-anal shields and two pairs flank the ventro-anal shield.

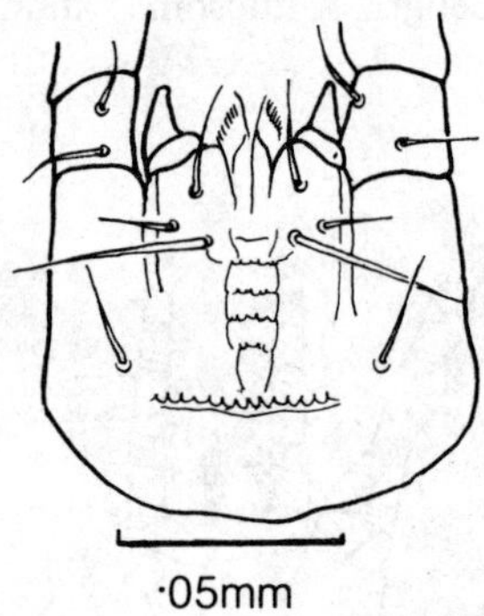

FIG. 397. *Digamasellus presepum* Berlese. Ventral view of gnathosoma.

The deutosternum (Fig. 397) has five rows of denticles, the proximal row extended laterally. The tectum (Fig. 395b) is tri-pronged, the centre prong attached to the internal surface of the central lobe. The chelicerae are chelate and dentate.

All the legs end in pretarsi and claws. Five small spines are present at the base of the pretarsus on all the legs.

MALE (Fig. 398). Length of idiosoma: about 375 μm.

The genital setae are on separate genital plates. The ventro-anal shield is fused laterally and posteriorly to the posterior dorsal shield. Four pairs of setae arise from the ventro-anal plate in addition to the anals.

The second pair of legs is not noticeably thickened (Fig. 399). On the ventral surface, the femur bears a stout apophysis and blunt projections replace setae on the genu, tibia and tarsus.

BIOLOGY

Hirschmann (1960) records *D. presepum* from manure and Brady (1970) found that it was one of the 10 species achieving the greatest density in litter and droppings of poultry in broiler houses. It was also collected from the edges of ceiling boards in the lower hold of a ship (G. Hosie coll.).

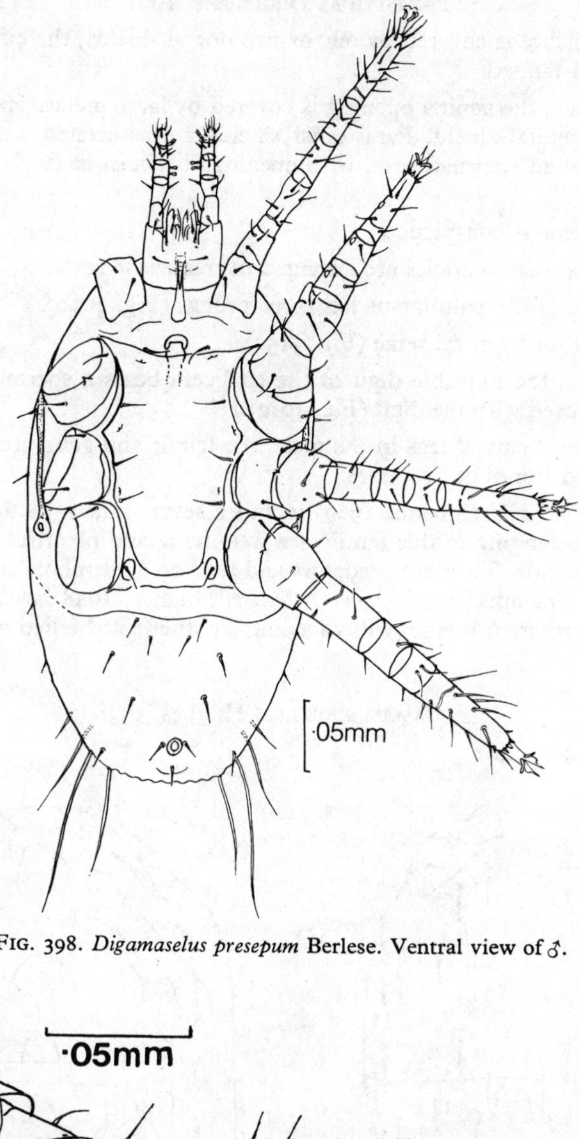

FIG. 398. *Digamaselus presepum* Berlese. Ventral view of ♂.

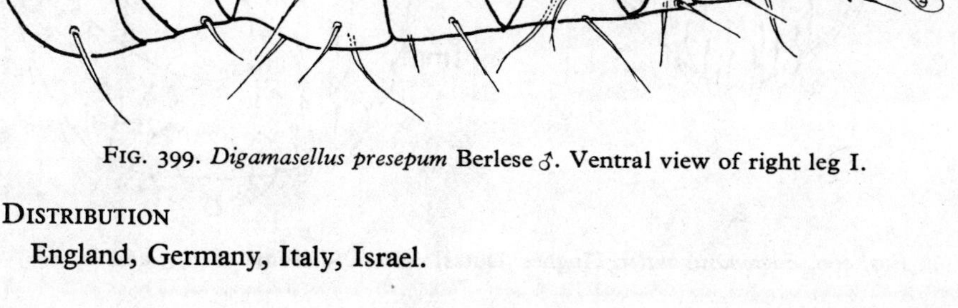

FIG. 399. *Digamasellus presepum* Berlese ♂. Ventral view of right leg I.

DISTRIBUTION

England, Germany, Italy, Israel.

PARASITIDAE Oudemans, 1901

The dorsal surface is covered by one or two dorsal shields, the cuticle of which is usually well-tanned.

In the female, the genital opening is covered by large metasternal shields and a triangular genital shield. Parasternal shields are associated with the genital (Fig. 407). Paired spermathecae, with openings between coxae III and IV, are absent.

The peritreme is conspicuous.

The deutosternal deuticles are arranged in transverse rows.

The apotele of the palp tarsus has three prongs (Fig. 402b).

Tibia I has four ventral setae (Fig. 315c).

In the male, the movable digit of the chelicera bears a spermadactyl whose distal end is fused with the digit (Fig. 404a).

On the second pair of legs in the male, the femur and genu are enlarged and bear thickened spines or processes.

Holzmann (in Hirschmann, 1969) includes seven genera in the Parasitidae. The species belonging to this family are free-living and plentiful in the surface litter of woodlands. They are predators and feed on Collembola and other small Arthropoda. One species—*Eugamasus butleri* Hughes, 1948 has been found in floor detritus where it was probably feeding on other soft-bodied mites.

EUGAMASUS BUTLERI Hughes, 1948

MALE (Figs. 400a, 401)

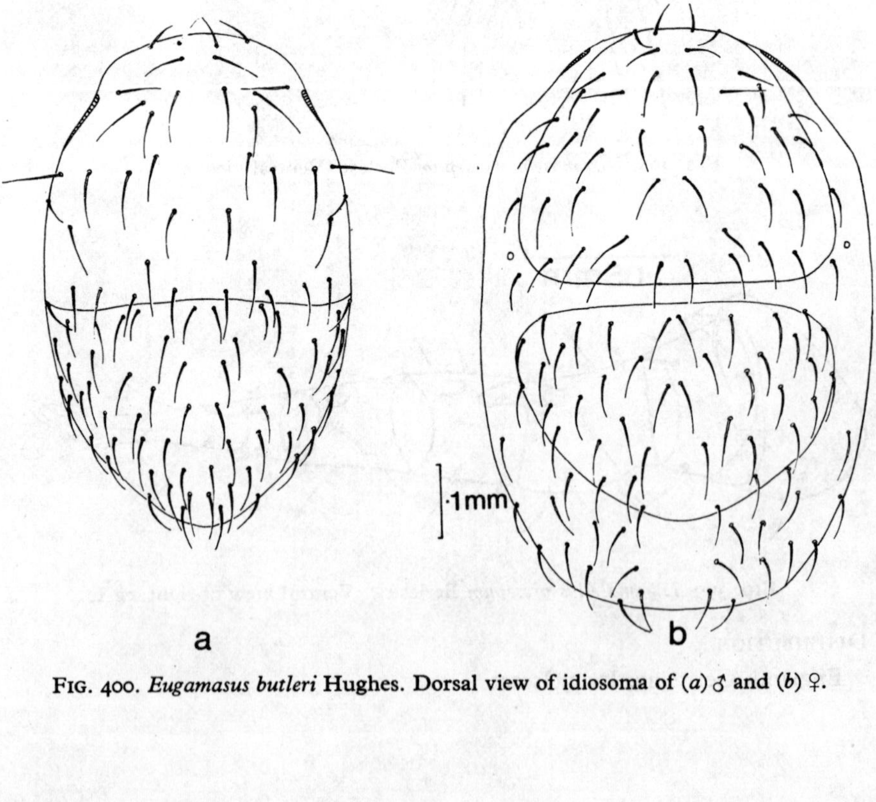

FIG. 400. *Eugamasus butleri* Hughes. Dorsal view of idiosoma of (*a*) ♂ and (*b*) ♀.

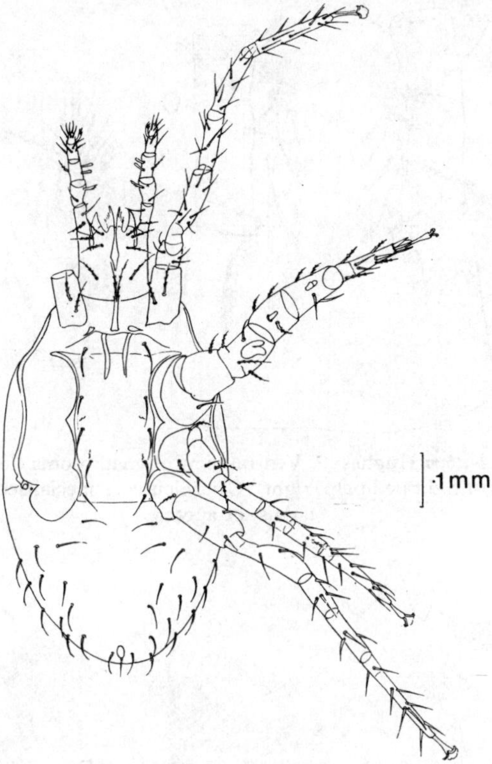

FIG. 401. *Eugamasus butleri* Hughes. Ventral view of ♂.

The idiosoma is reddish-brown in colour and has a slightly pointed posterior margin. Its dorsal surface is covered by two tightly-fitted plates ornamented with scale-like markings and bearing a number of smooth, pointed setae. These plates fuse with the peritrematal plate anteriorly and with the ventro-anal posteriorly, so that the whole of the body, with the exception of the coxal region, is covered with a rigid exoskeleton. A small pair of pre-endopodal plates flanks the tritosternum. The anterior margin of the sternal shield is not well-defined and the first pair of sternal setae arises from the interscutal membrane. The large sterno-genital shield extends from coxae II to IV, its anterior edges encircle coxae II, but do not articulate with the peritrematal or dorsal shields. In the region of coxae II to IV the sterno-genital shield is fused with the endopodal plates; it bears the usual five pairs of setae. The ventro-anal shield covers the rest of the opisthosoma and bears a number of fine setae. The stigma opens between coxae III and IV and the peritreme follows a rather undulating course up the side of the body.

On the gnathosoma, the corniculi (Fig. 402a) are well-defined and arise directly from the pedipalpal coxae. The tectum (Fig. 403a) is tri-pronged, the central prong being much longer than the two lateral. The chelicerae are large and well-sclerotized (Fig. 404a) the spermadactyl distally fused with the movable limb. The fixed digit bears a small tooth and an insignificant pilus dentilis.

The second pair of legs (Fig. 405) is thickened, its ventral surface bearing a number of irregular protuberances. The femur bears a large conical process,

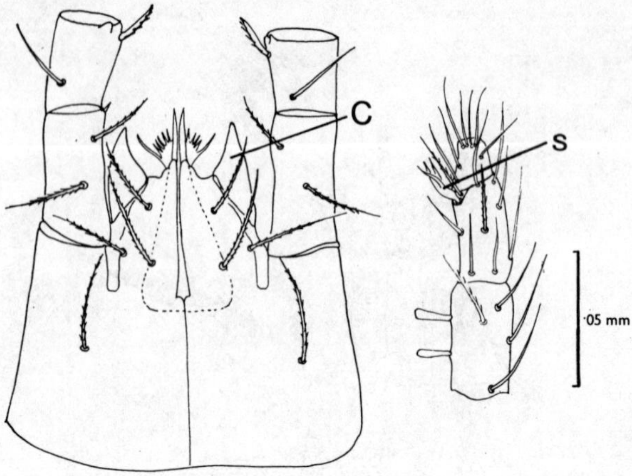

Fig. 402. *Eugamasus butleri* Hughes ♀. Ventral view of gnathosoma (left) and dorsal view of terminal segments of right pedipalp (right). C corniculus, s specialised seta on pediplapal tarsus, or apotele.

FIG. 403. *Eugamasus butleri* Hughes. Tectum of (*a*) ♂ and (*b*) ♀.

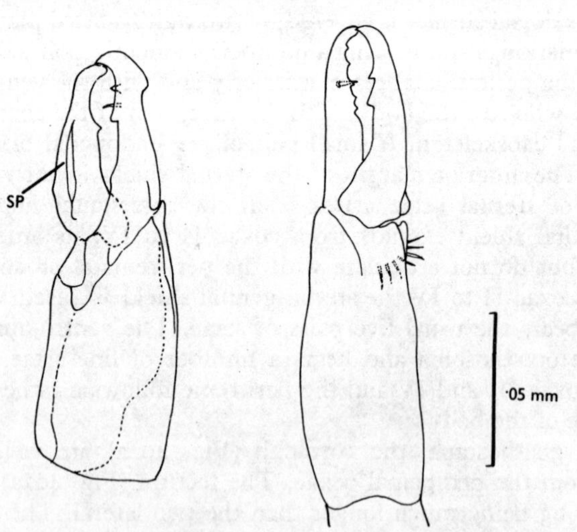

FIG. 404. *Eugamasus butleri* Hughes. Chelicera of (*a*) ♂ and (*b*) ♀. SP spermadactyl.

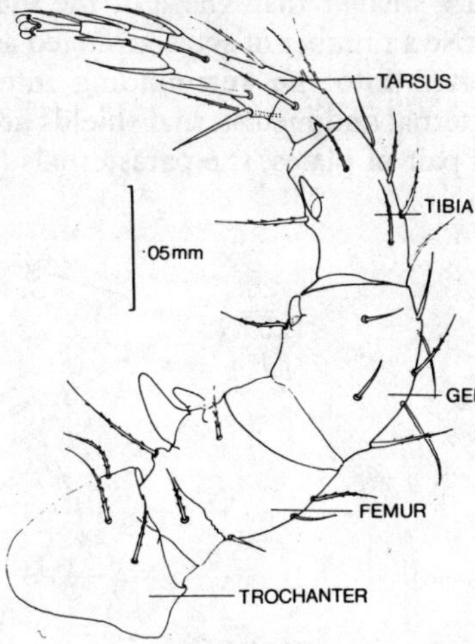

FIG. 405. *Eugamasus butleri* Hughes ♂. Internal view of right leg II.

distal to which is a small thumb-shaped structure. The genu and tibia each have similar projections in close association with a seta. The tarsus is relatively long, with two transverse fissures and numerous thickened setae.

FEMALE (Figs. 400b, 406).

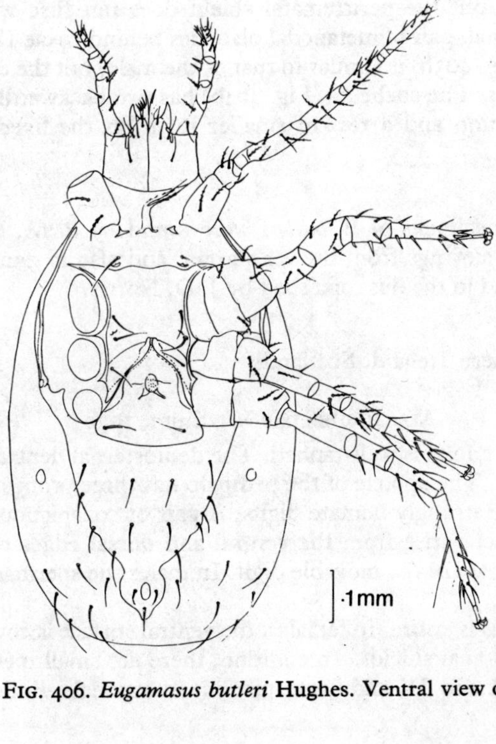

FIG. 406. *Eugamasus butleri* Hughes. Ventral view of ♀.

The dorsal shields are smaller than those of the male and almost triangular in shape. From them arise a number of setae arranged as in Fig. 400b. A number of setae are also inserted into the surrounding interscutal membrane. The lateral margins of the sternal and metasternal shields are fused with the endopodals, and an additional pair of plates, the parasternals (Fig. 407) are attached to

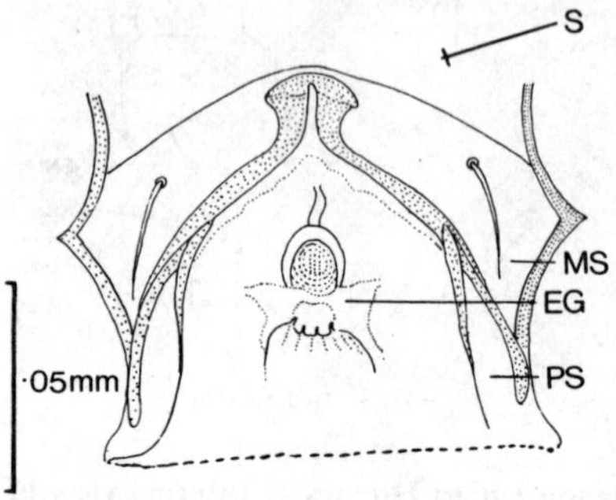

FIG. 407. *Eugamasus butleri* Hughes. Genital region of ♀. The genital plate has been removed. S sternal shield, MS metasternal shield, PS parasternal shield and EG endogynium.

the posterior margin of the metasternals and partially concealed by the hinged genital shield. The sclerites or endogynium supporting the wall of the vagina, are not conspicuous. The ventro-anal plate diminishes in width towards the posterior end of the body, and its anterior corners are extended to fuse with the peritrematal plate. Seven pairs of setae arise from it. The peritreme is similar to that of the male, but the peritrematal shield does not fuse with the anterior dorsal shield. A small pair of metapodal plate lies behind coxae IV.

The tectum (Fig. 403b) is similar to that of the male, but the central process is not so conspicuous. The chelicera (Fig. 404b) has two backwardly-directed teeth on the movable limb and a row of smaller teeth on the fixed. A small pilus dentilis is present.

BIOLOGY

A number of specimens of *E. butleri* were found by P. M. Butler and J. E. Cranham in floor sievings from brewer's grain, and also in damp floor dust. It has also been found in the dust of a shed by J. D. Norris.

DISTRIBUTION

England, Northern Ireland, Scotland.

MACROCHELIDAE Vitzthum, 1930

The cuticle of the adults is well-tanned. The deutosternal denticles are arranged in transverse rows. The apotele of the pedipalp has three prongs. The chelicerae in both sexes have strongly dentate digits. A pair of conspicuous plumose setae and one smooth seta arise from the ventral and dorsal edges of the arthrodial membrane at the base of the movable digit. In males the spermadactyl has a free distal end.

The dorsal shield is entire. In females, the ventral surface is covered by sternal, genital and ventro-anal shields. In addition, there are small metasternal shields between coxae III and IV and inconspicuous metapodals. Endopodal and exopodal plates are well-developed. The normal five pairs of setae arise from the sternal, metasternal and genital shields, but the ventro-anal bears from two to five pairs of setae in addition to the three associated with the anus. In males, the ventral surface may be covered by one holoventral shield or by sterno-genital and ventro-anal shields. The ventro-anal shield is not fused with the dorsal shield.

The stigmata are situated between coxae III and IV, the peritremes being well-developed and looped in the region of the stigma (Fig. 409). The peritrematal plate is fused anteriorly with the dorsal shield, but is free posteriorly.

The first pair of legs has no pretarsus, pulvillus or claws and terminates in a cluster of setae (Fig. 314b). Tibia I has three ventral, five dorsal and two antero-lateral setae. Genu I has two ventral and two antero-lateral setae.

The Macrochelidae contains a number of free-living mites commonly found in soil, compost heaps and dung, where they feed on living and dead insects, and possibly on fungi as well. The family is divided into three subfamilies, the British species all belonging to the Macrochelinae Vitzthum, 1930. This subfamily has been extensively revised by Evans and Browning, 1956 who recognize four genera. The two species which are sometimes found in stored food belong to the genus *Macrocheles* Latreille, 1829.

MACROCHELES Latreille, 1829

The tectum usually has three processes (Fig. 411).

The vertical setae j_1 (Fig. 408) arise from the edge of the dorsal shield, not on an outgrowth from it.

The ventro-anal shield has three pairs of setae, in addition to the three anal setae.

The femur of leg II is unarmed in the female.

Type species: *Acarus muscaedomesticae* Scopoli, 1772.

MACROCHELES MUSCAEDOMESTICAE (Scopoli, 1772)

Acarus muscaedomesticae Scopoli, 1772; also see Pereira and Castro, 1945.

FEMALE (Fig. 408). Length of idiosoma: 980 to 1,010 µm.

In life, the female is reddish brown in colour and moves very quickly. The dorsal surface is covered by a reticulate shield, from which arise 28 pairs of setae. It is perforated by the openings of about 20 dermal glands. The majority of the setae are short, straight and plumose in their distal third; only setae j_5, j_6, J_3, z_1, z_3 and z_6 are smooth. Setae J_5 are pectinate over more than half their length. The vertical setae j_1 are close together (Fig. 410) with their bases touching. The ventral surface of the body is covered by plates ornamented with characteristic punctures and ridges. All the setae arising from these plates are simple. The ventro-anal shield bears three pairs of setae in addition to those associated with the anus.

The gnathosoma is well-sclerotized, except for the tectum (Fig. 411) which remains colourless. This has three distinct prongs, the median one is forked distally, the lateral ones which are flat and thin with serrated margins are often asymmetrical.

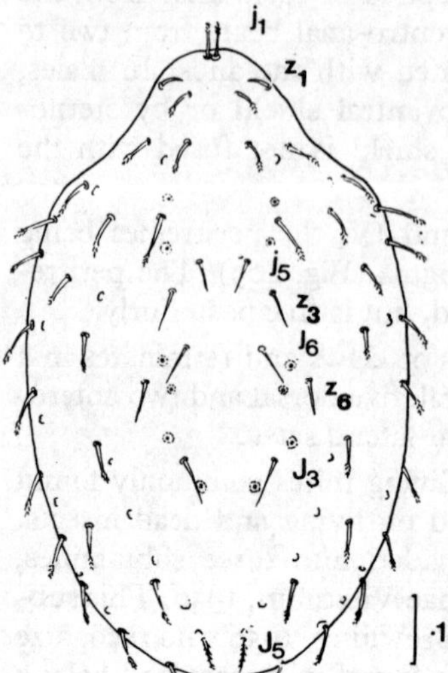

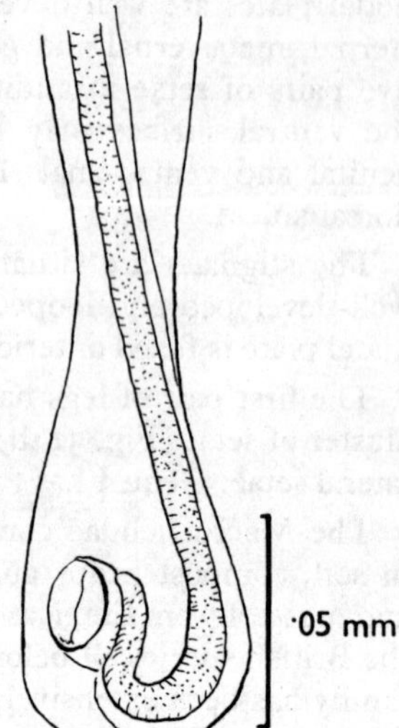

FIG. 408. *Macrocheles muscaedomesticae* (Scopoli) ♀. Dorsal view of idiosoma (left). Setae arising from dorsal shield: $j_1, j_5, j_6, J_3, J_5, z_1, z_3, z_6$.

FIG. 409. *Macrocheles muscaedomesticae* (Scopoli) ♀. Stigma and peritreme (right).

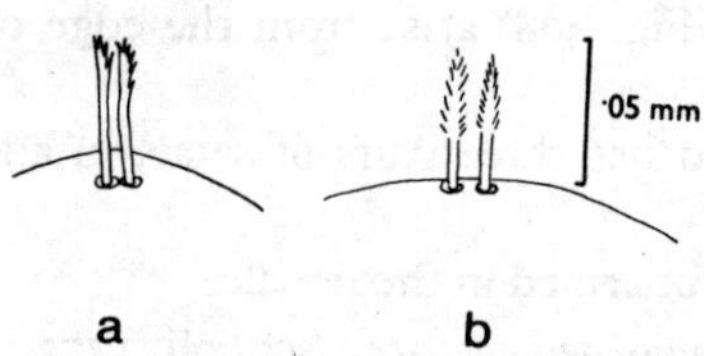

FIG. 410. Vertical setae (j_1) of (*a*) *Macrocheles muscaedomesticae* (Scopoli) ♀ and (*b*) *M matrius* (Hull) ♀.

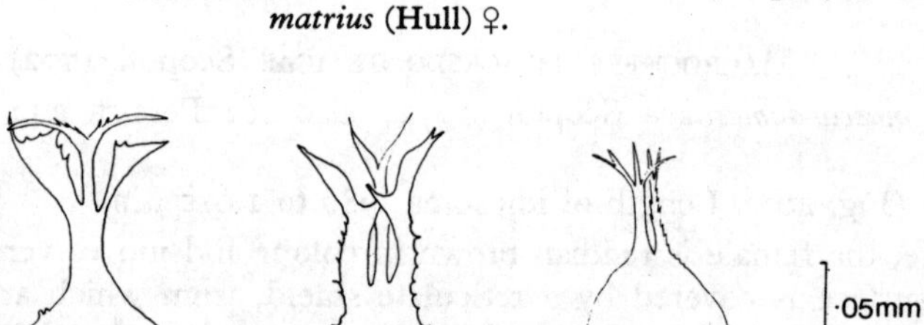

FIG. 411. Tectum of (*a*) *Macrocheles muscaedomesticae* (Scopoli) ♀, (*b*) *M. matrius* (Hull) ♀ and (*c*) *M. matrius* ♂.

BIOLOGY

This has been extensively investigated by Pereira and Castro, 1945 and 1947, Filipponi, 1955, Filipponi and Cervone, 1957, Wade and Rodriguez, Rodriguez and Wade, 1961, Rodriguez *et al.*, 1962.

M. muscaedomesticae is a predator of all stages, except the puparium, of house flies and allied species. It is found on substrates such as dung which provide a suitable habitat for the flies, and the adult fly and also dung beetles are used to transport the females from one breeding site to another. Eggs are laid by virgin and mated females, some attempt being made to conceal them in cracks or under clumps of substrate (Wade and Rodriguez, 1961), although this does not prevent the occurrence of cannibalism. The larvae and adults feed readily on fly eggs, but when they are available, proto- and deutonymphs prefer Nematodes such as *Rhabditella leptura* and, to a lesser extent, species of *Diplogaster* and *Panagrolaimus* (Rodriguez *et al.*, 1962). Two morphologically distinct deutonymphs are formed, one giving rise to males, the other to females. Virgin females always give rise to males, mated females to females and an occasional male (Pereira and Castro, 1945 and 1947). Rodriguez and his colleagues (1962) found that the nature of the substrate influenced egg production and this was found to be highest when a used standard house fly rearing medium was utilized, and Nematodes and fly eggs were added to serve as food.

In stored food, female specimens of *M. muscaedomesticae* are likely to be found anywhere that is open to the visit of flies. It has also been recorded from deep pit poultry manure in Essex (G. T. Bills coll.).

DISTRIBUTION

Cosmospolitan.

MACROCHELES MATRIUS (Hull, 1925)

Nothrholaspis matrius Hull, 1925; *Macrocheles carinatus* (Koch, 1839) sensu Hughes, 1948.

FEMALE (Figs. 313, 412). Length of idiosoma: 890 to 960 µm.

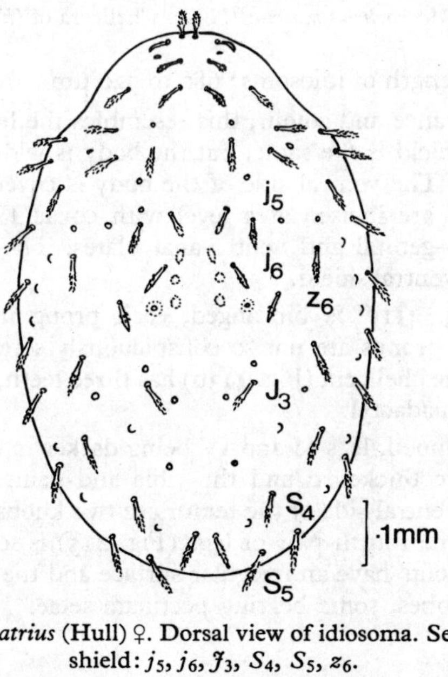

FIG. 412. *Macrocheles matrius* (Hull) ♀. Dorsal view of idiosoma. Setae arising from dorsal shield: j_5, j_6, J_3, S_4, S_5, z_6.

In general appearance and colour, this resembles the previous species. The dorsal shield has a rough surface and faint reticulate pattern, the reticulations being outlined in places by rounded punctuations. The arrangement of dorsal setae is the same as that of *M. muscaedomesticae*, but the setae themselves are plumose, although the pectinations are not so marked in setae j_5, j_6 and J_3. The remaining setae are pectinate over more than one third of their length. The vertical setae j_1 (Fig. 410b) are also plumose, their bases well separated from one another. Setae S_4 and S_5 are equal in length.

The sternal shield is ornamented with transverse lines and two conspicuous punctate areas lying internal to the third pair of sternal setae. The genital and ventro-anal shields are also ornamented with distinct punctate lines.

The tectum (Fig. 411b) has three prongs; the median one is bifid, the lateral often asymmetrical with irregularly serrated edges. The chelicera (Fig. 413a) has dentate blades.

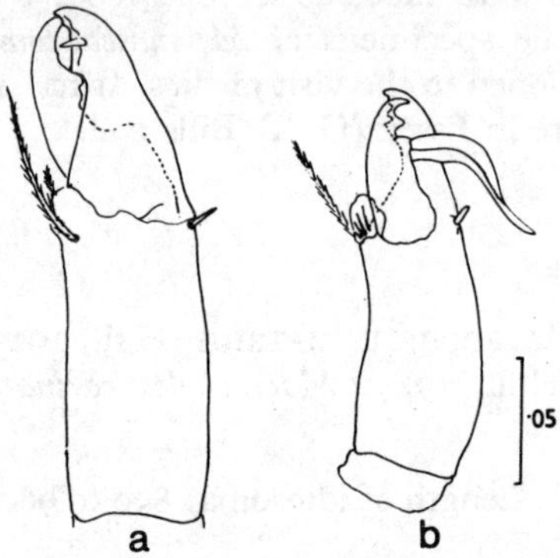

FIG. 413. *Macrocheles matrius* (Hull). Chelicera of (*a*) ♀ and (*b*) ♂.

MALE (Fig. 414). Length of idiosoma: 680 to 720 μm.

In general appearance and colour, this resembles the female, and the chaetotaxy of the dorsal shield is the same, but the body is wider in the region of the second pair of legs. The ventral side of the body is covered by a single shield, whose lateral edges are incised at a level with coxae IV, so as to indicate a division into sterno-genital and ventro-anal plates. The endopodal plates are also fused with this ventral shield.

The tectum (Fig. 411c) is bipronged, each prong irregularly forked; the lateral edges of the prongs are not so conspicuously serrated as in the female. The fixed digit of the chelicera (Fig. 413b) has three teeth, the movable only one and a recurved spermadactyl.

All the legs are tanned, legs II and IV being darker in colour than I and III. Legs II (Fig. 3) are thickened and the tibia and genu each bears one small projection. On the ventral side of the femur are two knobs, one of which is bent sharply on itself. The fourth pair of legs (Fig. 415) is longer than that of the female, all the segments have an irregular surface and the trochanter and femur are produced into lobes, some bearing pectinate setae.

BIOLOGY

M. matrius is widely distributed in poultry manure where the dung is mixed with large amounts of litter (Evans and Browning, 1956), and dung beetles such as *Geotrupes silvaticus* aid in its distribution (Gotz, 1952 quoted from Hirschmann, 1970). Evans and Browning, 1956 also found it in a nest of *Riparia riparia*.

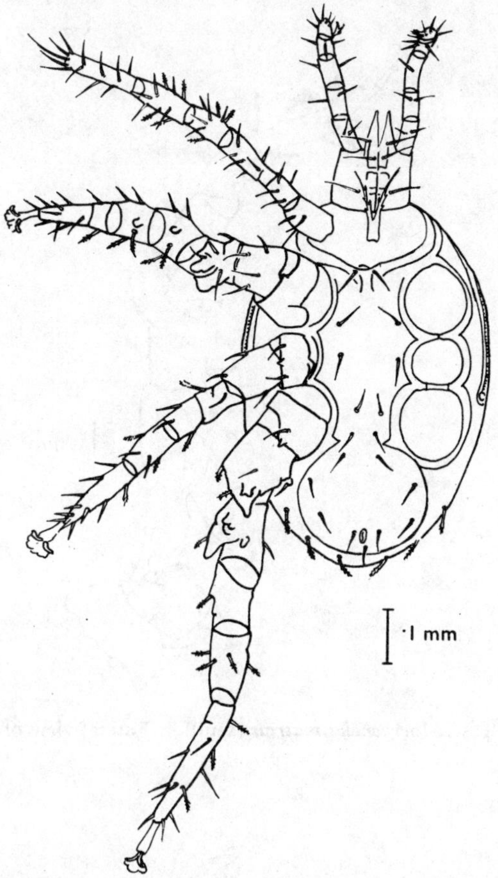

FIG. 414. *Macrocheles matrius* (Hull). Ventral view of ♂.

It has also been recorded from floor debris and sievings of grain. When males occur, several are found together.

DISTRIBUTION

England, Italy, Finland, Iceland, Germany, Austria.

UROPODINA Kramer, 1881

The bases of the hypostomal setae are almost in linear series (Fig. 416).

The stigmata are situated between coxae II and III and the peritremes are often displaced and convoluted owing to the development of leg grooves.

The tritosternum is often concealed by the enlarged, flattened coxae of the first pair of legs.

The palpgenu usually bears five setae, rarely four or six.

The male chelicera never has a spermadactyl.

The idiosoma usually has an entire or fragmented marginal shield (Fig. 417).

The anterior margin of the camerostome is usually produced into an upstanding ridge or scabellum (Fig. 426).

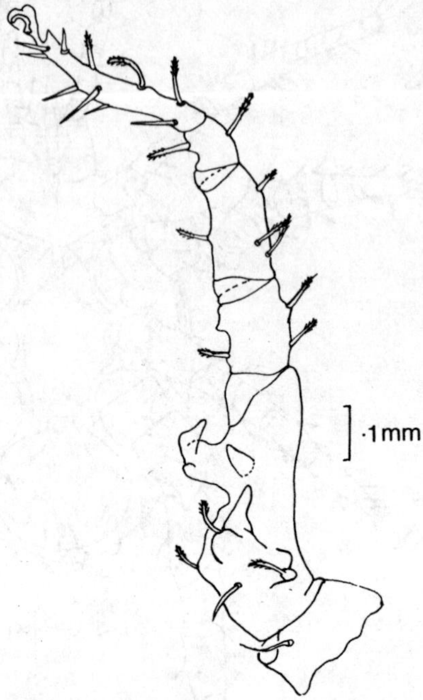

FIG. 415. *Macrocheles matrius* (Hull) ♂. Lateral view of leg IV.

FIG. 416. *Leiodinychus krameri* (Canestrini) ♂. Ventral view of gnathosoma. Hyp. 1 to 3 hypostomal setae 1 to 3; Cap C capitular seta.

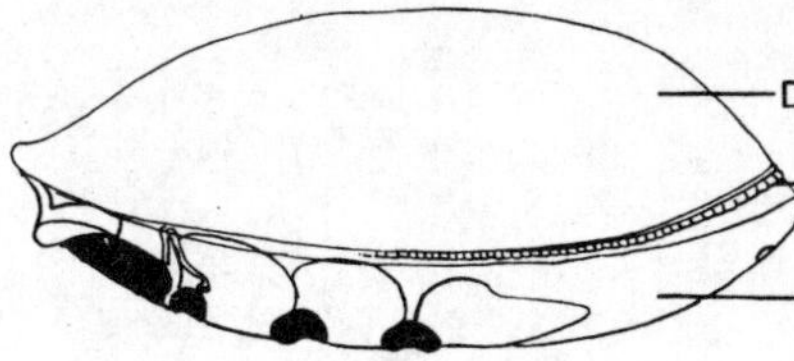

FIG. 417. *Leiodinychus krameri* (Canestrini). Lateral view of idiosoma.

Tibia I never has more than two ventral setae.

The Uropodina are free-living forms found in moss and under the bark of trees, where they feed on small fungi, lichens and other vegetable matter. The chelicerae tend to be unusually long, sometimes extending almost as far back as the posterior end of the body, and the cheliceral digits are small and feebly sclerotized. Many species are myrmecophilous and are found in various degrees of association in the nests or clinging to the bodies of ants. In some, a heteromorphic deutonymph may be formed which can be stimulated by touch to squirt a secretion from the anus, this hardens on exposure to air and serves to attach the mite to passing insects (the 'nympha pedunculata' of Berlese, wandernymph' of Faasch, see p. 368).

The basis of a classification of the Uropodina was laid by Berlese [1918] and adopted by subsequent authors. Between 1961 and 1969, it was extensively revised by Hirschmann and Zirngiebl-Nicol who used certain features of the gnathosoma as their main taxonomic characters and made considerable changes at the generic level. Evans (1972) discusses their new classification and emphasises the importance of studying a range of, rather than, few characters in any taxonomic survey of the Uropodina. I am therefore retaining the generic names used in my previous book rather than adopting those proposed by Hirschmann and Zirngiebl-Nicol.

Two species of Uropodina are often found in stored food and two others occur less frequently.

LEIODINYCHUS KRAMERI (G. and R. Canestrini, 1882)

Uropoda krameri G. and R. Canestrini, 1882; *Trichouropoda orbicularis* (C. L. Koch, 1839) sensu Hirschmann and Zirngiebl-Nicol (1961).

FEMALE (Figs. 418, 419). Length of idiosoma: about 760 μm.

The idiosoma is broadly oval and heavily tanned. The dorsal shield is pointed anteriorly and is joined by a strip of interscutal membrane to a marginal shield. The internal margin of this encircling shield is evenly crenulate and it bears a series of regularly spaced setae. The surface of the dorsal shield is slightly pitted and from it arise a number of bent setae arranged as in Fig. 418. The ventral surface of the body is gently rounded and is covered by a continuous plate. A transverse suture extends across the body in front of the anus. The genital opening is covered by a shield-shaped plate with a pointed apex which extends between coxae II and IV. Six pairs of setae encircle the anterior part of the shield and, by transparency, many fine projections can be seen in the wall of the oviduct. Setae arise from the posterior part of the ventral shield and these are arranged as in Fig. 419.

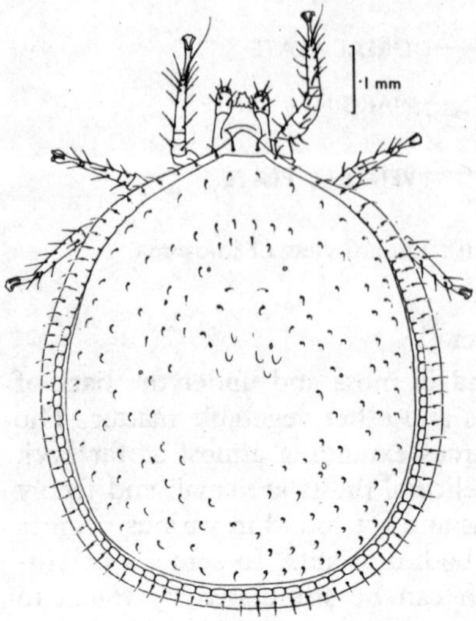

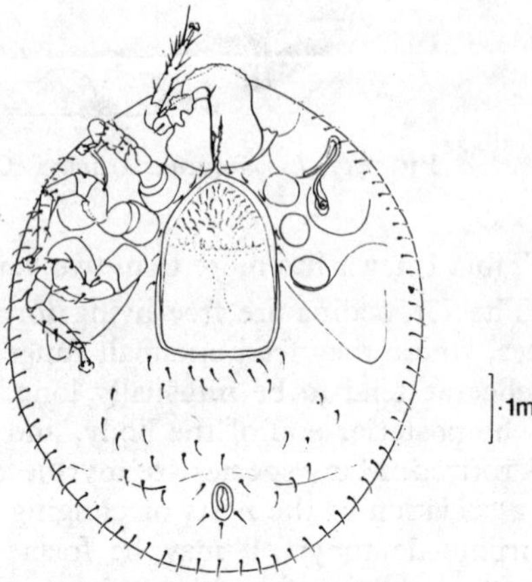

FIG. 418. *Leiodinychus krameri* (Canestrini). Dorsal view of ♀ (left).
FIG. 419. *Leiodinychus krameri* (Canestrini). Ventral view of ♀ with gnathosoma and left legs removed (right).

Leg grooves are well-developed and the margin of the posterior groove (the metapodal line) is obtusely pointed. The stigma opens between legs II and III and the peritreme outlines the ridge separating legs II and III.

The gnathosoma is feebly sclerotized. Hypostomal setae 2 and 3 and the capitular setae are branched, hypostomal setae I are smooth and flattened. The tectum is pointed, with a serrated edge. The pedipalps are short, the distal margins of the genu and femur serrated as in many uropodid mites. The chelicerae are long and thin, each digit bearing four or five teeth.

All the legs bear claws and pulvilli. The internal margins of coxae I (Fig. 420) are straight, the external produced into flat lobes with irregular edges. Similar flattened plates are attached to the internal edges of the trochanters and femora.

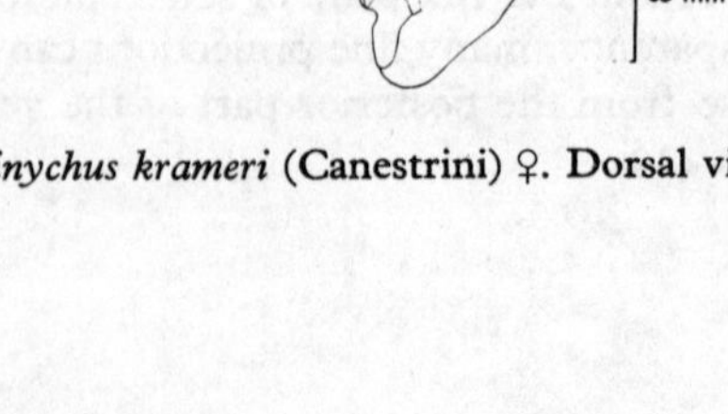

FIG. 420. *Leiodinychus krameri* (Canestrini) ♀. Dorsal view of left leg I.

MALE (Fig. 421). Length of idiosoma: about 680 μm.

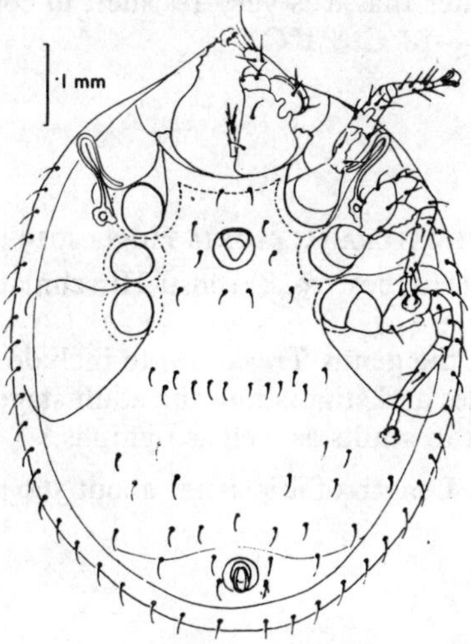

FIG. 421. *Leiodinychus krameri* (Canestrini). Ventral view of ♂ with gnathosoma and right legs removed.

This resembles the female. The genital opening, between coxae II and III, is covered by a small pear-shaped plate and encircled by five pairs of setae. A number of setae arise from the posterior region of the ventral shield and are arranged as in Fig. 421.

The gnathosoma (Fig. 416) differs from that of the female in having smooth hypostomal setae 2 and 3.

BIOLOGY

The life cycle takes from 54 to 80 days at 20°C and 95 per cent relative humidity with white bread as food. At 30°C and the same humidity it is completed in from 31 to 41 days. (Radinovsky, 1965a). Mites reared at 20°C were found to be larger than those reared at 30°C.

The male mates with the female by turning her over and, with the aid of chelicerae and legs, pushing a spermatophore under the apical region of the genital shield. The female then rights herself, but the male then climbs on to her back. The spermatophore remains in position and gradually becomes smaller as the spermatozoa are assimilated into the oviduct. Complete assimilation takes from 40 minutes to $7\frac{1}{4}$ hours and during this time, the male is carried around by the female. The spermatozoa are rod-shaped, 18.5 μm long, 1.5 μm wide (Radinovsky, 1965b).

Not more than one egg is usually laid per day and sometimes the young larvae escape from the egg at the moment of oviposition (Michael, 1889, Radinovsky, 1965b).

Sometimes the deutonymphal stage is prolonged (Radinovsky, 1965a, Hughes, 1961). This may be due to unfavourable conditions or it may be because the deutonymph is the stage at which dispersal normally occurs.

Very large numbers of these tortoise-like mites are often found in damp, mouldy sprouting barley, wheat or corn in ships and warehouses. It is also commonly found in moss. André (1934) records it as occurring in large numbers in a private house and J. D. Ethelston collected it from the litter of a broiler house. Sinha (1964b) states that it is very resistant to cold and will survive for more than 168 hours at —18°C ±1°C.

DISTRIBUTION

Cosmopolitan.

TREMATURA JACKSONIA Hughes, 1948

Trichouropoda jacksonia (Hughes, 1948) sensu Hirschmann and Zirngiebl-Nicol (1961).

Berlese (1918) erected the genus *Trematura* to include those species which he believed lacked peritremes and stigmata in the adult stage. Evans (1957) showed that stigmata are present in adults as well as nymphs.

FEMALE (Figs. 422, 423). Length of idiosoma: about 560 µm.

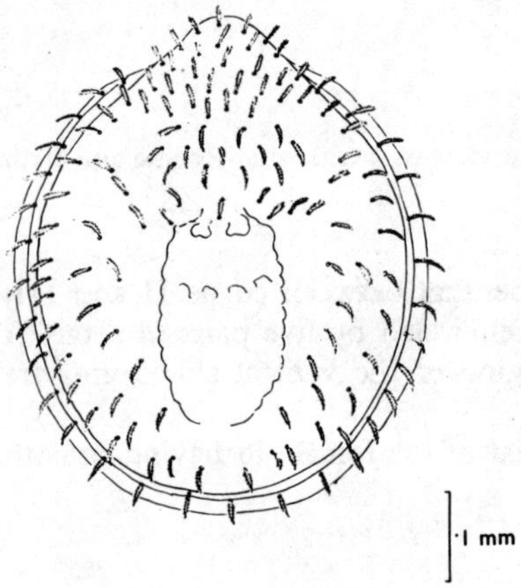

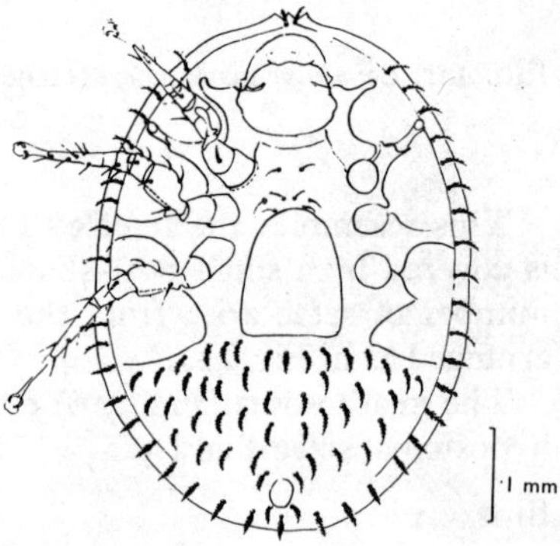

FIG. 422. *Trematura jacksonia* Hughes. Dorsal view of ♀ (left).

FIG. 423. *Trematura jacksonia* Hughes. Ventral view of ♀ (right). The gnathosoma, legs I and left legs II to IV have been removed.

The idiosoma is broadly oval in shape and dark brown in colour. The posterior margin of the body is distinctly rounded. The dorsal shield is almost smooth and from it and the marginal plate arise a number of colourless branched setae; in life, these give the mite an appearance of having been dusted with white powder. The ventral surface is flattened, its posterior half thickly covered with branched setae. The anterior half is bare, except for some smooth setae lying in front of the genital plate. The cuticle lining the leg grooves, as well as the marginal region of the body, are covered with reticulate and irregular thickenings (Fig. 425). The stigma is situated in the groove associated with leg III. The genital shield has a rounded anterior margin which is on a level with coxa III.

The gnathosoma is moderately well-sclerotized (Fig. 424b) and the tectum is a single lobe terminating in a thin, bifid prong (Fig. 424a). Both digits of the chelicerae (Fig. 424c) have forked tips and two to three feebly-developed teeth.

The coxae of legs I are flattened and square, their internal margins slightly concave, the external produced into membranous lobes. Similar lobes are also present on the trochanters and femora of all the legs.

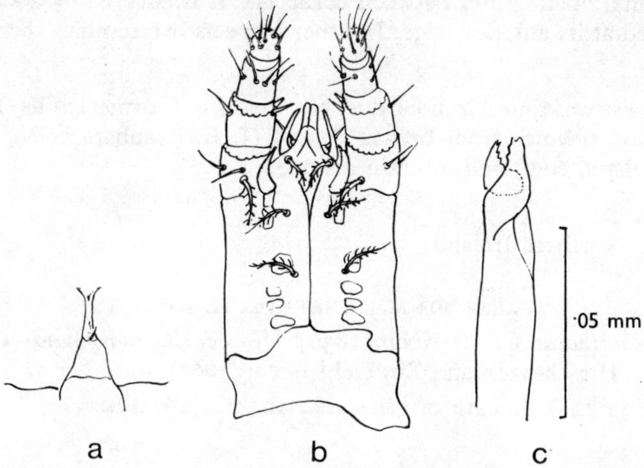

FIG. 424. *Trematura jacksonia* Hughes. (*a*) Tectum of ♂, (*b*) ventral view of gnathosoma of ♀ and (*c*) chelicera of ♀.

MALE (Fig. 425). Length of idiosoma: about 560 µm.

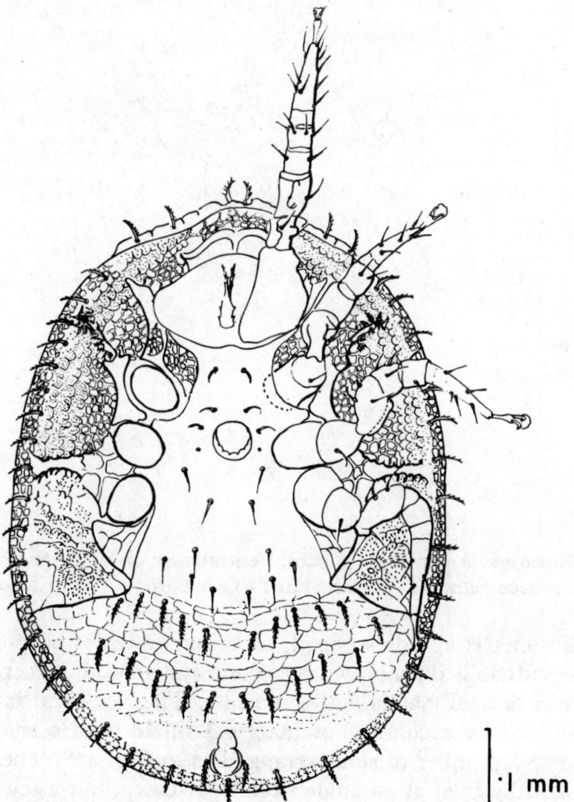

FIG. 425. *Trematura jacksonia* Hughes. Ventral view of ♂ with gnathosoma and right legs removed.

The genital opening lies between coxae III, it is covered by a small circular plate hinged at its anterior edge. In other respects it resembles the female.

BIOLOGY

T. jacksonia was found in floor dust obtained from oatmeal mills (P. M. Butler coll.), in fine sievings from brewer's yeast (J. E. Cranham coll.), in maltings (J. D. Ethelston coll.) and in grain residue.

DISTRIBUTION

England, Northern Ireland.

FUSCUROPODA MARGINATA (C. L. Koch, 1839)

Notaspis marginatus C. L. Koch, 1839; *Uroobovella marginata* (C. L. Koch, 1839) sensu Hirschmann and Zirngiebl-Nicol (1962).

FEMALE (Fig. 426). Length of idiosoma: about 1,210 μm.

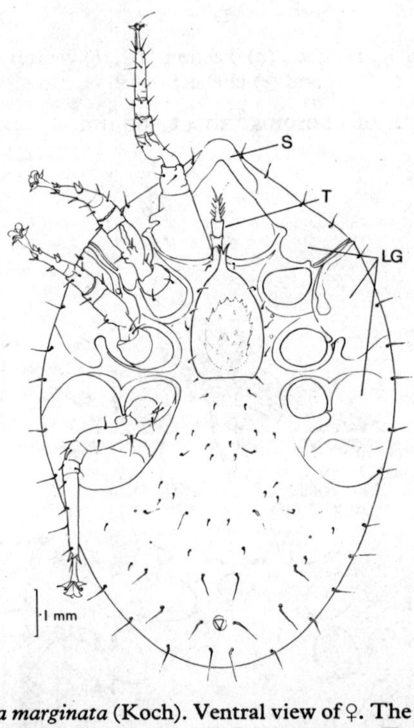

FIG. 426. *Fuscuropoda marginata* (Koch). Ventral view of ♀. The gnathosoma and legs have been removed. S scabellum; LG leg grooves; T tritosternum.

The idiosoma is spindle-shaped, tapering towards the posterior end of the body. The cuticle is dark brown in colour and smooth, except for the openings of numerous dermal glands and some pits. The dorsal surface is covered by a shield encircled by a continuous marginal shield with a smooth inner border. From it arise a number of setae arranged as in Fig. 428. These are smooth and characteristically bent at an angle near their bases, but a few terminal ones are straight and pectinate. A number of bent setae also arise from the ventral surface, as well as some much shorter ones. The genital opening is on a level with coxae II and III and is covered by a genital shield with a transparent bifid apex. The wall of the underlying oviduct is irregularly thickened. The exopodal region of the ventral surface is deeply grooved for the reception of the legs, the peritreme running from the stigma along the ridge separating legs II and III.

The gnathosoma is comparatively small, the hypostomal setae unequal in length and pectinate. The corniculi are short and the internal malae long, slender and pectinate. The chelicerae (Fig. 427) are long and slender, when

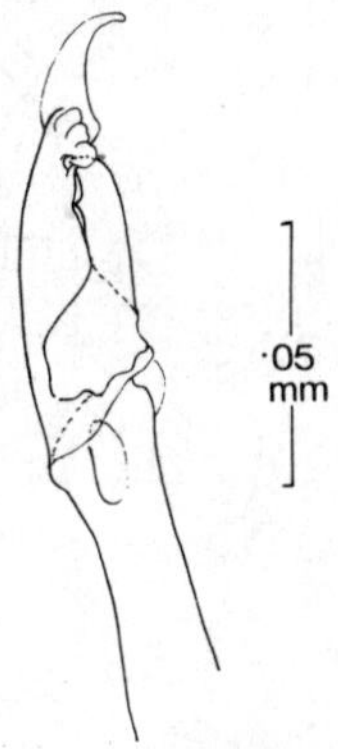

FIG. 427. *Fuscuropoda marginata* (Koch). Chelicera of ♀.

retracted, their shafts extend almost to the posterior margin of the body. The fixed digit terminates in a scalloped edge, to which is attached a transparent curved lobe. The movable limb extends to the base of this structure, it has a hooked end and a single small tooth in its distal half. The tritosternum has an oblong base and the proximal halves of the two laciniae are fused (Fig. 426).

The legs terminate in well-developed claws and pulvilli (Fig. 314a); the pulvilli on legs I are three-lobed, on legs II to IV five-lobed. The coxae of legs I are flattened, the internal margins are straight, but the external are produced into two backwardly-directed projections; dorsal to these is a membranous lobe with a fringed edge. Similar smaller lobes are also present on the internal margins of the trochanters and femora of legs II to IV. The second pair of legs bears stouter spines than the other legs; and these probably assist with egg-laying.

MALE (Figs. 428, 429). Length of idiosoma: about 1,120 μm.

The male closely resembles the female, but the idiosoma is more slender. The genital opening lies between coxae II and is covered by a small, pear-shaped plate hinged at its anterior end.

BIOLOGY

This has been extensively studied by H. Faasch (1967).

F. marginata is a scavenger, but it also feeds on fungal hyphae and spores and will attack soft-bodied animals such as nematodes, fly larvae and the immature stages of various mites, for example, *Caloglyphus mycophagus*. As in other uropodids, the spermatozoa are enclosed in a spermatophore which is manipulated by the male into the genital opening of the female with the help of his palps. Eggs are laid singly in a shallow hole excavated by the second pair of legs and the palps of the female.

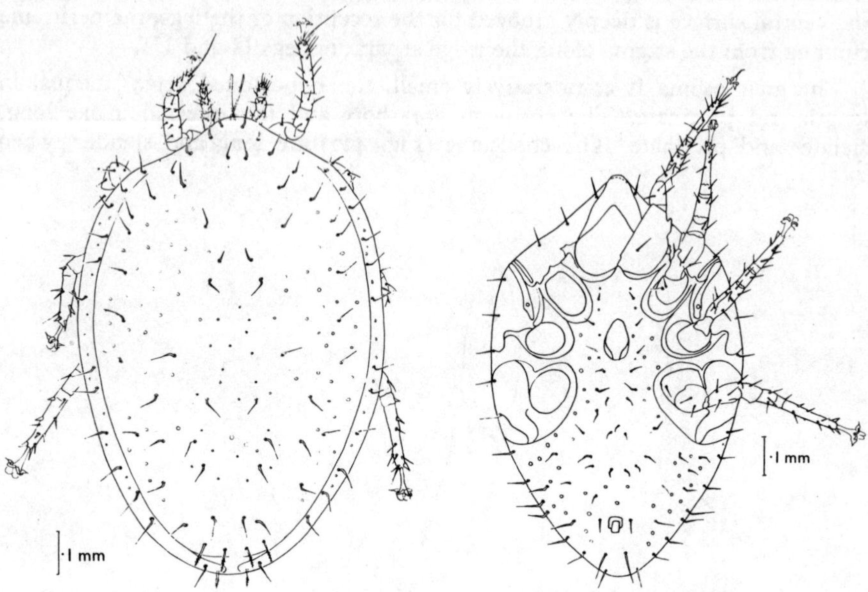

FIG. 428. *Fuscuropoda marginata* (Koch). Dorsal view of ♂ (left).
FIG. 429. *Fuscuropoda marginata* (Koch). Ventral view of ♂ (right).

In the course of the life cycle, one of two deutonymphs is formed: a 'dauernymph' (non-phoretic) or a 'wandernymph' which is phoretic. The wandernymph is also more resistant to drought and lack of food. The ambulacrum of the wandernymph differs from that of the dauernymph in having a more rounded pulvillus and proximally-directed claws, possibly better adapted for clinging to a smooth surface. In addition, the cuticle of the posterior region of the idiosoma is thickened and a gland lying above the hind gut is able to manufacture a secretion which is discharged through the anus and hardens in air to form a stalk. This remains attached to the anus and serves to fix the nymph to passing objects. Species of *Geotrupes* (dung beetles) are a favoured means of conveyance, but the nymphs will also attach themselves to Carabidae, Forficulidae and Myriapoda, in fact, to any smooth, water-repellant surface. When given a choice, the wandernymph will select species of *Geotrupes* and the presence of this beetle is detected by hair-like chemoreceptors situated on the apex of the tarsus of the first pair of legs. The stimulus provided by fresh manure induces the mite to end its wanderings. It then detaches itself from the beetle and in about two weeks moults into a male or female. According to Rohde (1959) the total life cycle ranges from 30 to 40 days.

F. marginata is commonly found in rotting plant material and in manure, and very large numbers of the wander or pedunculate nymphs were found on one occasion in a cucumber house by A. M. Brock. It is also sometimes found in damp, mouldy grain in warehouses and ships.

DISTRIBUTION

England, Holland, Germany, Italy, Jugoslavia.

UROSEIUS ACUMINATUS (C. L. Koch, 1847)

Uropoda acuminata C. L. Koch, 1847.

FEMALE (Figs. 430, 431). Length of idiosoma: about 1,160 μm.

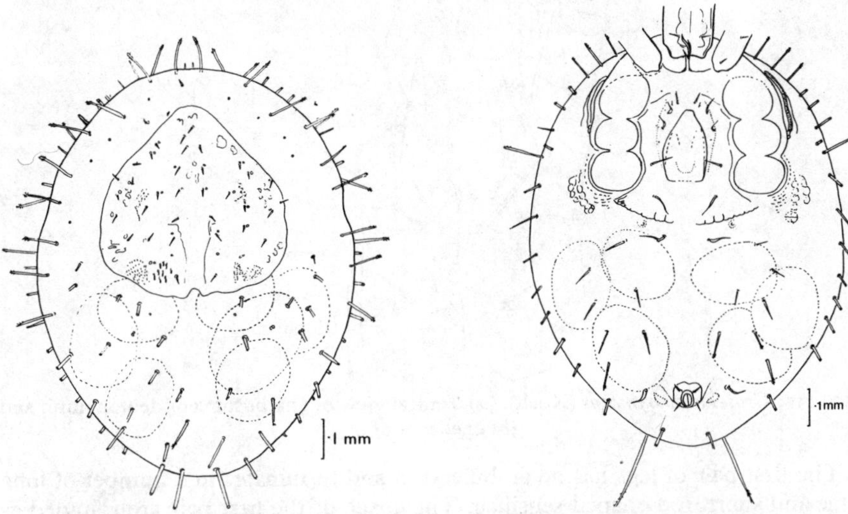

FIG. 430. *Uroseius acuminatus* (Koch) ♀. Dorsal view of idiosoma (left).
FIG. 431. *Uroseius acuminatus* (Koch) ♀. Ventral view of idiosoma (right).

The idiosoma is evenly sclerotized and brown in colour. The cuticle is rough, the dorsal shield almost triangular in shape and with an irregularly patterned surface, not very different in texture from that of the surrounding cuticle. From the shield arise a number of short, fine setae, not always symmetrically arranged. The setae arising from the surrounding cuticle are pectinate at their tips and brittle and therefore easily broken. Openings of dermal glands are scattered over the dorsal surface. The even sclerotization of the dorsum extends on to the ventral surface and forms a rigid covering, except in the region of the gnathosoma and legs. The genital opening which is midway between coxae II to IV, is covered by a shield and flanked by four pairs of setae, the fifth pair arising posterior to the genital shield.

The anus projects from the surface of the body and is surrounded by a distinct anal shield bearing two pairs of setae. The leg-bearing region of the body also projects, so as to partially conceal the peritreme; this extends from the anterior border of coxa II to the middle of coxa III and leads into a stigma and a small internal chamber. No leg grooves are present. An incomplete transverse line extends across the body behind coxae IV, and posterior to this, a number of setae arise from the surface of the body. The anterior ones are smooth, the posterior have pectinate ends. The tritosternum has an irregular serrated margin and arises from the anterior edge of the sternal plate.

The gnathosoma (Fig. 432a) has well-developed corniculi and the hypostomal and capitular setae are arranged in linear series, as is typical for the Uropodina. The tectum is very long, pointed and serrated. The apotele of the palp tarsus is bi-pronged. One of the two ventral setae of the palp trochanter (v_2) is long and irregularly forked. On the chelicera (Fig. 432b), the fixed digit is longer than the movable and ends in a colourless folded lobe.

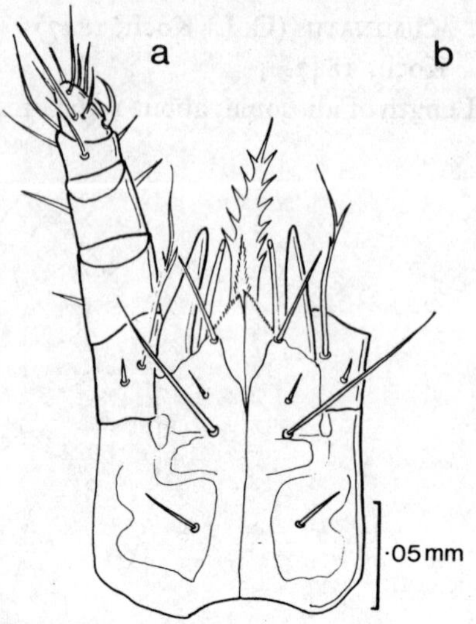

FIG. 432. *Uroseius acuminatus* (Koch). (*a*) Ventral view of gnathosoma of deutonymph and (*b*) chelicera of ♀.

The first pair of legs has no ambulacrum and terminates in a number of long setae and short, rod-shaped sensillae. The coxae of the first pair are rounded as in the Gamasina and widely spaced, so that the tritosternum is clearly visible.
MALE (Figs. 433, 434). Length of idiosoma: about 790 μm.

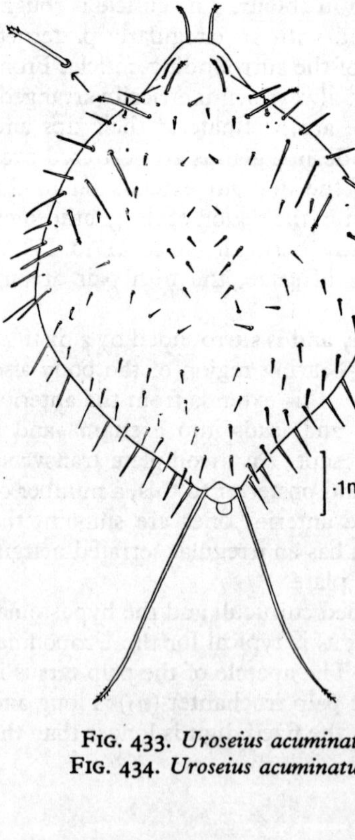

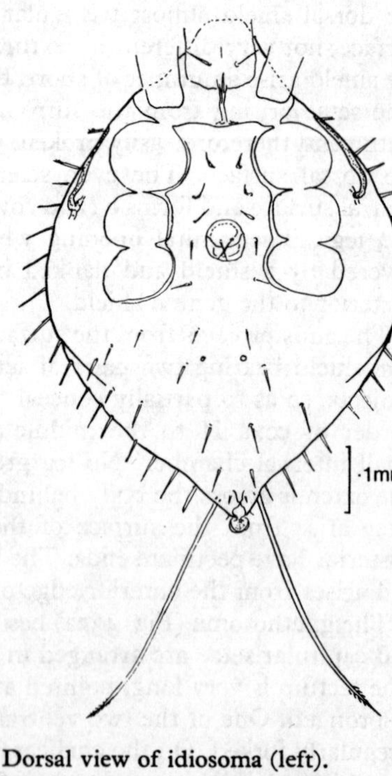

FIG. 433. *Uroseius acuminatus* (Koch) ♂. Dorsal view of idiosoma (left).
FIG. 434. *Uroseius acuminatus* (Koch) ♂. Ventral view of idiosoma (right).

The idiosoma is paler in colour than that of the female and more pointed anteriorly and posteriorly. The dorsal setae are arranged as in Fig. 433, the peripheral ones having pectinate ends, the central being short and smooth. A pair of very long setae arises from the bilobed posterior end of the body. The genital opening is between coxae III and IV and is covered by a small plate bearing two setae. Five pairs of setae are arranged on either side of it as in the female. The posterior region of the ventral surface bears a number of setae arranged as in Fig. 434. The anus is terminal and two pairs of setae lie close to it. A small membranous lobe is attached to the margin of the anus.

The gnathosoma is similar to that of the female. The second pair of legs (Fig. 435) has a large conical process arising from the ventral side of the femur, and smaller processes on genu and tibia.

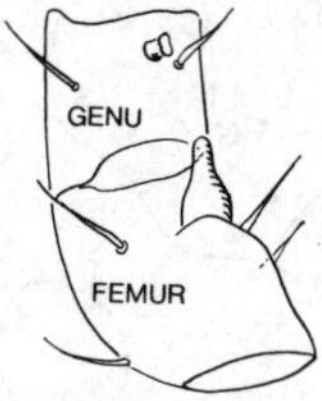

FIG. 435. *Uroseius acuminatus* (Koch) ♂. Femur and genu of leg II.

DEUTONYMPH (Fig. 436). Length of idiosoma: about 760 μm.

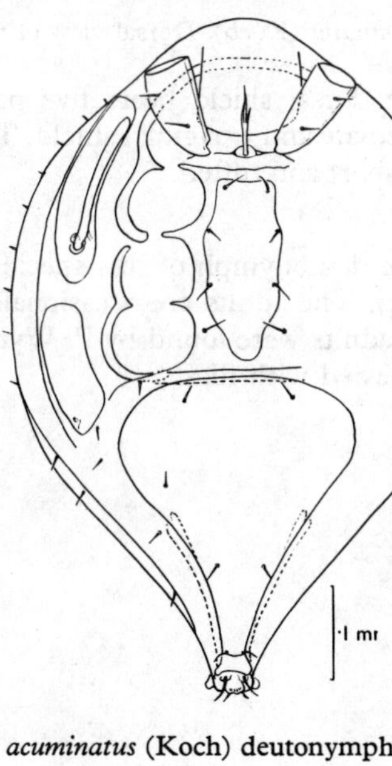

FIG. 436. *Uroseius acuminatus* (Koch) deutonymph. Ventral view of idiosoma.

The idiosoma is brown in colour and pear-shaped. The dorsal surface is covered by a single plate whose margins are reflected on to the ventral surface at the posterior end of the body. The ventral shields are arranged as in Fig. 436, the posterior one extending from coxae IV to the posterior margin of the body. The anus is terminal and provided with a number of membranous lobes and two elongated internal sclerites, the whole structure probably functioning as a sucker.

PROTONYMPH (Fig. 437). Length of idiosoma: about 590 µm.

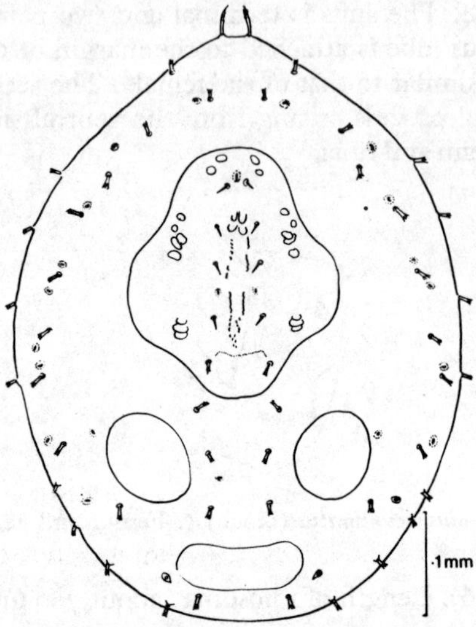

FIG. 437. *Uroseius acuminatus* (Koch). Dorsal view of protonymph.

This is oval in shape. The dorsal shield bears five pairs of fine setae. In addition, there are two intermediate and a pygidial shield. The setae arising from the interscutal membrane are short and tufted.

BIOLOGY

Vitzthum (1928) records the deutonymph of this species from bee hives and attached to *Aphiochaeta rufipes*. The adults are occasionally found in decaying plant material. Nymphs and adults were found by P. Wynter Cross on mouldy wheat in a ship which was infested with flies.

DISTRIBUTION

England, Germany.

Habitats of some species of mites

Inhabitants of grasslands, pastures, in and beneath haystacks, in soil and on leaves of plants

Astigmata
- *Acarus farris*
  - *A. immobilis*
- *Tyrophagus longior*
  - *T. neiswanderi*
  - *T. similis*
  - *T. palmarum*
- *Mycetoglyphus fungivorus*
- *Caloglyphus mycophagus*
  - *C. oudemansi*
  - *C. rhizoglyphoides*
- *Rhizoglyphus càllae*
  - *R. robini*
- *Glycyphagus destructor*
- *Ctenoglyphus plumiger*
  - *C. canestrinii*
- *Histiostoma feroniarum*
  - *H. sapromyzarum*

Cryptostigmata
- *Trichoribates trimaculatus*

Prostigmata
- *Cheyletus eruditus*
  - *C. trouessarti*
- *Cheletomorpha lepidopterorum*

Mesostigmata
- *Androlaelaps casalis casalis*
- *Hypoaspis aculeifer*
- *Proctolaelaps pygmaeus*
  - *P. scolyti*
- *Arctoseius butleri*
- *Kleemannia plumosus*
  - *K. plumigera*
- *Lasioseius penicilliger*
- *Fuscuropoda marginata*

Inhabitants of storage premises containing grain, flour, animal feeding material, etc.

Astigmata
- *Acarus siro*
  - *A. farris*
- *Tyrophagus putrescentiae*
  - *T. perniciosus*
- *Tyroborus lini*
- *Tyrolichus casei*
- *Aleuroglyphus ovatus*
- *Thyreophagus entomophagus*
- *Suidasia nesbitti*
  - *S. medanensis*
- *Caloglyphus berlesei*
- *Glycyphagus destructor*
  - *G. domesticus*
  - *G. privatus*
  - *G. michaeli*
  - *G. ornatus*
- *Blomia freemani*
- *Ctenoglyphus canestrinii*
  - *C. palmifer*
  - *C. plumiger*
- *Gohieria fusca*
- *Chortoglyphus arcuatus*
- *Euroglyphus (Gymnoglyphus) longior*
  - *E. (E.) maynéi*
- *Dermatophagoides farinae*

Prostigmata
- *Cheyletus eruditus*
  - *C. aversor*
  - *C. malaccensis*
  - *C. trouessarti*
- *Acaropsis sollers*
- *Eucheyletia taurica*
- *Cheletomorpha lepidopterorum*
- *Tarsonemus granarius*
- *Tydeus interruptus*

Mesostigmata
- *Haemogamasus pontiger*
- *Androlaelaps casalis casalis*
- *Eulaelaps stabularis*
- *Hypoaspis aculeifer*
  - *H. lubrica*
  - *H. miles*
  - *H. sardoa*
- *Proctolaelaps pomorum*
- *Lasioseius penicilliger*
- *Kleemannia plumigera*
  - *K. plumosus*
- *Blattisocius tarsalis*
  - *B. keegani*
  - *B. dentriticus*
  - *B. mali*

Macrocheles matrius
M. muscaedomesticae
Leiodinychus krameri
Tremature jacksonia
Fuscuropoda marginata

Inhabitants of dried fruit, christmas puddings, etc.

Astigmata
Carpoglyphus lactis

Mesostigmata
Melichares agilis
Blattisocius mali

Inhabitants of dried meat, bones, hides, dried fish and fishmeal.

Astigmata
Lardoglyphus konoi
L. zacheri
Glycyphagus domesticus
Pyroglyphus (Hughesiella) africanus

Inhabitants of cheese stores, dried eggs, copra, linseed, ground nuts, etc.

Astigmata
Acarus siro
A. farris
A. immobilis
Tyrophagus putrescentiae
T. longior
T. palmarum
T. perniciosus
T. tropicus

Astigmata—contd:
T. brevicrinatus
Tyrolichus casei
Suidasia medanensis
Caloglyphus berlesei
C. oudemansi
Glycyphagus domesticus
Euroglyphus (E.) maynéi

Inhabitants of houses

A. House dust, on walls and on window-sills

Astigmata
Glycyphagus domesticus
G. privatus
Acarus siro
Chortoglyphus arcuatus
Dermatophagoides pteronyssinus
D. farinae
D. microceras

Cryptostigmata
Phauloppia lucorum

Prostigmata
Cheyletus eruditus
Bryobia praetiosa

Mesostigmata
Androlaelaps casalis casalis
Haemogamasus pontiger
Kleemannia plumigera
K. plumosus
Proctolaelaps pomorum
P. pygmaeus
P. pomorum
Melichares agilis
Lasioseius penicilliger
Dermanyssus gallinae
Neoseiulus barkeri

B. Mattress dust

Astigmata
Dermatophagoides pteronyssinus
D. farinae
D. microceras
Euroglyphus (E.) maynéi
Chortoglyphus arcuatus

Astigmata—contd:
Gohieria fusca
Glycyphagus domesticus

Prostigmata
Cheyletus eruditus

C. Inside feather pillows

Dermatophagoides evansi
  D. microceras

Inhabitants of small mammal nests

Astigmata
- *Acarus farris*
- *Mycetoglyphus fungivorus*
- *Aleuroglyphus ovatus*
- *Tyrolichus casei*
- *Caloglyphus rhizoglyphoides*
- *Glycyphagus destructor*
  - *G. bicaudatus*
  - *G. michaeli*
  - *G. ornatus*

Prostigmata
- *Cheyletus eruditus*
  - *C. trouessarti*
- *Eucheyletia taurica*

Mesostigmata
- *Androlaelaps casalis casalis*
- *Hypoaspis lubrica*
  - *H. miles*
  - *H. sardoa*
- *Haemogamasus pontiger*
- *Eulaelaps stabularis*
- *Proctolaelaps pygmaeus*

Inhabitants of birds' nests

Astigmata
- *Acarus farris*
  - *A. gracilis*
  - *A. immobilis*
- *Tyrophagus longior*
  - *T. palmarum*
  - *T. similis*
- *Thyreophagus entomophagus*
- *Chortoglyphus arcuatus*
- *Glycyphagus domesticus*
  - *G. ornatus*
  - *G. privatus*

Astigmata—contd.
- *Austroglycyphagus geniculatus*
- *Diamesoglyphus intermedius*

Cryptostigmata
- *Trichoribates trimaculatus*

Prostigmata
- *Cheyletus eruditus*
- *Cheletomorpha lepidopterorum*

Mesostigmata
- *Androlaelaps casalis casalis*
- *Hypoaspis lubrica*
- *Blattisocius keegani*
- *Dermanyssus gallinae*

Inhabitants of bat roosts

Astigmata
- *Acarus gracilis*
- *Glycyphagus domesticus*
- *Coproglyphus stammeri*
- *Carpoglyphus munroi*

Mesostigmata
- *Androlaelaps casalis casalis*
- *Haemogamasus pontiger*

Inhabitants of deep litter of broiler houses

Astigmata
- *Acarus siro*
  - *A. farris*
- *Tyrophagus longior*
- *Tyroborus lini*
- *Aleuroglyphus ovatus*

Astigmata—contd.
- *Caloglyphus berlesei*
  - *C. oudemansi*
- *Chortoglyphus arcuatus*
- *Glycyphagus destructor*

Prostigmata
Cheyletus aversor
Mesostigmata
Androlaelaps casalis casalis
Hypoaspis lubrica
Eulaelaps stabularis
Haemogamasus pontiger
Mesostigmata—contd.
Proctolaelaps pygmaeus
P. scolyti
Kleemannia plumigera
Digamasellus presepum
Macrocheles muscaedomesticae
M. matrius

Preparation and storage of mites

EXAMINATION OF LIVE MATERIAL

When possible, it is important to examine living mites under a dissecting binocular microscope. Much useful information can be obtained regarding shape of body (always distorted on mounting); colour: either due to the presence of pigment or degree of tanning; gait and also relative sizes of different individuals of the same or unrelated species. For an account of the different techniques used in the collection of mites, see Evans, Sheals and Macfarlane (1961).

EXAMINATION OF DEAD MATERIAL

For any detailed examination, microscopic preparations must be made of the smaller mites. For this use ordinary glass slides and small coverslips (about ⅜ or ½ in. diameter) as it is easier to find the mites and the more delicate mites are not so easily crushed.

Temporary preparations

1. Mount the mites direct into 50 to 100 per cent *lactic acid*, using the lower concentrations for the weakly sclerotized forms. Warm the slide over a electric light bulb or on a hot plate kept at about 60°C. In time, the mites will clear if left in the cold. The effect of the lactic acid is gently to dissolve the soft parts of the body without causing shrinkage of the cuticle; it will not quickly dissolve guanin (often present in the body as an opaque white mass) or blood present in the gut of blood-sucking species. Mites can be re-orientated by (a) gently pushing the coverslip so that the mite can be rolled over; (b) using a grooved piece of plastic which permits a needle to be inserted under the coverslip (see Evans and Browning, 1955): (c) using a cavity slide with the depression half-covered by a coverslip (see Grandjean, 1949).
2. Use *Lactophenol* in place of lactic acid. This is liable to cause shrinkage in soft-bodied mites.
3. Use *lactic acid* coloured with lignin pink. This is often useful for poorly-tanned species as lignin pink stains the cuticle. *Note:* Guanin can be removed from the body by soaking for *as short a time as possible* in 1 per cent caustic potash solution.

Permanent preparations

Canada balsam is unsuitable for permanent preparations. The most suitable medium is *Faure's modification of Berlese medium.* Mount the mites direct into this and warm to clear them. The preparations need ringing after a few days, either with *glyceel* (obtainable from Gurrs), *gold size*, or ordinary *nail varnish.* The disadvantages of Berlese medium are (1) it can cause shrinkage of some more delicate mites; (2) minute structures such as setae, are not so clearly visible. Specimens mounted in Berlese medium are clearer when a phase-contrast microscope is used.

Permanent preparations can also be made by using C-M medium which contains a high proportion of lactic acid. Mites can be mounted direct into this medium, or else they can be cleared first of all in lactophenol. Material treated with potassium hydroxide should be rinsed in acid alcohol before mounting. Polyvinyl-alcohol mixtures are not suitable as they cause excessive shrinkage.

STORAGE OF UNMOUNTED SPECIMENS

When possible, it is desirable to keep a number of unmounted specimens as well as mounted ones. Mites can be preserved indefinitely in (a) 70 *to* 80 *per cent ethyl alcohol*; (b) *Oudemans' fluid*. The Oudemans' fluid is better for soft-bodies species as there is less shrinkage.

In either case, the mites can be placed in a small tube 25 x 6 mm (1 x ¼ in.) plugged with elder pith, and this is then put into a 51 x 13 mm (2 x ½ in.) tube plugged with cotton wool and also filled with preservative. Labels can be placed inside the larger tube and will be clearly visible. Many of these tubes can be stored together in larger containers and there is little danger of drying-up.

CULTURING OF MITES

Many free-living mites can be kept alive in individual cells. A convenient and simple cell can be made by boring a 3 mm (⅜ in.) diameter hole with sloping sides in 2 mm (1/16 in.) perspex. The base of the cell can be made by fixing a piece of taffeta (or any other material of similar weave) or filter paper with 'Durofix'; the lid is a 22 mm (⅞ in.) coverslip held in place by rubber bands. Use dark coloured material for colourless mites and vice versa. Mites are very sensitive to desiccation and high humidity can be maintained by enclosing the cells in desiccators containing a saturated salt solution. Sodium chloride gives a relative humidity of 76 per cent, potassium chloride 87 per cent, potassium sulphate 98 per cent. Condensation can be prevented by keeping the mites at a constant temperature: 20-25°C is suitable for many mites. As food, yeast, wheat germ or dried heart muscle is acceptable to many free-living mites, and predatory mites will often feed on the common flour mite *Acarus siro* or its eggs. The amount of food given should bear some relation to the size of the mite under observation—too much yeast or wheat germ rapidly develops mould which smothers the mites. These cells have been found suitable for culturing most astigmatid and some mesostigmatid mites.

1 Temporary mountants

(a) 50 to 100 per cent lactic acid.
Use lower concentrations for weakly sclerotized forms.

(b) Lactophenol

Phenol	20 g
Lactic acid	20 g (=16.5 ml)
Glycerol	40 g (=32.0 ml)
Distilled H_2O	20 ml

(c) Lactic acid and lignin pink (Evans and Browning, 1955)

Lactic acid	60 parts
Glycerine	40 parts
Lignin pink	trace

2 Permanent mountants

(a) Faure's modification of Berlese medium.

Distilled water	50 ml
Chloral hydrate	50 g
Glycerine	20 ml
Gum arabic	30 g

Mix in order given above and filter through bolting silk.

(b) C-M medium (Clark and Morishita, 1950).

Methocellulose (Methocel)	5 g (from B.D.H.)
Carbowax	2 g (from B.D.H.)
Diethylene glycol (Digol)	1 ml (from B.D.H.)
Ethyl alcohol (95 per cent)	25 ml
Lactic acid	100 ml
Distilled water	75 ml

The methocellulose and alcohol are mixed, added to remaining ingredients and filtered through glass wool. The medium is then placed in an oven at 40-45°C for three to five days, or until it has reached the desired consistency. The viscosity can be reduced by adding 95 per cent ethyl alcohol.

3 Preservation

(a) Oudemans' fluid

Ethyl alcohol (70 per cent)	87 parts
Glycerine	5 parts
Glacial acetic acid	8 parts

(b) 70-80 per cent ethyl alcohol.

Bibliography

ANDRÉ, M. (1934). Note sur un Acarien (*Leiodinychus krameri* G. and R. Can.) parasite des habitations humaines. *Bull. Soc. zool Fr.* **59,** No. 2, 112–22.

ATHIAS-HENRIOT, C. (1961). Mésostigmates (Urop. exel.) édaphiques méditerranéens (Acaromorpha, Anactinotrichida) (Collect. Prof. H. Franz et C. Athias-Henriot). Première Série. *Acarologia* 2, 159–174.

ATHIAS-HENRIOT, C. (1966). Contribution à l'étude des *Amblyseius* paléarctiques (Acariens, Anactinotriches, Phytoseiidae). *Bull. sci. Bourgogne* **24,** 181–230.

ATHIAS-HENRIOT, C. (1968). L'appareil d'insemination laelapoide (Acariens Anactinotriches: Laelapoidea, ♀♀). Premières observations possibilité d'emploi à des fins taxonomiques. *Bull. scient. Bourgogne* **25,** 229–74.

ATTIAH, H.H. (1969). Tyroglyphoid mites associated with stored food in U.A.R. (Sarcoptiformes–Acarina). *United Arab Republic. Ministry of Agriculture, Plant Protection Department.* Tech. Bull. No. 10, 1–51.

ATYEO, W.T. (1960). A revision of the mite family Bdellidae in North and Central America (Acarina: Prostigmata). *Univ.Kansas Sci. Bull.* **40,** 345–499.

BAKER, R.A. (1964a). Some preliminary observations on the mite fauna of sewage bacteria beds. *Ann. Mag. nat. Hist.* (13) **7,** 707–9.

BAKER, R.A. (1964b). The further development of the hypopus of *Histiostoma feroniarum* (Dufour, 1839) (Acari). *Ann. Mag. nat. Hist.* (13) **7,** 693–5.

BAKER, E.W. (1949). A review of the mites of the family Cheyletidae in the United States National Museum. *Proc. U.S. Nat. Mus.* **99,** 267–320.

BAKER, E.W. and BALOCK, J.W. (1944). Mites of the family Bdellidae. *Proc. ent. Soc. Washington* **46,** 7, 176–84.

BAKER, E.W. and HOFFMANN, A. (1948). Acaros de la familia Cunaxidae. *Anales de la Escuela Nacional de Ciencias Biologicas* V (3–4), 229–73.

BAKER, E.W. and WHARTON, G.W. (1952). An Introduction to Acarology. *The Macmillan Company, New York.*

BAKER, E.W., EVANS, T.M., GOULD, D.J., HULL, W.B. and KEEGAN, H.L. (1956). A manual of parasitic mites of medical or economic importance. *Nat. Pest Control Assoc., New York,* 1–170.

BARKER, P.S. (1967a). The effects of high humidity and different temperature on the biology of *Tyrophagus putrescentiae* (Schrank) (Acarina: Tyroglyphidae). *Can. J. Zool.* **45,** 91–6.

BARKER, P.S. (1967b). Bionomics of *Blattisocius keegani* (Fox) (Acarina: Ascidae), a predator on eggs of pests of stored grain. *Can. J. Zool.* **45,** 1,093–9.

BARKER, P.S. (1967c). Effect of humidity and temperature on the biology of *Aëroglyphus robustus* (Banks) (Acarina: Glycyphagidae). *Can. J. Zool.* **45,** 479–83

BARKER, P.S. (1968a). Effect of food quality on the reproduction of Mesostigmata: a review. *Manitoba Ent.* **2,** 46–8.

BARKER, P.S. (1968b). Bionomics of *Androlaelaps casalis* (Berlese) (Acarina: Laelapidae) a predator on mite pests of stored grain. *Can. J. Zool.* **46,** (6), 1,099–1,102.

BARKER, P.S. (1968c). Notes on the bionomics of *Haemogamasus pontiger* (Berlese) (Acarina: Mesostigmata), a predator on *Glycyphagus domesticus* (De Geer). *Manitoba Ent.* **2,** 85–7.

BARKER, P.S. (1969). The response of a predator, *Hypoaspis aculeifer* (Canestrini) (Acarina: Laelapidae), to two species of prey. *Can. J. Zool.* **47,** 343–5.

BEER, R.E. (1954). A revision of the Tarsonemidae of the Western hemisphere (order Acarina). *Kans. Univ. Sci. Bull.* **36,** 1,091–1,387.

BEER, R.E. and DAILEY, D.T. (1956). Biological and systematic studies of two species of cheyletid mites, with a description of a new species (Acarina, Cheyletidae). *Kans. Univ. Sci. Bull.* **38,** 393–437.

BERLESE, A. (1887). *Acari, Myr. Scorp. Ital.* fasc. **38,** No. 8.

BERLESE, A. (1897). *Acari, Myr. Scorp. Ital.* fasc. **89,** No. 10.

BERLESE, A. (1899). Gli acari agrarii. *Riv. Pat. veg.* **7,** 312–44.

BERLESE, A. (1918). Intorno agli Uropodidae. *Redia* **13,** 7–16.

BOCZCK, J. (1959). Biology and ecology of *Cheyletus eruditus* (Schrank). *Prace Naukowe, Inst. Ochrony Róslin, Warszawa* **1,** 175–230.

BOCZCK, J., JURA, C. and KRZYSZTOFOWIEZ, A. (1969). The comparison of the structure of the internal organs of post-embryonic stages of *Acarus farris* (Oud.) with special reference to the hypopus. In G.O. Evans (Ed.) Proc. 2nd. int. Congr. Acarology 1967. (*Akad. Kaidó*) *Bpest*, 265–71.

BOGDANOV, A. (1864). Deux acariens trouvés par M. Scheremetewsky sur homme. *Bull. Soc. Nat. Moscou* **37,** 341–5.

BOLLAERTS, D. and BRENY, R. (1951). Les acariens nuisibles aux matières entreposées. *Rev. Agric. Brux.* **4,** 738–64.

BOOTH, B.H. and JONES, R.W. (1952). Epidemiological and clinical study of grain itch. *J. Amer. Med. Ass.* **150,** 1,575–9.

BOULANOVA, E. (1937). Notes morphologique et biologique sur *Ferminia fusca* Ouds. *Bull. Soc. Nat. Moscow Biol. N.S.* **46,** 6, 304–16.

BRADY, J. (1970). The mites of poultry litter. Observations on the bionomics of common species, with a species list for England and Wales. *J. appl. Ecol.* **7,** 331–48.

BREGETOVA, N. (1956). Gamasid mites (Gamasoidea)—concise determinations. *Tabl. anal. Faune. U.S.S.R.* No. 61, 1–563.

BRODY, A.R. and WHARTON, G.W. (1970). *Dermatophagoides farinae*: ultra-structure of lateral opisthosomal dermal glands. *Trans. Amer. Microsc. Soc.* **89**(4), 499–513.

CANESTRINI, G. (1888). Prospetto dell' Acarofauna Italiana, III Padova.

CANESTRINI, G. (1891). Abbozzo del sistema Acarologico. *Atti. Ist. veneto*, **38,** 699–725.

CHABASSE, Y. and GENTHON, H. (1966). Les acariens detricoles. Parasitisme d'un jambon par *Glycyphagus domesticus*. *Recl. Méd. vét. Ec. Alfort* 142, 1,049–54.

CHANT, D.A. (1959). Phytoseiid mites (Acarina: Phytoseiidae). Part 1. Bionomics of seven species in South-Eastern England. Part 2. A taxonomic review of the family Phytoseiidae, with descriptions of 38 new species. *Canad. Ent.* **91,** Suppl. 12, 1–164.

CHANT, D.A. (1961). The effect of prey density on prey consumption and oviposition in adults of *Typhlodromus* (*T.*) *occidentalis* Nesbitt (Acarina: Phytoseiidae) in the laboratory. *Canad. J. Zool.* **39,** 311–15.

CHANT, D.A. (1963). The subfamily Blattisocinae Garman (=Aceosejinae Evans) (Acarina: Blattisocidae Garman) (=Aceosejidae Baker and Wharton) in North America, with descriptions of new species. *Canad. J. Zool.* **41,** 243–305.

CHMIELEWSKI, W. (1967a). The question of applied terminology for development stages of mites of the suborder Sarcoptiformes. *Polskie Pismo ent.* **37,** 603–9.

CHMIELEWSKI, W. (1967b). Stadium hypopus *Carpoglyphus lactis* (L., 1758) Glycyphagidae, Acarina). *Prace. nauk. Inst. Ochr. Róslin* **9,** 161–73.

CHMIELEWSKI, W. (1969). Fauna of mites in stored seeds of sugar beet. *Polskie Pismo ent.* **39,** 619–28.

CHMIELEWSKI, W. (1970). The passage of mites through the alimentary canal of vertebrates. *Ekologia Polska* **35,** 741–56.

CLARK, E.W. and MORISHITA, F. (1950). C-M medium: a mounting medium for small insects, mites and other whole mounts. *Science* 112 2922: 789–90.

CLEAT, N.D. (1952). Growth in the laboratory of economically important Oribatid mites. *Nature, Lond.* **169,** 280–81.

COOMBS, C.W. and WOODROFFE, G.E. (1968). Changes in the arthropod fauna of an experimental bulk of stored wheat. *J. appl. Ecol.* **5,** 563–74.

COOREMAN, J. (1942). Notes et observations sur les Acariens, 1. *Bull. Mus. Hist. Nat. Belg.* **18,** 33, 1–7.

COOREMAN, J. (1944). Notes et observations sur les Acariens. *Bull. Mus. Hist. Nat. Belg.* **20,** 8, 1–16.

COOREMAN, J. (1950). sur un Acarien nouveau, préjudiciable sur matières alimentaires entreposées: *Mealia maynéi* n.sp. *Bull. Ann. Soc. ent. Belg.* **86,** 164–8.

COOREMAN, J. (1959). Note sur le genre *Aëroglyphus* Zachvatkine, 1941. (Acaridae, Glycyphagidae). *Inst. Roy. Sci. Nat. Belgique* Bull. **35,** (33), 1–19.

COOREMAN, J. (1962). Invasion massive des habitations par *Haemolaelaps casalis* (Berlese) (Acari, Mesostigmata). *Bull. Ann. Soc. ent. Belg.* **98,** 388–91.

COSTA, M. (1961). Mites recovered from the nests of the Levant vole (*Microtus guentheri*) in Israel. *Ann. Mag. nat. Hist.* (13) **4,** 257–82.

CRUMB, S.E. (1934). A classification of some noctuid larvae of the subfamily Hypeninae. *Entomologica Americana* **14,** 159–60.

CUNNINGTON, A.M. (1965). Physical limits for complete development of the Grain mite, *Acarus siro* (Acarina, Acaridae), in relation to its world distribution. *J. appl. Ecol.* **2,** 295–306.

CUNNINGTON, A.M. (1969). Physical limits for complete development of the copra mite, *Tyrophagus putrescentiae* (Schrank) (Acarina: Acaridae). In G.O. Evans (Ed.) Proc. 2nd. int. Congr. Acarology 1967 (*Akad. Kaidó*) *Bpest.*, 241–8.

CUNNINGTON, A.M. (1971). House-dust mites and respiratory allergy: a note on the identity of *Dermatophagoides farinae* Hughes, 1961. *Clinical Allergy* **1,** 4, 447–9.

CUTCHER, J.J. and WOODRING, J.P. (1969). Environmental regulation of hypopial apolysis of the mite, *Caloglyphus boharti*. *J. Insect Physiol.*, **15,** 2045–57.

DAVIES, C.J. (1926). Vent disease in rabbits. *Veterin. J. London*, **83,** 320–322.

DE LEON, D. (1963). A new *Dermatophagoides*: it prevents the rising of self-rising flour (Acarina: Epidermoptidae). *Florida Ent.* **46,** 247–50.

DESCHIENS, R. (1951). L'acariase de l'appareil respiratrice chez les primates et chez l'homme. *Ann. Inst. Pasteur Paris*, **80,** 107–47.

DOSSE, G. (1958). Die Spermathecae, ein zusätzliches Bestimmungsmerkmal bei Raubmilben (Acar., Phytoseiidae). *PflSchBer. Ber.* **20,** 1–11.

DUBININ, V.B. (1953). Arachnoidea—Feather mites (Analgesoidae). Part 2. Families Epidermoptidae and Freyanidae. *Fauna U.S.S.R.* 6.6, 1–412.

EDWARDS, A.R. (1952). Support for the view that *Cheyletus eruditus* Schrank (Acarina, Trombidiformes) has heteromorphic males. *Ent. Mon. Mag.* **83,** 107.

EHARA, S. (1964). Some mites of the families Phytoseiidae and Blattisocidae from Japan (Acarina: Mesostigmata). *J. Fac. Sci. Hokkaido Univ.* (6) *Zool.* **15,** 378–94.

EVANS, G.O. and BROWNING, E. (1956). British mites of the subfamily Macrochelinae Trägårdh (Mesostigmata-Macrochelidae). *Brit. Mus.* (*nat. Hist.*) *Zool.* **4,** 1–54.

EVANS, G.O. (1957). An introduction to the British Mesostigmata (Acarina) with keys to families and genera. *J. Linn. Soc. Lond. Zool.* **43,** No. 291, 203–59.

EVANS, G.O. (1958). A revision of the British Aceosejinae (Acarina: Mesostigmata). *Proc. zool. Soc. Lond.* **131,** 177–229.

EVANS, G.O. (1963). Observations on the chaetotaxy of the legs in the free-living Gamasina (Acari: Mesostigmata). *Bull. Brit. Mus.* (*nat. Hist.*) *Zool.* **10,** 275–303.

EVANS, G.O. (1972). Leg chaetotaxy and the classification of the Uropodina (Acari: Mesostigmata). *J. Zool., Lond.* **167,** 193–206.

EVANS, G.O., and BROWNING, E. (1955). Techniques for the preparation of mites for study. *Ann. Mag. nat. Hist.* (12) **8,** 631–5

EVANS, G.O., SHEALS, J.G. and MACFARLANE, D. (1961). The terrestrial Acari of the British Isles. An introduction to their morphology, biology and classification. Vol. 1. Introduction and biology. *London* (*British Museum*) *nat. Hist.*, 1–219.

EVANS, G.O. and TILL, W.M. (1965). Studies on the British Dermanyssidae (Acari: Mesostigmata). Part 1. External morphology. *Bull. Br. Mus.* (*nat. Hist.*) (*Zool.*). **13,** 247–94.

EVANS, G.O. and TILL, W.M. (1966). Studies on the British Dermanyssidae (Acari: Mesostigmata). Part 2. Classification. *Bull. Br. Mus.* (*nat. Hist.*) (*Zool.*) **14,** 109–370.

EYNDHOVEN, G.L. VAN. (1960). Artunterschiede beim genus *Rhizoglyphus* (Acar.). *Verh. XI intern. Kongr. Ent. Wien.* 17–25 August 1960 1, 274–276.

EYNDHOVEN, G.L. VAN. (1963). The *Rhizoglyphus echinopus* of Fumouze and Robin. *Mitt. schweiz. ent. Ges.* **36,** 48–9.

EYNDHOVEN, G.L. VAN. (1964) *Cheletomorpha lepidopterorum* (Shaw, 1794) (=*Ch. venustissima*) (Acari, Cheyletidae) on Lepidoptera. *Beaufortia* **11,** 53–60.

EYNDHOVEN, G.L. VAN. (1968). *Rhizoglyphus engeli* nov. spec. with notes on the genus *Rhizoglyphus* (Acari, Acaridae). *Beaufortia* **15,** 95–103.

FAASCH, H. (1967). Beitrag zur Biologie der einheimischen Uropodiden *Uroobovella marginata* (C.L. Koch, 1839) und *Uropoda orbicularis* (O.F. Müller, 1776) und experimentelle Analyse ihres Phoresieverhaltens. *Zool. Jb.* (*Syst.*) **94,** 521–608.

FAIN, A. (1963a). Les tyroglyphides commensaux des chauves-souris insectivores. Description de cinq espèces nouvelles. *Rev. Zool. Bot. afr.* **67,** 33–58.

FAIN, A. (1963b). La spermathèque et ses canaux adducteurs chez les acariens mésostigmatiques parasites des voies respiratoires. *Acarologia* **5** (4), 463–79.

FAIN, A. (1965). Les acariens nidicoles et detriticoles de la famille Pyroglyphidae Cunliffe. (Sarcoptiformes). *Revue. Zool. Bot. afr.* **72,** 257–288.

FAIN, A. (1966). Nouvelle description de *Dermatophagoides pteronyssinus* (Trouessart, 1897). Importance de cet acarien en pathologie humaine (Psoroptidae: Sarcoptiformes). *Acarologia* **8,** 302–27.

FAIN, A. (1967a). Les hypopes parasites des tissues cellulaires des oiseaux (Hypodectidae. Sarcoptiformes). *Bull. Inst. r. Sci. nat. Belg.* **43,** 1–139.

FAIN, A. (1967b). Deux nouvelles espèces de Dermatophagoidinae. Rattachement de cette sous-familie aux Pyroglyphidae (Sarcoptiformes). *Acarologia* **9,** 870–881.

FAIN, A. and LOWRY, W.J. (1974). A new genus and two new species of Glycyphaginae from Australia. *Bull. Ann. Soc. r. belge Ent.*, **110,** 215–24.

FILIPPONI, A. (1955). Sulla natura dell' associazione tra *Macrocheles muscaedomesticae e Musca domestica. Riv. Parassit.* **16,** 83–102.

FILIPPONI, A. and CERVONE, L. (1957). Isolamento sessuale tra due specie di *Macrocheles,* foretiche et predatrici di *Musca domestica. Riv. Parassit.* **18,** 17–26.

FLANDERS, S.E. and BADGLEY, M.E. (1963). Prey-predator interactions in self-balanced laboratory populations. *Hilgardia* **35,** (8), 145–83.

FOX, I. (1947). Seven new mites from rats in Puerto Rico. *Ann. ent. Soc. Amer.* **40,** 4, 598–603.

FOX, I. (1950). A new tyroglyphid mite from Puerto Rico. *Proc. Biol. Soc. Wash.* **63,** 205–8.

FUMOUZE, A. and ROBIN, C. (1868). Observations sur une nouvelle espèce d'Acariens du genre *Tyroglyphus. J. Anat. Phys.*, **5** (3), 287–306.

GARMAN, P. (1937). A study of the bulb mite. *Conn. Agr. Exp. Sta. Bull.* **402,** 889–907.

GIGJA, G. (1964). Pests in stockfish. *University Research Institute, Reykjavik, Iceland. Department of Agricultural reports,* Series B. No. 18, 5–41.

GOULD, H.J., HUSSEY, N.W. and PARR, W.J. (1969). Large scale commercial control of *Tetranychus urticae* Koch on cucumbers by the predator *Phytoseiulus persimilis* A.H. in G.O. Evans (Ed.) Proc. 2nd int. Congr. Acarology 1967 (*Akad. Kaidó*) *Bpest,* 383–8.

GRANDJEAN, F. (1935a). Observations sur les Acariens (1 ère sèrie). *Bull. Mus. Hist. nat. Paris.* (2) **7,** No. 2, 119–26.

GRANDJEAN, F. (1935b). Les poils et les organes sensitifs portés par les pattes et le palpe chez les Oribates. *Bull. Soc. zool. Fr.* **60,** 6–39.

GRANDJEAN, F. (1938b). Observations sur les Bdelles. *Ann. Soc. ent. Fr.* **107,** 1–24.

GRANDJEAN, F. (1939a). La chaetotaxie des pattes chez les Acaridiae. *Bull. Soc. zool. Fr.* **64,** 50–60.

GRANDJEAN, F. (1939b). Observations sur les Acariens, Vme. sèrie. *Bull. Mus. Hist. nat. Paris* (2), **11,** 394–401.

GRANDJEAN, F. (1949). Observation et conservation des très petits Arthropodes. *Bull. Mus. Hist. nat. Paris* (2) **21,** 3: 363–70.

GRIFFITHS, D.A. (1958). A study of the inter-relationships between some Grain mites in the families Acaridae and Glycyphagidae and the storage mites of grain. Thesis submitted to the Graduate Faculty of the University of Minnesota.

GRIFFITHS, D.A. (1960). Some field habitats of mites of stored food products. *Ann. appl. Biol.* **48,** 134–44.

GRIFFITHS, D.A. (1964a). A revision of the genus *Acarus* (Acaridae, Acarina). *Bull. Brit. Mus.* (*nat. Hist.*) (Zool.) **11,** 413–64.

GRIFFITHS, D.A. (1964b). Experimental studies on the systematics of the genus *Acarus* Linnaeus 1758. (Sarcoptiformes, Acarina). Proc—1st—Congr., Acarology 1963—*Acarologia* **6** (fasc. hors sér.), 101–16.

GRIFFITHS, D.A. (1966). Nutrition as a factor influencing hypopus formation in *Acarus siro* species complex (Acarina: Acaridae). *J. Stored Prod. Res.* **1,** 325–40.

GRIFFITHS, D.A. (1969). The influence of dietary factors on hypopus formation in *Acarus immobilis* Griffiths (Acari, Acaridae), in G.O. Evans (Ed.) Proc. 2nd int. Congr. Acarology 1967. (*Akad. Kaidó*) *Bpest*, 419–32.

GRIFFITHS, D.A. (1970). A further systematic study of the genus *Acarus* L., 1758 (Acaridae, Acarina), with a key to species. *Bull. Brit. Mus.* (*nat. Hist.*) (Zool.) **19,** 89–120.

GRIFFITHS, D.A., HODSON, A.C. and CHRISTENSEN, C.M. (1959). Grain storage fungi associated with mites. *J. econ. Ent.* **52,** 514–18.

GRIFFITHS, D.A. and CUNNINGTON, A.M. (1971). *Dermatophagoides microceras* sp. n.: A description and comparison with its sibling species, *D. farinae* Hughes, 1961. *J. stored Prod. Res.* **7,** 1–14.

HAARLOV, N. and ALANI, M. (1970). House-dust Mites *Dermatophagoides pteronyssinus* (Trt.), *D. farinae* Hughes, *Euroglyphus maynei* (Cooreman) (Fain) in Denmark (Acarina). *Ent. scand.* **1,** 301–6.

HALBERT, N.J. (1923). Notes on Acari, with descriptions of new species. *J. Linn. Soc. Lond.* (Zool.) **35,** 363–92.

HALL, C.C. (1959). A dispersal mechanism in mites. *Journ. Kansas Ent. Soc.* **32,** 45–6.

HAMMEN, L. VAN DER. (1952). The Oribatei (Acari) of the Netherlands. *Zool. Verh. Leiden* No. 17, 1–139.

HAMMEN, L. VAN DER (1968). Stray notes on Acarida (Arachnida). *Zool. Meded., Leiden* **42,** 261–80.

HEINEMANN, R.L. and HUGHES, R.D. (1970). Reproduction, Reproductive Organs, and Meiosis in the Bisexual Non-parthenogenetic Mite *Caloglyphus mycophagus*, with Reference to Oocyte Degeneration in Virgins (Sarcoptiformes: Acaridae). *J. Morph.* **130,** 1, 93–102.

HENSCHEL, J. (1929). Reizphysiologische Untersuchung der Käsemilbe *Tyrolichus casei* (Oudemans). *Z. vergl. Physiol.* **9,** 802–37.

HIRSCHMANN, W. (1957). Acarologie—Gangsystematik der Parasitiformes Teil 1. Rumpfbehaarung und Rüchenflächen. Schriftenreihe für vergleichende Milbenkunde. *Fürth (Bayern)*, 1–20.

HIRSCHMANN, W. (1960). Acarologie. Gangsystematik der Parasitiformes. Teil 3. —Die Gattung *Dendrolaelaps* Halbert, 1915. *Fürth (Bayern)*, 1–27.

HIRSCHMANN, W. (1969). Acarologie, Schriftenreihe für vergleichende milbenkunde Folge 13. *Fürth (Bayern)*, 1–55.

HIRSCHMANN, W. (1970). Acarologie—Schriftenreihe für vergleichende milbenkunde. Folge 14. *Fürth (Bayern)*, 1–60.

HIRSCHMANN, W. and ZIRNGIEBL-NICOL, I. (1961). Acarologie—Gangsystematik der Parasitiformes, Teil 4. *Fürth/(Bayern)* 1–41.

HIRSCHMANN, W. and ZIRNGIEBL-NICOL, I. (1962). Die Gattung *Uroobovella* Berlese 1903 nov. comb. Acarologie—Gangsystematik der Parasitiformes Teil 6. *Fürth (Bayern)*, 57–80.

HORA, A.M. (1934). On the biology of the mite, *Glycyphagus domesticus* de Geer (Tyroglyphidae, Acarina). *Ann. app. Biol., Cambridge,* **21,** No. 3, 483–94.

HOSAYA, H. and KUGOH, T. (1954). On the new records of *Gohieria fusca* (Oudemans, 1903) from packed bean flour and sugar in Japan. *Jap. J. Sanit. Zool.* **5,** 154–6.

HSIN, K. -1. and CHEN, C.-p. (1964). Studies on the life history of *Aleuroglyphus ovatus* Troupeau (Acarina: Acaridae). *Act. ent. Sinica* **13,** 428–35.

HUGHES, A.M. (1954). On *Suidasia medanensis* Oudemans (Acaridae). A mite new to Great Britain. *Proc. zool. Soc. Lond.*, **125,** 41–7.

HUGHES, A.M. (1955). On the inert hypopial form of *Acarus siro* L. (=*Tyroglyphus farinae* L.). (Acarina). *Ent. mon. Mag.* **91,** 99–102.

HUGHES, A.M. (1956). The mite genus *Lardoglyphus* Oudemans, 1927. (*Hoshikadania* Sasa and Asanuma, 1951). *zool. Meded.* **34,** 271–85.

HUGHES, A.M. (1957). Silk production in the Cheyletidae (Acarina: Arachnida). *Nature, Lond.* **180,** 815–16.

HUGHES, A.M. (1961). Terrestrial acarina III. Acaridiae in Fridriksson, A, and Tuxen, S.L. *The Zoology of Iceland* **3,** 57C, 1–12.

HUGHES, A.M. (1961). The Mites of Stored food. *Tech. Bull. Minist. Agric. Lond.* No. 9, 1–287.

HUGHES, A.M., and HUGHES, T.E. (1939). The internal anatomy and post-embryonic development of *Glycyphagus domesticus* De Geer. *Proc. zool. Soc. Lond.*, **108,** 715–33.

HUGHES, A.M. and MAUNSELL, K. (1973). A study of a population of house dust mite in its natural environment. *Clinical Allergy*, **3:** 245–9.

HUGHES, T.E. (1943). The respiration of *Tyroglyphus farinae*. *J. exp. Biol. Lond*, **20,** 1, 1–5.

HUGHES, T. (1949). The functional morphology of the mouth parts of *Liponyssus bacoti*. *Ann. Trop. Med. Parasit. L'pool.* **43,** 3–4, 349–60.

HUGHES, T.E. (1953). The functional morphology of the mouth parts of the mite *Anoetus sapromyzarum* Dufour 1839, compared with those of the more typical Sarcoptiformes. *Proc. Acad. Sci. Amst.* **56C,** 2, 278–87.

HUGHES, T.E. (1954). The internal anatomy of the mite *Listrophorus leuckarti* (Pagenstecher, 1861). *Proc. zool. Soc. Lond.* **124,** 239–56.

HUGHES, T.E. (1958). The respiratory system of the mite *Cheyletus eruditus* (Schrank, 1781). *Proc. zool. Soc. Lond.* **130,** 231–9.

HUGHES, T.E. (1959). The Acarina, or the Mites. *Athlone Press, London*, 1–210.

HUGHES, T.E. (1964). Neurosecretion, ecdysis and hypopus formation in the Acaridei. Proc. 1st int. Congr. Acarology 1963—*Acarologia* **6** (fasc. hors sér.), 338–42.

HUGHES, R.D. and JACKSON, C.G. (1958). A review of the Anoetidae (Acari). *Virginia J. Sci.* 9. *New Series*, 1–198.

JOHNSTON, D.E. and BRUCE, W.A. (1965). *Tyrophagus neiswanderi*, a new acarid mite of agricultural importance. *Res. Bull. Ohio agric. Exp. Stn.* No. 977, 1–17.

JOYEUX, C. and BAER, G. (1945). Morphologie, evolution et position systématique de *Catenotaenia*. *Rev. Suisse. Zool.*, **52,** 13–51.

KARG, W. (1971). Zur Kenntnis der gattung *Tyrolichus* Oudemans, 1924. *Zool. Anz., Leipzig* **187,** 423–32.

KATES, K.C. and RUNKEL, C.E. (1948). Observations on oribated mite vectors of *Moniezia expansa* on pastures, with a report of several new vectors from the United States. *Proc. Helminth Soc. Wash.* **15,** (1), 18–33.

KENT-JONES, D.W. ET AL. (1948). The micro-analytical test for purity in food, with special reference to cereals. *Analyst* **73,** 128–40.

KEVAN, D.K.MC.E. and SHARMA, G.D. (1964). Observations on the biology of *Hypoaspis aculeifer* (Canestrini 1884), apparently new to North America (Acarina: Mesostigmata: Laelaptidae). *Acarologia* **6,** 647–58.

KILPIO, O. and PIRILA, V. (1952). A new tyroglyphid mite causing dermatitis. *Acta Derm.-Ven.* **32,** 197–200.

KNÜLLE, W. and WHARTON, G.W. (1964). Equilibrium humidities in arthropods and their ecological significance. Proc. 1st int. Congr. Acarology 1963—*Acarologia* **6** (*fasc. hors. sér.*), 299–306.

KRAMER, P. (1877). Grundzüge zur Systematik der Milben. *Arch. Naturg.* **2,** 215–47.

KRANTZ, G.W. (1970). A manual of Acarology. *O.S.U. Book stores Ltd., Corvallis, Oregon*, 1–335.

KREZAL, H. (1959). Systemik und Ökologie der Pyemotiden. In H.J. Stammer, *Beiträge zur Systematik und Ökologie Mitteleuropäischer Acarina* Bd. 1, Teil 2, 385–625.

LARSON, D.G., MITCHELL, W.F. and WHARTON G.W. (1968). A second species of house dust mite, *Dermatophagoides farinae* Hughes involved in house dust allergies. *Proc. N. cent. Bch. Ent. Soc. Am.* **23,** 74 (*Abstr.*)—*47 Ann. Congr. N. cent. States Entomologists.*

LARSON, D.G., MITCHELL, W.F. and WHARTON, G.W. (1969). Preliminary studies on *Dermatophagoides farinae* Hughes, 1961 (Acari) and house dust allergy. *J. med. Ent.* **6,** 295–9.

LIGNIÈRES, J. (1893). Étude zoologique et anatomique du *Tyroglyphus malus* et de sa nymphe hypopiale. *Mém. Soc. zool. France,* **6,** 5–15.

LINDQUIST, E.E. (1961). Taxonomic and biological studies of mites of the genus *Arctoseius* Thor from Barrow, Alaska. *Hilgardia* **30,** 301–50.

LINDQUIST, E.E. (1972). A new species of *Tarsonemus* from stored grain (Acarina: Tarsonemidae). *Can. Ent.* **104:** 1699–708.

LINDQUIST, E.E. and EVANS, G.O. (1965). Taxonomic concept in the Ascidae, with a modified setal nomenclature for the idiosoma of the Gamasina (Acarina: Mesostigmata). *Mem. ent. Soc. Can.* No. 47, 1–64.

MACHACEK, J.E., CHEREWICK, W.J., MEAD, H.W. and BROADFOOT, W.C. (1951). A study of some seed-borne diseases of cereals in Canada II. Kinds of fungi and prevalence of disease in cereal seeds. *Sci. Agr.* **31,** 193–206.

MAHUNKA, S. (1961). Contributions to the tyroglyphid fauna of Hungary (Acari). *Ann. Univ. Sci. Budapest, Biol.* **4,** 113–18.

MANSON, D.C.M. (1972). A contribution to the study of the genus *Rhizoglyphus* Claparède, 1869 (Acarina: Acaridae). *Acarologia* **13,** 4, 621–50.

MARIANI, G. (1952). Die episodi d'infestasione da *Pediculoides ventricosus* nel Bresciano. *Riv. Ital Igiene* **12,** 1–2, 76–9.

MATHYS, G. (1961). Note sur l'espece *Bryobia praetiosa* Koch (acarien, Tetranychidae). *Annu. agric. Suisse* **62,** 295–304.

MATHYS, G. and TENCALLA, Y. (1959). Note preliminaire sur la biologie et la valeur prédatrice de *Proctolaelaps hypudaei* Oudms. (Acarien, Mesostigmata, Aceosejidae). *Publ. Sta. féd. Essais Agric. Lausanne* No. 600, 645–654, also *Ann. agric. Suisse* **60,** 645–54.

MAUNSELL, K. WRAITH, D.G. and CUNNINGTON, A.M. (1968). Mites and house-dust allergy in bronchial asthma. *Lancet* (15 June), 1,267–1,270.

McKINLEY, D.J. (1963). The morphology and biology of *Haemolaelaps casalis* Berlese (Acarina: Mesostigmata). *Ann. Mag. nat. Hist.* **6** (62), 65–76.

MICHAEL, A.D. (1878). On a species of *Acarus* of the genus *Cheyletus* believed to be new. *Trans. Roy. Micr. Soc.* **1,** 135.

MICHAEL, A.D. (1883, 1887). British Oribatidae, vols I, II. *Ray Society, London.*

MICHAEL, A.D. (1889). Observations on the special internal anatomy of *Uropoda krameri. J. Roy. Microscop. Soc.,* 1–15.

MICHAEL, A.D. (1892). On the variations in the internal anatomy of the Gamasinae especially in that of the genital organs, and of their mode of coition. *Trans. Linn. Soc. London Zool.* 2nd Ser. **5** (9), 281–324.

MICHAEL, A.D. (1896). The internal anatomy of *Bdella. Trans. Linn. Soc. London Zoology* **6,** 477–528.

MICHAEL, A.D. (1901, 1903). British Tyroglyphidae I, II. *Ray Society, London.*

MITROFANOV, V.I. (1971). On the system of mites of the family Bryobiidae (Acariformes, Tetranychoidea). Third International Congress of Acarology, *Prague, Czechoslovakia. Abstracts*, 130.

MUMA, M.H. (1955). Phytoseiidae (Acarina) associated with *Citrus* in Florida. *Ann. Ent. Soc. Amer.* **48,** 262–72.

MUMA, M.H. and DENMARK, H.A. (1971). Phytoseiidae of Florida. *Florida Department of Agriculture and Consumer Services*. Contribution No. 148 *Bureau of Entomology*, 1–150.

NARAYANAN, E.S. and GHAI, S. (1961). A new species of *Melichares* (*Blattisocius*) Keegan (Aceosejinae). *Proc. nat. Inst. Sci. India* **27B:** 18–20.

NESBITT, H.H.J. (1945). A revision of the family Acaridae (Tyroglyphidae) order Acari, based on comparative morphological studies. *Canada J. Res.* D, **23,** 139–88.

NESBITT, H.H.J. (1951). A taxonomic study of the Phytoseiinae (family Laelaptidae) predaceous upon Tetranychidae of economic importance. *Zool. Verhandl.* **12,** 1–64.

NEWSTEAD, R. and DUVALL, H.M. (1918). Bionomic, morphological and economic report of acarids of stored grain and flour—Part 1. *Roy. Soc. Rep. Grain Pests* (*War*) *Committee* No. 2, 1–48.

NEWSTEAD, R. and DUVALL, H.M. (1920). Bionomic, morphological and economic report on acarids of stored grain and flour—Part II. *Roy. Soc. Rep. Grain Pests* (*War*) *Committee* No. 8, 1–25.

NIXON, J.W. (1944). Cheese 'itch' and 'itchy' cargoes in reference to workmen's compensation. *Proc. Roy. Soc. Med. London* **87,** 405–10.

NORRIS, J.D. (1958). Observations on the control of mite infestations in stored wheat by *Cheyletus* sp. (Acarina). *Ann. appl. Biol.* **46,** 3, 411–22.

OBOUSSIER, H. (1939). Beiträg zur Biologie und Anatomie der Wohnungsmilben. *Z. angew. Ent. Berlin* **26,** 2, 253–96.

O'FARRELL, A.F. and BUTLER, P.M. (1948). Insects and Mites associated with the storage and manufacture of foodstuffs in Northern Ireland. *Econ. Proc. Roy. Dublin Soc.* **3,** 22, 343–407.

OSHIMA, S. (1968). Redescription of three species of *Mealia* Trouessart, 1897 (Acarina: Pyroglyphidae) from house dust in Japan. *Jap. J. sanit. Zool.* **19,** 165–91.

OUDEMANS, A.C. (1906a). Nieuwe classificatie der Acari. *Ent. Ber.* **2,** 43–6.

OUDEMANS, A.C. (1906b). Revision des Chéletinés. *Mem. Soc. zool. France* **19,** 36–219.

OUDEMANS, A.C. (1910). Acarologische Aanteekeningen XXXII. *Ent. Ber.* III, **53,** 67–74.

OUDEMANS, A.C. (1924a). Analytical key for the classification of families and genera of Diacotricha Oud., 1906. *Ent. Ber.* VI, 226–35.

OUDEMANS, A.C. (1924b). Acarologische Aanteekeningen LXXVII. *Ent. Ber.* VI, 317–36.

OUDEMANS, A.C. (1929). Acarologische Aanteekeningen XCIX. *Ent. Ber.* VIII, 11–20.

PARAMESWARAN PILLAI, P.R. (1957). Pests of stored fish and prawns. *Bull Res. Inst. Univ. Travancore, V*, **3,** 1–79.

PEREIRA, C. and CASTRO, M.P. DE (1945). Contribuicao para o conhecimento da espécie tipo de '*Macrocheles* Latr'. ('Acarina'): '*M. muscaedomesticae* (Scopoli, 1772)'. emend. *Arq. Inst. Biol., Sao Paulo* **16,** 153–86.

PEREIRA, C. and CASTRO, M.P. DE (1947). Forese e partenogenese arrenótoca em 'Macrocheles muscaedomesticae' (Scopoli). ('Acarina: Macrochelidae'). e sua signifacao ecólogica. *Arq. Inst. Biol., S. Paulo*, **18,** 71–89.

PERRON, R. (1954). Untersuchungen über Bau, Entwicklung und Physiologie der Milbe *Histiostoma laboratorium* Hughes. *Acta zool., Stockh.* **35,** 71–176.

PIMENTAL, D., RUMSEY, M.W. and STREAMS, F.A. (1960). Rearing tyroglyphid mites on *Neurospora*. *Ann. ent. Soc. Amer.* **53,** 549.

POLEZHAEV, W.G. (1938). Influence de la faim sur la formation des hypopes enkystes chez *Glycyphagus destructor* (Schrk.). *Zool. J. Moscow*, **17,** 617–21.

POLEZHAEV, W.G. (1940). Effect of atmospheric humidity and temperature on the appearance of hypopial stage in *Glycyphagus destructor* Shrk. and *Tyroglyphus farinae* L. *Wiss. Ber. moskauer Staats. Univ. Zool.* **42,** 185–96.

PULPÁN, J. and VERNER, P.H. (1965). Control of tyroglyphoid mites in stored grain by the predatory mite *Cheyletus eruditus* (Schrank). *Can. J. Zool.* **43,** 417–32.

RACK, G. (1962). Milben aus Taubennestern mit Beschreibung einer neuen Art, *Acheles gracilis* (Acarina, Raphignathidae). *Zool. Anz.* **168,** 275–92.

RACK, G. (1968) Milben als Hausungeziefer u. Vorratsschädlinge in Nordwestdeutschland. *Ent. Mitt. Zool. St. Inst. Zool. Mus Hamburg*, **3,** No. 62, 1–19.

RACK, G. (1970). Massenvorkommen von *Neoseiulus barkeri*, Hughes, 1948 (Acarina, Phytoseiidae) und sechs weiterer Milbenarten in einem Neubau. *Anz. für Schädlingskunde und Pflanzenschutz* **43,** Heft 12, 184–6.

RACK, G. (1971). Milben in Neubauten. *Der praktische Schädlingsbekämpfer* **23,** 11: 149–52.

RADINOVSKY, S. (1965a). The biology and ecology of granary mites of the Pacific Northwest. 3. Life history and development of *Leiodinychus krameri* (Acarina: Uropodidae). *Ann. ent. Soc. Am.* **58,** 259–67.

RADINOVSKY, S. (1965b). The biology and ecology of granary mites of the Pacific Northwest. 4. Various aspects of behaviour of *Leiodinychus krameri* (Acarina: Uropodidae). *Ann. ent. Soc. Am.* **58,** 267–72.

RIVARD, I. (1958a). Influence of humidity on mortality and rate of development of immature stages of the grain-infesting mite *Tyrophagus castellanii* (Hirst) (Acarina: Acaridae) reared on mould cultures. *Canada. Ent.* **90,** 721–24.

RIVARD, I. (1958b). A technique for rearing tyroglyphid mites on mould cultures. *Canada. Ent.* **90,** 146–7.

RIVARD, I. (1960). A technique for individual rearing of the predacious mite *Melichares dentriticus* (Berlese) (Acarina: Aceosejidae) with notes on its life history and behaviour. *Canad. Ent.* **92,** 834–9.

ROBERTSON, P.L. (1946). Tyroglyphid mites in stored products in New Zealand. *Trans. Proc. Roy. Soc. N.Z.* **76,** 2, 185–207.

ROBERTSON, P.L. (1959). Revision of the genus *Tyrophagus* with a discussion on its taxonomic position in the Acarina. *Aust. J. Zool.* **7,** 146–81.

ROBERTSON, P.L. (1961). A morphological study of variation in *Tyrophagus* (Acarina), with particular reference to populations infesting cheese. *Bull. ent. Res.* **52,** 501–29.

ROBINSON, I. (1953). The hypopus of *Hericia hericia* (Kramer), Acarina, Tyroglyphidae. *Proc. zool. Soc. London,* **123,** 267–72.

RODRIGUEZ, J.G. and WADE, C.F. (1961). The nutrition of *Macrocheles muscaedomesticae* (Acarina: Macrochelidae) in relation to its predatory action on the house fly. *Ann. ent. Soc. Amer.* **54,** 782–88.

RODRIGUEZ, J.G., WADE, C.F. and WELLS, C.N. (1962). Nematodes as a natural food for *Macrocheles muscaedomesticae* (Acarina: Macrochelidae) a predator of the house fly egg. *Ann. ent. Soc. Amer.* **55,** 507–11.

ROHDE, C.J. (1959). Studies on the biologies of two mite species, predator and prey, including some effects of gamma radiation on selected developmental stages. *Ecology* **40,** 572–9.

ROHDENDORF, B.P. (1940). Mites of the families Cheyletidae and Pediculoididae. *Moscow Univ. Uchenye zapiski, Wissench. Ber.* **42,** 69–98.

RYKE, P.A.J. (1964) Acarina associated with *Protea* flowers in the Cape Province. *J. ent. Soc. S. Afr.* **26,** 337–54.

SAMŠIŇÁK, K. (1960). Ueber einige myrmekophile Milben aus der Familie Acaridae. *Acta. Soc. ent. Čech.* **57,** 185–92.

SAMŠIŇÁK, K. (1966). Die Neuerrichtung der Gattung *Cosmoglyphus* Oudemans, 1932, gleichzeitig ein Beitrag zum Problem der 'Copra itch'. *Zool. Anz.* **176,** 27–42.

SASA, M. (1964). Special problems of mites in stored food and drugs in Japan. Proc. 1st int. Congr. Acarology 1963 *Acarologia* **6** (fasc. hors sér), 390–91.

SCHEUCHER, R.I. (1957). Systematik und Ökologie der deutschen Anoetinen. *Beitr. Syst. Ökol.* **1,** 1, 233–84.

SCHULZE, H. (1923). Über die Widerstandsfahigkeit der Dauerformen von wirtschaftlich winchtigen milben. *Naturwissenschaften* XI, 763–5.

SCHULZE, H. (1924). Zur kenntnis der dauerformen (Hypopi) der Mehlmilbe *Tyroglyphus farinae* (L.). *Centralbl-Bakt. Paras. Infect.* **60,** 536–49.

SELLNICK, M. (1929). Formenkreis: Hornmilben, Oribatei. *Die Tierwelt Mitteleuropas* Band II, Lieferung **4,** 1–42.

SELLNICK, M. (1958). Milben aus landwirtschaftlichen Betrieben Nordschwedens. *Medd. Vaxtskyddsanst. Stockh.* **11,** 71, 9-59.

SHEALS, J.G. (1956). Notes on a collection of soil Acari. *Ent. Mon. Mag.* **92,** 99–103.

SINHA, R.N. (1963). Stored product Acarology in Canada, in Naegele *Adv. in Acarol.* **1,** 70–88.

SINHA, R.N. (1964a). Ecological relationships of stored product mites and seedborne fungi. Proc. 1st int. Congr. Acarology 1963—*Acarologia* **6** (fasc. hors sér.), 372–89.

Sinha, R.N. (1964b). Effect of low temperature on the survival of some stored products mites. *Acarologia* **6,** 336–41.

Sinha, R.N. (1966a). Feeding and reproduction of some stored-product mites on seed-borne fungi. *J. econ. Ent.* **59,** 5, 1,227–32.

Sinha, R.N. (1966b). *Aëroglyphus robustus*, a pest of stored grain. *J. econ. Ent.* **59**(3), 686–8.

Sinha, R.N. (1968a). Adaptive significance of mycophagy in stored-product Arthropoda. *Evolution* **22,** 4, 785–98.

Sinha, R.N. (1968b). Seasonal changes in Mite Populations in Rural Granaries in Japan. *Ann. ent. Soc. Am.* **61,** 938–49.

Sinha, R.N., Liscombe, E.A.R. and Wallace, H.A.H. (1962). Infestation of mites, insects and micro-organisms in a large wheat bulk after prolonged storage. *Can. Entomol.* **94** (5), 542–55.

Sinha, R.N. and Wallace, H.A.H. (1966). Association of granary mites and seed-borne fungi in stored grain and in outdoor and indoor habitats. *Ann. ent. Soc. Am.* **59,** 1,170–81.

Sinha, R.N. and Mills, J.T. (1968). Feeding and reproduction of the grain mite and the mushroom mite on some species of *Penicillium. J. econ. Ent.* **61,** 6, 1548–52.

Sinha, R.N., Wallace, H.A.H. and Chebib, F.S. (1969a). Principal-component analysis of interrelations among fungi, mites and insects in grain bulk ecosystems. *Ecology* **50,** 536–47.

Sinha, R.N., Wallace, H.A.H. and Chebib, F.S. (1969b). Canonical correlation between groups of acarine, fungal and environmental variables in bulk grain ecosystems. *Res. Popul. Ecol.* **11,** 92–104.

Smirnov, E. and Polezaeff, W. (1936). Uber die Einwirkung von Blausäure und chlorpikrin auf das Hypopial-stadium von *Glycyphagus destructor* Schr. *Zool. J. Moscow* **15,** 340–48.

Smith, K.G. (1969). Infestation by mites as a factor in the long term storage of flour, in G.O. Evans (Ed.) Proc. 2nd int. Congr. Acarology 1967 (*Akad. Kaidó*) *Bpest*, 249–53.

Snodgrass, R.E. (1948). The feeding organs of Arachnida, including mites and ticks. *Smithsonian misc. Coll.* **110,** 10, 1–93.

Solomon, M.E. (1946). Tyroglyphid mites in stored products. Ecological studies. *Ann. appl. Biol. London*, **33,** 1, 82–97.

Solomon, M.E. (1962). Ecology of the flour mite, *Acarus siro* L. (=*Tyroglyphus farinae* De G.). *Ann. appl. Biol.* **50,** 178–84.

Solomon, M.E. (1966). Moisture gains, losses and equilibria of flour mites, *Acarus siro* L., in comparison with larger arthropods. *Ent. exp. and appl.* **9,** 24–41.

Solomon, M.E., Hill, S.T., Cunnington, A.M. and Ayerst, G. (1964). Storage fungi antagonistic to the flour mite (*Acarus siro* L.). *J. appl. Ecol.* **1,** 119–25.

Sorokin, S.V. (1951). The causes of the formation of the hypopi in cereal mites. *Zool. zh., Moscow*, **30,** 523–39.

SPIEKSMA, F.Th. (1967). The house-dust mite *Dermatophagoides pteronyssinus* (Trouessart, 1897), producer of the house-dust allergen (Acari: Psoroptidae). (*Bateljee and Terpstra*) *Leiden*, 1–65.

SPIEKSMA, F.Th.M. and SPIEKSMA-BOEZEMAN, M.I.A. (1967). The mite fauna of house dust with particular reference to the house-dust mite *Dermatophagoides pteronyssinus* (Trouessart, 1897) (Psoroptidae: Sarcoptiformes). *Acarologia* **9,** 226–41.

STRANDTMANN, R.W. (1963). Some previously unpublished drawings of gamasid mites by the late Dr. A. C. Oudemans. *J. Kansas ent. Soc.* **36,** 1–24.

STRANDTMANN, R.W. and MENZIES, G.C. (1948). A new species of mite *Hypoaspis murinus,* frequently taken from *Rattus* sp. (Laelaptidae: Hypoaspinae). *Ann. ent. Soc. Amer. Columbus,* **41,** 4, 479–82.

STRANDTMANN, R.W. and WHARTON, G.W. (1958). Manual of mesostigmatid mites parasitic on vertebrates. *Inst. Acarology, Maryland, U.S.A.* Contribution No. 4, 1–330.

SUMMERS, F.M. and PRICE, D.W. (1970). Review of the mite family Cheyletidae. *University of California publications in Entomology* **61,** 1–153.

SZWABOWICZ, A. and MIEDZOBRODSKI, K. (1957). Toxicity of the meal mite *Tyroglyphus farinae.* I. Experiments in mice, guinea pigs, pigeons and fowls. *Vet. Bull.* **28,** 475–8.

SZWABOWICZ, A., MIEDZOBRODSKI, K. and DONIGIEWICK, K. (1957). Toxicity of *Tyroglyphus farinae* for animals. II Experimental research in horses and sheep. *Med. vet. Varsovie* **13,** 722–4.

SZWABOWICZ, A., MIEDZOBRODSKI, K. and SCHMIDT, W. (1958). Toxicity of *Tyroglyphus farinae* for animals. III Experiments in swine. *Vet. Bull.* **29,** 344–6.

SZWABOWICZ, A., MIEDZOBRODSKI, K., PANKOWA, J. and HOLNICKA, B. (1958). Toxicity of *Tyroglyphus farinae* to animals. IV Experiments on chickens and ducks. *Med. vet. Varsovie* **14,** 556–8.

TABOADA, M. DE F. (1954). Pulmonary acariasis in Spain. An illustrative case report. *Brit. Med. J.* No. 4859, 437–8.

THOR, S. (1931). Bdellidae, Nicoletiellidae, Cryptognathidae. *Das Tierreich* **56,** 1–65.

THOR, S. and WILLMANN, C. (1941). Eupodidae, Penthalodidae, Penthaleidae, Rhagidiidae, Pachygnathidae, Cunaxidae. *Das Tierreich* **71,** 1–186.

TRÄGÅRDH, I. (1932). Palaeacariformes, a new sub-order of Acari. *Ark. Zool.* **24B** (2), 1–6.

TRAVÉ, J. (1963). Écologie et biologie des Oribates (Acariens) saxicoles et arboricoles. *Vie et Milieu Suppl.* **14,** 1–267.

TREAT, A. E. (1967). Mites from noctuid moths. *J. Lepidopterists Soc.* **21,** 169–79.

TREAT, A.E. (1969a). Behavioural aspects of the association of mites with noctuid moths. in G.O. Evans (Ed.) Proc. 2nd int. Congr. Acarology 1967 (*Akad. Kaidó*) *Bpest.* 275–86.

TREAT, A.E. (1969b). Association of the mite *Blattisocius tarsalis* with the moth *Epizeuxis aemula. Journ. New York Ent. Soc.* **77,** 171–75.

TROUESSART, E. (1901). Sur deux espèces, formant un genre nouveau, de Sarcoptides détriticoles parasites des fourrures. *Bull. Soc. zool. France* **26,** 82–4.

TURK, F.A. and TURK, S.M. (1952). Studies on Acari—7th Series. Records and descriptions of mites new to the British fauna, together with short notes on the biology of sundry species. *Ann. Mag. nat. Hist.* (12) **5,** 475–506.

TÜRK, E. and TÜRK F. (1957). Systematik und Ökologie der Tyroglyphiden Mitteleuropas. *Beitr. Syst. Ökol.* **1,** 1–233.

USHATINSKAYA, R.S. (1954). The biological bases for the use of low temperature in the struggle against pests of stored grain (Insects and Mites). *Acad. Sci. U.S.S.R., Moscow,* 88pp.

VITZTHUM, H. GRAF (1919). Acarologische Beobachtungen. 3 Reihe. *Arch. Natg. Berlin* **85A,** 1–62.

VITZTHUM, H.G. (1928). Die Milben-seuche unserer Hönigbiene. *Wiener Tierärztl. Monatsschr.* **15,** 861–2.

VITZTHUM, H. GRAF (1929). Ordnung: Milben, Acari. *Tierwelt Mitteleuropas Leipzig* **3.3,** 1–112.

VITZTHUM, H. GRAF (1940). Die Deutonympha von *Carpoglyphus lactis* (L., 1763) (Acari: Tyroglyphidae). *Zool. Anz. Leipzig.* **129,** 197–201.

VITZTHUM, H. GRAF (1943). Acarina. *Brons Tierreich,* 5, Abt. 4, Buch 5, Lf. 5, 1–1,011.

VOLGIN, V.I. (1961). New species of the mites of the family Glycyphagidae. *Parazit. Sborn., Moscow* **20,** 257–66.

VOLGIN, V.I. (1969). Acarina of the family Cheyletidae from the world fauna. *Opred. Faune SSSR* **101,** 1–432.

VOORHORST, R., SPIEKSMA-BOEZEMAN, M.I.A. and SPIEKSMA, F.Th.M., (1964). Is a mite (*Dermatophagoides* sp.) the producer of the house-dust allergen? *Allergie und Asthma* **10** (6), 329–34.

VYSOTZKAYA, S.O. (1961). The tyroglyphoid mites (Sarcoptiformes) from the nests of rodents and insectivorous mammals of the Leningrad region. *Parazit. Sborn., Moscow* **20,** 267–82.

WADE, C.F. and RODRIGUEZ, J.G. (1961). Life history of *Macrocheles muscaedomesticae* (Acarina: Macrochelidae), a predator of the house fly. *Ann. ent. Soc. Amer.* **54,** 776–81.

WASYLIK, A. (1959). Mite fauna (Tyroglyphoidea) in the nests of the common sparrow (*Passer domesticus* L.) *Ekol. polska* **5,** 187–90.

WASYLIK, A. (1963). A method of continuous analyse of the Acarina of birds' nests. *Ekol. polska* **B9,** 219–24.

WASYLIK, A. (1964). Remarks on the dispersion of certain mites. *Ekol. polska* **10,** 189–92.

WALLACE, D.R.J. (1960). Observations on hypopus development in the Acarina. *J. Insect Physiol.,* **5,** 212–29.

WHARTON, G.W. (1970). Mites and Commercial Extracts of House Dust. *Science* **167,** 1,382–3.

WILLMANN, C. (1931). Moosmilben oder Oribatiden (Cryptostigmata) in Dahl *Die Tierwelt Deutschlands* Teil **22,** 79–200.

WOMERSLEY, H. (1933). A preliminary account of the Bdellidae (Snout mites) of Australia. *Trans. Roy. Soc. South Australia* **57,** 97–107.

WOODRING, J.P. (1963). The nutrition and biology of saprophytic Sarcoptiformes. in Naegele *Adv. in Acarol.* **1,** 89–111.

WOODRING, J.P. (1969a). Environmental regulation of andropolymorphism in tyroglyphids (Acari). In G.O Evans (Ed) Proc. 2nd int. Congr. Acarology 1967. (*Akad. Kaidó*) *Bpest*, 433–40.

WOODRING, J.P. (1969b). Observations on the biology of six species of acarid mites. *Ann. ent. Soc. Am.* **62,** 102–8.

WOODRING, J.P. and COOK, E.F. (1962). The biology of *Ceratozetes cisalpinus* Berlese, *Scheloribates laevigatus* Koch, and *Oppia neerlandica* Oudemans (Oribatei), with a description of all stages. *Acarologia* **4,** 101–37.

WOODROFFE, G.E. (1953). An ecological study of the insects and mites in the nests of certain birds in Britain. *Bull. ent. Res.* **44,** 739–72.

WOODROFFE, G.E. (1954). An additional note on the fauna of birds' nests in Britain. *Bull. ent. Res.* **45,** 135–6.

WOODROFFE, G.E. (1956). Some insects and mites associated with bat-roosts, with a discussion of the feeding habits of the cheyletids (Acarina). *Ent. mon. Mag.* **92,** 138–41.

YUNKER, C.E. (1955). A proposed classification of the Acaridiae (Acarina: Sarcoptiformes). *Proc. Helminth Soc. Wash.* **22,** 98–105.

ZACHVATKIN, A.A. (1936). Notes systématiques sur les Acariens habitant les greniers. *Bull Soc. Nat. Moscow Sect. Biol. N.S.* **45,** 263–70.

ZACHVATKIN, A.A. (1937). Études sur les Tyroglyphides. 1. Le groupe Caloglyphini. *Wiss Ber. moskauer Staats Univ.* **3,** 169–202.

ZACHVATKIN, A.A. (1941). Arachnoidea, Acariens, Tyroglyphoides. Fauna de l'U.R.S.S. 6. 1. *Inst. zool. Acad. Sci. Moscow N.S.* No. 28, 1–475.

ZACHVATKIN, A.A. (1949). New species of granary mites from Azerbaijan S.S.R. *Ent. Obozr., Moscow* **30,** 287–90.

ŽĎÁRKOVÁ, E. (1967). Stored food mites in Czechoslovakia. *J. Stored Prod. Res.* **3,** 155–71.

ŽĎÁRKOVÁ, E. (1969). Mites of stored food in Czechoslovakia, in G.O. Evans (Ed.) Proc. 2nd int. Congr. Acarology 1967 (*Akad. Kaidó*) *Bpest*, 261–4.

ŽĎÁRKOVÁ, E. (1971). Orientation of *Tyrophagus putrescentiae* (Schrank) towards olfactory stimuli. Third International Congress of Acarology, 1971, *Prague, Czechoslovakia. Abstracts*, 206.

ZHELEVA, M. and ANGELKOVA, E. (1964). Contribution to the study of the tyroglyphid fauna in Bulgaria. *God. Sof. Univ. Biol.* **56,** 141–9.

Index

Printed in England for Her Majesty's Stationery Office by Perivan Press Williams Lea Group
Dd 506748 K20 12/75